Assembly

Robot (reprogrammable) vs Posit[illegible]

- typically 6 DoF

4 DOF robot

- Don't need 6 DoF robot for screws

Robots: Gripper uses Gear & rack >> slider crank (pg 390)

↳ used in welding

↳ rivet robot

Brazing → Furnace – preplaced filler, into a furnace

→ Dip – complete immersion into vat

Soldering → Wave – [illegible] PCB, but sticks only to [illegible]

→ Vapor phase – most popular

- remelting of preplaced solder via convection

Manufacturing

Design, Production, Automation, and Integration

MANUFACTURING ENGINEERING AND MATERIALS PROCESSING
A Series of Reference Books and Textbooks

1. Computers in Manufacturing, *U. Rembold, M. Seth, and J. S. Weinstein*
2. Cold Rolling of Steel, *William L. Roberts*
3. Strengthening of Ceramics: Treatments, Tests, and Design Applications, *Harry P. Kirchner*
4. Metal Forming: The Application of Limit Analysis, *Betzalel Avitzur*
5. Improving Productivity by Classification, Coding, and Data Base Standardization: The Key to Maximizing CAD/CAM and Group Technology, *William F. Hyde*
6. Automatic Assembly, *Geoffrey Boothroyd, Corrado Poli, and Laurence E. Murch*
7. Manufacturing Engineering Processes, *Leo Alting*
8. Modern Ceramic Engineering: Properties, Processing, and Use in Design, *David W. Richerson*
9. Interface Technology for Computer-Controlled Manufacturing Processes, *Ulrich Rembold, Karl Armbruster, and Wolfgang Ülzmann*
10. Hot Rolling of Steel, *William L. Roberts*
11. Adhesives in Manufacturing, *edited by Gerald L. Schneberger*
12. Understanding the Manufacturing Process: Key to Successful CAD/CAM Implementation, *Joseph Harrington, Jr.*
13. Industrial Materials Science and Engineering, *edited by Lawrence E. Murr*
14. Lubricants and Lubrication in Metalworking Operations, *Elliot S. Nachtman and Serope Kalpakjian*
15. Manufacturing Engineering: An Introduction to the Basic Functions, *John P. Tanner*
16. Computer-Integrated Manufacturing Technology and Systems, *Ulrich Rembold, Christian Blume, and Ruediger Dillman*
17. Connections in Electronic Assemblies, *Anthony J. Bilotta*
18. Automation for Press Feed Operations: Applications and Economics, *Edward Walker*
19. Nontraditional Manufacturing Processes, *Gary F. Benedict*
20. Programmable Controllers for Factory Automation, *David G. Johnson*
21. Printed Circuit Assembly Manufacturing, *Fred W. Kear*

22. Manufacturing High Technology Handbook, *edited by Donatas Tijunelis and Keith E. McKee*
23. Factory Information Systems: Design and Implementation for CIM Management and Control, *John Gaylord*
24. Flat Processing of Steel, *William L. Roberts*
25. Soldering for Electronic Assemblies, *Leo P. Lambert*
26. Flexible Manufacturing Systems in Practice: Applications, Design, and Simulation, *Joseph Talavage and Roger G. Hannam*
27. Flexible Manufacturing Systems: Benefits for the Low Inventory Factory, *John E. Lenz*
28. Fundamentals of Machining and Machine Tools: Second Edition, *Geoffrey Boothroyd and Winston A. Knight*
29. Computer-Automated Process Planning for World-Class Manufacturing, *James Nolen*
30. Steel-Rolling Technology: Theory and Practice, *Vladimir B. Ginzburg*
31. Computer Integrated Electronics Manufacturing and Testing, *Jack Arabian*
32. In-Process Measurement and Control, *Stephan D. Murphy*
33. Assembly Line Design: Methodology and Applications, *We-Min Chow*
34. Robot Technology and Applications, *edited by Ulrich Rembold*
35. Mechanical Deburring and Surface Finishing Technology, *Alfred F. Scheider*
36. Manufacturing Engineering: An Introduction to the Basic Functions, Second Edition, Revised and Expanded, *John P. Tanner*
37. Assembly Automation and Product Design, *Geoffrey Boothroyd*
38. Hybrid Assemblies and Multichip Modules, *Fred W. Kear*
39. High-Quality Steel Rolling: Theory and Practice, *Vladimir B. Ginzburg*
40. Manufacturing Engineering Processes: Second Edition, Revised and Expanded, *Leo Alting*
41. Metalworking Fluids, *edited by Jerry P. Byers*
42. Coordinate Measuring Machines and Systems, *edited by John A. Bosch*
43. Arc Welding Automation, *Howard B. Cary*
44. Facilities Planning and Materials Handling: Methods and Requirements, *Vijay S. Sheth*
45. Continuous Flow Manufacturing: Quality in Design and Processes, *Pierre C. Guerindon*
46. Laser Materials Processing, *edited by Leonard Migliore*
47. Re-Engineering the Manufacturing System: Applying the Theory of Constraints, *Robert E. Stein*
48. Handbook of Manufacturing Engineering, *edited by Jack M. Walker*
49. Metal Cutting Theory and Practice, *David A. Stephenson and John S. Agapiou*
50. Manufacturing Process Design and Optimization, *Robert F. Rhyder*
51. Statistical Process Control in Manufacturing Practice, *Fred W. Kear*
52. Measurement of Geometric Tolerances in Manufacturing, *James D. Meadows*
53. Machining of Ceramics and Composites, *edited by Said Jahanmir, M. Ramulu, and Philip Koshy*
54. Introduction to Manufacturing Processes and Materials, *Robert C. Creese*
55. Computer-Aided Fixture Design, *Yiming (Kevin) Rong and Yaoxiang (Stephens) Zhu*
56. Understanding and Applying Machine Vision: Second Edition, Revised and Expanded, *Nello Zuech*
57. Flat Rolling Fundamentals, *Vladimir B. Ginzburg and Robert Ballas*

58. Product Design for Manufacture and Assembly: Second Edition, Revised and Expanded, *Geoffrey Boothroyd, Peter Dewhurst, and Winston Knight*
59. Process Modeling in Composites Manufacturing, Suresh G. Advani and E. Murat Sozer
60. Integrated Product Design and Manufacturing Using Geometric Dimensioning and Tolerancing, Robert G. Campbell and Edward S. Roth
61. Handbook of Induction Heating, Valery Rudnev, Don Loveless, Raymond Cook, and Micah Black
62. Re-Engineering the Manufacturing System: Applying the Theory of Constraints, Second Edition, Revised and Expanded, Robert E. Stein
63. Manufacturing: Design, Production, Automation, and Integration, *Beno Benhabib*

Additional Volumes in Preparation

Manufacturing

Design, Production, Automation, and Integration

Beno Benhabib
University of Toronto
Toronto, Ontario, Canada

MARCEL DEKKER, INC.   NEW YORK • BASEL

Transferred to Digital Printing 2005

Although great care has been taken to provide accurate and current information, neither the author(s) nor the publisher, nor anyone else associated with this publication, shall be liable for any loss, damage, or liability directly or indirectly caused or alleged to be caused by this book. The material contained herein is not intended to provide specific advice or recommendations for any specific situation.

Trademark notice: Product or corporate names may be trademarks or registered trademarks and are used only for identification and explanation without intent to infringe.

Library of Congress Cataloging-in-Publication Data
A catalog record for this book is available from the Library of Congress.

ISBN: 0-8247-4273-7

Headquarters
Marcel Dekker, Inc., 270 Madison Avenue, New York, NY 10016, U.S.A.
tel: 212-696-9000; fax: 212-685-4540

Distribution and Customer Service
Marcel Dekker, Inc., Cimarron Road, Monticello, New York 12701, U.S.A.
tel: 800-228-1160; fax: 845-796-1772

Eastern Hemisphere Distribution
Marcel Dekker AG, Hutgasse 4, Postfach 812, CH-4001 Basel, Switzerland
tel: 41-61-260-6300; fax: 41-61-260-6333

World Wide Web
http://www.dekker.com

The publisher offers discounts on this book when ordered in bulk quantities. For more information, write to Special Sales/Professional Marketing at the headquarters address above.

Current printing (last digit):

10 9 8 7 6 5 4 3 2

Preface

This book is a comprehensive, integrated treatise on manufacturing engineering in the modern age. By addressing the three important aspects of manufacturing—namely, design, production processes, and automation—it presents the state of the art in manufacturing as well as a careful treatment of the fundamentals. All topics have been carefully selected for completeness, researched, and discussed as accurately as possible, with an emphasis on *computer integration*. *Design* is discussed from concept development to the engineering analysis of the final product, with frequent reference to the various processes of fabrication. Numerous common fabrication processes (traditional and modern) are subsequently detailed and contextualized in terms of product design and automation. In the third part of the book, manufacturing control is discussed at the machine level as well as the system level (namely, material flow control in flexible manufacturing systems).

Although the book does discuss the totality of the *design* cycle, it does not present an exhaustive discussion of all manufacturing *processes* in existence. It emphasizes the most common types of metal processing, plastics processing, and powder processing, including modern processes such as laser cutting and numerous lithography-based methods. In the third part of the book, continuous control is not discussed in detail; students interested in automation are expected to have a basic knowledge of the topic. Discrete-event control—a topic rarely introduced in manufacturing books—is addressed because of its vital importance in system control.

Although this book was written mainly for undergraduate and graduate students in mechanical and industrial engineering programs, its integrated treatment of the subject makes it a suitable reference for practicing engineers and other professionals interested in manufacturing. For the classroom setting, the book offers the following benefits: (1) providing the undergraduate-level instructor with the flexibility to include several advanced topics in a course on manufacturing fundamentals and (2) providing graduate students with a background of manufacturing fundamentals, which they may not have fully studied as undergraduates.

TEACHING MANUFACTURING ENGINEERING USING THIS BOOK

Although manufacturing practice in industry has evolved significantly over the past two decades, existing textbooks rarely reflect these changes, thus severely restricting the way manufacturing courses are taught. Most textbooks are still compartmentalized in the manner that manufacturing practice was in the distant past; namely, there are design books, process books, and automation books. In practice, manufacturing is a concurrent, integrated process that requires engineers to think simultaneously of all issues and their impact on one another. This book attempts to advance the teaching of manufacturing engineering, keeping pace with practice in industry while providing instructors with options for course development. Instructors can configure the book to be suitable for two consecutive (one-term) courses: one at an introductory undergraduate level (Fundamentals of Manufacturing Engineering) and one at an advanced level (Manufacturing Automation):

Fundamentals of Manufacturing Engineering

Chapter 1: Competitive Manufacturing
Chapter 2: Conceptual Design
Chapter 3: Design Methodologies
(*Optional*) Chapter 4: Computer-Aided Design
Chapter 6: Metal Casting, Powder Processing, and Plastics Molding
Chapter 7: Metal Forming
Chapter 8: Machining
Chapter 9: Modern Manufacturing Techniques
Chapter 10: Assembly
(*Optional*) Chapter 11: Workholding—Fixtures and Jigs

(*Optional*) Chapter 12: Materials Handling
(*Optional*) Chapter 16: Control of Manufacturing Quality

Manufacturing Automation

Chapter 1: Competitive Manufacturing
(*Optional*) Chapter 2: Conceptual Design
(*Optional*) Chapter 3: Design Methodologies
Chapter 4: Computer-Aided Design
Chapter 5: Computer-Aided Engineering Analysis and Prototyping
(*Optional*) Chapter 9: Modern Manufacturing Techniques
(*Optional*) Chapter 10: Assembly
(*Optional*) Chapter 11: Workholding—Fixtures and Jigs
(*Optional*) Chapter 12: Materials Handling
Chapter 13: Instrumentation for Manufacturing Control
Chapter 14: Control of Production and Assembly Machines
Chapter 15: Supervisory Control of Manufacturing Systems
Chapter 16: Control of Manufacturing Quality

CHAPTER HIGHLIGHTS

Chapter 1 focuses on major historical developments in the manufacturing industry in the past two centuries. The emergence of machine tools and industrial robots is discussed as prelude to a more in-depth review of the automotive manufacturing industry. Technological advancements in this industry have significantly benefited other manufacturing industries over the past century. Various manufacturing strategies adopted in different countries are reviewed as prelude to a discussion on the expected future of the manufacturing industry—namely, information technology–based manufacturing.

Chapter 2 emphasizes the first stage of the engineering design process: development of viable concepts. Concurrent engineering (CE) is defined as a systematic approach to the integrated design of products and their manufacturing and support processes. Identification of customer need is described as the first step in this process, followed by concept generation and selection. The importance of industrial design (including human factors) in engineering design is also highlighted. The chapter concludes with a review of modular product design practices and the mass manufacturing of such customized products.

Chapter 3 describes four primary design methodologies. Although these methodologies have commonly been targeted for the post–conceptual design phase, some can also be of significant benefit during the conceptual design phase of a product—for example, *axiomatic design* and *group technology* during the conceptual design phase, *design for manufacturing/assembly/environment* during configuration and detailed design, and the *Taguchi method* during parametric design.

Chapter 4 begins with a review of geometric-modeling principles and then addresses several topics in computer-aided design (CAD), such as solid-modeling techniques, feature-based design, and product-data-exchange standards.

In Chapter 5 a discussion of prototyping (physical versus virtual) serves as introduction to a thorough review of the most common computer-aided engineering (CAE) analysis tool used in mechanical engineering: finite-element modeling and analysis. Subsequently, several optimization techniques are discussed.

Chapter 6 describes three distinct fusion-based production processes for the net-shape fabrication of three primary engineering materials: casting for metals, powder processing for ceramics and high-melting-point metals and their alloys (e.g., cermets), and molding for plastics.

Chapter 7 describes several common metal-forming processes, focusing on two processes targeted for discrete-parts manufacturing: forging and sheet-metal forming. Quick die exchange, which is at the heart of productivity improvement through elimination of "waste," is also briefly addressed.

Chapter 8 surveys nonabrasive machining techniques (e.g., turning and milling) and discusses critical variables for finding material removal rate, such as cutting velocity and feed rate. The economics of machining—which is based on the utilization of these variables in the derivation of the necessary optimization models—is also discussed in terms of the relationship of cutting-tool wear to machining-process parameters. A discussion of representative abrasive-machining methods is also included.

In Chapter 9, several (nontraditional) processes for material removal are reviewed in separate sections devoted to non–laser-based and laser-based fabrication. This leads to a discussion of several modern material-additive techniques commonly used in the rapid fabrication of layered physical prototypes.

Chapter 10 describes various methods used for joining operations in the fabrication of multicomponent products. These include mechanical fastening, adhesive bonding, welding, brazing, and soldering. The chapter concludes with a detailed review of two specific assembly applications: automatic assembly of electronic parts and automatic assembly of small mechanical parts.

In Chapter 11, following the description of general workholding principles and basic design guidelines for jigs and fixtures, the use of such devices in manufacturing is discussed, in the form of dedicated or modular configurations. CAD techniques for fixture/jig development are briefly described.

In Chapter 12, the focus is on the handling of individual goods (i.e., "unit loads") with a primary emphasis on material-handling equipment, as opposed to facility planning and movement control. Industrial trucks (including automated guided vehicles), conveyors and industrial robots are reviewed as the primary mechanized/automated material-handling equipment. The automated storage and retrieval of goods in high-density warehouses, as well as the important issue of automatic part identification (including bar codes), are also discussed. The chapter ends with a discussion of automobile assembly.

Chapter 13 describes the various sensors that can be used for automatic control in manufacturing environments. A brief introduction to the control of devices in the continuous-time domain precedes a discussion of pertinent manufacturing sensors: motion sensors, force sensors, and machine vision. A brief discussion of actuators concludes the chapter.

Chapter 14 focuses on the automatic control of two representative classes of production and assembly machines: material-removal machine tools and industrial robotic manipulators, respectively.

Chapter 15 describes two of the most successful discrete-event-system (DES) control theories developed by the academic community: Ramadge–Wonham automata theory and Petri nets theory. The chapter ends with a description of programmable logic controllers (PLCs), which are used for the autonomous DES-based supervisory control of parts flow in flexible manufacturing workcells.

Chapter 16 addresses quality control with an emphasis on on-line control (as opposed to postprocess sampling), focusing on measurement technologies and statistical process-control tools. Inspection is defined and some common metrological techniques are presented. An overview of probability and statistics theories are presented as prelude to a discussion of statistical process capability and control. A discussion of ISO 9000:2000 concludes the chapter.

Beno Benhabib

Acknowledgments

Most books on manufacturing engineering—and this one is no exception—reflect the worldwide efforts of thousands of engineers and scientists who have, over the past century and even earlier, advanced the state of the art. Herein, that cumulative achievement is augmented by the knowledge and experience I have gathered over the past two decades. Although many individuals have helped me in my endeavors, the primary contributors are the dedicated graduate students and postdoctoral fellows whose work I have had the privilege of supervising over the past two decades at the University of Toronto:

Ph.D. students: A. Bonen, E. A. Croft, H. R. Golmakani, X. He, M. Mehrandezh, M. Naish, G. Nejat, W. Owen, A. Qamhiyah, A. Ramirez-Serrano, R. Saad, R. Safaee-Rad, E. Tabarah, G. Zak, and D. Zlatanov

M.A.Sc. students: F. Agah, A. Bahktari, M. Bonert, J. Borg, K. C. Chan, C. Charette, P. Chen, H. Chiu, M. Eskandari, M. Ficocelli, M. Haberer, D. He, I. Heerah, D. Hujic, Z. Jiang, S. Lauzon, M. Lipton, O. Partaatmadja, R. Ristic, S. Rooks, A. Sun, R. Williams, F. Wong, and V. Yevko

M.Eng. students: K. H. Chan, S. W. Chan, Y. F. Chan, V. Cheung, A. Cupillari, M. Doiron, T. Kolovos, O. Kornienko, K. Leung, A. Ma,

H. Maatouk, I. Naguib, B. Nouri, W. Nasser, M. Tam, I. Tropak, and D. Valliere

Postdoctoral fellows: R. Cohen, P. Han, G. Hexner, S. Kaizerman, N. Sela, H.-Y. Sun, and X. Wang

Throughout my career in academia, I have also had the pleasure to collaborate and interact with many colleagues, frequently through the work of our graduate students. These individuals have also dedicated their professional lives to the advancement of manufacturing practices, and have therefore indirectly contributed to this work.

University of Toronto: R. Ben-Mrad, R. G. Fenton, A. A. Goldenberg, J. K. Mills, M. Paraschivoiu, J. Paradi, C. B. Park, L. Shu, K. C. Smith, I. B. Turksen, A. N. Venetsanopoulos, and R. D. Venter
University of British Columbia: E. A. Croft, Y. Altintas, and F. Sassani
Queen's University: G. Zak
University of Montpellier: E. Dombre
National University of Singapore: Y. H. Fuh and A. Y. C. Nee

Two colleagues I thank especially are A. Ber (Technion) and R. G. Fenton (University of Toronto). During my early years in academia, they acted as invaluable advisors and mentors to me and to many others.

Publication of this book would not have been possible without the contributions of W. Smith (text preparation), J. Kolba (artwork), and M. Bienenstock (artwork) at the University of Toronto, and John Corrigan (Acquisitions Editor) and Michael Deters (Production Editor) at Marcel Dekker, Inc.

Finally, I would like to thank my family (Sylvie, Neama, and Hadas) for their unconditional love, patience, encouragement, and support. Thank you all. *This book is dedicated to you!*

Contents

1

Competitive Manufacturing

1.1 MANUFACTURING MATTERS

In the earlier part of the 20th century, manufacturing became a capital-intensive activity. A rigid mode of mass production replaced mostly small-batch and make-to-order fabrication of products. A turning point was the 1920s. With increased household incomes in North America and Europe came large-scale production of household appliances and motor vehicles. These products steadily increased in complexity, thus requiring design standardization on the one hand and labor specialization on the other. Product complexity combined with manufacturing inflexibility led to long product life cycles (up to 5 to 7 years, as opposed to as low as 6 months to 1 year in today's communication and computation industries), thus slowing down the introduction of innovative products.

In the post–World War II (WWII) era we saw a second boom in the manufacturing industries in Western Europe, the U.S.A., and Japan, with many domestic companies competing for their respective market shares. In the early 1950s, most of these countries imposed heavy tariffs on imports in order to protect local companies. Some national governments went a step further by either acquiring large equities in numerous strategic companies or providing them with substantial subsidies. Today, however, we witness the fall of many of these domestic barriers and the emergence of multinational companies attempting to gain international competitive advantage via distributed design and manufacturing across a number of countries (sometimes several continents), though it is important to note that most such successful companies are normally those that encountered and survived intense domestic competition, such as Toyota, General Motors, Northern Telecom (Nortel), Sony, and Siemens. Rapid expansion of foreign investment opportunities continue to require these companies to be innovative and maintain a competitive edge via a highly productive manufacturing base. In the absence of continuous improvement, any company can experience a rapid drop in investor confidence that may lead to severe market share loss.

Another important current trend is conglomeration via mergers or acquisitions of companies who need to be financially strong and productive in order to be internationally competitive. This trend is in total contrast to the 1970s and 1980s, when large companies (sometimes having a monopoly in a domestic market) broke into smaller companies voluntarily or via government intervention in the name of increased productivity, consumer protection, etc. A similar trend in political and economic conglomeration is the creation of free-trade commercial zones such as NAFTA (the North American Free Trade Agreement), EEC (the European Economic Community), and APEC (the Asia-Pacific Economic Cooperation).

One can thus conclude that the manufacturing company of the future will be multinational, capital as well as knowledge intensive, with a high level of production automation, whose competitiveness will heavily depend on the effective utilization of information technology (IT). This company will design products in virtual space, manufacture them in a number of countries with the minimum possible (hands-on) labor force, and compete by offering customers as much flexibility as possible in choices. Furthermore, such a company will specialize in a minimal number of products with low life cycles and high variety; mass customization will be the order of the day.

In the above context, computer integrated manufacturing (CIM) must be seen as the utilization of computing and automation technologies across the enterprise (from marketing to design to production) for achieving the most effective and highest quality service of customer needs. CIM is no longer simply a business strategy; it is a required utilization of state-of-the-art technology (software and hardware) for maintaining a competitive edge.

In this chapter, our focus will be on major historical developments in the manufacturing industry in the past two centuries. In Sec. 1.2, the beginnings of machine tools and industrial robots will be briefly discussed as a prelude to a more in-depth review of the automotive manufacturing industry. Advancements made in this industry (technological, or even marketing) have benefited significantly other manufacturing industries over the past century. In Sec. 1.3, we review the historical developments in computing technologies. In Secs. 1.4 and 1.5, we review a variety of "manufacturing strategies" adopted in different countries as a prelude to a discussion on the expected future of the manufacturing industry, namely, "information-technology–based manufacturing," Sec. 1.6.

1.2 POST–INDUSTRIAL-REVOLUTION HISTORY OF MANUFACTURING TECHNOLOGIES

The industrial revolution (1770–1830) was marked by the introduction of steam power to replace waterpower (for industrial purposes) as well as animal-muscle power. The first successful uses for such power in the U.K. and U.S.A. were for river and rail transport. Subsequently, steam power began to be widely used in mechanization for manufacturing (textile, metal forming, woodworking, etc.). The use of steam power in factories peaked around the 1900s with the start of the wide adoption of electric power. Factory electrification was a primary contributor to significant productivity improvements in 1920s and 1930s.

Due to factory mechanization and social changes over the past century, yearly hours worked per person has declined from almost 3000 hours to 1500 hours across Europe and to 1600 hours in North America. However, these decreases have been accompanied by significant increases in labor productivity. Notable advances occurred in the standard of living of the population in these continents. Gross Domestic Product (GDP) per worker increased seven fold in the U.S., 10-fold in Germany, and more than 20-fold in Japan between 1870s and the 1980s.

1.2.1 Machine Tools

Material-removal machines are commonly referred to as "machine tools." Such machines are utilized extensively in the manufacturing industry for a variety of material-removal tasks, ranging from simple hole making (e.g., via drilling and boring) to producing complex contoured surfaces on rotational or prismatic parts (e.g., via turning and milling).

J. Wilkinson's (U.K.) boring machine in 1774 is considered to be the first real machine tool. D. Wilkinson's (U.S.A.) (not related to J. Wilkinson) screw-cutting machine patented in 1798 is the first lathe. There exists some disagreement as to who the credit should go to for the first milling machine. R. Johnson (U.S.A.) reported in 1818 about a milling machine, but probably this machine was invented by S. North well before then. Further developments on the milling machine were reported by E. Whitney and J. Hall (U.S.A.) around 1823 to 1826. F. W. Howe (U.K.) is credited with the design of the first universal milling machine in 1852, manufactured in the U.S.A. in large numbers by 1855. The first company to produce machine tools, 1851, Gage, Warner and Whitney, produced lathes, boring machines, and drills, though it went out of business in the 1870s.

As one would expect, metal cutting and forming has been a major manufacturing challenge since the late 1700s. Although modern machine tools and presses tend to be similar to their early versions, current machines are more powerful and effective. A primary reason for up to 100-fold improvements is the advancement in materials used in cutting tools and dies. Tougher titanium carbide tools followed by the ceramic and boron-nitride (artificial diamond) tools of today provide many orders of magnitude improvement in cutting speeds. Naturally, with the introduction of automatic-control technologies in 1950s, these machines became easier to utilize in the production of complex-geometry workpieces, while providing excellent repeatability.

Due to the worldwide extensive utilization of machine tools by small, medium, and large manufacturing enterprises and the longevity of these machines, it is impossible to tell with certainty their current numbers (which may be as high as 3 to 4 million worldwide). Some recent statistics, however, quote sales of machine tools in the U.S.A. to be in the range of 3 to 5 billion dollars annually during the period of 1995 to 2000 (in contrast to $300–500 million annually for metal-forming machines). It has also been stated that up to 30% of existing machine tools in Europe, Japan, and the U.S. are of the numerical control (NC) type. This percentage of NC machines has been steadily growing since the mid-1980s, when the percentage was below 10%, due to rapid advancements in computing technologies. In Sec. 1.3 we will further address the history of automation in machine-tool control during the 1950s and 1960s.

1.2.2 Industrial Robots

A manipulating industrial robot is defined by the International Organization for Standardization (ISO) as "An automatically controlled, re-programmable, multi-purpose, manipulative machine with several degrees of freedom, which may be either fixed in place or mobile for use in industrial application" (ISO/TR 8373). This definition excludes automated guided vehicles, AGVs, and dedicated automatic assembly machines.

The 1960s were marked by the introduction of industrial robots (in addition to automatic machine tools). Their initial utilization on factory floors were for simple repetitive tasks in either handling bulky and heavy workpieces or heavy welding guns in point-to-point motion. With significant improvements in computing technologies, their application spectrum was later widened to include arc welding and spray painting in continuous-path motion. Although the commercial use of robots in the manufacturing industry can be traced back to the early 1960s, their widespread use only started in the 1970s and peaked in the 1980s. The 1990s saw a marked decline in the use of industrial robots due to the lack of technological support these robots needed in terms of coping with uncertainties in their environments. The high expectations of industries to replace the human labor force with a robotic one did not materialize. The robots lacked artificial perception ability and could not operate in autonomous environments without external decision-making support to deal with diagnosis and error recovery issues. In many instances, robots replaced human operators for manipulative tasks only to be monitored by the same operators in order to cope with uncertainties.

In late 1980s, Japan clearly led in the number of industrial robots. However, most of these were manipulators with reduced degrees of freedom (2 to 4); they were pneumatic and utilized in a playback mode. Actually, only about 10% of the (over 200,000) robot population could be classified as "intelligent" robots complying with the ISO/TR 8373 definition. The percentage would be as high as 80%, though, if one were to count the playback manipulators mostly used in the automotive industry. Table 1 shows that the primary user of industrial robots has been indeed the automotive industry worldwide (approximately 25–30%) with the electronics industry being a distant second (approximately 10–15%).

Today, industrial robots can be found in many high-precision and high-speed applications. They come in various geometries: serial (anthropomorphic, cylindrical, and gantry) as well as parallel (Stewart platform and hexapod). However, still, due to the lack of effective sensors, industrial robots cannot be utilized to their full capacity in an integrated sense with other production machines. They are mostly restricted to repetitive tasks, whose pick and place locations or trajectories are a priori known; they are not robust to positional deviations of workpiece locations (Figure 1).

TABLE 1 Industrial Robot Population in 1989

	France	Germany	Italy	Japan	U.K.	U.S.A.	World
Total population (1000s)	7	22	10	220	6	37	387
Automotive industry (%)	33	N/A	30[a]	26	33	N/A	N/A

[a] Calculated based on installations during the past 5 years.

FIGURE 1 A FANUC Mate 50:L welding robot welding a part.

1.2.3 Automotive Manufacturing Industry

The automotive industry still plays a major economic role in many countries where it directly and indirectly employs 5 to 15% of the workforce (Tables 2 to 4). Based on its history of successful mass production that spans a century, many valuable lessons learned in this industry can be extrapolated to other manufacturing industries. The Ford Motor Co., in this respect, has been the most studied and documented car manufacturing enterprise.

Prior to the introduction of its world-famous 1909 Model T car, Ford produced and marketed eight earlier models (A, C, B, F, K, N, R, and S). However, the price of this easy-to-operate and easy-to-maintain car (sold for under $600) was indeed what revolutionized the industry, leading to great demand and thus the introduction of the moving assembly line in 1913. By 1920, Ford was building half the cars in the world (more than 500,000 per year) at a cost of less than $300 each. A total of 15 million Model T cars were made before the end of the product line in 1927 (Figure 2).

TABLE 2 Motor Vehicle[a] Production Numbers per Year per Country (1000s)

	1899	1905	1910	1925	1950	1968	1993	1999
U.S.A.	3	25	187	4,265	8,005	10,206	10,864	13,024
Canada				161	388	1,353	2,237	3,056
France	2	20	38	177	357	2,459	3,155	3,032
Germany[b]	1	4	13	55	304	3,739	3,990	5,687
Italy	N/A	1	5	40	127	1,592	1,267	1,701
Japan				N/A	32	4,674	11,227	9,905
S. Korea[c]						45	2,050	2,832
U.K.	1.6	3	14	176	785	2,183	1,569	1,972
World	N/A	55	256	4,800	10,577	29,745	46,856	54,947

[a] "Motor vehicle" includes passenger cars, trucks, and buses.
[b] Federal Republic of Germany only prior to 1980.
[c] South Korean motor vehicle industry started in 1962 (3000 vehicles).

The first automobile, however, is attributed to N. J. Cugnot, a French artillery officer, who made a steam-powered three-wheeled vehicle in 1769. The first internal-combustion–based vehicle is credited to two inventors: the Belgian E. Lenoir (1860) and the Austrian S. Marcus (1864). The first ancestors of modern cars, however, were the separate designs of C. Benz (1885) and G. Daimler (1886). The first American car was built by J. W. Lambert in 1890–1891.

Since the beginnings of the industry, productivity has been primarily achieved via product standardization and mass production at the expense of competitiveness via innovation. Competitors have mostly provided customers with a price advantage over an innovative advantage. Almost 70

TABLE 3 Motor Vehicle Registration by Country by Year (1000s)

	1925	1950	1953	1992	1998
U.S.A.	19,954	49,177	101,039	190,362	210,901
Canada	718	2,537	7,539	17,010	17,581
France	735	2,422	13,220	29,060	32,300
Germany	323	998[a]	14,289	42,009	46,030
Italy	114	758	8,976	32,260	30,000
Japan	33	337	12,482	61,658	71,209
S. Korea	—	15	58	5,231	10,739
U.K.	902	3,306	12,786	26,651	25,283
World	24,564	70,400	216,608	613,530	663,038

[a] Federal Republic of Germany.

TABLE 4 Employment in U.S. Automobile Industry Plants (1000s)

1925	1950	1976	1999
474	839	881	1,000

automotive companies early on provided customers with substantial innovative differences in their products, but today there remain only three major U.S. car companies that provide technologically very similar products.

From 1909 to 1926, Ford's policy of making a single, but best-priced, car allowed its competitors slowly to gain market share, as mentioned above, via technologically similar but broader product lines. By 1925, General Motors (GM) held approximately 40% of the market versus 25% of Ford and 22% of Chrysler. In 1927, although Ford discontinued its production of the Model T, its strategy remained unchanged. It introduced a second generation of its Model A with an even a lower price. (Ford discontinued production for 9 months in order to switch from Model T to Model A). However, once again, the competitiveness-via-price strategy of Ford did not survive long. It was completely abandoned in the early 1930s (primarily owing to the introduction of the V-8 engine), finally leading to some variability in Ford's product line.

In 1923–1924, industrial design became a mainstream issue in the automobile industry. The focus was on internal design as well as external

FIGURE 2 The Ford Model T car.

styling and color choices. In contrast to Ford's strategy, GM, under the general management of A. P. Sloan (an MIT graduate), decided to develop a line of cars in multiple pricing categories, from the lowest to the highest. Sloan insisted on making GM cars different from the competition's, different from each other, and different from year to year, naturally at the expense of technological innovation. The objective was not a radical innovation but an offer of variety in frequent intervals, namely incremental changes in design as well as in production processes. Sloan rationalized product variety by introducing several platforms as well as frequent model changes within each platform. His approach to increased productivity was however very similar to Ford's in that each platform was manufactured in a different plant and yearly model changes were only minor owing to prohibitive costs in radically changing tooling and fixturing more than once every 4 to 6 years. The approach of manufacturing multiple platforms in the same plant in a mixed manufacturing environment was only introduced in the late 1970s by Toyota (Table 5). The question at hand is, naturally, How many platforms does a company need today to be competitive in the decades to come?

Chrysler followed GM's lead and offered four basic car lines in 1929; Chrysler, DeSoto, Dodge, and Plymouth. Unlike GM and Ford, however, Chrysler was less vertically integrated and thus more open to innovation introduced by its past suppliers. (This policy allowed Chrysler to gain market share through design flexibility in the pre-WWII era).

The automobile's widespread introduction in the 1920s as a non luxury consumer good benefited other industries, first through the spin-off of manufacturing technologies (e.g., sheet-metal rolling used in home appliances) and second through stimulation of purchases by credit. Annual production of washing machines doubled between 1919 and 1929, while annual refrigerator production rose from 5000 to 890,000 during the same period. Concurrently, the spillover effect of utilization of styling and color as a marketing tool became very apparent. The market was flooded with purple bathroom fixtures, red cookware, and enamelled furniture. One can draw parallels to the period of 1997–2000, when numerous companies, including Apple and Epson, adopted marketing strategies that led to the production of colorful personal computers, printers, disk drives, and so forth.

1.3 RECENT HISTORY OF COMPUTING TECHNOLOGIES

The first electronic computer was built by a team led by P. Eckert and J. Mauchley, University of Pennsylvania, from 1944 to 1947 under the auspices of the U.S. Defense Department. The result was the Electronic

TABLE 5 Platforms/Models for Some Automotive Manufacturers During the Period 1964–1993

	Ford		GM		Chrysler		Fiat		Renault		Volkswagen		Nissan		Toyota	
	1960/64	1990/93	60/64	90/93	60/64	90/93	60/64	90/93	60/64	90/93	60/64	90/94	60/64	90/93	60/64	90/93
Platforms	5.8	7.5	10.0	15.8	5.6	7.8	4.8	7.0	2.2	6.8	1.4	5.0	2.4	17.5	2.0	13.8
Models	7.2	12.5	20.8	31.8	10.0	13.3	8.4	13.5	3.8	8.0	2.6	9.0	2	22.3	2.0	24.3

Numerical Integrator and Computer (ENIAC); the subsequent commercial version, UNIVAC I, became available in 1950.

The first breakthrough toward the development of modern computers came, however, with the fabrication of semiconductor switching elements (transistors) in 1948. What followed was the rapid miniaturization of the transistors and their combination with capacitors, resistors, etc. in multi-layered silicon-based integrated circuits (ICs). Today, millions of such elements are configured within extremely small areas to produce processor, memory, and other types of ICs commonly found in our personal computers and other devices (such as calculators, portable phones, and personal organizers).

Until the late 1970s, a typical computer network included a centralized processing unit ("main-frame"), most probably an IBM make (such as IBM-360), which was accessed by users first by punched cards (1950–1965) and then by "dumb" terminals (1965–1980). The 1970s can be considered as the decade when the computing industry went through a revolution, first with the introduction of "smart" graphic terminals and then with the development of smaller main-frame computers, such as the DEC-PDP minicomputer. Finally came the personal (micro) computers that allowed distributed computing and sophisticated graphical user interfaces (GUIs).

In the late 1980s, the impact of revolutionary advances in computer development on manufacturing was twofold. First, with the introduction of computer-aided design (CAD) software (and "smart" graphic terminals), engineers could now easily develop the geometric models of products, which they wanted to analyze via existing engineering analysis software (such as ANSYS). One must, however, not forget that computers (hardware and software) were long being utilized for computer-aided engineering (CAE) before the introduction of CAD software. The second major impact of computing technology was naturally in automatic and intelligent control of production machines. But we must yet again remember that numerical control (NC) was conceived of long before the first computer, at the beginning of the 20th century, though the widespread implementation of automatic-control technology did not start before the 1950s. An MIT team is recognized with the development of the NC machine-tool concept in 1951 and its first commercial application in 1955.

The evolution of computer hardware and software has been mirrored by corresponding advances in manufacturing control strategies on factory floors. In late 1960s, the strategy of direct numerical control (DNC) resulted in large numbers of NC machines being brought under the control of a central main-frame computer. A major drawback with such a centralized control architecture was the total stoppage of manufacturing activities when the main-frame computer failed. As one would expect, even short periods of downtime on factory floors are not acceptable. Thus the DNC strategy was quickly abandoned until the introduction of computer numerical control (CNC) machines.

In the early 1970s, with the development of microprocessors and their widespread use in the automatic control of machine tools, the era of CNC started. These were stand-alone machines with (software-based) local processing computing units that could be networked to other computers. However, owing to negative experience that manufacturers had with earlier DNC strategies and the lack of enterprise-wide CIM-implementation strategies, companies refrained from networking the CNC machines until the 1990s. That decade witnessed the introduction of a new strategy, distributed computer numerical control (DCNC), in which CNC machines were networked and connected to a central computer. Unlike in a DNC environment, the role of a main-frame computer here is one of distributing tasks and collecting vital operational information, as opposed to direct control.

1.3.1 CAD Software and Hardware

Research and development activities during the 1960s to 1980s resulted in proprietary CAD software running on proprietary computer platforms. In 1963, a 2-D CAD software SKETCHPAD was developed at M.I.T. CADAM by Lockheed in 1969, CADD by Unigraphics, and FASTDRAW by McDonell-Douglas followed this initial development. The 1970s were dominated by two major players, Computer Vision and Intergraph. IBM significantly penetrated the CAD market during the late 1970s and early 1980s with its CATIA software, which was originally developed by Dessault Industries in France, which naturally ran on IBM's main-frame (4300) computer, providing a time-sharing environment to multiple concurrent users.

With the introduction of minicomputers (SUN, DEC, HP) in the late 1970s and early 1980s, the linkage of CAD software and proprietary hardware was finally broken, allowing software developers to market their products on multiple platforms. Today, the market leaders in CAD software (ProEngineer and I-DEAS) even sell scaled-down versions of their packages for engineering students (for $300 to 400) that run on personal computers.

1.4 MANUFACTURING MANAGEMENT STRATEGIES

It has been said many times, especially during the early 1980s, that a nation can prosper without a manufacturing base and survive solely on its service industry. Fortunately, this opinion was soundly rejected during the 1990s, and manufacturing once again enjoys the close attention of engineers, managers, and academics. It is now agreed that an enterprise must have a competitive manufacturing strategy, setting a clear vision for the company and a set of achievable objectives.

A manufacturing strategy must deal with a variety of issues from operational to tactical to strategic levels. These include decisions on the level of vertical integration, facilities and capacity, technology and workforce, and of course organizational structure.

The successful (multinational) manufacturing enterprise of today is normally divided into a number of business units for effective and streamlined decision making for the successful launch of products and their production management as they reach maturity and eventually the end of life. A business unit is expected continually and semi-independently to make decisions on marketing and sales, research and development, procurement, manufacturing and support, and financial matters. Naturally, a manufacturing strategy must be robust and evolve concurrently with the product.

As the history of manufacturing shows us, companies will have to make difficult decisions during their lives (which can be as short as a few years if managed unsuccessfully) in regard to remaining competitive via marketing efforts or innovative designs. As one would expect, innovation requires investment (time and capital): it is risky, and return on investment can span several years. Thus the majority of products introduced into the market are only marginally different from their competitors and rarely survive beyond an initial period.

No manufacturing enterprise can afford the ultraflexibility continually to introduce new and innovative products into the market place. Most, instead, only devote limited resources to risky endeavors. A successful manufacturing company must strike a balance between design innovation and process innovation. The enterprise must maintain a niche and a dominant product line, in which incremental improvements must be compatible with existing manufacturing capability, i.e., fit within the operational flexibility of the plant. It is expected that a portion of profits and cost reductions achieved via process innovations on mature product lines today will be invested in the R&D of the innovative product of tomorrow. One must remember that these innovative products of the future can achieve up to 50 to 70% market-share penetration within a short period from their introduction.

1.4.1 Manufacturing Flexibility

Manufacturing flexibility has been described as the ability of an enterprise to cope with environmental uncertainties: "upstream" uncertainties, such as production problems (e.g., machine failures and process-quality problems) and supplier-delivery problems, as well as "downstream" uncertainties due to customer-demand volatility and competitors' behavior. Rapid technological shifts, declining product life cycles, greater customization, and increased globalization have all put increased pressure on manufacturing companies significantly to increase their flexibility. Thus a competitive company must today have the ability to respond to customer and market demands in a timely and profitable manner. Sony is such a company, that has introduced hundreds of variations of its original Walkman in the past decade.

Manufacturing flexibility is a continuous medium spanning from operational to strategic flexibilities on each end of the spectrum: *operational flexibility* (equipment versatility in terms of reconfigurability and reprogrammability), *tactical flexibility* (mix, volume, and product-modification robustness), and *strategic flexibility* (new product introduction ability). One can rarely achieve strategic flexibility without having already achieved the previous two. However, as widely discussed in the literature, tactical flexibility can be facilitated through in-house (advanced-technology-based) flexible manufacturing systems or by outsourcing, namely, through the development of an effective supply chain.

It has been argued that as an alternative to a vertically integrated manufacturing company, strategic outsourcing can be utilized to reduce uncertainties and thus to build competitive advantage without capital investment. As has been the case for several decades in Germany and Japan, early supplier involvement in product engineering allows sharing of ideas and technology, for product as well as process improvements. Naturally, with the ever-increasing effectiveness of current communication technologies and transportation means, supply chains do not have to be local or domestic. Globalization in outsourcing is here to stay.

1.4.2 Vertical Integration Versus Outsourcing

Every company at some time faces the simple question of "make or buy." As discussed above, there exists a school of thought in which one maintains tactical or even strategic flexibility through outsourcing. But it is also common manufacturing wisdom that production adds value to a product, whereas assembly and distribution simply add cost. Thus outsourcing must be viewed in the light of establishing strategic alliances while companies join together with a common objective and admit that two hands sometimes can do better than one. Naturally, one can argue that such alliances are in fact a form of vertical integration.

The American auto industry, in its early stages, comprised companies that were totally vertically integrated. They started their production with the raw material (for most of the vehicle components) and concluded their organizational structure with controlling distribution and retail sales. Chrysler was one of the first American companies to break this organizational structure and adopt the utilization of (closely allied) supply chains. IBM was one of the latecomers in reducing its vertical integration and forming alliances with chip makers and software developers for its PC product line.

Managers argue in favor of vertical integration by pointing to potential lower costs through savings on overall product design and process optimization, better coordination and concurrency among the activities of different manufacturing functions (financial, marketing, logistics), and finally by maintaining directly their hand on the pulse of their customers. Another strong argument is the reduction of uncertainties via better control over the environment (product quality, lead times, pricing strategies, and of course intellectual property).

A common argument against vertical integration has been that once a company crosses an optimal size, it becomes difficult to manage, and it loses its innovative edge over its competitors. Many such companies quickly (and sometimes not so quickly) realize that expected cost reductions do not materialize and they may even increase. Vertical integration may also lead a company to have less control over its own departments. While it is easier to let an under-performing supplier go, the same simple strategy cannot be easily pursued in-house.

1.4.3 Taylor/Ford Versus Multitalented Labor

Prior to discussing the role of labor in manufacturing, it would be appropriate briefly to review production scales. Goods produced for the population at large are manufactured on a larger scale than the machines used to produce them. Cars, bicycles, personal computers, phones, and household appliances are manufactured on the largest scale possible. Normally, these are manufactured in dedicated plants where production flexibility refers to a family of minor variations. Machine tools, presses, aircraft engines, buses, and military vehicles on the other hand are manufactured in small batches and over long periods of time. Naturally, one cannot expect a uniform labor force suitable for both scales of manufacturing.

While operators in a job-shop environment are expected to be multitalented ("flexible"), the labor force in the mass production environment is a collection of specialists. The latter is a direct product of the labor profile advocated by F. Taylor (an engineer by training) at the turn of the 20th century and perfected on the assembly lines of Ford Motor Company.

In the pre-mass-production era of the late 1880s, manufacturing companies emphasized "piece rates" in order to increase productivity, while floor management was left to the foremen. However, labor was not cooperative in driving up productivity, fearing possible reductions in piece rates. In response to this gridlock, Taylor introduced the "scientific management" concept and claimed that both productivity and salaries (based on piece rates) could be significantly improved. The basis of the claim was optimization of work methods through a detailed study of the process as well as of the ergonomic capability of the workers. (Some trace the beginning of the discipline of industrial engineering to these studies.)

Taylor advocated the breaking down of processes into their smallest possible units to determine the optimal way (i.e., the minimum of time) of accomplishing the individual tasks. Naturally at first implementation depended on the workers' willingness to specialize on doing a repetitive task daily, which did not require much skill, in order to receive increased financial compensation. (Some claim that these well-paying blue-collar jobs significantly reduced motivation to gain knowledge and skills in the subsequent generations of labor.)

In order to reduce wasted time, Taylor required companies to shorten material-handling routes and accurately to time the deliveries of the subassemblies to their next destination, which led to in-depth studies of routing and scheduling, and furthermore of plant layouts. Despite significant productivity increases, however, Taylor's ideas could not be implemented in job shops, where the work involved the utilization of complex processes that required skilled machinists to make decisions about process planning. Lack of mathematical modeling of such processes, even today, is a major factor in this failure, restricting Taylor's scientific management ideas to simple assembly tasks that could be timed with a stopwatch.

Taylor's work, though developed during 1880 to 1900, was only implemented on a larger scale by H. Ford on his assembly lines during 1900 to 1920 (and much later in Europe). The result was synchronous production lines, where operators (treated like machines) performed specialized tasks during their shifts for months. They were often subjected to time analyses in order to save, sometimes, just a few seconds. On a larger scale, companies extrapolated this specialization to the level of factories, where plants were designed to produce a single car model, whose discontinued production often resulted in the economic collapse of small towns.

The standarization of products combined with specialized labor increased efficiency and labor productivity at the expense of flexibility. Ford Motor Company's response to growing demands for product variety was "They can have any color Model T car, so long as it is black." This attitude almost caused its collapse in the face of competition from GM under the management of A. Sloan, which started to market four different models by 1926. GM managed to remain competitive by maintaining standarization at the fundamental component and subassembly level, while permitting customers to have some choice in other areas.

Following the era of the Taylor/Ford paradigm of inflexibility, flexible manufacturing was developed as a strategy, among others, in response to increased demand for customization of products, significantly reduced lead-times, and a need for cost savings through in-process and post-process inventory reductions. The strategy has become a viable alternative for large-batch manufacturing because of (1) increases in in-process quality control (product and process), (2) technological advancements spearheaded via innovations in computing hardware and software, and (3) changes in production strategies (cellular manufacturing, just-in-time production, quick setup changes, etc.).

One can note a marked increased in customer inflexibility over the past two decades and their lack of willingness to compromise on quality and lead-time. Furthermore, today companies find it increasingly hard to maintain a steady base of loyal customers as global competitiveness provides customers with a large selection of goods. In response, manufacturing enterprises must now have the ability to cope with the production of a variety of designs within a family of products, to change or to increase existing product families and be innovative.

Due to almost revolutionary changes in computing and industrial-automation technologies, shop-floor workers must be continually educated and trained on the state of the art. The above described "factory of the future" requires labor skilled not only in specific manufacturing processes but as well in general computing and control technologies. Naturally, operators will be helped with monitoring and decision-making hardware and software integrated across the factory. A paramount task for labor in manufacturing will be maintenance of highly complex mechatronic systems. Thus these people will be continuously facing intellectual challenges, in contrast to the boredom that faced the specialists of the Taylor Ford factories.

1.4.4 MRP Versus JIT

A follow-up to Taylor's paradigm of minimizing waste due to poor scheduling was the development of the material requirements planning (MRP) technique in the 1960s. MRP is time-phased scheduling of a product's components based on the required delivery deadline of the product itself. An accurate bill of materials (BOM) is a necessity for the successful implementation of MRP. The objective is to minimize in-process inventory via precise scheduling carried out on computers.

Just-in-time (JIT) manufacturing, as pioneered in Japan by the Toyota Motor Company in early 1970s and known as the *kanban* or card system, requires operators to place orders to an earlier operation, normally by passing cards. As with MRP, the objective is inventory minimization by delaying production of components until the very last moment.

Although often contrasted, MRP and JIT strategies can be seen as complementary inventory management strategies. JIT emphasizes that production of any component should not be initiated until a firm order has been placed—a pull system. MRP complements this strategy by back-scheduling the start of the production of this part in order to avoid potential delays for lengthy production activities. MRP anticipates a pull command in advance of its occurrence and triggers the start of production for timely completion and meeting a future demand for the product in a timely manner.

U.S. manufacturers, prior to their encounter with JIT manufacturing, expected MRP magically to solve their complex scheduling problems in the early 1970s, they quickly abandoned it while failing to understand its potential. Although the modest gains of MRP were to be strengthened by the development of manufacturing resource planning (also known as MRP II) in the 1980s, with the introduction of JIT at the same time period, many manufacturing managers opted out from implementing MRP II in favor of JIT, only to recognize later that the two were not competitive but actually complementary techniques for inventory management. A key factor in this was the common but false belief that MRP requires large-batch production owing to the long periods of time needed to retool the machines.

Naturally, JIT was quickly noted to be not as a simple technique as it appeared to be but very challenging to implement. JIT had arrived to the U.S.A. from Japan, where the concept of single-minute die exchange (SMDE) allowed manufacturers to have small batches and product mix on the same line. SMDE, when combined with in-process quality control, was a winning strategy. It took almost a decade for the U.S. manufacturers to meet the triple-headed challenge of JIT, SMDE, and quality control.

Today one can easily see the natural place of JIT in manufacturing enterprises, where orders are received via the internet and passed on to the factory floor as they arrive. JIT eliminates large in-process (or even post production) inventories and allows companies to pass on the significant cost savings to the customers. However, with reduced in-process inventories, a plant is required to have eliminated all potential problems in production in regard to machine failures and product quality. For example, it is not unusual for an automotive parts manufacturing company to work with half-a-day inventory. Industrial customers expect multiple daily deliveries from their suppliers, with potentially severe penalties imposed on delivery delays.

1.5 INTERNATIONAL MANUFACTURING MANAGEMENT STRATEGIES

The 20th century witnessed the development of manufacturing strategies typical to certain continents, countries, and even some specific regions within federalist countries. Current multinational companies, however, must develop manufacturing strategies tailored to local markets as well as have an overall business strategy to compete globally. Prior to a brief review of several key economic engines in the world, it would be appropriate to define manufacturing strategy as a plan to design, produce, and market a well-engineered product with a long-range vision. Competitive priorities in this context can be identified as quality (highest ranked), service, cost, delivery, and product variety. Thus a comprehensive strategy would require design and manufacture of a superior product (backed by an excellent service team) produced at lower costs than the competitor's and delivered in a timely manner.

1.5.1 The U.S. Approach

The U.S.A. has always been the leader in product innovation but not very adept at converting basic R&D into viable commercial products. An exception is software design and marketing, where the U.S.A. maintains three quarters of the world's software market with an excellent information network.

The 1980s and early 1990s were typified in U.S.A. by significant downsizing, where companies tried to achieve lean manufacturing machines capable of producing products of superior quality (as good as Japanese). Reengineering became a key word for change in the way managers thought about their manufacturing processes, though the results were far from revolutionary. Often external consultants were brought in to propose management strategies that were not followed up after their departure.

The late 1990s, however, saw a dramatic shift in U.S. productivity, building on innovation in the philosophy of product design. This combined with the economic (mostly financial) problems that came about in Japan resulted in an unprecedented manufacturing boom in the U.S.A. Hewlett-Packard (HP) was a typical U.S. company capturing a large share of the world's color ink-jet printers and scanners. HP went from no printer manufacturing in 1984 to nearly $8 billion in sales in the mid-1990s. A primary factor in this success was HP's strategic flexibility.

It is important, however, to note that although the U.S.A. currently has a quarter of the world gross domestic product (GDP), the European Economic Community (EEC) is now the world's largest market, with the U.S.A. in second place. U.S. manufacturing companies are partially responsible for this drop, primarily because of their short-term vision and concentration on domestic markets. Despite the economic good times, most still continue to emphasize the objective of quarterly profits by maximizing the utilization of their current capacity (technological and workforce).

The following selective objectives are representative of the current (not-so-competitive) state of the U.S. manufacturing industry:

Customer responsiveness: Deliver what is ordered, in contrast to working with customers to provide solutions that fit their current product's life-cycle requirements and furthermore anticipate their future requirements.

Manufacturing process responsiveness: Dependence on hard tooling, fixed capacity and processes that lag product needs, in contrast to having a reconfigurable and scalable manufacturing plant that implements cost-effective processes that lead product needs and can react to rapidly changing customer requirements. One must not confuse automated machines with truly autonomous systems that have closed-loop processing capability for self-diagnosis and error recovery. Variable capacity must be seen as a strategic weapon to be used for competitiveness and not something to be simply solved by outsourcing or leasing equipment based on the latest received orders.

Human resource responsiveness: Encouragement of company loyalty in exchange for lifetime employment promise, in contrast to hiring of "knowledge individuals" who plan their own careers and expect to be supported in their continuing education efforts. The current U.S. workforce is in a high state of flux, where a company's equity is constantly evaluated by the knowledge and skills of its employees as opposed to only by the value of their capital. In the future, companies will be forced invest not only in capacity and technology but also in training that will increase the value of their employees, without a fear of possible greater turnover.

Global market responsiveness: Dependence on local companies run by locals but that are led by business strategies developed in the U.S.A., in contrast to operating globally (including distributed R&D efforts) and aiming to achieve high world market share. Globalization requires understanding of local markets and cultures for rapid responsiveness with no particular loyalty to any domestic politics.

1.5.2 Germany's Approach

Germany's industrial strength has been in the manufacturing of high-performance products of excellent quality. A common virtue to all German companies is to get things right the first time in well-ordered plants. The workforce is highly skilled, drawn from a population of young people who have passed through a traditional apprenticeship system. Their in-depth knowledge of manufacturing processes lets them more easily adapt to new technologies. At the upper echelon of management, one finds managers with Ph.D. degrees in engineering who are well-versed in economics. Engineering is a degree held in the highest esteem among all professional degrees.

Most German companies have long had reliable supply chains that they utilize for the joint design of well-engineered products. Long-term business objectives mandate strategic management decisions with lower intervention levels from stockholders. However, with rapid globalization of companies and their markets, the German approach to manufacturing management may have to evolve as well.

One must note that, as is the case with their Japanese counterparts, German companies tend to improve on their products and manufacturing processes, as opposed to emphasizing innovation as the U.S. companies attempt to do. Their long history of very high labor costs forced German manufacturers to invest heavily in plant renewals through advanced production machines and in the process achieve at least tactical flexibility levels in many of their companies.

1.5.3 Japan's Approach

Japanese engineering has long concentrated on incremental innovation and commercialization of economically viable inventions. Television, the VCR, and the CD player are a few products developed offshore (by RCA, Ampex, and Phillips, respectively) but successfully commercialized by Sony.

In the 1960s, the Made-in-Japan stamp on products was seen as a symbol of unsuccessful imitations of their American and European counterparts, attempting to penetrate foreign markets based solely on a price advantage. The following two decades caught the world by surprise when (once again) low-price but (this time) superior quality (strategically selected) Japanese household products flooded the world markets. First came televisions, then audio equipment, and finally cars. Although the Japanese companies easily penetrated the U.S. and British markets (and in some instances completely eliminated local competitors), the European continent mostly shut these products out by protectionist actions. In the U.S.A., the local and federal governments joined forces in the 1980s to help the American auto industry survive and not suffer the fate of the television industry for example.

In the 1980s, numerous Japanese automobile makers opened assembly plants in the U.S.A., the U.K., and Canada in order to deal with the increasing local criticism that imports took jobs away from local people. Though they were strictly assembly plants at the start, most of their valuable components being imported from Japan (for maintaining a high level of quality), these plants now have their own local supply chains as a true step toward globalization.

Like Germany, Japan must also heavily rely on exports of manufactured goods to owing the lack of local raw materials as and energy sources. Most such export companies have developed their competitive edge through intense local competitions in attempting to satisfy the domestic population's demand for high quality and timely delivery of goods. The just-in-time production strategy developed in Japan could not be implemented unless manufacturing processes were totally predictable. Another factor adding to the low uncertainty environment was the concept of *keiretsu* (family) based supply chains, which in most cases included large financial institutions. These institutions provided local manufacturing companies with large sums for investment, for capital improvements that did not come with any strings attached, thus, letting companies develop long-term strategies. With the globalization of the world's financial markets, it is now difficult for Japanese companies to secure such low-risk investments.

Like their German counterparts, most Japanese companies have developed operational and tactical flexibility which they rely on for stable, repetitive mass production of goods. However, unlike their European competitors, the Japanese companies have developed a fundamental advantage, significantly shorter product development cycles. This advantage is now being challenged on several fronts by European and American competitors in markets such as telecommunications, automotive, computing, and lately even household electronics.

Japanese companies are currently being forced to shift to innovative product development and marketing as see witness several phenomena occurring worldwide: (1) competition catching up with their productivity (including quality) and tactical flexibility levels, (2) financial globalization eroding their long-enjoyed unconditional investment support, and (3) penetration of information technology into all areas of manufacturing. It did not take long before for companies such as Sony rapidly shifted paradigms and stopped the economic slide.

Keiretsu

The Japanese term *keiretsu*, as used outside Japan, has normally referred to a horizontal group of companies that revolve around a large financial core (a bank plus a trading company—*shosha*). Most horizontal keiretsus also include a large manufacturing company in the center of the group. On the periphery, there is a large number of smaller companies (local banks, insurance companies, manufacturers, etc.) that add up to hundreds of firms associated with an individual keiretsu. Occasionally, there is also a large manufacturing company (for example, Toyota) on the periphery with a loose connection to the horizontal keiretsu, but having a vertical keiretsu itself.

Many vertical keiretsus (supply chains), such as Toyota, Toshiba, and Nissan, belong to one or another horizontal keiretsu. However, some may belong to several horizontal keiretsus (for example, Hitachi), while others maintain a (relative) independence (for example, Sony and Bridgestone). It has been estimated that several thousands of smaller companies form a pyramid to supply the flagship company that bears the name of a vertical keiretsu.

Most horizontal keiretsus have started as businesses owned by individual families at the turn of the 20th century (some even earlier). The four largest families were Mitsui (one of the largest conglomerates in the world), Mitsubishi, Sumitomo, and Yasuda. All these groups prospered throughout the century, but they lost their family control after WWII owing to political pressures and antimonopoly laws. The 1950s were a decade of intense efforts by the Japanese government for the formation of strong and competitive keiretsus. The result was the birth of many clusters, including the big six: Mitsui, Mitsubishi, Sumitomo, Fuyo, Sanwa, and Dai-Ichi Kangyo (DKB).

The Mitsui keiretsu (founded in 1961) has at its core the Sakura Bank, the Mitsui Bussan trading company, and the Mitsui Fudosan real-estate company. Toyota and Toshiba are peripheral vertical keiretsus aligned with the Mitsui Group. The Fuyo keiretsu (founded in 1966) has at its core the Fuji Bank supported by the Marubeni trading company and Canon. Other large manufacturing companies on the periphery of this group include Nissan Motors and Hitachi, the latter belonging to the Sanwa keiretsu as well.

The vertical keiretsus in Japan can be classified into either of manufacturing or trading/distribution. From the start, companies within a vertical keiretsu supplied exclusively those above them in the pyramid, thus developing and maintaining a total social loyalty to the parent company—unlike in the U.S.A., where subcontractors could provide competitors with similar or the same components. Since the 1980s, these keiretsu ties are slowly loosening, especially owing to the establishment of many satellite Japanese plants across the world that supply other local competitors.

The leading vertical keiretsus include the Toyota Group and the Sony Group. The parent flagship company of the former group is Toyota Motors (an automobile manufacturer), which totally dominates the local vehicle market in Japan (as high as 40 to 50%) and whose sales were near $70 billion in the mid 1990s. There are ten core companies at the top level of the Toyota group, and there are several thousands of companies at the lowest level, which generate sales also at a comparable to level that of Toyota Motors ($50 billion in the mid-1990s).

Although the Matsushita Group is the largest vertical keiretsu in the Japanese electronics industry (comprising companies such as, Panasonic, Technics, and JVC), the Sony Group has the most widely recognized name in the world. While Matsushita made almost $60 billion in yearly sales in the mid-1990s, Sony's sales were less than half of these and primarily targeted for export. Lately, the Sony group has made several acquisitions around the world (outside the audio-visual industry), primarily in the entertainment industry (music and movies).

1.5.4 Italy's Approach

Italy is one of the world's most industrialized (top seven) countries, and the northern part of the country enjoys a historical manufacturing base. Owing to cultural attitudes, there are only a very few large companies, most of which were government-owned or government-dominated for many decades. The many thousands of small companies have been owned by individuals and compete in niche markets. Unlike in Germany, most manufacturing managers in Italy have a sales or finance background, and there are few engineers. Being primarily an export-oriented country, domestic competition is underdeveloped. As a company that was forced to deal with this issue, Fiat had to adopt a completely new manufacturing strategy in the late 1980s to maintain its share of the European car market (12–15%). Computer integrated manufacturing (CIM) was adopted in Fiat at a huge financial burden and coupled with a merger of automation and a highly skilled labor force.

1.5.5 Sweden's Approach

The Scandinavian countries of Sweden, Finland, and Norway are culturally similar in putting an emphasis on their populations' welfare. The 1990s in Sweden were a period of increased productivity and of the revitalization of private companies having a strong interest in exports, which led to a reevaluation of the countrys' social infrastructure. Companies such as Volvo and Ericsson have decentralized structures and emphasize teamwork and the utilization of multiskilled operators frequently working in manufacturing workcells (in contrast to flow lines). Lead-time is an important issue in their supply chains that include a very large number of (non-Swedish) European companies.

1.5.6 The U.K. Approach

In the 1890s, the U.K. was the largest manufacturing economy, and its output dominated 25% of the world's market. A century later, in the 1990s, this number shrank to less than 5%, and the U.K. has been overtaken by Germany, France, Japan, and the U.S.A. The U.K. experienced a deindustrialization since the 1960s, and major manufacturing industries (including the automotive) were significantly weakened. As expected, dominance in the world's financial services did not contribute to the U.K.'s development to compensate for the major deindustrialization. Since the 1990s, there has been a reversal in government policies that emphasize once again the importance of manufacturing to the U.K.'s well being. However, most companies investing in manufacturing are foreign multinationals (German, Japanese, and American). It is expected that these companies will lead the U.K. out of deindustrialization and teach the local people the importance of global competitiveness.

1.6 INFORMATION-TECHNOLOGY-BASED MANUFACTURING

The transition from the agrarian society of the 1700s to the industrial society of the 1900s resulted in the industrialization of agriculture, and not its disappearance. Today, only 3% of Americans are engaged in agricultural activities in contrast to the 90% of the workforce in the 1700s. Similarly, in the past century, we did not witness the disappearance of manufacturing, but only its automation (Tables 6 to 8). By 1999, the manufacturing sectors in the U.S.A. constituted only 18% of overall employment, while the number for Japan was down to 21%. At the same time, the services industry grew to 72% in the U.S.A. and to 63.7% in Japan. As we progress through the first decades of the information age, it is expected that globalization will cause the total entanglement of the world's economies as never before.

1.6.1 The Internet and the World Wide Web

The start of the World Wide Web (WWW), or simply the web, can be traced to the work of T. Berners-Lee at the European Particle Physics Laboratory (CERN) in Switzerland around 1989. Although the internet was already around since the 1970s, the difficulty of transferring information between locations restricted its use primarily to academic institutions. It took more than two decades and tens of dedicated computer scientists in Europe and the U.S.A. to bring the web into the forefront. The first version of the hypertext application software only ran on one platform (NEXT, developed

TABLE 6 Employment Percentage by Sector

	U.S.A.			Japan			Germany			Canada		
	1930	1970	1999	1930	1970	1990	1933	1970	1990	1931	1970	1990
Agriculture	22.9	4.5	2.9	49.9	19.9	6.9	2.9	8.5	3.5	35.2	7.6	4.3
Manufacturing	24.5	26.4	18.0	16.1	27.4	23.4	31.6	39.5	31.6	16.4	22.3	15.7
Social services/govt.	9.2	22.0	25.5	5.5	10.3	14.3	6.8	15.7	24.3	8.9	22.0	22.6

TABLE 7 Employment Percentage by Sector (Excluding Agriculture)

	U.S.A.			Japan			Germany			Canada		
	1930	1970	1999	1930	1970	1990	1933	1970	1990	1931	1970	1990
Industry[a]	43.3	33.1	25.1	40.7	35.7	33.8	56.6	48.7	38.9	37.2	29.8	23.4
Services[b]	33.8	62.3	72.0	9.4	47.4	59.2	14.4	42.8	57.6	27.6	62.6	72.3

[a] Mining, construction, manufacturing.
[b] Financial, social, entertainment, communications, government.

TABLE 8 Employment by Profession (Percentage)

	U.S.A.		Japan		Germany		France		Italy		U.K.		Canada	
	1970	1990	1970	1990	1970	1987	1970	1989	1971	1990	1970	1990	1971	1992
Goods handling	61.1	52.6	73.2	65.9	71.6	60.8	66.8	54.9	76.1	62.2	67.6	54.2	58.6	54.3
Information handling	38.9	47.4	26.8	34.1	28.4	39.2	33.2	45.1	23.9	37.8	32.4	45.8	41.4	45.7

by S. Jobs, cofounder of Apple Computer) and was released to a limited number of users in 1991.

P. Wiu, a Berkeley university student, released a graphical browser in the U.S.A. in 1992 that was capable of displaying HTML graphics, doing animation, and downloading embedded applications off the internet. The two following browsers were Mosaic, developed in 1993, and Gopher, developed at the University of Minnesota at about the same time. However, when the University of Minnesota announced that they would consider a licensing fee for Gopher, it was disowned by the academic community and died quickly. The principle at stake was the threat to academic sharing of knowledge in the most open way.

In 1994, the general public was for the first time given access to the web through several internet service providers via modem connections. The year was also marked by the release of Netscape's first version of Navigator, originally named Mozilla, free of charge. Finally, late in the year, the WWW Consortium (W3C) was established to oversee all future developments and set standards. Microsoft's version of their browser, Internet Explorer, was released bundled with their Windows 95 version after a failed attempt to reach a deal with Netscape. By 1996, millions of people around the world were accessing the web, an activity that finally caught the attention of many manufacturing companies and started the transformation of the whole industry into information technology (IT)-based supply chains (spanning from customers at one end to component suppliers at the other).

1.6.2 IT-Based Manufacturing

As mentioned above, the transformation to an IT-based economy began in the 1970s with rapid advances in computing and the continued spirit of academics who believe in the free spread of knowledge. The 1990s were marked by the emergence of the web as a commercial vehicle. Today, highly competitive markets force manufacturing enterprises to network; they must place the customer at the center of their business while continuing to improve on their relationships with suppliers. This transformation will, however, only come easy to companies that spent the past two decades trying to achieve manufacturing flexibility via advanced technologies (for design, production, and overall integration of knowledge sharing) and implementation of quality-control measures.

IT-based manufacturing requires rapid response to meet personalized customer demands. A common trend for manufacturing enterprises is to establish reliable interconnected supply chains by pursuing connectivity and coordination. A critical factor to the success of these companies will be the managing of (almost instantaneous) shared information within the company through intranets and with the outside world through extranets. The task becomes increasingly more difficult with large product-variation offerings.

Information sharing is an important tool in reducing uncertainties in forecasting and in thus providing manufacturers with accurate production orders. In the next decade, we should move toward total collaboration between the companies within a supply chain, as opposed to current underutilization of the web through simple information exchange on demand via extranets. True collaboration requires the real-time sharing of operational information between two supply-chain partners, in which each has a window to the other's latest operational status. In a retail market supply environment this could involve individual suppliers having real-time knowledge of inventories as well as sales patterns and make autonomous decisions on when and what quantity to resupply. Similarly, in supplying assemblers, component manufacturers can access the formers' production plans and shop status to decide on their orders and timing.

Whether the web has been the missing link in the advancement of manufacturing beyond the utilization of the latest autonomous technologies will be answered in the upcoming decade by manufacturing strategy analysts. In the meantime, enterprises should strive to achieve high productivity and offer their employees intellectually challenging working environments via the utilization of what we know now as opposed to reluctantly waiting for the future to arrive.

REVIEW QUESTIONS

1. Discuss recent trends in the structuring of manufacturing companies and comment on their expected operational strategies in the future, including the issue of computer-integrated manufacturing (CIM).
2. What specific advancement has contributed to significant improvements in the efficiency of modern machine tools when compared to their very early versions?
3. Discuss problems experienced with the commercial use of robots in the manufacturing industry during the period 1960 to 1990.
4. Why have vehicle manufacturing practices been very closely studied and implemented in other industries?
5. Compare Ford's passenger car manufacturing and marketing strategies in the earlier part of the 20th century to those implemented at GM by A. P. Sloan. Elaborate on the continuing use of these competitive practices in today's manufacturing industries.

6. State several benefits of the common use of computers in the manufacturing industry. In your discussion compare the state of manufacturing before and after the development and widespread use of information technologies (IT).
7. Computer users long resisted paying for software products and expected hardware makers to provide these at no cost and bundled with the hardware. What led to changes in consumer sentiments in regard to this issue, now that users are willing to pay even for the operating system and not only for specific application software?
8. What is a manufacturing strategy? Why should companies attempt to strike a balance between design innovation and process innovation?
9. Discuss manufacturing flexibility. Address the issues of vertical versus horizontal (including outsourcing) integration of manufacturing enterprises.
10. Discuss the Taylor/Ford paradigm of inflexibility and list its potential advantages.
11. State one positive and one critical aspect of manufacturing management strategies typically adopted by companies in the following world regions: U.S.A./Canada, U.K., continental Europe (excluding the U.K.), and Japan.
12. Discuss the use of IT (hardware and software) in the next two decades, where customers can effectively and in a transparent manner access the computers of the suppliers for order placement, tracking, etc.

DISCUSSION QUESTIONS

1. Most engineering products are based on innovative design rather than on fundamental inventions. They are developed in response to a common customer demand, enabled by new materials and/or technologies. Review the development of a recently marketed product that fits the above description from its conception to its manufacturing and marketing: for example, portable CD players, portable wireless phones, microwave ovens.
2. Computers may be seen as machines that automatically process information, as do automated production machines process materials. Discuss a possible definition of manufacturing as "the processing of information and materials for the efficient (profitable) fabrication and assembly of products."
3. Explain the importance of investigating the following factors in the establishment of a manufacturing facility: availability of skilled labor, availability and closeness of raw materials and suppliers, closeness of

customers/market, and availability of logistical means for the effective distribution of products.

4. Manufacturing flexibility can be achieved at three levels: operational flexibility, tactical flexibility, and strategic flexibility. Discuss operational flexibility. Is automation a necessary or a desirable tool in achieving this level of flexibility?
5. When IBM's subsidiary Lexmark moved from producing manual typewriters to electrical typewriters and, eventually, to computer-input based printers, what level of manufacturing flexibility did they have to have and why? Discuss the three levels of flexibility prior to your answer to the above question: strategic, tactical, and operational.
6. Discuss strategies for retrofitting an existing manufacturing enterprise with automation tools for material as well as information processing. Among others, consider issues such as buying turn-key solutions versus developing in-house solutions and carrying out consultations in a bottom-up approach, starting on the factory floor, versus a top-to-bottom approach, starting on the executive board of the company and progressing downward to the factory floor, etc.
7. The period of 1980 to 2000 has witnessed the dismantling of the vertical integration of many large manufacturers and rapid movement toward supply-chain relationships. Discuss the impact of recent technological and management developments on this movement: short product lives, concurrent engineering carried out in the virtual domain (i.e., distributed design), minimization of in-process inventories, etc.
8. There have been numerous significant approaches proposed during the period of 1980 to 2000 to the reduction of lead times in the production of multiprocess, multicomponent products. However, companies still face tight lead times in an economic environment of short product lives. Discuss the following options and others when faced with a possibility of not being able to meet a customer-expected lead time: expanding/improving the manufacturing facility, subcontracting parts of the work, refusing the order.
9. Fast-food outlets have been often managed via familiar manufacturing strategies that have evolved over the past century, moving from a mass-production environment to mass customization. Discuss the manufacturing strategies of several popular food chains that fabricate/assemble hamburgers, deli/cheese sandwiches, and pizza-based products. In your discussion, compare the manufacturing of these food products to other engineering products, such a personal computers and wireless phones. Do universities also employ such manufacturing strategies in educating students?

10. Computers and other information-management technologies have been commonly accepted as facilitators for the integration of various manufacturing activities. Define/discuss integrated manufacturing in the modern manufacturing enterprise and address the role of computers in this respect. Furthermore, discuss the use of intranets and extranets as they pertain to the linking of suppliers, manufacturers, and customers.
11. The widespread use of the internet, owing to significant increases in the numbers of household computers around the world, has forced companies to provide customers with an on-line shopping capability, creating e-business. Discuss the benefits of e-business as it is expected to allow customers to place/modify orders and access up-to-date information on the status of these orders via the internet, and as they progress through the manufacturing process. Briefly expand your discussion to relationships between suppliers and manufacturers in the context of e-business.
12. The 20th century has witnessed an historical trend in the strong reduction of manual labor in the agricultural industry with the introduction of a variety of (mechanized) vehicles, irrigation systems, crop-treatment techniques, etc. Discuss the current trend of the continuing reduction in the (manual) labor force involved in materials-processing activities versus increases in information-processing activities in manufacturing enterprises. Identify similarities to what has happened in the agricultural industry (and even in the book-publishing, textile, and other industries in earlier centuries) to what may happen in the manufacturing industry in the 21st century.

BIBLIOGRAPHY

Abernathy, William J. (1978). *The Productivity Dilemma: Roadblock to Innovation in The Automobile Industry*. Baltimore: Johns Hopkins University Press.

Agility Forum, Leaders for Manufacturing, and Technologies Enabling Agile Manufacturing. Next-Generation Manufacturing, A Framework for Action. http://www.dp.doe.gov/dp_web/documents/ngm.pdf, 1997.

American Automobile Manufacturers Association (1993–2000). *Automobile Facts and Figures, 1920–75 and 1993–2000*. New York: AAMA.

American Machine Tool Distributors' Association. http://www.amtda.org/usmtc/history.htm. Rockville, MD, 2001.

American Machinist (1978). *Metalworking: Yesterday and Tomorrow: The 100th Anniversary Issue of American Machinist*. New York: American Machinist.

Ayres, Robert U. (1991–1992). *Computer Integrated Manufacturing. 4 Vol.* New York: Chapman and Hall.

Berners-Lee, Tim, and Fischetti, Mark (1999). *Weaving the Web: The Original Design and Ultimate Destiny of the World Wide Web by Its Inventor*. San Francisco: Harper.
Bushnell, P. Timothy (1994). *Transformation of the American Manufacturing Paradigm*. New York: Garland.
Castells, Manuel (1996). *The Rise of the Network Society. Vol. 1. The Information Age: Economy, Society and Culture*. Malden, MA: Blackwell.
Council for Asian Manpower Studies (ed. Konosuke Odaka) (1983). The Motor Vehicle Industry in Asia: A Study of Ancillary Firm Development. Singapore University Press, Singapore.
Foreman-Peck, James, Bowden, Sue, and McKinlay, Alan (1995). *The British Motor Industry*. Manchester, UK: Manchester University Press.
Fulkerson, W. (2000). Information-based manufacturing in the informational age. *International Journal of Flexible Manufacturing Systems* 12:131–143.
Hayes, Robert H., and Wheelwright, Steven C. (1984). *Restoring Our Competitive Edge: Competing Through Manufacturing*. New York: John Wiley.
Hill, Terry (1989). *Manufacturing Strategy: Text and Cases*. Homewood, IL: Irwin.
Instrument Society of America, ISA Ad Hoc Committee (1992). *The Computer Control Pioneers: A History of the Innovators and Their Work*. NC: Research Triangle Park.
Karlsson, Jan M. (1991). A Decade of Robotics. *Mekanförbundets Förlag*. Sweden: Stockholm.
Laux, James Michael (1992). *The European Automobile Industry*. Toronto, Canada: Maxwell Macmillan.
Lindberg, Per, Voss, A. Christopher, and Blackmon, Kathryn L. (1998). *International Manufacturing Strategies: Context, Content, and Change*. Boston: Kluwer.
Lung, Yannick (1999). *Coping With Variety: Flexible Productive Systems for Product Variety in the Auto Industry*. Brookfield, VT: Ashgate.
Meikle, Jeffrey L. (1979). *Twentieth Century Limited: Industrial Design in America, 1925–1939*. Philadelphia: Temple University Press.
Miyashita, Kenichi, and Russell, David (1996). *Keiretsu: Inside the Hidden Japanese Conglomerates*. New York: McGraw-Hill.
Narasimhan, R., and Das, A. (1999). An Empirical Investigation of the Contribution of Strategic Sourcing to Manufacturing Flexibilities and Performance. *Decision Sciences* 30(3):683–718.
Nishiguchi, Toshihiro (1994). *Strategic Industrial Sourcing: the Japanese Advantage*. New York: Oxford University Press.
Porter, Michael E. (1986). *Competition in Global Industries*. Boston: Harvard Business School Press.
Porter, Michael E. (1990). *The Competitive Advantage of Nations*. Free Press of New York.
Porter, Michael E., Takeuchi, Hirotaka, and Sakakibara, Mariko (2000). *Can Japan Compete?* Cambridge, MA: Perseus.
Rolt, Lionel T. C. (1965). *A Short History of Machine Tools*. Cambridge, MA: M.I.T. Press.

Steeds, William (1969). *A History of Machine tools, 1700–1910*. Oxford: Clarendon Press.
Shaw, J. M. (2000). Information-based manufacturing with the web. *International Journal of Flexible Manufacturing Systems* 12:115–129.
Shingo, Shigeo (1985). *A Revolution in Manufacturing: The SMED System*. Stamford, CT: Productivity Press.
Shingo, Shigeo (1986). *Zero Quality Control: Source Inspection and the Poka-Yoke System*. Stamford, CT: Productivity Press.
Shingo, Shigeo (1989). *A Study of the Toyota Production System from an Industrial Engineering Viewpoint*. Stamford, CT: Productivity Press.
Underwood, Lynn (1994). *Intelligent Manufacturing*. Reading, MA: Addison-Wesley.
Viswanadham, N. (2000). *Analysis of Manufacturing Enterprises: An Approach to Leveraging Value Delivery Processes for Competitive Advantage*. Boston: Kluwer.
Woodbury, Robert S. (1972). *Studies in the History of Machine Tools*. Cambridge, MA: M.I.T. Press.

Part I

Engineering Design

The Accreditation Board for Engineering and Technology (ABET)* defines engineering design as "the process of devising a system, component or process to meet desired needs." ABET emphasizes that design is an iterative decision-making process, in which natural sciences, mathematics, and applied sciences (engineering) are applied to meet a stated objective in an optimal manner. The schematic of this process is as follows:

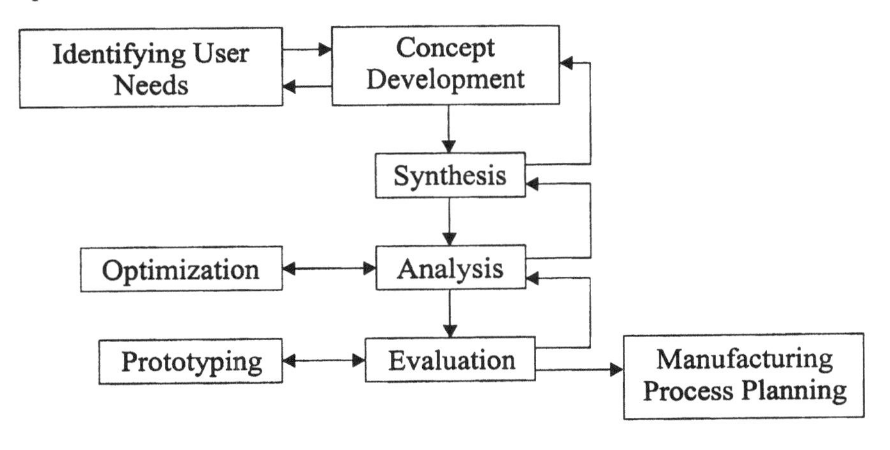

* Accreditation Board for Engineering and Technology, Inc. http://www.abet.org. Baltimore, MD, 2001.

One can rightfully argue that design is not such a neat sequential process as is shown in the figure. Today, product development teams have multidisciplinary members, who concurrently work on several aspects of design without being totally restricted by any sequential approach. Thus our focus in this Part I will be on concurrent design and engineering analysis.

In Chap. 2, conceptual design is discussed as the first step in the engineering design process. Customer-needs evaluation, concept development (including industrial design), and identification of a viable product architecture are the three primary phases of this stage of design. Several engineering design methodologies are discussed in Chap. 3 as common techniques utilized in the synthesis stage of the design process. They include the axiomatic design methodology developed by N. Suh (M.I.T.), the Taguchi method for parameter design, as well as the group-technology (GT)-based approach, originally developed in Europe in the first half of the 20th century, for efficient engineering data management.

In Chapter 4, computer-aided solid modeling techniques such as constructive solid geometry and boundary representation methods, are presented as necessary tools for downstream engineering analysis applications. Feature-based computer-aided design is also discussed in this chapter. In Chap. 5, the focus is on the computer-aided engineering (CAE) analysis and prototyping of products in "virtual space." Finite-element analysis is highlighted in this context. Parameter optimization is also discussed for choosing the "best" design.

2

Conceptual Design

Engineering design starts with a need directly communicated by the customer or with an innovative idea developed by a research team that would lead to an incremental improvement on the state of the art, or to a totally new product. One can, naturally, claim that there have been only a very few inventions in the 20th century and that most products have been incrementally innovated. The Walkman by Sony certainly falls into this second category, while the telephone can be classified as one of the true inventions. In this chapter, the emphasis is on the first stage of the engineering design process, namely development of viable concepts.

2.1 CONCURRENT ENGINEERING

The need for accelerated product launch in the face of significantly shortened product life cycles, especially in the communications and computing industries, has forced today's manufacturing companies to assemble multidisciplinary product design teams and ask them for concurrent input into the design process. In 1987, a U.S. Defense Advanced Research Projects Agency (DARPA) working group proposed the following definition: "Concurrent Engineering (CE) is a systematic approach to the integrated (concurrent) design of products and their manufacturing and support process." The product development team must consider all elements of the product life-cycle from the outset, including safety, quality, cost, and disposal (Fig. 1). Boeing was one of the first large manufacturing companies to utilize CE in their development of the Boeing 777, widely utilizing computer-aided design and engineering (CAD/CAE) tools for this purpose.

It has been advocated that CE could benefit from moving away from a function-based manufacturing structure toward a team-based approach. Lately, however, companies have been adopting a hybrid approach: they maintain product-based business units as well as function units that comprise highly skilled people who work (and help) across product business units. In this context, CE-based companies (1) use CAD/CAE tools for analysis of design concepts and their effective communication to others, (2) employ people with specialties but who can work in team environments, (3) allow teams to have wide memberships but also a high degree of autonomy, (4) encourage their teams to follow structured and disciplined (but parallel) design processes, and finally (5) review the progress of designs via milestones, deliverables, and cost.

The following partial list of concurrent design guidelines will be addressed in more detail in Chap. 3, where different product design methodologies are presented.

The conceptual design phase should receive input from individuals with diverse (but complementary) backgrounds.

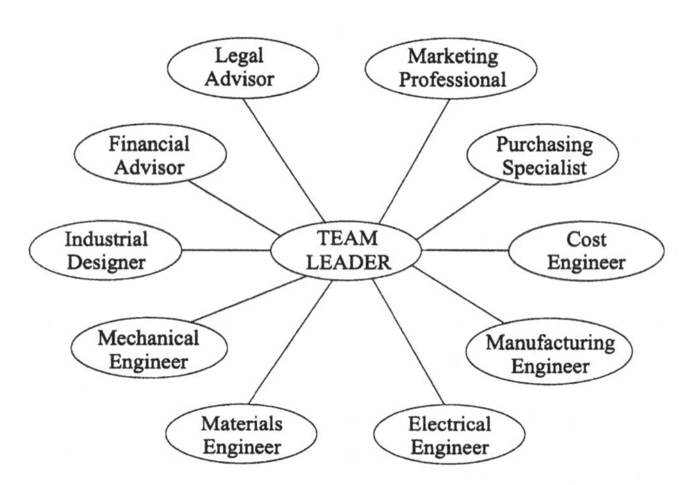

FIGURE 1 Example of structure of product development team.

Irreversible decisions should be delayed as much as possible, if they cannot altogether be avoided.
Designs should allow "continuous improvement" based on potential future feedback.
Product features should be analyzed with respect to manufacturability, assembly, and human factors.
Product modularity, standardization, and interchangeability should be maximized where profitable.
Product parameters should be designed in anticipation of imperfect use—"design for robustness."
Production processes should be finalized concurrently with product design selection.
Production plans and capacities should be in sync with marketing efforts and aim for short lead times (for delivery).

2.2 CONCEPT DEVELOPMENT PROCESS

Conceptual design encompasses many activities carried out by people with a variety of backgrounds with the ultimate objective of a profitable product launch. Industrial designers and human-factor engineers are normally involved at this stage of (conceptual) design and preliminary prototyping in order to provide timely input to the product design team. The first step in this process is customer-need identification and the second is concept generation and selection.

The process of customer-need identification must be carried without attempting to develop product specifications. The latter can only be decided upon once a concept is chosen and preliminarily tested to be technologically feasible and economically viable. Gathering useful data from the customer may include interviews with a select (representative) group in order to identify all their requirements, preferably in a ranked order. Naturally, need identification is an iterative process that involves returning to the "focus group" with more questions following the analysis of earlier collected data.

Concept generation follows the step of customer-need identification and development of some functional (target) specifications based on the experience and know-how of the product design team members. As will be discussed in Chap. 3, it is expected that the team will follow one (or a mixture) of the design methodologies developed in the past three decades in order to decompose the problem into its manageable parts and provide decoupled solutions. Assuming that the product design problem at hand is an incremental-innovation type, the team members are expected to search through existing similar products, technologies, and tools for "clues." At this stage, it is natural to develop (in an unrestricted way) as many concepts as possible and not dismiss any ideas—"brainstorming." This stage can be concluded, however, with a methodical review of all data/ideas/proposals in order to narrow the field of options to a few conceptual design alternatives. Figure 2 shows two alternative scooter designs patented in the U.S.A., Patents US D438,911 S and US D433,718.

The final "winning" concept selection process is a critical stage in product design and does not necessarily imply the rejection of all in favor of one. This stage seeks a wider input from manufacturing engineers and (future) product-support group members in order to rank all proposals (or even subsystems within each proposal). Preliminary prototyping (physical or virtual) may be necessary in order to consult with potential customers and evaluate usability (or even quality) of the selected product design concept.

As discussed in Chap. 1, the multinational manufacturing company of the future will need to develop and design products for global markets. Several key issues will have to be addressed in this respect: industrial designs for different domestic markets and cultures, ergonomic designs for different populations and segments of these populations, and utilization of modular design concepts for remaining competitive in several domestic markets.

2.3 INDUSTRIAL DESIGN

The Industrial Designers Society of America (IDSA) defines industrial design as "the professional service of creating and developing concepts and specifications that optimize the function, value and appearance of products and systems for the mutual benefit of both user and manufacturer." The following objectives have been commonly accepted by the industrial design community:

Appearance: The form, styling and colors of the product must convey a pleasing feeling to the user.

Human factors: The ergonomic and human-interface design of the product should facilitate its utilization in a safe manner.

Maintenance: Design features should not hinder maintenance and repair.

Other important factors include minimization of manufacturing costs through the utilization of appropriate materials and easy-to-produce form features. Most companies would also prefer to convey a corporate identity that is easily recognizable by the customer, through the product's design.

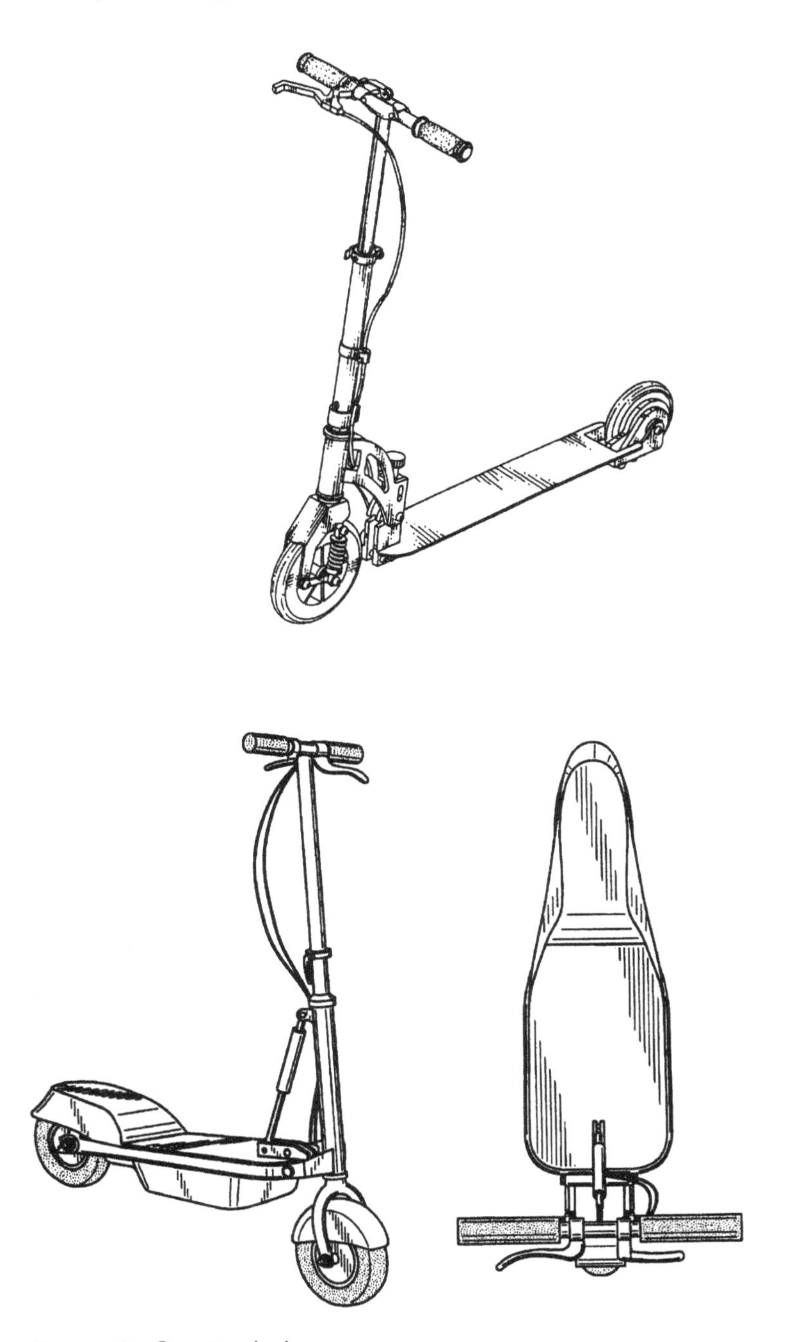

FIGURE 2 Scooter designs.

2.3.1 Industrial Design History

The beginnings of industrial design can be traced to the start of the mass production of household items (especially automobiles) in the early 1900s. While most European designers of the time were drawn from the ranks of engineers, their U.S. counterparts were primarily individuals with arts backgrounds, including marketing people. The latter group advocated utilization of nonfunctional features on the exterior of the product for maximum appeal with little emphasis on the interior of the product. Thus, while the European products were simple, precise, and economical, the American products were colorful and fancy looking (aerodynamically designed, even when the aerodynamic features were totally nonfunctional, for example, on furniture and refrigerators) (Figure 3).

In the U.S.A., the use of industrial design in the automotive industry started in around 1924 or 1925 and was due to the personal efforts of GM's manager A. P. Sloan. Sloan insisted on styling and color variations in GM cars, which were competing at the time with the single model of Ford. For a period of time, color was the answer to the demand of beauty by the U.S. public. Around 1926, the market was flooded with colorful products (automotive and other household and office products), including the successful Corona typewriters.

In the late 1920s, large manufacturers started to hire designers and create appropriate departments within their corporations, though in parallel many designers formed consulting firms and maintained their independence. The latter group, however, spent most of their efforts on package design. The early 1930s witnessed the birth of streamlining, which employed sweeping horizontal lines, rounded corners and projected frictionless motion, for the design of many different products (chairs, refrigerators, cars, etc.). The number of industrial designers in the U.S.A. rose from 5,500 in 1931 to 9,500 in 1936 as industrial design became an (accepted) standard practice among manufacturers.

Naturally, industrial design trends were quickly brought into line with engineering, as most designers became internal members of larger design teams in manufacturing enterprises. Consequently, today, industrial designers actively participate in the conceptual design stage of the product as opposed to simply being consulted for marketing purposes once the product has reached the premanufacturing stage.

2.3.2 Industrial Design Process

The intensity of industrial design in the development of a product is a direct function of its future utilization, namely, the characteristics of the customer.

Aerodynamic truck

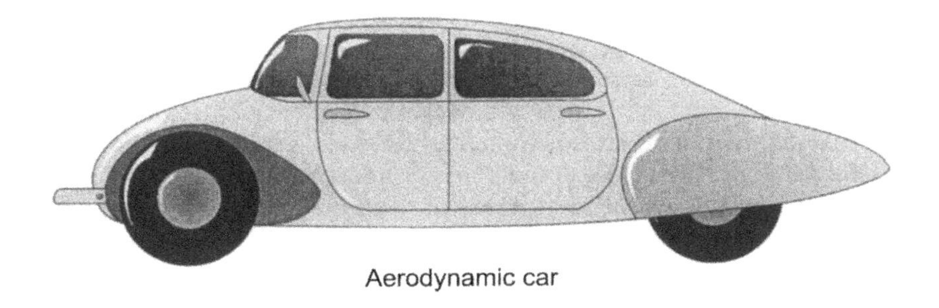

Aerodynamic car

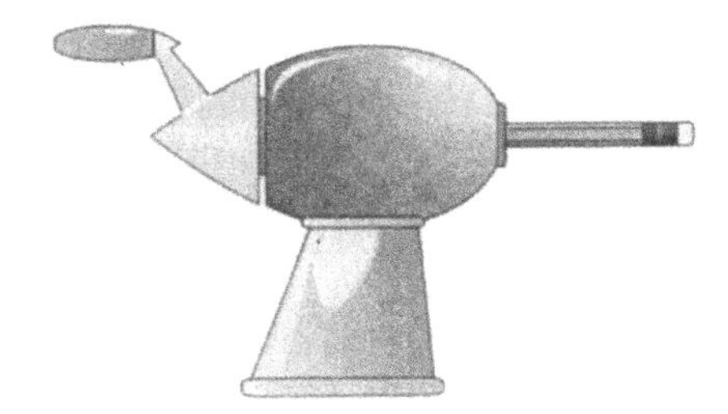

Aerodynamic pencil sharpener

FIGURE 3 Aerodynamic design.

An ink-jet printer, an office photocopier, and a lathe all require different emphases on the importance of certain design features. An office photocopier must allow users to understand its operational features with minimum mental effort but must also be designed for ease of maintenance by the service people. Users of a lathe, however, are expected to be well qualified to use the machinery, whose paramount concern is safety and of course ease of daily maintenance tasks. Household appliances would require to have aesthetic appeal and convey some brand-name identity.

As a vital member of the overall product design team, an industrial designer's first task is the evaluation of customer needs during the concept development phase. Industrial designers are expected to have the skills necessary to interview customers and research the market for identifying the needs clearly and communicate them to the engineers. At the concept generation phase, they concentrate on the form and human interfaces of the product, while engineers are mostly pre-occupied with addressing the functional requirements. Having an artistic background, industrial designers can also take a hands-on approach in generating alternative prototype models for conveying form and aesthetic requirements.

Once the field of design alternatives has been narrowed down, industrial designers return to their interactions with customers for collecting vital information on the customers' views and preferences regarding the individual concepts. At the final stages of the industrial design process, the role of the design engineers can vary from actually selecting the winning design and dictating the terms of manufacturing (mostly for consumer products, such as phones, wrist watches, and furniture) to simply participating in the marketing effort (mostly for products used by manufacturers, such as lathes, presses, and robots).

2.4 HUMAN FACTORS IN DESIGN

Interactions between people and products can be classified into three categories: occupying common space, acting as a source of input power, and acting as a supervisory controller. Human factors must be considered for every possible interaction, whether it being simply the operation of the product or its manufacture. Designers must analyze their products for evaluation of hazards, preferably for their subsequent elimination, and when impossible, for their avoidance. The following hazards could be noted in most mechanical systems: kinematic (moving parts), electrical, energy (potential, kinematic, and thermal), ergonomic/human factors (human–machine interface) and environmental (noise, chemicals, and radiation).

Safety of the person and quality of the product are the two paramount concerns. As noted above, if a hazard cannot be eliminated through design, the human users of the product should be provided with sufficient defense for hazard avoidance and with clear feedback, via signs, instruction, or warning sensors, to indicate the potential for a future hazard.

Of the three mentioned above, for interactions of the first type (i.e., occupying the same place), designers must carefully analyze available statistical data on the human metrics (anthropometric data) in order to determine the optimal product dimensions and to decide where to introduce reconfigurability (for example, different car-seat positions). The multinational company that aims to compete in different domestic markets must allow for this parametric variability in their design.

People often interact with their environment through touch: they have to apply force in opening a car door, twisting a bottle cap, or carrying boxes or parts on the shop floor. As with available human metrics for height, weight, reach, etc., there also exist empirical data on the capability of people in applying forces while they have different postures. Safety is also a major concern here. Products and processes must be designed ergonomically in order to prevent unnecessary injuries to the human body, especially in the case of operators who carry out repetitive tasks.

Supervisory control of machines and systems is the most common human/machine interaction. In such environments, people monitor ongoing machine activities through their senses and exercise supervisory control (when necessary) based on their decisions. (It has been estimated that 80% of human interactions with the environment is visual. Hearing is the next most important sense for information gathering.) The issues of sensing and control, thus, should be first individually examined. (With significant advances in artificial sensing and computing technologies, today, many supervisory control activities are carried out by computer controlled mechanical systems when economically viable, or when people cannot perform these tasks effectively, for example, automatic landing of aircraft.)

The following are only some representative issues that a designer must consider in designing human/machine interfaces (Fig. 4).

Clear and unambiguous display of sensory data: Displays should be clear, visible, and large. Analog displays are easier for quick analysis of a phenomenon, whereas digital displays provide precise information. Additionally, we must note that (1) the number of colors easily distinguishable by the human eye is less than ten, (2) the visual field extends 130° vertically and about 200° horizontally, (3) it takes about half a second to change focus, (4) a moving object's velocity and acceleration greatly reduce its accurate positioning, (5) the hearing range is between 20 and 20,000 Hz, and (6) noise above 120 dB (for example, generated by a jet

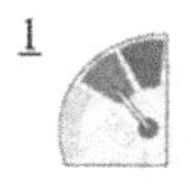

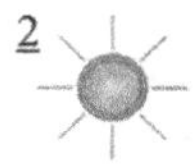

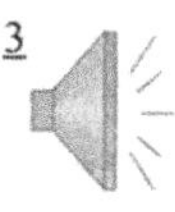

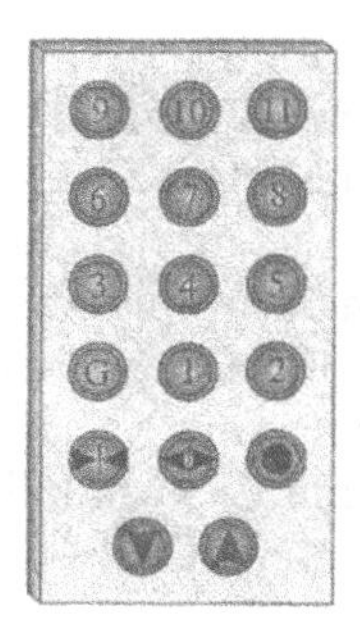

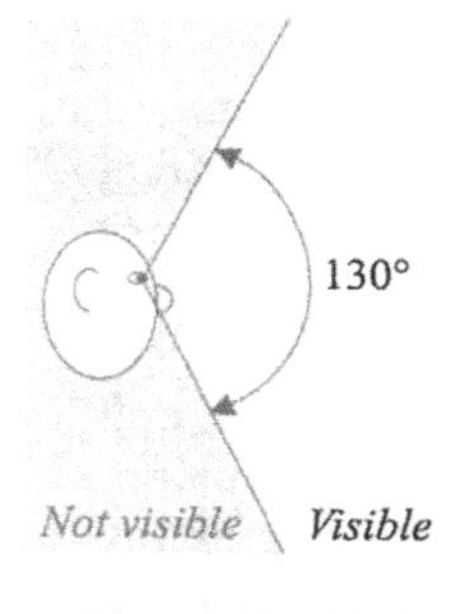

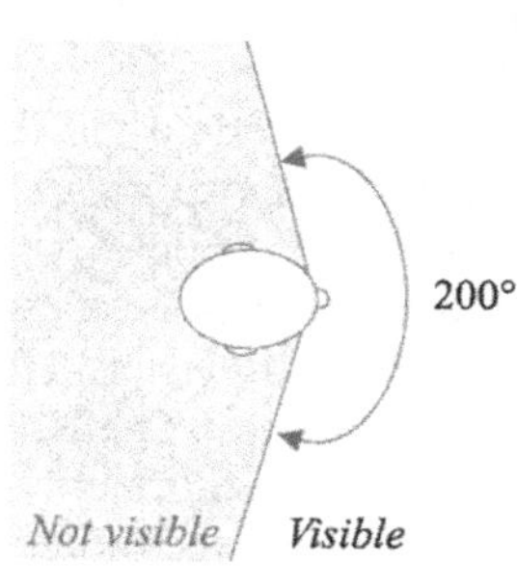

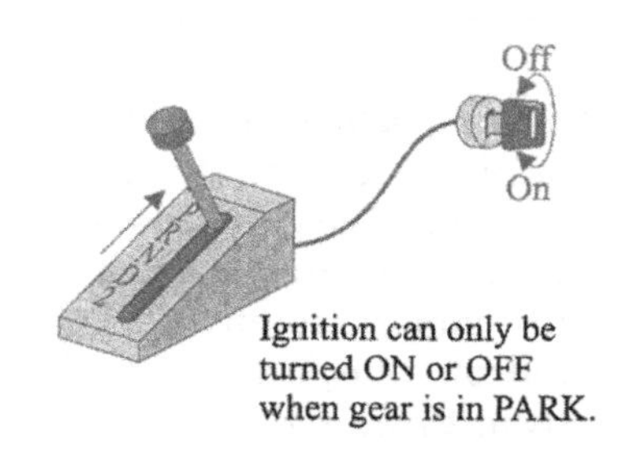

FIGURE 4 Human factors issues.

airplane) would cause discomfort if listened to by more than a few minutes.

Simplification and constraining of tasks: Control operations must involve a minimum number of actions. Where the possibility of incorrect actions exists, they should be prevented by clever design.

Suitable placement of input devices: Control devices, such as levers and buttons, should be placed for intuitive access and be easy to notice and to differentiate.

Providing feedback of control actions: The operator should be provided with a clear feedback (light or sound) in response to a control action undertaken, especially in anticipation of unwanted actions (where these cannot be physically prevented).

In order to cope with emergencies, we must also be aware of the information-processing limitations of human operators. As expected, people's information processing efficiency is significantly degraded when performing repetitive, boring tasks. General tiredness and personal stress further degrades this efficiency. The human operator is restricted in recalling from memory a task to be done within the next very short period of time. This phenomenon is further complicated if the necessary operation requires multiple subtasks. Thus, at emergencies, people react according to expected stereotypical actions. For example, they expect an increased reading with a clockwise dial display, push a switch upward for "on," and press down a brake pedal for stopping.

2.5 CONCEPTUAL DESIGN

Design projects can be classified broadly as (1) varying a product by modifying one or more of its parameters, but maintaining its overall functionality and performance, (2) redesigning a product by improving upon its performance, a large number of its characteristics, and/or its quality, (3) development of a new product, whose development process (design and materials) is affected by the expected production level (batch versus large volume), and (4) made-to-order design of a product. No matter what category the project falls into, however, the first step in the conceptual design process is problem formulation, followed by concept generation and concept evaluation phases.

2.5.1 Problem Formulation

Problem formulation must not be treated as an intuitive and trivial step of the engineering design process. One can never overemphasize this stage of identifying a customer's needs, which should be carried out with great care. Since design must yield an optimal solution to the problem at hand, the objective and constraints of the problem must be defined (preferably in a tangible manner). Let us, for example, consider the development of Sony's (portable) Walkman, where the overall goal could have been stated as "Provide individuals with a device capable of replaying tape/CD-recorded music, while they are mobile, with a carry-on power source and in a private listening mode." From an engineering perspective, there would exist several objectives (that could be interrelated) to satisfy this overall goal. The unit must have its own (preferably integral) power source, provide earphone connection, and of course be affordable. Typical constraints for this example product would be size, weight, and durability. (Naturally, some objectives can be formulated as equality, i.e., "must-have," constraints).

The technical literature provides us with numerous empirical and heuristic techniques for analyzing the problem at hand (as defined by the customer) and relate the customer requirements to engineering design parameters. Quality function development (QFD) is such a technique, first developed in Japan, that utilizes a chart representation of these relationships (Fig. 5). The primary elements of the QFD chart are

Customer requirements: A list of the characteristics of the design as explained by the customer.
Engineering requirements: A list that is generated by the engineers in response to the customer requirements. The list should be as comprehensive as possible.
Benchmarking: A comparison process to competitors' similar products.
Engineering targets: A set of target values for engineering requirements.

Customer needs are normally expressed qualitatively or in fuzzy terms, whereas engineering characteristics are usually quantitative. Engineers are required to determine the functional requirements of the product that influence the needs expressed by the customer. These requirements can be qualitative (an acceptable form at the conceptual design phase) or expressed

		Engineering requirements							
		Self-powered	Size	Weight	Shape	Ease of operation	Ruggedness	Cost	Benchmark (Competitor A)
Customer requirements	Portable	X	X	X	X				W
	Reliable					X	X		M
	Appealing		X		X	X			W
	Inexpensive	X					X	X	S
		6 V		100 g				$30	
		Engineering targets							

FIGURE 5 An exemplary QFD chart for Sony's Walkman.

as ranges (with possible extreme limits), for example, "the cost should be between $35 and $45."

Functional requirements can express goals and constraints in the following categories: performance, (geometrical) form, and aesthetics, environmental and life cycle, and manufacturability. Performance requirements would include goals on output (rate, accuracy, reliability, etc.), product life, maintenance, and safety. Form requirements refer to physical space and weight and industrial-design issues. Manufacturability requirements refer to the determination of fabrication and assembly methods that need to be employed for a profitable product line.

In the QFD chart shown in Fig. 5, the "X" mark indicates the existence of a relationship between the corresponding customer and engineering requirements. In the benchmarking column, "S" refers to a strong competitive position, whereas "M" and "W" refer to moderate and weak competitive positions, respectively.

2.5.2 Concept Generation

The conceptualization stage of design can benefit from uninhibited creative thinking combined with wide knowledge of engineering principles and of the state of the art in the specific product market. Creativity is not a (scientifically) well understood process, though it has been researched by numerous psychologists. The Creative Education Foundation model proposed in 1976 has five stages that form a sequential process: (1) fact finding, (2) problem formulation, (3) idea finding (narrowing of ideas toward feasible solutions), (4) evaluation, and (5) acceptance finding (premanufacturing stage of design).

2.5.3 Concept Evaluation

One can appreciate the difficulty a design team faces in decision making, during the concept evaluation phase, without having the engineering design specifications to compare the alternative concepts. Quantifying designs based mostly on intangible criteria is the task at hand.

Pugh's method of concept selection, for example, evaluates each concept relative to a "reference concept" and rates it (according to some criteria) as being better (+), about the same (S), or poor (−) (Table 1). The evaluation process starts by choosing the criteria based on the engineering requirements (as listed in the QFD chart), or, if these are underdeveloped, based on the customer requirements. The criteria can be ranked without attempting to assign specific weights. The next step would be choosing a reference concept (preferably the "best" perceived concept). The evaluation stage of the process, then, requires comparison of each

TABLE 1 An Exemplary Pugh Concept Comparison Table

	Ref. concept	Concept 1	Concept 2	Concept 3	Concept 4
Criterion 1	D	+	−	−	S
Criterion 2	A	S	+	−	−
Criterion 3	T	+	−	S	−
Criterion 4	U	−	S	+	−
	M				
Σ (+)		2	1	1	0
Σ (−)		1	2	2	3
Σ (S)		1	1	1	1

concept to the reference according to the criteria chosen by the product team and the assignment of the corresponding score (+, S, or −). Based on the assigned scores, one ranks all the concepts and redefines the reference concept as the best among the ranked. (For example, in Table 1, at the stage of comparison shown Concept 1 could be chosen as the next reference concept.) The procedure would then be repeated with the new reference concept as our new comparison concept and stopped, eventually, if the repeated evaluations yield the same reference concept. At that time, the design team may simply decide to proceed with one or with the top n concepts to the next product design stage.

2.6 MODULAR PRODUCT DESIGN

As defined by Ulrich and Eppinger, "product architecture is the assignment of the individual functional elements (duties/requirements) of a product to the physical building blocks (clusters) of the product." The functional elements of a product refer to the specific subtasks a product would perform (for example, feeding paper in a printer), whereas the physical building blocks are the clusters of components that allow implementation of these functions (for example, paper feed being achieved via a collection of rollers and a motor subassembly in a printer). In a modular product design, clusters implement functions in their *entirety* and independently, whereas in an integral design, a function may be implemented using more than one (physical) cluster.

Standardization has long been a cost-saving measure, normally implemented at the component level for integral designs. Modular product design elevates standardization to the level of functional elements, where they can be used in different product models to carry out the same functions, allow easy replacement, and provide expansion (add-on) capability. One can conclude that product (design) modularity is a necessary step in achieving tactical flexibility in a manufacturing environment and providing customers with economically viable variety (Sec. 1.4).

2.6.1 Modularity Levels

There exist six levels of modularity (Fig. 6):

Component sharing: This is the lowest level of standardization: the same components (e.g., motors and clutches) are used across many products (which may be modular or integral in design).

Component swapping: This is a component sharing modularity approach built around a single core product. Great numbers of variations can be presented to the customers (almost approaching a one-of-a-kind product line). The Swatch family of watches is a typical example.

Cut-to-fit: This is a parametric design variability achieved by customizing a small number of geometric features on the product. In the 1990s, Matsushita in Japan provided customers with personalized bicycles with a two-week delivery schedule once the order was received from the "fitting" store.

Mixing: The product is simply a mixture of components, in which the components lose their identity within the final product. An exemplary application area could be the mixture of chemicals according to a recipe.

Bus configuration: Similar to mixing, a mixture of components is assembled on a "mother" bus/board/platform. Typical examples include computers and automobiles. Naturally, modularity can only be achieved through a flexible design of the bus.

Sectional: This is the ultimate level of modularity, where the product's architecture is reconfigurable itself (as opposed to being fixed). Individual modules are configured to yield different products. The most common example is the reconfiguration of software modules to yield different application programs.

2.6.2 Modular Design Process

A three-step procedure has been commonly proposed for developing a modular product architecture:

1. Create a schematic representation of the product, which normally would comprise a set of functional objectives as opposed to physical building blocks (or their components).

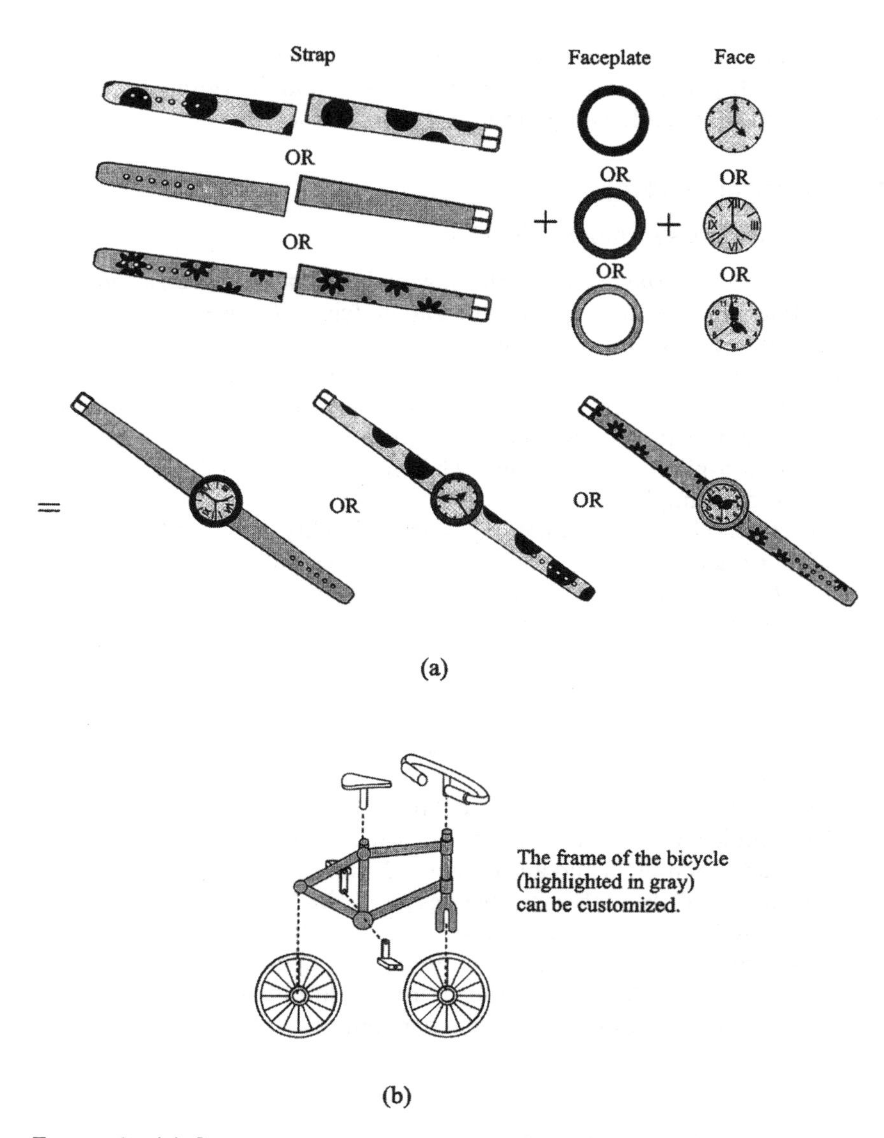

FIGURE 6 (a) Component swapping; (b) cut-to-fit modularity.

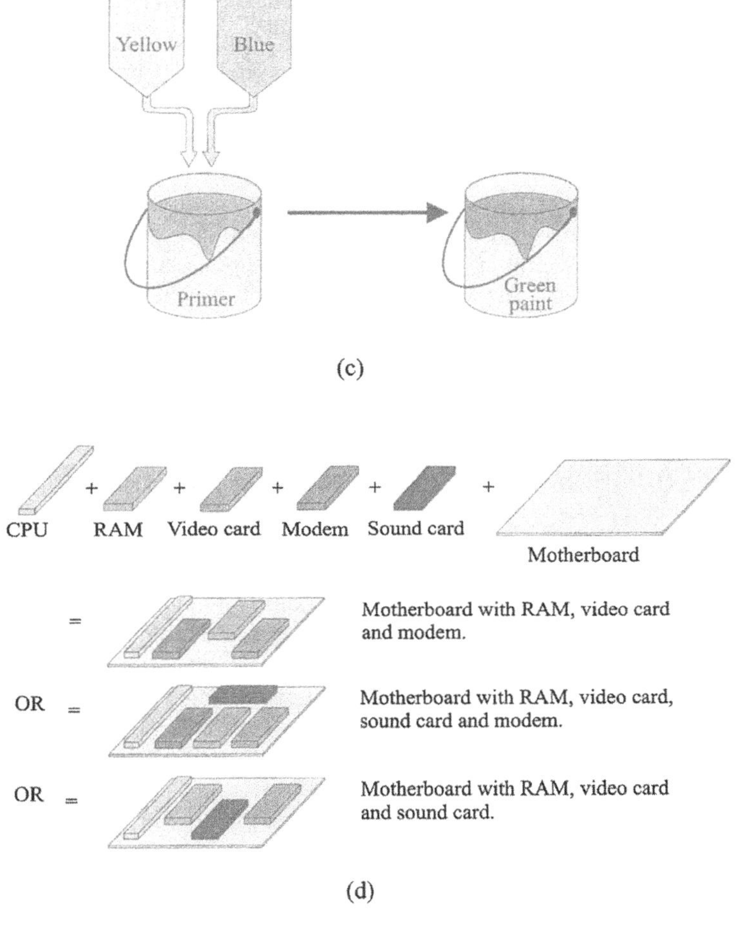

FIGURE 6 (c) Mixing; (d) bus configuration modularity.

2. Group the functional objectives into functional clusters, where possible. At this stage the designer can consider issues such as physical relationships (and proximity) between components, the potential for standardization, and even the capability of suppliers to provide clusters.
3. Create a rough geometric layout of the product in order to evaluate operational feasibility through analysis of the interactions between the clusters, as well as the feasibility of production and assembly while maintaining a high degree of quality and economic viability.

2.7 MASS CUSTOMIZATION VIA PRODUCT MODULARITY

Mass production adopted in the earlier part of the 20th century was based on the principles of interchangeable parts, specialized machines, and division of labor. The focus was primarily on improving productivity through process innovation. The primary objective was to reduce cost and thus cause an increase in demand. Most large companies ignored niche markets and customer desires, leaving them to the small companies. This manufacturing management paradigm started to loosen its grip on most consumer industries around the 1960s and 1970s in response to developing global competition pressures. A paradigm shift toward customization was full blown by the late 1980s in several industries, naturally, at different levels. The objective was set as "variety and customization through flexibility and quick responsiveness."

The key features of today's marketplace are (1) fragmented demand (the niches are the market) (2) low cost and high quality (customers are demanding high-quality products, not in direct relation to the cost of the product), (3) short product development cycles, and (4) short product cycles. The result is less demand for a specific product but increased demand for the overall product family of the company, whose strategy is to develop, produce, market, and deliver affordable goods with enough variety and customization that almost everyone purchases their own desired product.

The primary (fundamental) prerequisite to achieving mass customization can be noted as having customizable products with modularized components. Examples of customizable (reconfigurable) products include Braun's flex-control electric razor, which is self-adjusting to the user's facial profile, Reebok's Pump shoes that can be (air) pumped for better fit (similar to customizable "removable" casts for foot fractures), and finally Dell's personal computers, customized by the buyer and assembled specifically for them. In this context, standardization for customization is a competitive tool for companies marketing several related products, such as Black & Decker's line of power tools, which use a common set of standardized subassemblies (clusters, modules, etc.).

The primary steps for the design of a mass customizable product are

1. *Identifying customer needs*: This stage is similar to any product (concept) design stage with the exception of identifying potential personal differences in requirements for a common overall functional requirement for the product.
2. *Develop concepts*: Concepts (alternatives) should be developed and compared with a special emphasis for allowing modularity

in final engineering design. (QFD and Pugh's methods should be utilized.)

3. *Modularization of chosen concept*: The chosen design concept should be evaluated and iteratively modified with the objective of modularization (i.e., mass customization) and fit within the larger family of products, with which the proposed design will share modules.

REVIEW QUESTIONS

1. Define concurrent engineering (CE) and discuss its practical implementation in manufacturing enterprises.
2. Discuss the CE design guideline "design for robustness."
3. Discuss techniques for increasing the effectiveness of the customer-need-identification process.
4. Discuss the role of industrial design in the development of engineering products. Should industrial designers be consulted prior, during, or after a product has been designed and its manufacturing plans have been finalized?
5. Discuss some of the important issues that a human-factor engineer has to deal with for the design of products/systems that allow effective human/machine interface for supervisory control, maintenance, etc.
6. How can the quality function development (QFD) method be used in relating customer requirements to engineering design parameters?
7. The conceptualization stage of design can benefit from uninhibited creative thinking, eventually leading to several concepts for the solution of the problem at hand. Discuss Pugh's method of concept selection and difficulties associated with it.
8. Define product modularity and compare it to standardization. Provide several examples in your discussion, while classifying their level of modularity.
9. Discuss modularity for software products versus hardware products in the computing industry.
10. What is the key product design requirement for mass customization?

DISCUSSION QUESTIONS

1. Most engineering products are based on innovative design, rather than on fundamental inventions. They are developed in response to a common customer demand, enabled by new materials and/or technologies. Review the development of a recently marketed product that fits

the above description from its conception, to its manufacturing and marketing: for example, portable (personal Walkman type) CD players, portable wireless phones, microwave ovens, etc.

2. A common manufacturing strategy advocates assigning responsibility for a product to a team. This team designs the product, plans its fabrication, and remains responsible for it until the product reaches maturity while providing customer support. The team may grow or shrink in its membership during the life cycle of the product. Discuss this strategy versus a compartmentalized strategy, where different groups of people would take on responsibility for the product during the different periods of its life cycle without maintaining a tangible continuity.
3. Discuss the role of computers in the different stages of the (iterative) design process: concept development, synthesis, and analysis.
4. Product marketability is an important factor in the design and subsequent manufacturing of consumer products. Marketing efforts frequently concentrate on highlighting the non-functional, eye-pleasing design features of products in their promotion. Discuss the issue of incorporating such features (versus functional features) into product designs in the context of their impact on the manufacturing of the products. Include specific products and features in your discussion, such as furniture versus refrigerators versus passenger cars and aerodynamic geometry versus colors versus packaging (i.e., exterior of products).
5. Products can be designed for specific ranges of anthropometrics, for a targeted demographics, in two distinct modes: (1) Those that allow reconfiguration via continuous and/or discrete incremental changes, or even through modularity of certain subcomponents, or (2) those that have been already manufactured in different dimensions, etc., for different customer anthropometrics. Discuss these modes of design in terms of manufacturing difficulties, durability, safety, cost, customer response, etc. In your discussion, include specific products/features, for example car seats, bicycles, headphones, office chairs, personal clothing items.
6. Human factors (HF) studies encompass a range of issues from ergonomics to human-machine (including human–software) interfaces. Discuss the role of HF in the autonomous factory of the future, where the impact of human operators is significantly diminished and emphasis is switched from operating machines to supervision, planning, and maintenance.
7. Flexible manufacturing has often been proposed as a (tactical) production strategy. Discuss whether such a strategy can be justified economically for all products. In the same context, also discuss specific product features that would allow customization (e.g., geometry, material, fabrication process, etc.), which in turn requires manufacturing

flexibility. Consider products such as furniture, household appliances, bicycles, and personal clothing.

8. Discuss a design strategy for multicomponent products whose support and maintenance would not be negatively affected by significant variations in the life expectancy of their individual components, (i.e., large variations within the same batch of components). In your discussion, assume that these variations would occur for material or technological reasons, such as the absence of machines that can provide high levels of quality in terms of component life, and that they are unavoidable.
9. Discuss the concept of progressively increasing *cost of changes* to a product as it moves from the design stage to full production and distribution. How could you minimize necessary design changes to a product, especially for those that have very short development cycles, such as portable communication devices?

BIBLIOGRAPHY

Backhouse, Christopher J., Brookes, Naomi J., eds. (1996). *Concurrent Engineering: What's Working Where*. Brookfield, VT: Gower.

Burgess, John H. (1986). *Designing for Humans: The Human Factor in Engineering*. Princeton, NJ: Petrocelli.

Cohen, Lou (1995). *Quality Function Deployment: How to Make QFD Work for You*. Reading, MA: Addison-Wesley.

Dieter, George Ellwood (2000). *Engineering Design: A Materials and Processing Approach*. New York: McGraw-Hill.

Ericsson, Anna, Erixon, Gunnar (1999). *Controlling Design Variants: Modular Product Platforms*. Dearborn, MI: Society of Manufacturing Engineers.

Forty, Adrian (1986). *Objects of Desire*. New York: Pantheon Books.

Gu, P., Sosale, S. (1999). Product modularization for life cycle engineering. *Journal of Robotics and Computer Integrated Manufacturing* 15:387–401.

Hill, Terry (2000). *Manufacturing Strategy: Text and Cases*. Houndmills: Palgrave.

Hunter, Thomas A. (1992). *Engineering Design for Safety*. New York: McGraw-Hill.

Hyman, Barry I. (1998). *Fundamentals of Engineering Design*. Upper Saddle River, NJ: Prentice Hall.

Lorenz, Christopher (1987). *The Design Dimension: The New Competitive Weapon for Business*. Oxford: Blackwell.

Meikle, Jeffrey L. (1979). *Twentieth Century Limited: Industrial Design in America*. Philadelphia: Temple University Press.

Pahl, Gerhard, Beitz, W. (1996). *Engineering Design: A Systematic Approach*. New York: Springer-Verlag.

Pine, B. Joseph (1993). *Mass Customization: The New Frontier in Business Competition*. Boston: Harvard Business School Press.

Prasad, Biren (1996). *Concurrent Engineering Fundamentals.* Upper Saddle River, NJ: Prentice Hall PTR.

Pugh, Stuart (1996). *Creating Innovative Products Using Total Design.* Reading, MA: Addison Wesley.

Pugh, Stuart (1997). *Total Design: Integrating Methods for Successful Product Engineering.* Reading, MA: Addison Wesley.

Salhieh, S. M., Kamrani, A. K. (1999). Macro Product Development Using Design for Modularity. *Journal of Robotics and Computer Integrated Manufacturing* 15:319–329.

Stoll, Henry W. (1999). *Product Design Methods and Practices.* New York: Marcel Dekker.

Tilley, Alvin R. (1993). *The Measure of Man and Woman: Human Factors in Design.* New York: Whitney Library of Design.

Tseng, M. M., and Jiao, J. Design for mass customization by developing product family architecture. ASME Design Engineering Technical Conferences, CD Proceedings of Design for Manufacturing Conference, Atlanta, GA, Paper No. DETC98/DFM-5717, September 1998.

Ullman, David G. (1997). *The Mechanical Design Process.* New York: McGraw-Hill.

Ulrich, K. T., Eppinger, S. D. (2000). *Product Design and Development.* Boston: McGraw Hill.

Walsh, Vivien (1992). *Winning by Design: Technology, Product Design, and International Competitiveness.* Cambridge, MA: Blackwell.

3

Design Methodologies

This chapter presents four primary design methodologies developed in the past several decades for increased design productivity and the resultant product quality. Although these methodologies are suitable and have been commonly targeted for the post-conceptual-design phase, some can also be of significant benefit during the conceptual design phase of a product. Axiomatic design methodology, for example, falls into this category.

Designers should attempt to use as many established design methodologies as possible during product development: For example, Axiomatic design and group technology at the conceptual design phase, design for manufacturing/assembly/environment guidelines during configuration and detailed design, and the Taguchi Method during parametric design.

3.1 AXIOMATIC DESIGN METHODOLOGY

As discussed in Chap. 2, the conceptual design phase starts with examining and identifying the customer's needs, which subsequently must be related to engineering requirements. Axiomatic design methodology, developed in the late 1970s by N. P. Suh, but only widely implemented since the 1980s, is primarily an analysis technique for the evaluation of designs. Users of this methodology can utilize its two axioms and their numerous corollaries as guidelines for good design.

The two axioms of the theory advocate that for good design the functional requirements (FRs) of the product (as dictated by the customer) must be independently satisfied by the design parameters (DPs), onto which they would be mapped, in the simplest possible manner. Suh defines the FRs as "a minimum set of independent requirements that completely characterize the functional needs of the product in the functional domain:"

Axiom 1: The *independence axiom* states that a change in a DP should preferably only affect its corresponding FR.
Axiom 2: The *information axiom* states that among all alternatives considered, which satisfy Axiom 1, the simplest solution is the best design.

As a way to categorize designs, Suh introduced the following design categories:

Uncoupled design: A concept that satisfies Axiom 1.
Coupled design: A concept that violates Axiom 1, where a perturbation in a DP affects multiple FRs.
Decoupled design: A concept that is initially a coupled design due to lack of sufficient DPs, but one that can be decoupled with the use of extra DPs.

A decoupled design would naturally have an information content that is more than that of an uncoupled (competing) design.

As a simple example, let us consider a user need for a control device for hot-water supply—that is, the device must control the flow rate as well as the temperature of the water. These can be defined as FR_1 and FR_2, respectively. A possible design would be to have two knobs (DP_1, DP_2) individually controlling the flow rate of hot and cold water, respectively, prior to their mixing (Fig. 1a). As one can note, however, a user would have difficulty in achieving a desired output of a specific flow-rate set at a certain temperature using this design, necessitating numerous control interventions. This design concept can thus be classified as a coupled design. An uncoupled design would require that the individual DPs satisfy their corresponding FRs independently. Many such uncoupled design–based commercial products exist today, in which users can turn a lever (or a knob) left or right for a desired water temperature and tilt the lever up or down (or, pull or depress the knob) for flow-rate control (Fig. 1b).

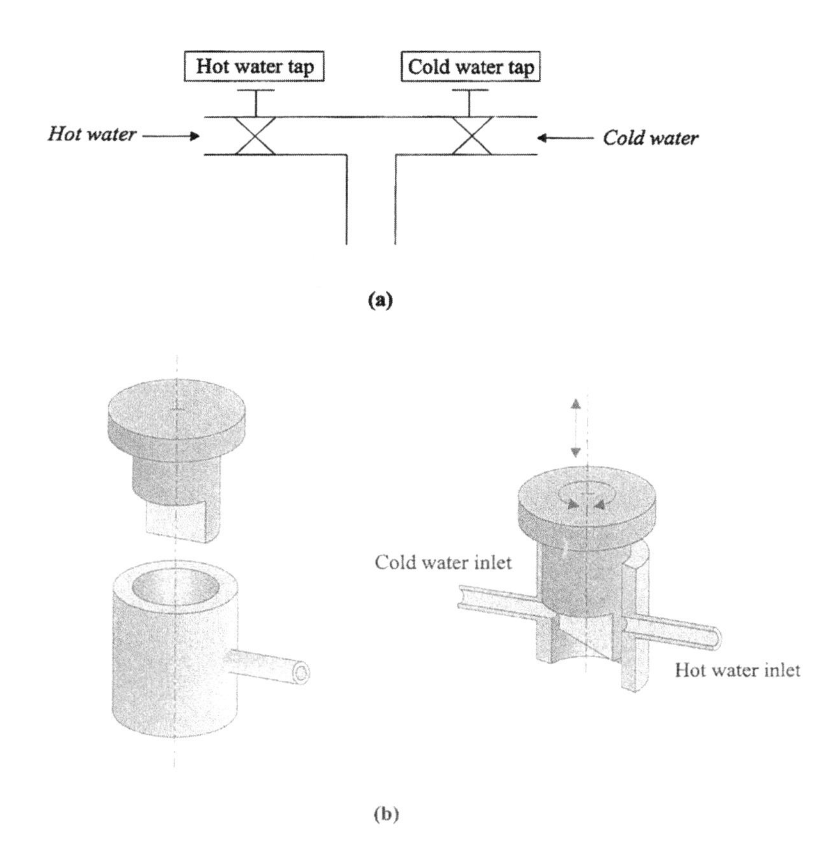

FIGURE 1 Axiomatic designs: (a) coupled and (b) uncoupled.

The mapping process between the FRs and the DPs can be expressed in a (linear) matrix form as

$$\{FR\} = [A]\{DP\} \tag{3.1}$$

where {FR} is the functional requirement vector and {DP} is the design parameter vector. The matrix [A] maps FRs into DPs.

An uncoupled design would have all the non diagonal elements of its [A] matrix as zero, thus satisfying the independence axiom. A coupled design, on the other hand, would have an [A] matrix with some nonzero nondiagonal elements. That is, some of the FRs will be functions of more than one DP.

For the first (coupled) water flow control device discussed above (Fig. 1a), the design matrix is

$$\begin{aligned} FR_1 &= A_{11}DP_1 + A_{12}DP_2 \\ FR_2 &= A_{21}DP_1 + A_{22}DP_2 \end{aligned} \tag{3.2}$$

Thus, change in the two design parameters (both knobs) affects simultaneously both the flow rate (FR_1) and the temperature (FR_2)—the latter by proportioning the amount of water coming from two sources (hot and cold).

A coupled design can be decoupled by redesign: the [A] matrix becomes triangular (all elements above or below the diagonal have zero values):

$$\begin{Bmatrix} FR_1 \\ FR_2 \end{Bmatrix} = \begin{bmatrix} A_{11} & 0 \\ A_{21} & A_{22} \end{bmatrix} \begin{Bmatrix} DP_1 \\ DP_2 \end{Bmatrix} \tag{3.3}$$

In the above equation, it is noted that FR_1 is only a function of DP_1. Once the value of DP_1 is set to correspond to a desired value of FR_1, subsequently, DP_2 can be appropriately adjusted so that the combination of DP_1 and DP_2 yields a desired (functional requirement) value for FR_2, in which

$$FR_2 = A_{21}DP_1 + A_{22}DP_2 \tag{3.4}$$

When the number of FRs, m, is different from the number of DPs, n, the design is either coupled, $m > n$, or it is redundant in nature, $m < n$. A coupled design can be decoupled, first by the use of additional DPs, so that m becomes equal to n, and, subsequently by varying them (if necessary) so that the (square) mapping matrix, [A], becomes diagonal, normally, through a trial and error process. A redundant design, on the other hand, may not be a coupled design necessarily.

Let us consider the following case of $m = 2$ and $n = 3$:

$$\begin{Bmatrix} FR_1 \\ FR_2 \end{Bmatrix} = \begin{bmatrix} A_{11} & A_{12} & A_{13} \\ A_{21} & A_{22} & A_{23} \end{bmatrix} \begin{Bmatrix} DP_1 \\ DP_2 \\ DP_3 \end{Bmatrix} \tag{3.5}$$

In Eq. (3.5), if A_{13}, A_{21}, and A_{22} are zero, FR_1 would be only a function of DP_1 and DP_2, whereas FR_2 would only be a function of DP_3, yielding functional independence. A preferable scenario might be to combine DP_1 and DP_2 into one design parameter to yield a more effective uncoupled design.

As one would expect, at the conceptual design phase, the objective of the designer is to note whether the elements of the mapping matrix [A] are zero or not. Once the design parameters (DPs) have been established, these will act as detailed design requirements and be mapped into specific variables for parametric design.

The minimization of information in a specific design, in order to satisfy Axiom 2, refers to its simplification process, for example, to increase its manufacturability. Design information may include geometric tolerances that are set realistically and material constraints/preferences and metallurgical treatments that should be chosen according to process availability and economic viability.

Suh provides a comprehensive list of design rules based on his two axioms, three of which are

Minimize the number of functional requirements and constraints.
Use standardized components.
Specify achievable tolerances.

The above should be seen as guidelines to be used in conjunction with numerous design rules to be specified in this chapter to satisfy objectives, such as manufacturability, ease of assembly, and environment friendliness.

3.2 DESIGN FOR X

Today it is commonly accepted that consideration of manufacturing and assembly issues during the design phase of a product is a fundamental part of concurrent engineering (CE). This was not the case in the first half of the 20th century, when CE was not a central manufacturing management policy, and designers were expected to be familiar with all manufacturing processes (and they actually were). In the latter half of the century (1950s to 1970s), though, manufacturing engineering was neglected as an undergraduate studies subject in the curricula of many North American universities; consequently most junior engineers lacked comprehensive knowledge of fabrication and assembly processes. Furthermore, these engineers became specialists in their fields (in the spirit of the Taylor/Ford paradigm) with little knowledge or appreciation of other disciplines. Thus the 1980s saw the necessary birth of CE-based design and the reintroduction of breadth into engineering curricula, so that engineers could communicate more effectively within their product design teams.

In this chapter, a limited number of design guidelines is presented for several manufacturing processes, for assembly, and for environmental considerations. The objective is to make the reader aware of the existence of such Design for X methodologies. The guidelines presented in the following subsections, though not comprehensive nor inclusive of all processes, are derivatives of the following general design guidelines:

Design parts for ease of (and profitable) manufacturing—select materials and corresponding fabrication processes suitably.
Specify tolerances, surface finish, and other dimensional constraints that are realistic.
Minimize the number of parts, and furthermore use as many as standard parts as possible.
Note that mechanical properties (and consequently a part's life) are affected by specific production process parameters, such as the location of parting lines in casting and molding.

3.2.1 Design for Manufacturing

Consideration of manufacturing (also termed as production or fabrication) processes during the design stage of a product is fundamental to successful design. Since selection of materials must precede consideration and analysis of manufacturability, herein it will be assumed that this stage is part of the definition of functional requirements (in response to customer needs) and thus will not be addressed. Primary issues that do arise during material selection include product life, environmental conditions, product features, and appearance factors.

In the following subsections, a select set of manufacturing processes will be reviewed, specifically from the perspective of design guidelines, in order to illustrate the importance of considering manufacturability during the product development stage. These and other design guidelines will be revisited when specific manufacturing processes are discussed in Part II of this book.

Design for Casting

In casting, a molten metal is poured (or injected at high pressure) into a mold with a single or multiple cavities (Fig. 2). The liquid metal solidifies within the cavity and is normally subject to shrinkage problems. A good mold design will thus have features to compensate appropriately for shrinkage and avoid potential defects.

Castings must be designed so that parts (and patterns for sand casting) can be removed easily from the cavities. Different sections of a part with varying thicknesses should have gradual transitions. Projecting details should be avoided. Ribs should not be allowed to cross each other but should be offsett. In die casting, parts should have thin-walled structures to

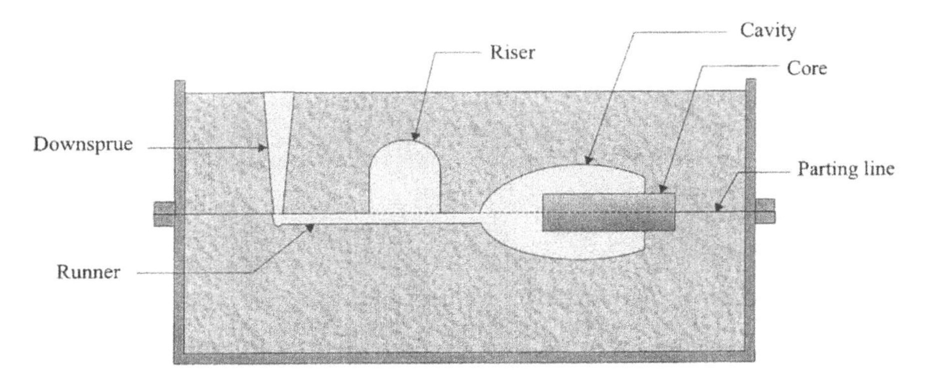

FIGURE 2 Sand casting.

ensure smooth metal flow and minimum distortion due to shrinkage. One should also note that die casting is normally limited to nonferrous metals. Furthermore, different casting techniques yield different dimensional accuracy and surface finish. For example, although sand casting can be used for any type of metal, poor surface finish and low dimensional accuracy are two of its disadvantages.

Design for Forging

Forging is the most common (discrete-part) metal forming process in which normally a heated workpiece is formed in a die cavity under great (impact) pressure. Owing to the high forces involved, generally a workpiece is formed in multiple iterations (Fig. 3). As with casting, vertical surfaces of a part

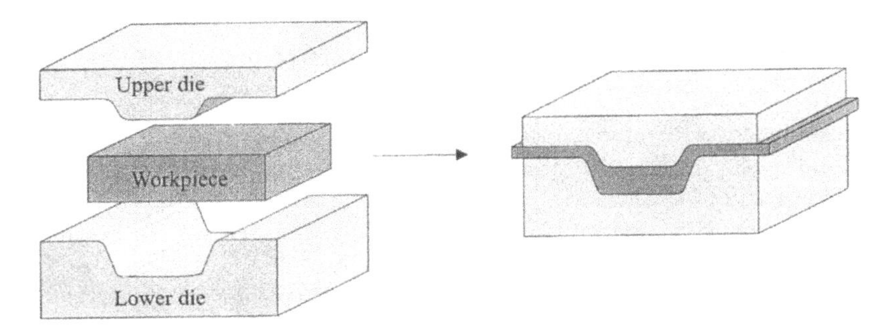

FIGURE 3 Closed-die forging.

must be tapered for ease of removal (normally done manually). Furthermore, rapid changes in section thicknesses should be avoided to prevent potential cracks. Finally, when designing a product that is to be forged, the location of the "parting line," where the two die halves meet, should be carefully chosen to influence positively the grain flow and thus the mechanical properties of the part.

Design for Machining

There is a variety of material-removal processes that are collectively called machining. Although both metals and plastics can be machined, for example, using turning, milling, and grinding operations, machining is primarily reserved for metal workpieces. Cylindrical (rotational) geometries can be obtained using a lathe or a boring machine (for internal turning), whereas prismatic (nonrotational) geometries can be obtained on a milling machine (Fig. 4). A drill press is reserved for making holes in prismatic objects. Although there are many other material-removal techniques, they will be discussed only in Part II of this book.

Machining is a flexible manufacturing operation in which metal-cutting parameters can be carefully controlled to produce almost any external detail on a (one-of-a-kind) part, including 3-D complex surfaces. Automated machine tools can be programmed to fabricate parts in large quantities as well, such as nuts, bolts, and gears. Being material-removal techniques, such processes can take long periods of time when high accuracies are required and/or material hardnesses are very high. Thus a designer must carefully consider configuring features on a product that would require several setup activities, to rotate and realign the part, and subsequently prolong manufacturing times.

A common error in designing parts for machining is placement of holes (or other details) on a workpiece that would not be accessible due to collision between the tool-holder and the part (or even the fixture that holds the part) (Fig. 5).

Numerous design guidelines for machining have been described by Boothroyd et al., some of which are

Preshape parts through casting or forging to minimize machining time.
Avoid specifying features or tolerances that your machine tools cannot profitably fabricate.
Ensure that the workpiece can be rigidly fixtured to withstand common high cutting forces.
Avoid internal features in long workpieces (including cylindrical bores).
Avoid dimensional ratios that are very high.

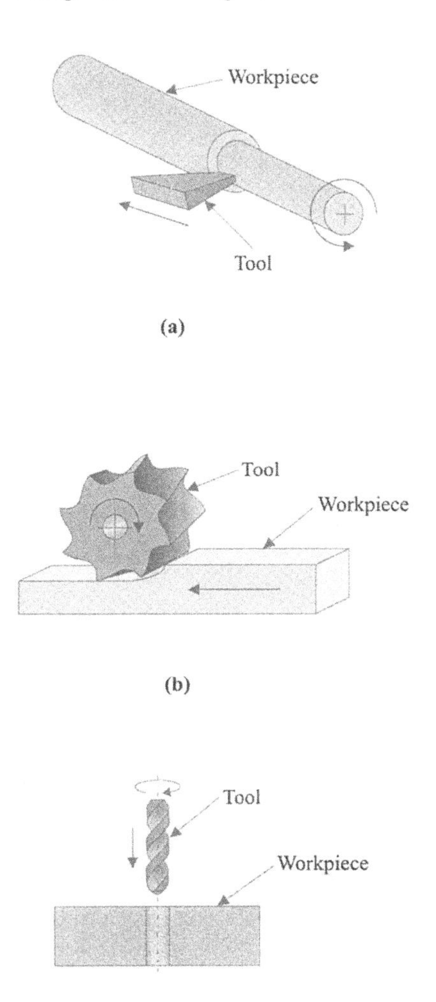

FIGURE 4 (a) Turning; (b) milling; (c) drilling.

Design for Injection Molding

Injection molding is the most common plastic-parts manufacturing process for thin-walled objects. It commonly utilizes (recyclable) thermoplastic polymer granules that are melted and forced into a mold cavity (Fig. 6). It is a very efficient process in which multicavity molds (up to 16 or more) can manufacture several thousands of parts per hour and last for several

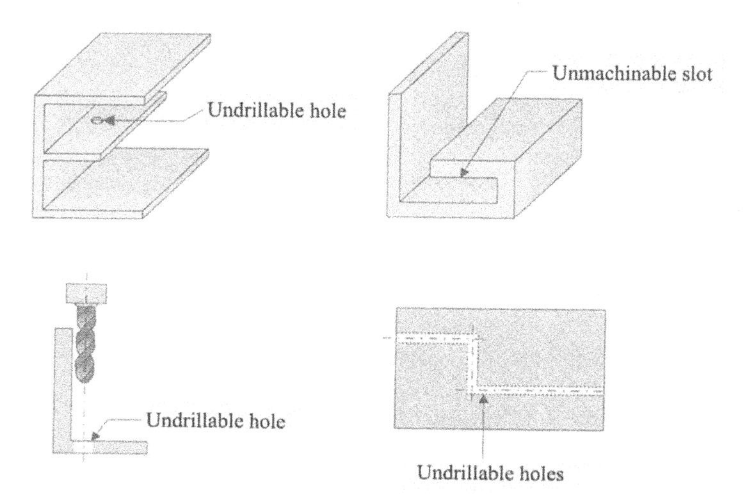

FIGURE 5 Unmachinable parts.

millions of parts. The three-step fabrication process comprises injection of molten plastic into the cavities, cooling (solidification) through the cavity walls, and forced ejection (for a typical total cycle time of 10–60 s).

As in casting and forging, the two most important design considerations in injection molding are wall thicknesses and parting lines. One always aims for gradual wall thickness changes through the part (typically, several mm) as well for incorporating as many features as possible into the design (snap fits, countersinks, holes, bosses, etc.) to avoid secondary operations. Other design guidelines proposed by Boothroyd et al. include

Configure your part geometry for adequate tapers for easy ejection from the cavities.

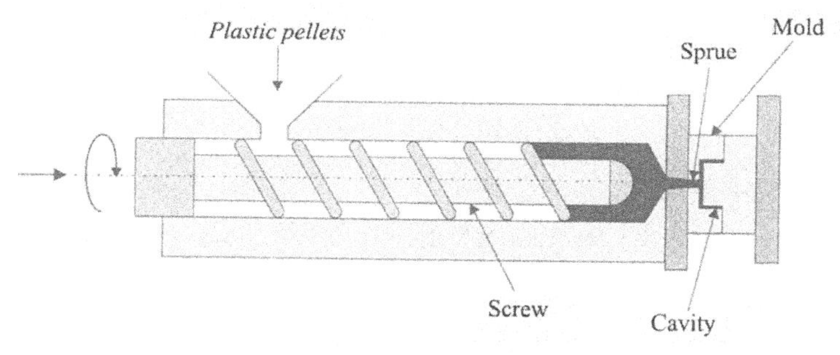

FIGURE 6 Injection molding.

Ensure proper proportioning of wall thicknesses for minimum distortion during cooling.
Minimize wall thicknesses (through the use of supporting elements) for fast cooling.
Avoid depressions on the inner side surfaces of the part to simplify mold design and minimize cost.

3.2.2 Design for Assembly

Assembly is a manufacturing process normally seen as an activity that does not add value to the final product. Thus every effort should be made to minimize assembly costs by minimization of the total number of parts, avoidance of several directions of assembly, and maximizing assemblability through the use of guidance features.

Most of the product design (for assembly) issues discussed below have been extensively reported in the pioneering works of G. Boothroyd, P. Dewhurst, and W. Knight since the 1970s. Although their work addressed the topics of manual and automatic assembly separately, such a distinction will not be made herein, since the emphasis of this book is on autonomous manufacturing systems; furthermore, most guidelines developed for the former case apply to the latter.

As providing a first level discussion to the three general design-for-assembly guidelines provided above, this list addresses the issues of parts manipulation and joining (Fig. 7):

Design parts with geometrical symmetry, and if not possible exaggerate the asymmetry.
Avoid part features that will cause jamming and entanglement, and if needed add nonfunctional features to achieve this objective.
Incorporate guidance features to part's geometry for ease of joining, such as chamfers; clearances should be configured for maximum guidance, but for minimum potential of jamming.
Design products for unidirectional vertical (layered) assembly in order to avoid securing the previous subassembly while turning it.
Incorporate joining elements into the parts (such as snap fits) in order to avoid holding them in place when utilizing additional joining elements (such as screws, bolts, nuts, or even rivets). Snap fits can be designed to allow future disassembly or be configured for permanent joining owing to potential safety hazards.

The major cost of assembly is determined by the number of parts in the product. Thus one should first and foremost attempt to eliminate as many parts as possible, primarily by combining them. Conditions for

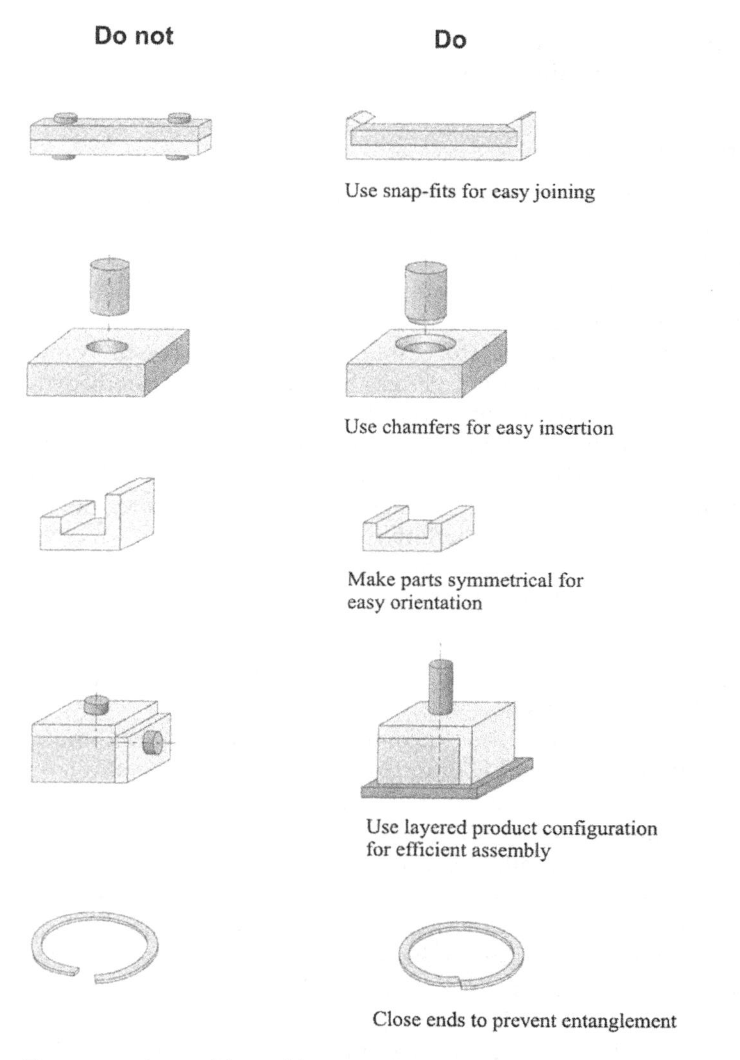

FIGURE 7 Assembly problems and solutions.

elimination can be recognized by examining the product: (1) Does the part move after the assembly, or simply remain static? (2) Why must the part be of a different material than the neighboring part? (3) Would the part prevent the assembly of other parts, by presenting an obstruction, if it were to be combined with a neighboring part?

As argued by the axiomatic design theory, eliminating parts from a product is in line with Axiom 2, which requires minimization of information (Sec. 1). Integration of parts (consolidation) may also reduce typical stress concentration points in parts owing to the use of external fasteners. (Naturally, the use of snap fits in product assembly also introduces stress concentration points, and thus their use should be carefully examined.) A beneficial side effect of part reduction is the elimination of future potential loosening in joints and subsequent vibration noise.

3.2.3 Design for the Environment

Human population growth is a major factor in the well-being of our environment, and when coupled with the complexity of our lifestyles it presents an enormous pressure on the world's precious resources. It is anticipated that in the 21st century, the world's population may peak at between 10 and 15 billion.

In the past century, the world's industrial production grew more than 100-fold. In the same period of time, the consumption of fossil fuel increased by a factor of more than 50. It has been eagerly argued that we cannot continue to use materials and resources at their current rates without experiencing severe shortages within the next 50 to 100 years.

No industrial activity today happens in isolation. It impacts the environment from the materials it uses to the products it manufactures, which have to be dealt with at the end of their life cycles. The approach to industrial-environmental interactions is commonly referred to as industrial ecology. It aims at designing industrial processes and products that minimize their impact on the environment, while maintaining manufacturing competitiveness. In that respect, one must be concerned about the following global issues: climate change, ozone depletion, loss of habitat and reductions in biodiversity, soil degradation, precipitation acidity, and degradation in water and air qualities.

A primary issue in industrial ecology is life cycle assessment (LCA), a formal approach to addressing the impact of a product on the environment as it is manufactured, used and finally disposed of. The first step of LCA is inventory analysis. That is, we need to determine the inputs (materials and energy) used in manufacturing and the outputs (the product itself, waste, and other pollutants) resulting during manufacturing and beyond. The second step of LCA is quantifying the impact of the outputs on the environment (a most contentious issue). The final step is the improvement analysis. Proposals are presented to manufacturers for reducing environmental impact—this stage is also referred as design for environment.

Design for Energy Efficiency

The manufacturing industry uses a considerable amount of energy. For example, manufacturing activities consume almost 20% of electricity in the U.S.A. The answer to energy-source selection is not a simple one, because uses of different resources impact the environment at varied levels. As far as the atmosphere is concerned, for example, fossil-fuel combustion is more harmful than energy produced by nuclear power. That is, energy-source efficiency must often be balanced with toxicity concerns. However, no matter what its source of energy is, a manufacturing company must always aim for energy conservation when evaluating product design and fabrication process alternatives.

Design for Minimum Residues

Numerous toxic chemicals are released to the environment during many of today's manufacturing processes. These residues can be solid, liquid, and/or gaseous. For example, in the U.S.A., municipal solid waste discarded in landfills is less than 2% of the amount of industrial waste (of which more than half comes from manufacturing activities). Solid residues come in several forms: product residues generated during processing (for example, small pieces of plastic trimmings), process residues (such as cutting tools disposed of at the end of their useful life), and packaging residues (packaging and transportation material brought to the factory (drums, pallets, cardboard, etc.). A manufacturing company should make every effort to minimize all waste (recycle waste material as well as utilize reusable packaging).

Design for Optimal Materials

Although it is a difficult issue to tackle, designers can make an effort in choosing product materials for minimal environmental impact, especially in relation to their extraction as well as processing. Naturally, an efficient recycling operation may provide manufacturers with adequate material supply with lower costs and minimal environmental impact. Thus recyclabilty of the product's material should be a factor in this selection. Today metals are recycled with reasonable efficiency, returning them to their original condition through rework (remanufacture) or at worst by melting them. No matter what materials are utilized, one should always strive toward minimization of their amount through suitable engineering design (for example, by using thinner walls supported by many ribs).

Design for Recycling

In the past decade, besides the obvious economic benefits, manufacturers have also had to consider various government regulations (i.e., punitive incentives) when employing design-for-recycling practices. The common hierarchy of preferences in recycling practices available to manufacturing companies has been as follows (Fig. 8):

Subassemblies (highest): Replacement of a subassembly in a product in order to restore the product to its original operational level; the failed subassembly can also be recycled in order to restore it to its original performance level for future use.

Components: Refurbishment of products first by their complete removal from operation, then by replacement of failed components with new or recycled components, and finally by their return to their normal operational level.

Materials (lowest): Removal of products from operation, recycling of materials that can be recovered, and use of these materials (most frequently mixed with virgin materials) for the manufacturing of new components/products.

It is important to minimize the number of different materials used in the manufacturing of a product and, where possible, to keep them separated for ease of joining and subsequent recycling.

Another important issue is "design for disassembly." There exist two methods for common disassembly: reversible (where screws are removed, snap-fits unsnapped, etc.) and destructive (where the joints are broken). Economic and safety issues play major roles in deciding which joining technique to use. A modular design will greatly simplify the task of disassembly, as we can quickly identify the part/component/subassembly to be replaced (Chap. 2).

The design guidelines for "green" products and processes can be summarized as

Increase efficiency of energy use, while considering environmental impact.

Minimize the amount of materials used.

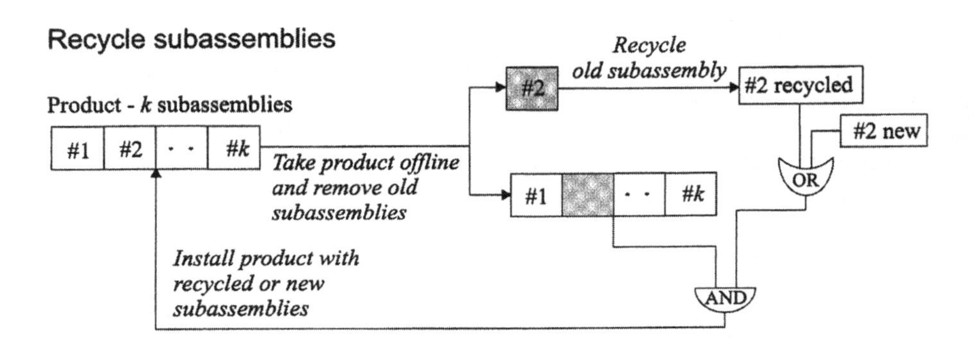

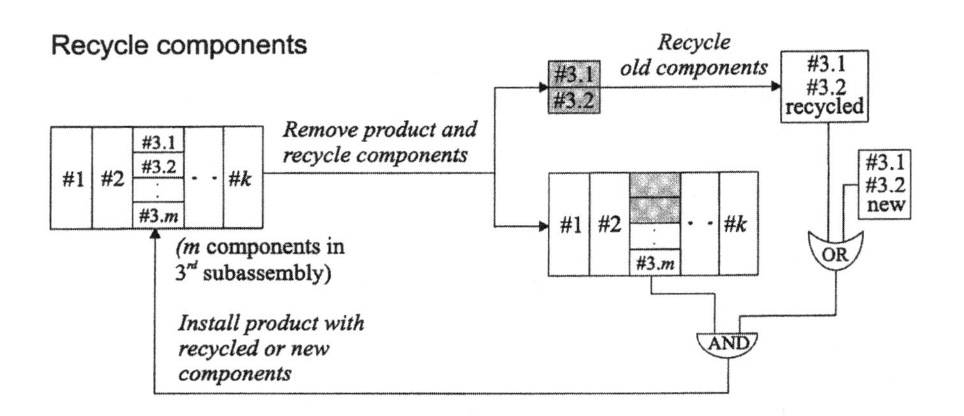

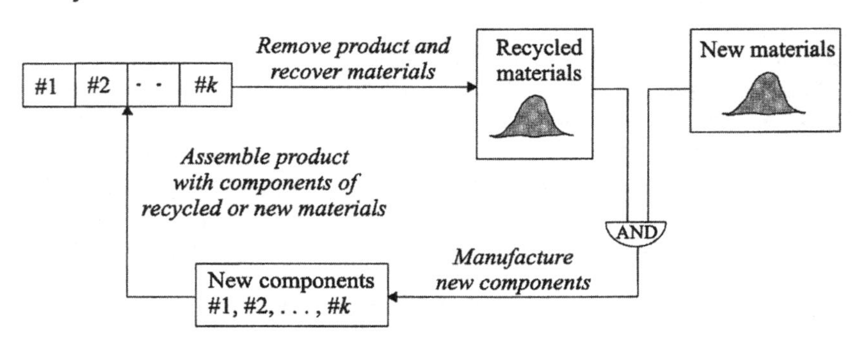

FIGURE 8 Design for recycling.

Use recyclable and biodegradable materials where possible.
Maximize the life expectancy of the product (in materials as well as technology).
Design a modular product for ease of disassembly and remanufacturing,

3.3 DESIGN OF EXPERIMENTS AND TAGUCHI'S METHOD

Parameter and tolerance design follows the conceptual design and engineering requirements determination phases of a product. At this stage, most engineers review functional requirements and decide on parameter and tolerance values based on experience, handbooks, etc. In the case of multi-parameter design, however, where the choice of one parameter affects the other, engineers are advised to run experiments and optimize their values. Experimentation (for optimization) can be in the physical domain or in virtual space, where numerical simulations are performed.

3.3.1 Parameter Design Using Design of Experiments and Response-Surface Optimization

It is strongly recommended that engineers take advantage of well-established statistical design of experiments (DOE) theories in order to minimize the search efforts for the optimal parameter values. The alternative would be to run a random (not well thought) set of experiments, from which one cannot easily infer meaningful conclusions. DOE theory advocates a factorial approach to experiments, that is, the controllable variables (parameters) of the experiment are discretized to a very limited number of levels (e.g., low, medium and high) and are randomized methodically in order to create a limited (but well thought) set of experiments. There exist a number of techniques for factorial design—the Latin square, the Youden square, and of course the Taguchi method.

As an example of a full-factorial design of experiments, let us assume that the fatigue failure level of a product depends on three (dimensional) parameters (A, B, and C). We design an experiment to evaluate these dependencies, and decide to test two levels (low and high) for every parameter, respectively. Table 1 illustrates the results of the experiments for a set of $2^3 = 8$ experiments. (Naturally, if it is economically viable, one may decide to repeat the experiments several times, if they are physical in nature, in order to minimize the effect of noise on the observations.)

Once the experiments have been completed a design engineer's objective would be to search for the optimal values of the parameters (not only

TABLE 1 A Design-of-Experiments Example

A	B	C	Failure (cycles $\times 10^6$)
L	L	L	1.27
H	L	L	1.29
L	H	L	1.31
L	L	H	1.28
H	H	L	1.58
H	L	H	1.29
L	H	H	1.41
H	H	H	1.39

among the specific levels tested during the particular set of a limited number of experiments run but also through the complete feasible search space). Prior to this stage one may examine the obtained results through an analysis-of-variance study, to determine whether some of the parameters have a low impact on the output. In the case of a large number of parameters, it would be wise to select quickly the appropriate values for these parameters and exclude them from future searches for the optimal values of the remaining parameters. Based on such an investigation, one would note that the variation of Parameter A in our example (Table 1) has a low impact on the value of the failure cycle and thus could be eliminated after choosing a suitable value for it.

Response surface (RS) methodology is a common technique that can be utilized to facilitate the search for the optimal parameter values. As the name implies, the first step of the RS methodology is to establish a (continuous-variable) relationship between the variables and the output (observation) through a surface fit (a hypersurface, if the number of variables is above two). Least-squares-based regression methods are commonly used for this purpose—namely, in fitting a response surface to experimentally obtained data. (Naturally, it is strongly advised that experimental data be collected using a DOE theory.) One must recall that, during the surface-fitting process, the actual numerical values of the variables that correspond to the two levels examined (e.g., L and H) are utilized. For our example above, Table 1, after eliminating Parameter A from the optimization, the outcome of the regression analysis would be a three-dimensional surface, if we assume a nonlinear relationship between the variables (B and C) and the output (failure cycles).

As the last step, a nonlinear constrained search method must be utilized in order to search effectively for the optimal values of the variables within the search space defined by the response surface.

Therefore in conclusion to the above discussion, we can summarize the three-step parameter-design phase as follows:

1. Use DOE theory for selecting a limited set of experiments (not necessarily full factorial).
2. Determine the relationship between the variables and the output using a RS methodology (based on the experimental data).
3. Employ an efficient optimization search technique for determining the best parameter values that minimize/maximize the output value.

3.3.2 The Taguchi Method

The use of statistical methods in engineering can be attributed to two mathematicians in the earlier part of the 20th century (1920s): Sir R. A. Fisher in the U.K. (who first developed the DOE technique) and W. A. Shewhart in the U.S.A. (who developed the process control charts used today in statistical process control—SPC). G. Taguchi's contribution to the field can be traced to the early 1950s during his employment period by Nippon Telephone and Telegraph. During this early period, Taguchi advocated the use of orthogonal arrays in order to reduce significantly the number of experiments dictated by a full-factorial experimental design. For example, in a design problem of 13 parameters each to be evaluated at 3 different levels, we would have to run 1,594,323 experiments, whereas Taguchi's orthogonal arrays would require only 27 trials.

In the above context, the Taguchi method for parameter design aims at choosing the levels of the control variables that are robust to environmental noise. As a complementary approach to this selection of parameter values, Taguchi also proposed a technique for choosing corresponding tolerance values aimed at maximizing the quality of the manufactured product.

Robust Parameter Design

In Taguchi's approach, design parameters of a product are referred to as "controllable" factors versus noise factors that refer to disturbances, which cannot be controlled. The objective is, thus, to select optimal design parameter values that are least affected by noise to be encountered during the future utilization of the product. The method is a simple approach to selecting parameter values that maximize a signal-to-noise ratio (a term borrowed from the communications engineering field) defined as

$$S/N = 10 \log \left(\frac{\mu^2}{\sigma^2}\right) \tag{3.6}$$

where the mean, (μ), and variance, (σ^2), values of the output, y (for a set of constant parameter values and n different noise levels) are defined as

$$\mu_i = \frac{1}{n}\sum_{j=1}^{n} y_{ij} \quad \text{and} \quad \sigma_i^2 = \frac{1}{n}\sum_{j=1}^{n}(y_{ij} - \mu_i)^2 \tag{3.7}$$

Based on the above definitions, Taguchi's S/N ratio encapsulates both the mean value and the variance of the output values in a single term that can be optimized by varying the design parameter values (controllable factors):

$$(S/N)_i = -10 \log \left(\frac{1}{n}\sum_{j=1}^{n} y_{ij}^2\right) \tag{3.8}$$

That is, instead of maximizing/minimizing the mean value of the output only, Taguchi's S/N value can be used simultaneously to minimize the effect of noise (variance) on this mean value. In a DOE process based on Taguchi's method, orthogonal arrays are utilized to vary both the parameter values and the noise factors.

Let us consider a new example, in which a product's desired output is affected by four parameters, which may in turn be influenced by three noise sources. The experimental design for this example is shown in Table 2.

TABLE 2 L_9 and L_4 Orthogonal Arrays

	Parameter set, i				Output @ noise set, j				$(S/N)_i$ ratio
Experiment # (i)	A	B	C	D	(1,1,1)	(1,2,2)	(2,2,1)	(2,1,2)	Eq. (3.8)
1	1	1	1	1	y_{11}	y_{12}	y_{13}	y_{14}	$(S/N)_1$
2	1	2	2	2	y_{21}	y_{22}	y_{23}	y_{24}	$(S/N)_2$
3	1	3	3	3	y_{31}	y_{32}	y_{33}	y_{34}	$(S/N)_3$
4	2	1	2	3	y_{41}	y_{42}	y_{43}	y_{44}	$(S/N)_4$
5	2	2	3	1	y_{51}	y_{52}	y_{53}	y_{54}	$(S/N)_5$
6	2	3	1	2	y_{61}	y_{62}	y_{63}	y_{64}	$(S/N)_6$
7	3	1	3	2	y_{71}	y_{72}	y_{73}	y_{74}	$(S/N)_7$
8	3	2	1	3	y_{81}	y_{82}	y_{83}	y_{84}	$(S/N)_8$
9	3	3	2	1	y_{91}	y_{92}	y_{93}	y_{94}	$(S/N)_9$

In Table 2, one can notice that, the nine experiments (L_9) selected based on the three levels of four parameter values must each be repeated for two levels of three noise values (L_4) yielding a total of 36 experiments, as opposed to a total of $3^4 \times 2^3 = 648$ (full-factorial) combinations. Once the nine S/N values have been determined, a response surface is fitted to these nine data points—(A, B, C, and D)$_i$ versus the $(S/N)_i$ values. A search through the five-dimensional response surface, then, determines the optimal parameter values for (A, B, C, and D), where the function value to be maximized is the S/N ratio.

Tolerance Design

Taguchi defines quality as an inverse function of a desired characteristic of a product and treats it as a loss. That is, every product has an associated quality loss, which could be zero if the product has the exact (expected) characteristic from the consumer's point of view. This quality "loss function" (for the output) is defined by $L(y)$. If one assumes that process noise is normally distributed and that the mean value of this distribution is the expected output value by the customer—nominal is the best—the loss function can be defined as

$$L(y) = k(y - \mu)^2 \tag{3.9}$$

where k is a cost coefficient defined by

$$k = \frac{\text{consumer's loss}}{\text{functional tolerance}} = \frac{A_0}{\Delta_0^2} \tag{3.10}$$

The above concept of loss experienced by a customer, who expects the product to yield an output equal in magnitude to the mean, but which is actually a distance away from the mean $(y - \mu)$, is shown in Fig. 9. The product's output value, y, although it could be within its functional tolerance limits, defined by $\mu \pm \Delta_0$, still represents a loss to the customer, since it is not exactly equal to the mean value, where $L(y = m) = 0$. Naturally, the coefficient k in Eq. (3.9) differs from product to product and would include cost elements such as replacement, repair, service, customer loyalty, etc. Thus it is difficult to measure.

Taguchi proposes further tightening of the functional tolerance interval, specified for the desired output of the product according to the cost coefficient, k, by the cost of "fixing" the "problem" before it is shipped to the customer, as

$$\Delta = \left(\frac{A}{A_0}\right)^{1/2} \Delta_0 \tag{3.11}$$

where Δ is the new tolerance tightened according to the cost of fixing the deviation in-house, A. If one would assume that fixing a problem before the

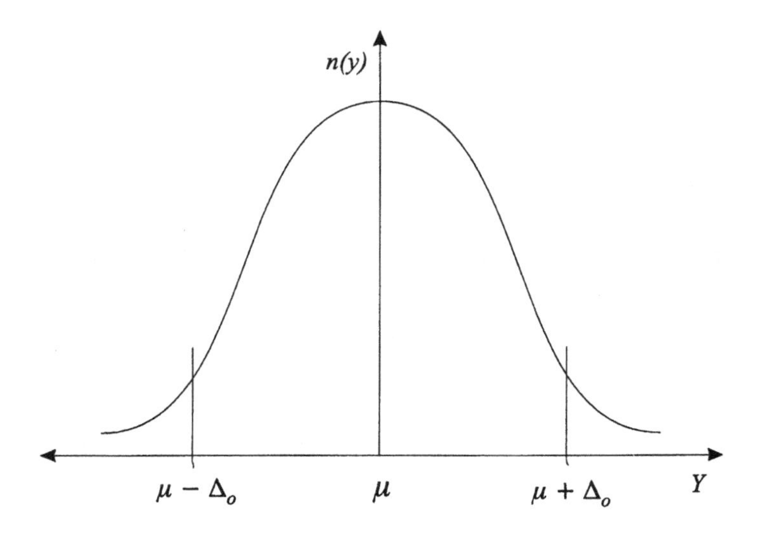

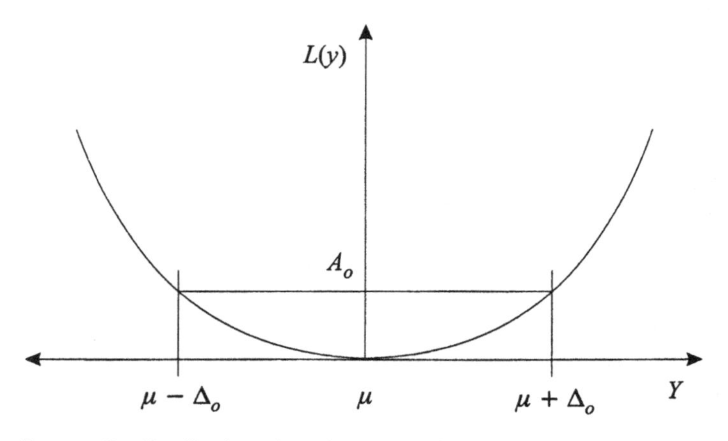

FIGURE 9 Quality loss function, $L(y)$, for a normally distributed, $n(y)$, output.

product is shipped would cost the manufacturer less than the consequence (cost of lost quality to the customer), i.e., $A < A_0$, then $\Delta < \Delta_0$. Thus, manufacturers should choose their process variability (variance of $n(y)$ Fig. 9) according to the tightened tolerance limits, Δ, and not according to Δ_0.

In parameter design, once the tightened tolerance level, Δ, is determined for the product's output, y, it must be propagated downward for specifying tolerances on individual (optimal) parameter values, so that their combination yields the expected Δ on the product's output.

The overall conclusion of Taguchi's studies has been that manufacturers must minimize the variability of their process as much as is economically viable, since customers do experience a loss in quality when they do not receive the mean output value that they expect. The tighter the process variability is, the higher the percentage of products within the (engineering) tolerance limits would be, and furthermore, the less the total cost of quality loss, which can be defined as the integral of $L(y)$ from $y = \mu - \Delta$ to $\mu + \Delta$. The product quality topic is further discussed in Chapter 16 of this book.

3.4 GROUP-TECHNOLOGY-BASED DESIGN

Group technology (GT) was first proposed and developed in Europe (prior to WWII) and exported to North America in the later decades of the 20th century with the start of widespread implementation of flexible manufacturing systems (FMSs). Although it has primarily been proposed for the increased efficiency of manufacturing activities, GT can be very effectively applied to engineering design. The premise of GT philosophy is fast access to pertinent (similar) historical data available within the enterprise and its modification for the design and fabrication of new parts within the same family. In this chapter, we will concentrate on the benefits of GT in the design of products.

3.4.1 History of Group Technology

It has been commonly agreed upon that S. P. Mitrofanov (of the former USSR) is the originator of GT. It is also accepted that this development was based on the earlier work of A. P. Sokolovski (also of the former USSR) in the 1930s, who argued that "parts of similar geometry and materials should be manufactured in the same way by standardized technological processes." Mitrofanov elaborated on this definition by advocating the use of physical cells (machines placed in closed proximity).

The initial work of Mitrofanov was adopted in the U.K. by E. G. Brisch in the 1950s, who later with Birn developed the Brisch and Birn coding and classification method. Next, in the 1960s, came the work of Prof. H. Opitz (of the former Federal Republic of Germany), who developed the most commonly used GT system (OPITZ) in Europe. Opitz's work was originally targeted for the investigation of part statistics in the machine-tool industry. However, since then it has been used for design retrieval, process planning, and cell formation. Other GT developments in Europe included the VUOSO system developed in the former Czechoslovakia for the optimization of machine tool design, the PGM system developed in Sweden for design retrieval, and The IAMA system developed in the former Yugoslavia for manufacturing (very similar to OPITZ). All these efforts occurred in the 1960s.

The widespread utilization of GT in the U.S.A. started first with the adaptation of the BRISCH–BIRN system named CODE for specifying geometry and function. The commercialization of the MICLASS (Metal Institute classification system) GT system, originally developed in the Netherlands by the Organization for Applied Scientific Research (TNO) in the 1970s, followed.

3.4.2 Classification

Classification is the most important element of GT—it refers to a logical and systematic way of grouping things based on their similarities but then subgrouping them according to their differences. The four principles developed by Brisch for the classification of a population of parts are as follows:

All-embracing: The adopted classification system must be inclusive. It must classify all current parts within the population at hand and also allow for future product features.

Mutually exclusive: Once the classification structure has been developed, a part should have only one class to be included within. The system must be mutually exclusive for achieving an unambiguous distribution of parts.

Based on permanent features: The classification system must utilize only the final geometrical features of the part and not any intermediate shapes.

From a user's point of view: The rules of classification must be obvious to the users, and thus should be developed based on extensive interviews with all designers within the company.

The first step in implementing a classification system is a detailed review of past products and identification of primary similarities according to, for example, overall geometry (rotational versus prismatic), presence of external features (grooves, key slots, etc.) or internal features (holes, threads, etc.). Uniformity of class sizes is desirable, but owing to increased speeds of current computers, which can search databases very quickly, it is no longer a necessity. Once a representative set of historical data has been examined, and the overall classes have been determined, the next step is examining each class for differences. This step is the most critical task in classification—one must look for representative features

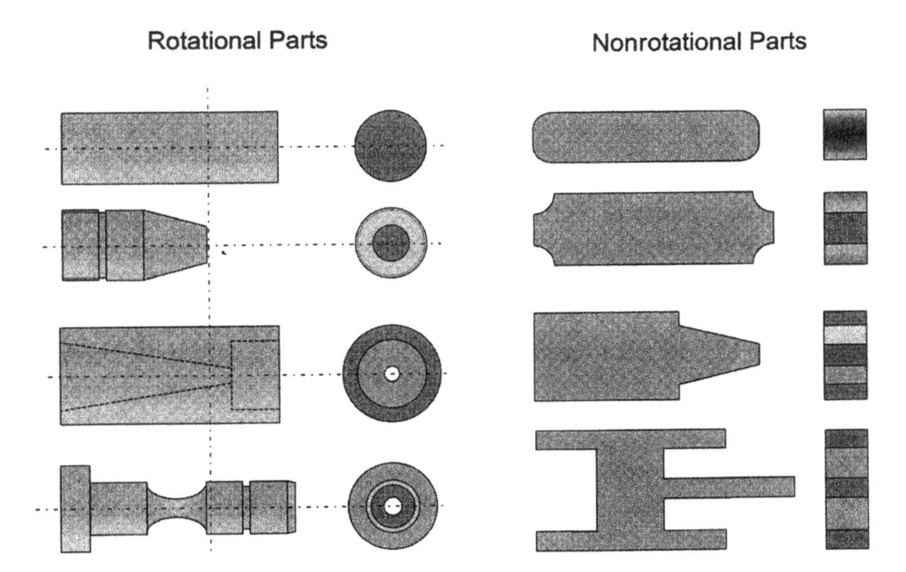

FIGURE 10 Geometrically similar parts.

that will differentiate parts and not for unique features that may never be encountered in other parts. That is, one would, actually, expect these features to be found on other past or future parts, so that when we eventually search our database we would discover past parts with similar characteristics and start our new design based on the utilization of a most similar past part—one that has the maximum number of similar features (Fig. 10).

A second level of features in a GT system would include ratios of diameter to length for rotational parts or ratios of maximum-dimension-to-minimum-dimension for prismatic parts (but rarely actual dimensions). Other features could be the presence of external or internal steps, specific shapes of external or internal features, presence of threads/teeth, etc. One should recall that classification at the first or subsequent levels of features may consider characteristics, such as material type, surface finish, and tolerances, which would not be very useful to geometric modeling of a part, but critical for the use of GT in process planning and assignment of parts to certain manufacturing workcells.

In conclusion to the above classification discussion, one must note that future users of GT can easily develop their own classification system after a careful review of the literature or past developments. There exist only a very few available commercial GT systems, and these should never be treated as turnkey systems. Classification is best achieved by expert, in-house designers.

3.4.3 Coding

Coding in the context of GT refers to the utilization of an alphanumeric system that will allow us to access past data with maximum efficiency. A GT coding procedure must be logical and concise. Below is a list of guidelines developed by Brisch and partners:

A code should not exceed five characters in length without a break in the string.
A code should be of fixed length and pattern.
All-numeric codes are preferable—causing fewer errors.
Alphanumeric (mixed) codes must have fixed fields for alphabetic and numeric codes, respectively—though they should be avoided if possible.

As examples, consider the following postal codes: a five-digit code followed by an additional 4-digit code for the U.S. (e.g., 17123-9254) versus the two-part, three-digit alphanumeric code in Canada (e.g., M5S 3G8).

All GT-based systems would have one of the following coding structures (Fig. 11):

Monocode: It can store a large amount of information within a short (length) code due to its hierarchical structure. That is, the meaning of any digit in the code is dependent on the value of the preceding digit, resulting in a tree-structure representation of a product's characteristics. Monocodes cannot be easily interpreted by people by simply examining the long code of a part. However, such codes could be easily decoded by computers.
Polycode: This can only store a limited amount of information, since each character is of fixed meaning—i.e., reserved (fixed) attribute. Although easily recognizable by people (in meaning), such codes can be excessive in length.
Hybrid: This is a mixture of mono- and polycodes. That is, it has a mixed structure, in which some fields have reserved (fixed) attribute meanings, regardless of the meaning of the preceding digits. Hybrid codes can be utilized for classification systems that yield group sizes of nonuniform size.

Let us now briefly review the (hybrid) MICLASS system as an example: it comprises two primary sections, a 12-digit first part and an

Monocode

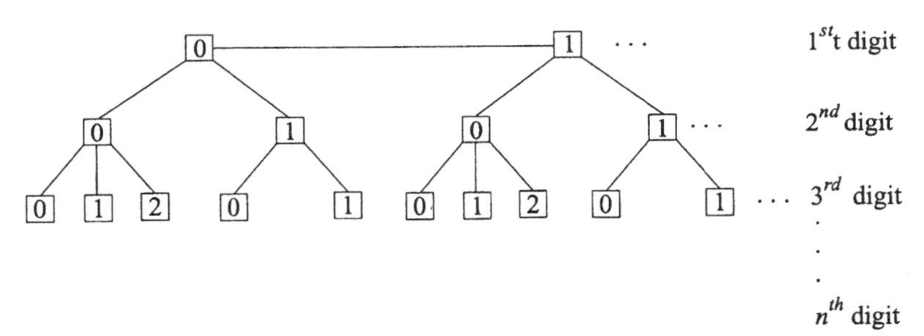

Polycode

2	4	7	3	. . . .	5
1st	2nd	3rd	4th		n^{th} digit

Hybrid Code

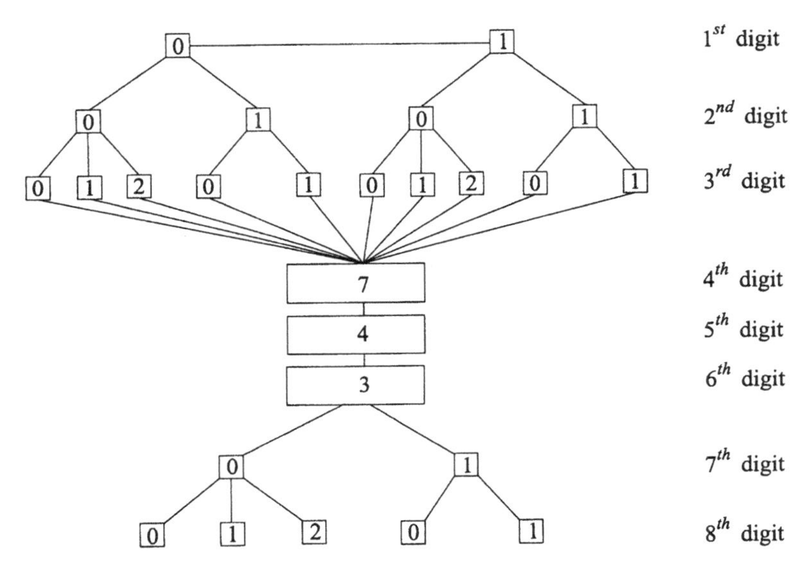

FIGURE 11 Coding systems.

additional 18-digit second part. The first four digits represent the overall geometry of a part:

1st digit	Main shape (rotational, nonrotational, special)
2nd and 3rd digits	Shape elements (stepped, with threads)
4th digit	Position of shape elements
5th and 6th digits	Main dimensions (maximum dimension less than X)
7th digit	Dimension ratio
8th digit	Auxiliary dimension
9th and 10th digits	Tolerance codes (range of tolerance on dimensions and surface finish)
11th and 12th digits	Material codes (nonferrous versus composite)

3.4.4 Implementation

Anyone who has visited design departments of small-batch manufacturing companies, up to the early 1980s, would have noticed large numbers of cabinets full of part drawings stored for potential future use, but later completely forgotten about (or "lost in the pile") due to the lack of a logical classification and coding system. With the introduction of CAD systems, after the 1970s, the filing cabinets have been complemented, first with large numbers of magnetic tapes, later with (soft) magnetic discs, and finally with today's common hard drives. When proposing GT systems to these companies, in early 1970s, a common reply received was that they indeed did have a coding system that assigned numbers and names to these product designs, which were recorded on a log book! GT, however, is based on the utilization of a classification and coding system that would allow designers to access earlier product designs based on a select set of similarities and not simply sequentially numbering these or naming them according to their functions. Thus, today, one must still emphasize this principle in attempting to sell GT to manufacturing engineers.

Once a company has developed and installed a GT system, the first decision at hand is how to start. If it is economically feasible, a large set of past products should be coded (this step can take 1 to 6 person-months, depending on the availability of a menu-driven computer-based coding system as well as on the amount of information to be stored). An alternative would be simply to code only new parts—which would postpone a meaningful usage of the GT system by at least one year.

With the availability of an effective database of past designs, a designer can code a new part, based on available sketchy information, and request the GT system to identify and retrieve the most similar part model from the database. The designer must subsequently decide whether it would be more economical to modify this past model rather than starting from scratch. The worst case scenario is the time wasted on the search that would, normally, take less than a few minutes. The time spent on coding the new (future) part is not wasted, since this code will be used when storing the new part in the database.

Several other points can be made at this time of discussing the use of GT in design efforts.

Classification of a population of similar products can help manufacturers in standardizing parts or even deciding on how to modularize their products.

If the GT system does include a component for process planning, fixture selection, and other manufacturing issues, at the time of information retrieval for the most similar past product, the product development team can concurrently review these pieces of information as well and make more educated design decisions on manufacturability, etc.

GT classification and coding systems could be used in conjunction with other methods developed for feature-based design, where CAD-based solid models are automatically analyzed for similar geometric (form) features, as will be discussed in Chapter 4.

REVIEW QUESTIONS

1. In the context of axiomatic design, define the functional requirements (FRs) and design parameters (DPs) of a product. As examples, consider two household/office products and define their FRs and DPs.
2. Review the two design axioms proposed by Suh and attempt to propose a third axiom that would be independent of and add value to the original two.
3. In the context of axiomatic design, describe the following three design classifications: uncoupled design, coupled design, and decoupled design. Which would you prefer and why?
4. What is the primary purpose of a design-for-manufacturing approach? Describe one design-for-manufacturing guideline each for casting, injection molding, forging, and machining.
5. What is the primary purpose of a design-for-assembly approach? As examples, consider two multicomponent household/office products and comment on whether they were designed for assembly efficiency.

6. What is the primary purpose of a design-for-environment approach? As an example consider the use of disposable coffee/tea cups versus the use of washable (long-term, reusable) mugs. Some argue that the latter consume more resources to manufacture and maintain than the former. Discuss both sides of the argument.
7. Multicomponent products should be designed for recycling. Describe the three most common forms of recycling.
8. Describe the data-mining activity in analyzing failed products and its benefit as a feedback tool for the design of future products.
9. What is parameter and tolerance design? What is design of experiments (DOE)? What is response-surface optimization? How can product design benefit from DOE?
10. Describe the Taguchi method/approach to DOE and review his proposed parameter-design and tolerance-design activities.
11. What is the primary purpose of a group-technology (GT) based design approach?
12. Describe the typical steps of implementing a GT-based design strategy in a job-shop type manufacturing enterprise that designs and fabricates similar-geometry, make-to-order products, for example injection molds for thin-walled plastic containers for the food-packaging industry.

DISCUSSION QUESTIONS

1. Axiomatic design rightfully argues that products/systems must be designed based on two fundamental axioms for efficient manufacturing and effective utilization by the customer. The functional requirements of the product/system should be addressed as independently as possible, and simplicity of design should be an important objective. Review several products/systems that have been, or could have been, designed and manufactured in accordance with these axioms (e.g., light dimmers that are also used as on/off switches, auto-focus cameras, etc.). Discuss whether these axioms might not always necessarily lead to better (profitable) designs for all products/systems. Provide an example if you agree with this statement.
2. Quality improvement is a manufacturing strategy that should be adopted by all enterprises. Although quality control is a primary concern for any manufacturing company, engineers should attempt to improve quality. In statistical terms, all variances should be minimized, and furthermore, where applicable, the mean values should be increased (e.g., product life, strength, etc.) or decreased (e.g.,

weight) appropriately. Discuss the quality-improvement issue and suggest ways of achieving continual improvements. Discuss also whether companies should concentrate on gaining market share through improved product performance or/and quality or only through cost/price.

3. Composite materials have been increasingly developed and used widely owing to their improved mechanical/electrical/chemical properties when compared to their base (matrix) material. For example, the use of glass, carbon, and Kevlar fibers in polymer base composites has significantly increased their employment in the automotive products and sports products industries. The concept of composite materials, however, may be in direct conflict with environmental and other concerns, which advocate that products should be designed so that material mix is minimized or totally avoided for ease of manufacturing and/or recycling (including decomposition) purposes. Discuss the above issues in favor of continuing to use composite materials; otherwise, propose alternatives.
4. Design of experiments (DOE) is a statistical approach that can be used in the design of physical or simulation-based experiments for the determination of optimal variable values. Such factorial-based experiments help engineers in the narrowing of the field of search to those parameters that have the greatest impact on the performance of the product as well as limiting the combinatoric number of variations of these variables. Discuss the role of DOE in the overall design (synthesis/analysis) of a product, while stating advantages, benefits, etc.
5. Discuss the need of developing a GT-based classification and coding system in-house as opposed to purchasing an already available (generic) commercial package.
6. Would several different GT-based classification and coding systems be needed in a company for different objectives? That is, would one system be needed for design, one system for manufacturing planning and yet another for cost engineering?
7. Group technology (GT) relies on the availability of an efficient coding system that could identify past similar product codes. Naturally, the corresponding search engine must be designed so that it closely follows the classification method employed. Two codes that are identical except for a single digit that differs by one value (e.g., 24708 versus 23708) could refer to totally dissimilar parts. Discuss the process of (tentatively) GT coding a new part under development (not in existence yet), searching the company database for the most similar past product, and proceeding from that point. Recall that two parts having identical GT codes are just "most similar" and not necessarily identical.

8. Failed products may provide very valuable information to manufacturers for immediate corrective actions on the design and manufacturing of current and/or future lines of products. Discuss how would you collect and analyze product failure (or survival) data for industries such as passenger vehicles, children's toys, and computer software.
9. Discuss the concept of progressively increasing cost of changes to a product as it moves from the design stage to full production and distribution. How could you minimize necessary design changes to a product, especially for those that have very short development cycles, such as portable communication devices?

BIBLIOGRAPHY

Allenby, Braden R. (1999). *Industrial Ecology: Policy Framework and Implementation*. Upper Saddle River, NJ: Prentice Hall.

Benhabib, B., Charania, M., Cupillari, A., Fenton, R. G., Turksen, I. B. (1990). CIM Implementation for Job Shop Environments—Phase I: Computer Aided Design, Production Planning and Shop Floor Control. *Int. J. of Production Planning and Control* 1(4):235–250.

Boothroyd, Geoffrey, Dewhurst, Peter, Knight, Winston (1994). *Product Design for Manufacture and Assembly*. New York: Marcel Dekker.

Dieter, George Ellwood (1991). *Engineering Design: A Materials and Processing Approach*. New York: McGraw-Hill.

Dixon, John R., Poli, Corrado (1995). *Engineering Design and Design for Manufacturing: A Structured Approach*. Conway, MA: Field Stone Publishers.

Ealey, Lance A. (1988). *Quality by Design: Taguchi Methods and U.S. Industry*. Dearborn, MI: ASI Press.

Graedel, T. E., Allenby, B. R. (1995). *Industrial Ecology*. Englewood Cliffs, NJ: Prentice Hall.

Hildebrand, Francis B. (1956). *Introduction to Numerical Analysis*. New York: McGraw-Hill.

Huang, G. Q., ed. (1996). *Design for X: Concurrent Engineering Imperatives*. New York: Chapman and Hall.

Hyde, William F. (1981). *Improving Productivity by Classification, Coding and Database Standardization: The Key to Maximizing CAD/CAM and Group technology*. New York: Marcel Dekker.

Jiang, Z., Shu, L. H., Benhabib, B. (June 2000). Reliability analysis of non-constant-size populations of parts in design for remanufacture. *ASME Trans., J. of Mechanical Design* 122(2):172–178.

Kolovos, Thomas. A Classification and Coding System for Injection Mold Parts. M.Eng. thesis, Department of Mechanical Engineering, University of Toronto, 1988.

Lindbeck, John R., Wygant, Robert M. (1995). *Product Design and Manufacture*. Englewood Cliff, NJ: Prentice Hall.

Maffin, D. (1998). Engineering design models: context, theory and practice. *J. of Engineering Design* 9(4):315–327.

Mitrofanov, Sergei Petrovich (1966). *Scientific Principles of Group Technology*. Boston: National Lending Library for Science and Technology.

Opitz, H. (1970). *A Classification System to Describe Workpieces*. Elsford, NY: Pergamon Press.

Stoll, Henry W. (1999). *Product Design Methods and Practices*. New York: Marcel Dekker.

Suh, Nam P. (1990). *The Principles of Design*. New York: Oxford University Press.

Suresh, Nallan C., and Kay, John M., eds. (1998). *Group Technology and Cellular Manufacturing: A State-of-the-art Synthesis of Research and Practice*. Boston, MA: Kluwer Academic.

Taguchi, Genichi (1986). *Introduction to Quality Engineering: Designing Quality into Products and Processes*. Tokyo: Asian Productivity Organization.

Taguchi, Genichi (1993). *Taguchi on Robust Technology Development: Bringing Quality Engineering Upstream*. New York: ASME Press.

Walpole, Ronald E., Myers, Raymond H., Myers, Sharon L. (2002). *Probability and Statistics for Engineers and Scientists*. Upper Saddle River, NJ: Prentice Hall.

Zak, G., Fenton, R. G., Benhabib, B. (March 1994). Determination of the optimum cost-residual error trade-off in robot calibration. *ASME J. of Mechanical Design* 116(1):28–35.

4

Computer-Aided Design

Geometric modeling is the first step in the computer-aided engineering (CAE) analysis of a designed product. The objective is to encapsulate all geometric data pertaining to the part in a single model and specify all necessary material properties as additional information. In this context, solid modeling, as a branch of geometric modeling, refers to the geometric description of solid objects in their entirety. Solid models (1) must be complete: the graphical model must not be an ambiguous representation, (2) must have integrity: operation on geometric models must preserve integrity, such as maintaining the connection of edges at a point when it is moved, and (3) provide accuracy in modeling of complex shapes.

Solid modeling is a multifaceted operation. At the forefront, a user describes a geometric model, through a graphical representation, to the computer, which in turn stores this representation, in one format or another, and furthermore allows the manipulation of this representation through a set of mathematical transformations/operators/etc. Thus a user of a computer-aided design (CAD) system for solid modeling purposes should have a basic knowledge of computer graphics principles needed for the manipulation and storage of graphical data.

As a preamble to solid modeling, this chapter will first review geometric modeling principles and concepts in Sec. 2 and then address the topics of solid modeling techniques, feature-based design, and product-data exchange standards in Sec. 3 to 5.

4.1 GEOMETRIC MODELING—HISTORICAL DEVELOPMENT

Sketchpad is known as the first graphical user interface (GUI), developed at M.I.T. by I. E. Sutherland, capable of interpreting information sketched on a computer display monitor. The software was developed during the period 1960 to 1962 on a TX-2 computer and primarily utilized a light pen (in conjunction with a push button) for data input (points, straight lines, circles, etc.). (It is interesting to note that the period was also *marked* by the development of the APT, automatically programmed tool, computer language, also developed at MIT, for the programming of numerical-control machine tools, the former in the Electrical Engineering department and the latter in the Mechanical Engineering department.)

Topological data related to an object model was stored in the computer as a "ring" structure, novel to sketchpad. When the user moved a vertex, the object geometry was be self-adjusted accordingly by the movements of the attached edges. The software was also used for basic engineering analysis operations, such as computing distribution of forces on the member links of a truss bridge.

The sketchpad system was followed by the development of DAC-1 (design augmented by computers) by General Motors in 1964 and CADAM (computer-aided design and manufacturing) by Lockheed Aircraft in 1965. The 1970s and early 1980s were marked by the development of numerous CAD systems, such as Computervision's Designer series that ran on proprietary hardware—however, only a handful of these systems survived beyond the late 1990s. Today, Pro/Engineer by Parametric Technology Corporation and I-DEAS by Structural Dynamics Research Corporation (SDRC) are the two primary CAD software packages that hold a large share of the CAD market. Both packages run on microcomputer (SUN, HP, etc.) as well personal computer platforms (IBM, Dell, etc.).

4.2 BASICS OF GEOMETRIC MODELING

4.2.1 Points and Curves

Points are the simplest geometric entities normally represented in Cartesian space by three coordinates (x, y, z). Points are also referred to as vertices when discussed in the context of bounding a line (or an edge of a surface).

Three-dimensional curves, in turn, can be represented in a parametric form, as a function of a single variable $u \in [0, 1]$:

$$x = x(u) \qquad y = y(u) \qquad \text{and} \qquad z = z(u) \tag{4.1}$$

Any point on such a parametric curve is defined by the components of the vector $\mathbf{p}(u)$. Thus the boundary conditions of a parametric curve are defined by the vectors $[\mathbf{p}(0), \mathbf{p}(1), \mathbf{p}'(0), \mathbf{p}'(1)]$, where

$$\mathbf{p}'(u) = \frac{d\mathbf{p}(u)}{du} \tag{4.2}$$

In parametric form, a straight line would be represented as

$$x = a + ku \qquad y = b + lu \qquad \text{and} \qquad z = c + mu \tag{4.3}$$

where (a, b, c) and (k, l, m) are constants. Similarly, a planar circle would be represented as,

$$x = x_c + r\cos 2\pi u \qquad y = y_c + r\sin 2\pi u \qquad \text{and} \qquad z = z_c \tag{4.4}$$

where r is the radius of the circle and (x_c, y_c, z_c) are constants. A circular arc, in turn, is represented as

$$x = x_c + r\cos u \qquad y = y_c + r\sin u \qquad \text{and} \qquad z = z_c \tag{4.5}$$

where $u \in [u_s, u_e]$—u_s and u_e represent the start and end points of the arc.

Although any curve can be represented by a corresponding parametric set of equations, in practice, several curves might have to be joined in order to achieve a specific part geometry. For such an objective, the two curves s_1 and s_2 can be manipulated in Cartesian space and joined end to end while satisfying the continuity constraint. That is,

$$\mathbf{p}_1(1) = \mathbf{p}_2(0) \quad \mathbf{p}_1'(1) = \mathbf{p}_2'(0) \quad \text{and} \quad \mathbf{p}_1''(1) = \mathbf{p}_2''(0) \tag{4.6}$$

where $\mathbf{p}'$ and $\mathbf{p}''$ are the first and second parametric derivatives, respectively. In Eq. (4.6), the first two constraints simply ensure continuity of end-to-end meeting and having identical slopes at this point, respectively. The third constraint (i.e., continuity of second derivatives), on the other hand, further ensures that the two curves have equal curvature at the joining point.

Curve Fitting

On many occasions a designer faces the task of curve fitting to a set of data points collected through experimentation. In industrial design, for example, this task would correspond to approximating a handcrafted surface by a mathematical representation, where a coordinate-measuring

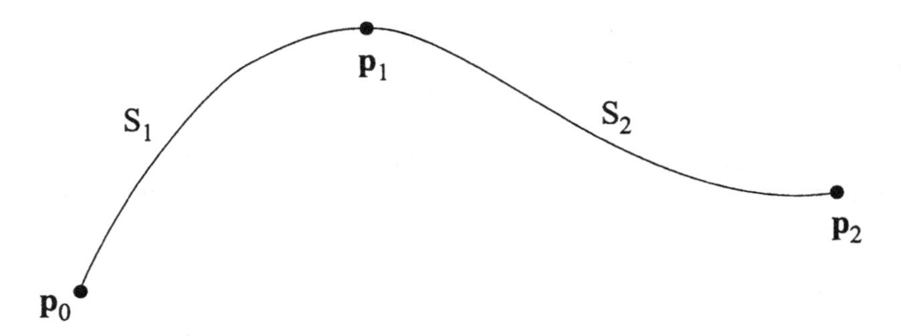

FIGURE 1 Cubic spline fit to three points.

machine (CMM) would be used to determine a sufficiently large number of points on the actual surface.

Two possible solutions to the curve-fitting problem would be the least-squares fit, where the best curve would most likely not pass through any one of the points, and the spline fit, where a set of curves would be determined that pass through all the given points and furthermore provide the designer with any desired degree of continuity at meeting points (i.e., matching higher-order derivatives), as in Eq. (4.6). In both cases, the mathematical problem at hand is the determination of the coefficients of the equations.

As an example, let us consider a cubic spline fit to three points, ($\mathbf{p}_0$, $\mathbf{p}_1$, $\mathbf{p}_2$). The designer is required to find the coefficients of two curves (both third-degree polynomials), one from $\mathbf{p}_0$ to $\mathbf{p}_1$ and another from $\mathbf{p}_1$ to $\mathbf{p}_2$. The constraints imposed on this problem (i.e., finding simultaneously the coefficients of both curve representations) are (1) the coordinates of all the three points and (2) the desired first and second derivative values at the first and last points, $\mathbf{p}_0$ and $\mathbf{p}_2$, respectively. Additionally, the solution algorithm is required to determine the curves' coefficients such that the first and second derivatives of both match at the joining point, $\mathbf{p}_1$ (Fig. 1).

The coefficients of both sets of equations, c_{ijk}, $k = 1, 2$, can be described in a matrix form as

$$\begin{pmatrix} x_1 \\ y_1 \\ z_1 \end{pmatrix} = \begin{bmatrix} c_{111} & c_{121} & c_{131} & c_{141} \\ c_{211} & c_{221} & c_{231} & c_{241} \\ c_{311} & c_{321} & c_{331} & c_{341} \\ c_{411} & c_{421} & c_{431} & c_{441} \end{bmatrix} \begin{pmatrix} u^3 \\ u^2 \\ u \\ 1 \end{pmatrix} \tag{4.7}$$

$$\begin{pmatrix} x_2 \\ y_2 \\ z_2 \end{pmatrix} = \begin{bmatrix} c_{112} & c_{122} & c_{132} & c_{142} \\ c_{212} & c_{222} & c_{232} & c_{242} \\ c_{312} & c_{322} & c_{332} & c_{342} \\ c_{412} & c_{422} & c_{432} & c_{442} \end{bmatrix} \begin{pmatrix} u^3 \\ u^2 \\ u \\ 1 \end{pmatrix} \tag{4.8}$$

The above spline fit technique, though ensuring that the curves pass through all the given points and satisfy the boundary conditions, may yield curves with undesirable inflection points, especially when overly constrained (Fig. 2). In response to this problem, P. Bézier (a mechanical engineer) of the French automobile firm Renault developed the curve now known as the Bézier curve in the late 1960s.

A Bézier curve satisfies the following four conditions while attempting to approximate the given points (but not passing through all of them) (Fig. 3a). For $(n+1)$ points,

1. The curve must only interpolate the first and last control points $(\mathbf{p}_0, \mathbf{p}_n)$.
2. The order of the polynomial is defined by the number of control points considered, where

$$\mathbf{p}(u) = \sum_{i=0}^{n} \mathbf{p}_i \, B_{i,n}(u) \tag{4.9}$$

and

$$B_{i,n}(u) = \left[\frac{n!}{i!(n-i)!}\right] u^i (1-u)^{n-i} \tag{4.10}$$

For example, for four control points, $n+1=4$,

$$\mathbf{p}(u) = (1-u)^3 \mathbf{p}_0 + 3u(1-u)^2 \mathbf{p}_1 + 3u^2(1-u)\mathbf{p}_2 + u^3 \mathbf{p}_3 \tag{4.11}$$

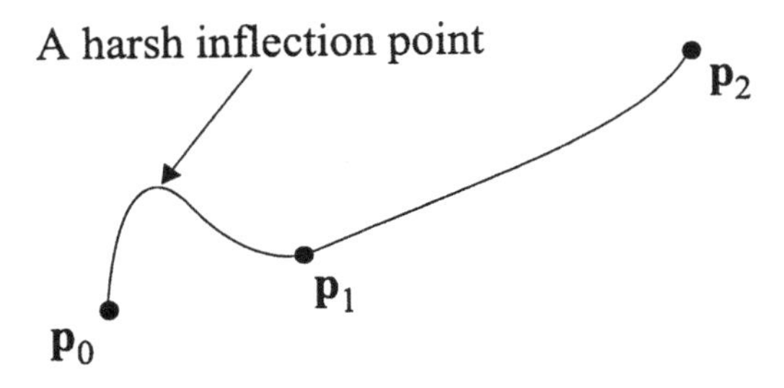

FIGURE 2 An undesirable spline fit.

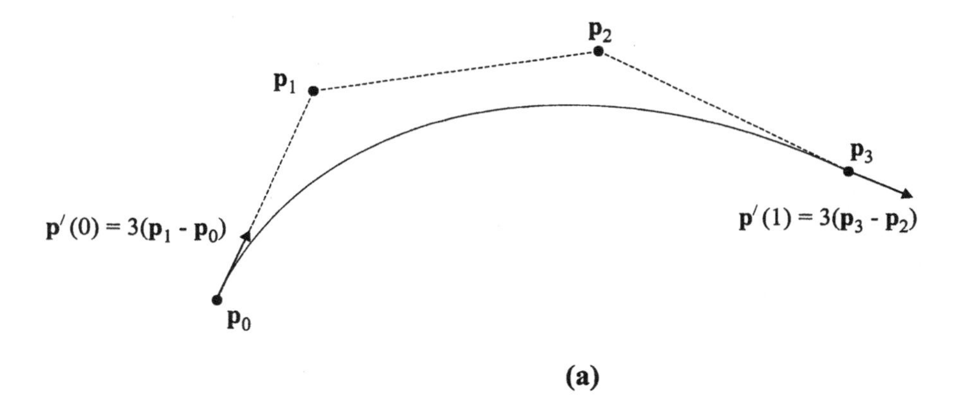

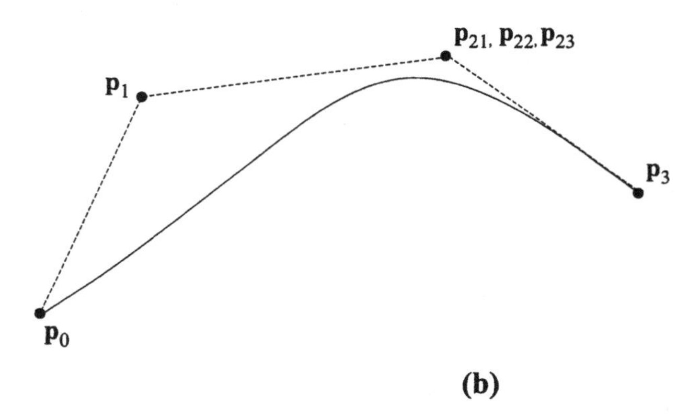

FIGURE 3 (a) Unweighted and (b) weighted Bézier curves.

where at $u = 0$, $\mathbf{p}(0) = \mathbf{p}_0$, and at $u = 1$, $\mathbf{p}(1) = \mathbf{p}_3$ satisfying Condition 1 above.

3. The curve satisfies r^{th} order derivatives at the first and last points only, where $r \leq n$ (for $n+1$ control points):

$$\mathbf{p}^r(0) = \frac{n!}{(n-r)!} \sum_{i=0}^{r} (-1)^{r-i} C(r,i)\mathbf{p}_i \tag{4.12}$$

and

$$\mathbf{p}^{r}(1) = \frac{n!}{(n-r)!}\sum_{i=0}^{r}(-1)^{i}\ C(r,i)\mathbf{p}_{n-i} \tag{4.13}$$

where

$$C(r,i) = \left[\frac{r!}{i!(r-i)!}\right]$$

The first two derivatives for a Bézier curve with four control points would be

$$\mathbf{p}'(0) = 3(\mathbf{p}_1 - \mathbf{p}_0) \qquad \mathbf{p}''(0) = 6(\mathbf{p}_2 - 2\mathbf{p}_1 + \mathbf{p}_0)$$
$$\mathbf{p}'(1) = 3(\mathbf{p}_3 - \mathbf{p}_2) \qquad \mathbf{p}''(1) = 6(\mathbf{p}_3 - 2\mathbf{p}_2 + \mathbf{p}_1)$$

4. The shape of the curve can be changed by emphasizing certain desired points by creating pseudopoints coinciding at the same location. For example, for the curve shown in Fig. 3b, we fit a Bézier curve to six points, three of which coincide, thus emphasizing the importance of that specific location.

4.2.2 Surfaces

Surface modeling is a natural extension of curve representation and an important step toward solid modeling. In three-dimensional space, a surface has the following parametric description:

$$x = x(u,w) \qquad y = y(u,w) \qquad \text{and} \qquad z = z(u,w) \tag{4.14}$$

where a point on this surface is defined by $\mathbf{p}(u,w)$, and u, $w \in [0, 1]$.

If one considers a patch of surface, the four vertices of this patch, $(\mathbf{p}_{00}, \mathbf{p}_{01}, \mathbf{p}_{10}, \mathbf{p}_{11})$, are defined by their respective coordinate values as well as by the two first-order derivatives at each vertex:

$$\begin{array}{ll} \mathbf{p}_{00}^{u} = \left.\dfrac{\partial \mathbf{p}}{\partial u}\right|_{u=0,w=0} & \mathbf{p}_{00}^{w} = \left.\dfrac{\partial \mathbf{p}}{\partial w}\right|_{u=0,w=0} \\ \vdots & \vdots \\ \mathbf{p}_{11}^{u} = \left.\dfrac{\partial \mathbf{p}}{\partial u}\right|_{u=1,w=1} & \mathbf{p}_{11}^{w} = \left.\dfrac{\partial \mathbf{p}}{\partial w}\right|_{u=1,w=1} \end{array} \tag{4.15}$$

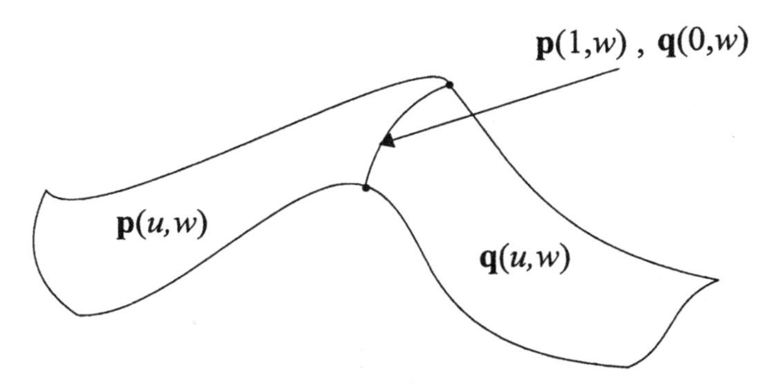

FIGURE 4 Patching of surfaces.

A unit normal vector at any point on this surface can be defined as

$$\mathbf{n}(u,w) = \frac{\left(\frac{\partial \mathbf{p}}{\partial u}\right) \times \left(\frac{\partial \mathbf{p}}{\partial w}\right)}{\left|\frac{\partial \mathbf{p}}{\partial u} \times \frac{\partial \mathbf{p}}{\partial w}\right|} \tag{4.16}$$

The unit normal is an important tool to be utilized in the geometric modeling of solids, usually required to point outward.

As in the case of curves, multiple surfaces can be patched together at their edges—that is two patches, $\mathbf{p}(u,w)$ and $\mathbf{q}(u,w)$, share a curve on each patch, for example $\mathbf{p}(1,w)$ and $\mathbf{q}(0,w)$ (Fig. 4)

Surface Fitting

In fitting a surface to a set of points, one can choose to carry out this operation via a number of spline-fitted, patched surfaces or by using one single "approximate surface," such as a Bézier surface. No matter what the method is, one needs to consider the first-order (and even second-order) order derivatives of the surfaces' boundary conditions.

The Bézier surface equation is defined as

$$\mathbf{p}(u,w) = \sum_{i=0}^{m} \sum_{j=0}^{n} \mathbf{p}_{ij} B_{i,m}(u) B_{j,n}(w) \tag{4.17}$$

where $\mathbf{p}_{ij}$ are the $(m+1)\times(n+1)$ control points, $B_{i,m}$ and $B_{j,n}$ are defined as in Eq. (4.10), and u, $w \in [0, 1]$. As in the Bézier curve case, only a limited number of control points actually lie on the Bézier surface [(e.g., the four points in Fig. 5: $(u,w)=(0,0)$, $(0,1)$, $(1,0)$ and $(1,1)$]. The remaining points control the curvature of the Bézier surface. Furthermore, as in the case of

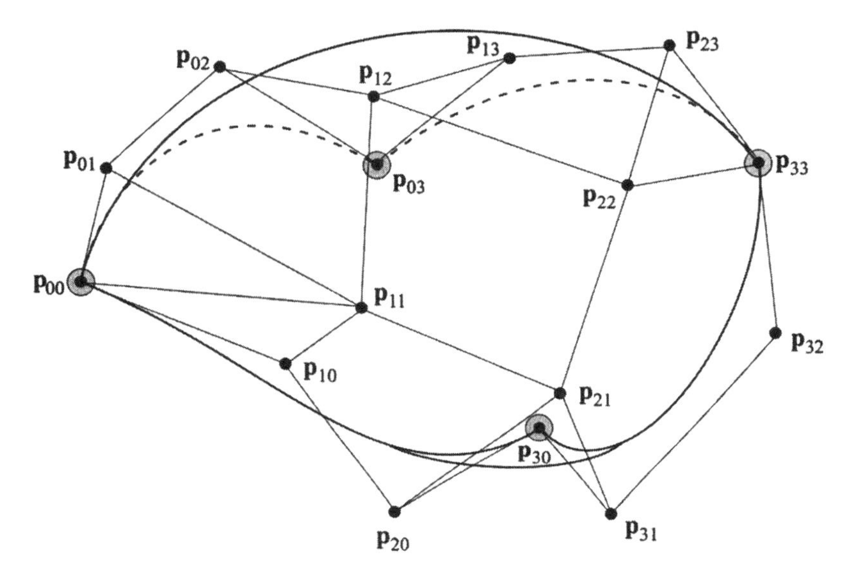

FIGURE 5 4×4 Bézier surface.

the Bézier curve, certain control points can be emphasized by creating a larger number of coinciding pseudopoints at a specific desired location.

4.2.3 Solids

Several solid modeling techniques were developed over the past two decades, three of which will be detailed below in Sec. 4.3. In this subsection, however, a brief review of pertinent issues will be addressed to provide a transition from the above discussion on surface modeling to these solid-modeling techniques.

A solid can be described as a "hyperpatch" by the parametric representation

$$x = x(u, v, w) \qquad y = y(u, v, w) \qquad \text{and} \qquad z = z(u, v, w) \tag{4.18}$$

where $u, v, w \in [0,1]$ (Fig. 6). In Eq. (4.18), fixing the value of any one of the three parameters would result in the definition of a surface that can be on or within the solid.

The simplest example of a solid is a rectangular prism obtained by substituting the proper constraints into Eq. (4.18) to yield

$$\begin{aligned} x &= a + (b-a)u \\ y &= c + (d-c)v \\ z &= e + (f-e)w \end{aligned} \tag{4.19}$$

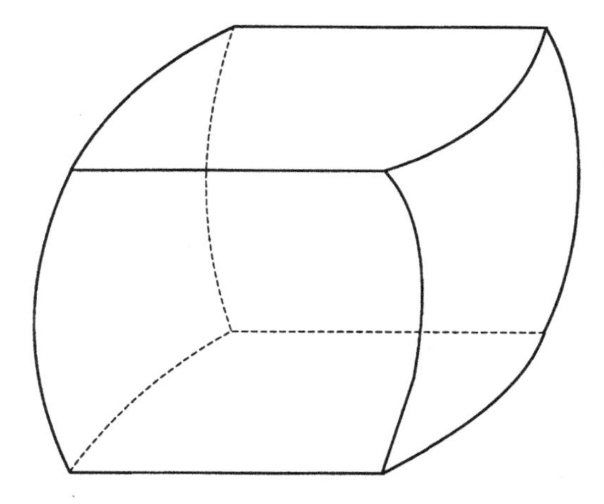

FIGURE 6 A solid.

One can note that the above equation describes points that are on the surface as well as inside the prism.

Solid models of objects must satisfy the following criteria:

Rigidity: The shape of the object remains fixed as it is manipulated in Cartesian space (i.e., translated and/or rotated).

Homogeneity: All boundaries of the model must be in contact with and enclosing the volume of the solid.

Finiteness: No dimension of the model can be infinite in magnitude.

Divisibility: The solid model must yield valid subvolumes when divided by Boolean operations.

4.3 SOLID MODELING

Computer-aided design (CAD) software packages are based on the mathematical principles of geometric modeling, some of which were discussed above in Sec. 4.2. Prior to the discussion of solid modeling techniques commonly employed by CAD systems, it will be beneficial to list briefly some of the tools that these systems utilize in manipulating curves, surfaces, and solids:

Segmentation: This is a division of a curve or a surface into several segments, while preserving the characteristics of the original entity in every one of the segments. This objective is achieved through reparameterization of the original entity.

Intersection: The intersection of two curves in three-dimensional space is a root-finding problem (for determining the coordinates of the intersection point). It is a nonlinear problem, for which numerical methods must be utilized. The complexity of the problem is increased for surface-with-curve and surface-with-surface intersections. Numerical methods developed for this purpose may follow a procedure such as the one developed by H. G. Timmer: Select one of the surfaces and create a grid structure; examine all grids for possible intersection points; trace individual intersection segments within each grid; order and connect the individual segments; and parameterize the intersection curve.

Transformation: Geometric transformation of an object may involve translation, rotation, or even scaling of its shape. Homogeneous transformation is the most efficient way of carrying out translation and rotation simultaneously—it defines the transformation of a coordinate frame attached to an entity with respect to a fixed "world" coordinate frame. It is defined by a (4 × 4) matrix,

$$T = \begin{bmatrix} R_{3\times3} & d_{3\times1} \\ 0\ 0\ 0 & 1 \end{bmatrix}_{4\times4} \tag{4.20}$$

where $R_{3\times3}$ is a square rotational matrix defining three successive rotations with respect to the world coordinate frame and $d_{3\times1}$ is the translation vector defining three simultaneous translations along the three orthogonal axes of the world coordinate frame.

Scaling: The size of a geometric entity (curve or surface) may be changed by scaling its geometric coefficients pointwise. The elements of the scaling matrix can be chosen to scale down the entity (with positive element values less than 1) or scale it up (with element values greater than 1). (Negative scaling factors cause reflection.)

Boolean operations: Set theory is an important tool in combining solid geometries (usually, simple shapes, "primitives"). The term set refers to a collection of (well-defined) objects—points in geometric modeling. Different sets can be combined, through Boolean operators, to create new sets. The three common Boolean operators are union, intersection, and complement (Fig. 7):

Union $\quad C = A \cup B,$
Intersection $\quad D = A \cap B,$
Complement $\quad E = (A \cup B)' = S - (A \cup B)$

The new set E above includes all the elements in the universal set, S, which are not included in A or B.

The three most common solid modeling techniques used by CAD systems are primitive instancing and sweeping, construction, and boundary representation. Decomposition models that describe solids based on a

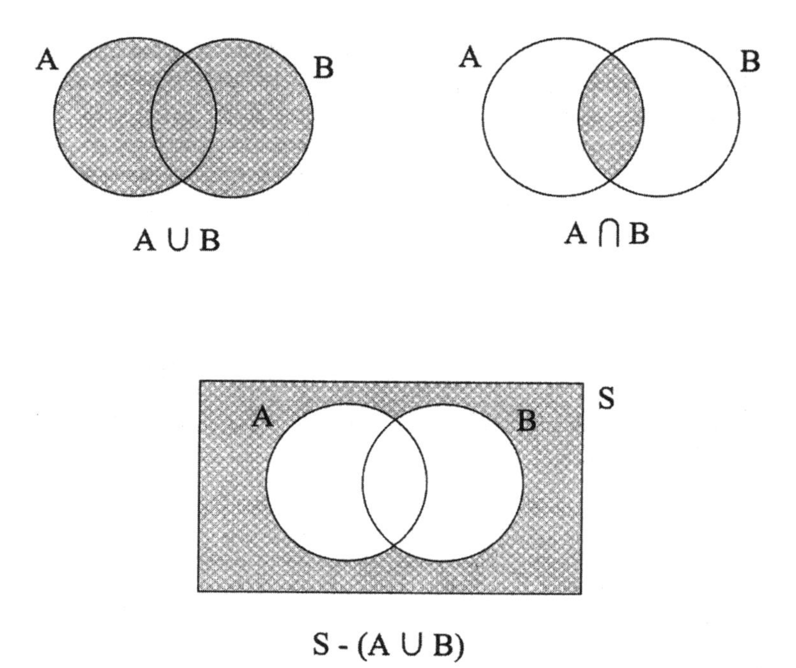

FIGURE 7 Venn diagrams of Boolean operations.

combination of geometric blocks will not be discussed in this chapter. We will, however, discuss briefly the issue of conversion of a solid representation from one model to another, for example from a constructive solid geometry model to a boundary representation model.

4.3.1 Primitive Instancing and Sweeping

Primitive instancing refers to the scaling of simple geometrical models (primitives) by manipulating one or more of their descriptive parameters, for example, elongating a cylinder, changing the dimensions of a rectangular prism, etc. As will be discussed below in Sec. 4.4, geometric primitives can play an integral role in feature-based design, where a set of (form) features are combined to generate a more complex model. It will also be shown that such primitives can be combined through Boolean operators for constructive solid geometry modeling.

Due to their simplicity, most geometric primitives can be generated by a sweeping ("extrusion") process, where a surface is either translated along spatial curve or rotated about it (Fig. 8). (The designer must be careful that

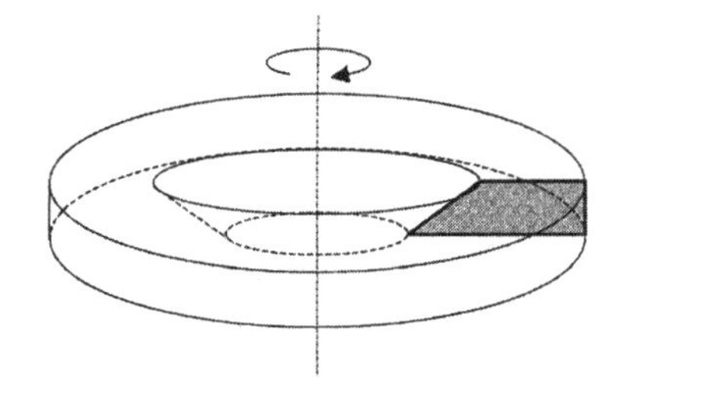

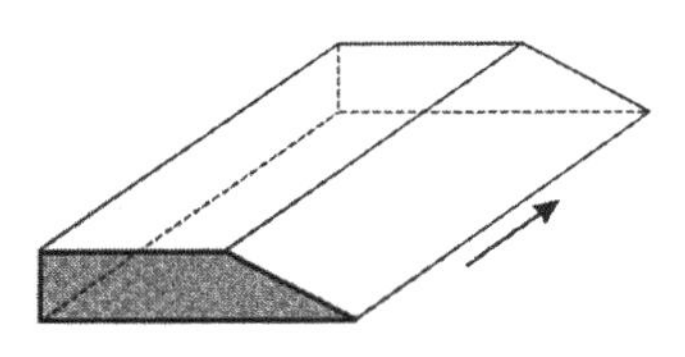

FIGURE 8 Sweeping of surfaces.

the end result is a valid solid.) In most cases, solid geometric models generated by a sweeping operation can be converted to construction and boundary representation models.

4.3.2 Constructive Solid Geometry

Constructive solid geometry (CSG) modelers allow designers to combine a set of primitives through Boolean operations. In the background (transparent to the user), these modelers represent and store the primitives as "half-space" models—these are simple geometric models comprising point sets bounded by a surface, i.e., points in three-dimensional space are defined as belonging to the half-space or being excluded. (An example half-space model would be that bounded by a cylindrical surface extending to infinity—points thus would be on and within the volume enveloped by the surface or be on the outside.) There do exist some CAD systems, however, that allow designers to work with bounded primitives, which are indeed a collection of patched half spaces themselves.

CSG-based solid models are represented as tree (or graph) structures. The leaves of the graph are the primitives, while the nodes that connect the branches are the Boolean operations applied on the individual (leaves) primitives (Fig. 9).

Naturally, CSG modelers rely on several geometric modeling tools discussed in this chapter: properly scaled primitives must be transformed (positioned and oriented) prior to their combinations; the modeler must determine exact intersection curves between the surfaces of the two primitives to be combined, and finally the modeler must use set theory to determine the new solid model obtained.

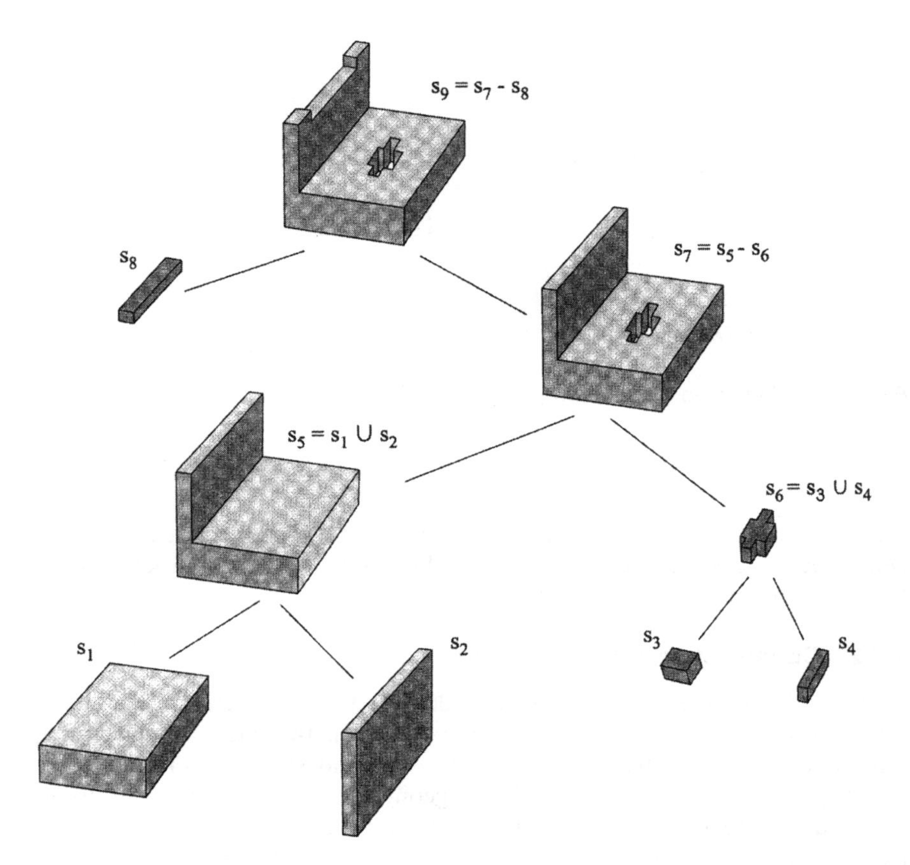

FIGURE 9 CSG model.

4.3.3 Boundary Representation

Boundary representation (B-Rep) models describe solids "topologically." That is, they rely on the notion that all solids are bounded by surfaces. Based on this surface-oriented view, a B-Rep model comprises faces, edges, and vertices, and each face has an unambiguous mathematical representation. A face may have several inner bounding loops in addition to the outer bounding curve. For example, a surface may have the bounding loops of holes/cavities included within it. Although B-Rep is a surface-oriented model, one can easily calculate the volumetric properties of the enclosed solid through integration.

Most engineering objects have either polyhedral or curved (cylindrical or spherical) surfaces. The former are easier and more intuitive to represent via their (finite in number) vertices and connected (linear) edges (Fig. 10a). For a cylindrical object, on the other hand, the side curved surface can be represented by one edge and two vertices, whereas the two opposite (circular) planar surfaces can be each represented by one edge and one vertex (coinciding with one of the vertices of the side surface) (Fig. 10b). A sphere can be represented by one face, one vertex, but no edges.

In the formal sense, a vertex is a unique point in Cartesian space defined by three coordinates. An edge is a finite-length curve bounded by two vertices—it must be non-self-intersecting. A loop is an ordered, directed collection of vertices and edges—i.e., a boundary. A face is a finite-size surface, non-self intersecting and bounded by one or more loops. The most common B-Rep modelers structure geometric data based on edge information, where a face is represented in terms of its loops. One can go a step

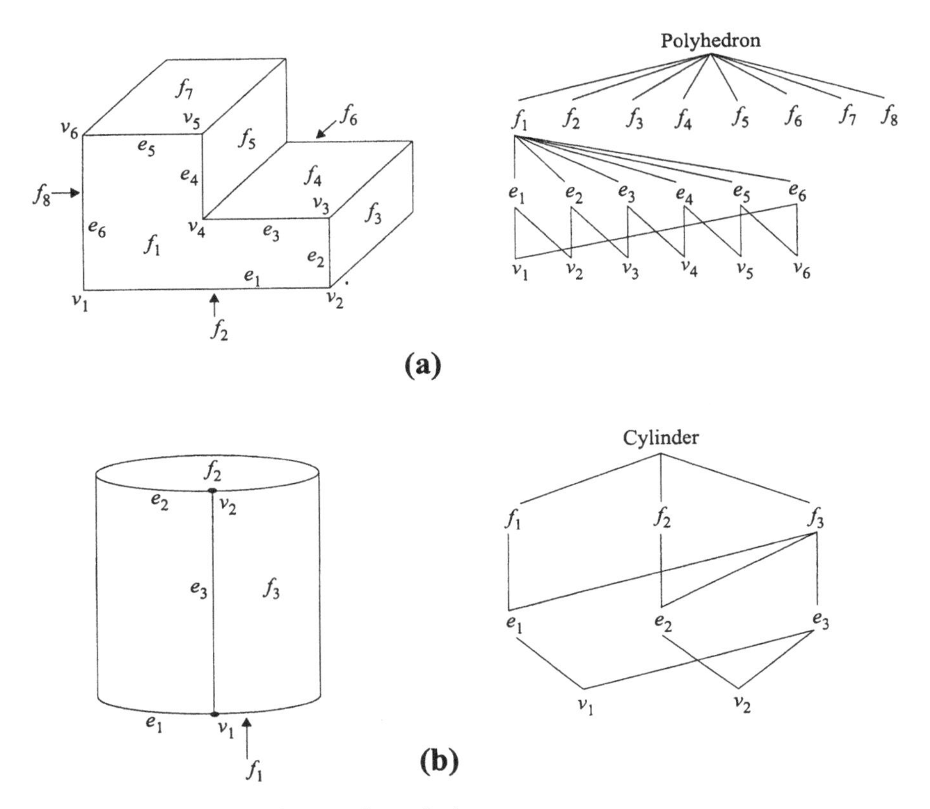

FIGURE 10 A polyhedron and a cylinder.

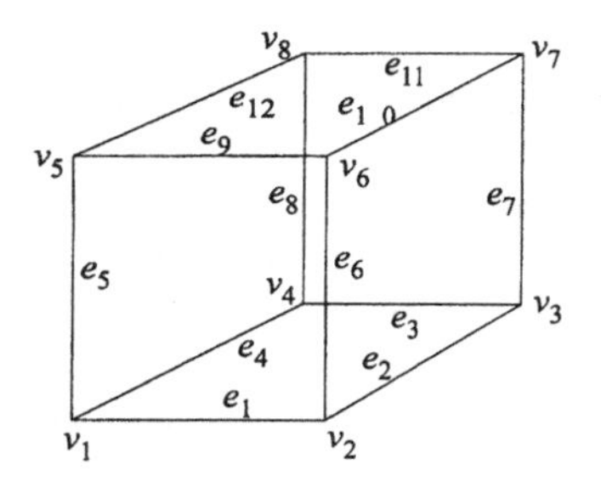

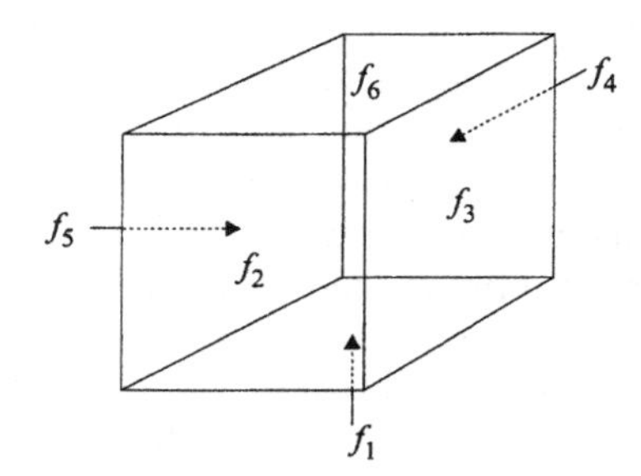

FIGURE 11 A polyhedron.

further by describing the adjacency of the edges through a directed search through the loops. The "winged-edge" data structure, first introduced by B. Baumgart, is commonly used for this purpose. It identifies a "first" edge for every face and a (loop) transverse direction for every edge, thus identifying the "next" edge on the loop. For example, let us consider the polyhedron in Fig. 11 and its partial winged-edge structure in Table 1, where cw is clockwise and ccw is counter-clockwise, ncw is next clockwise edge, pcw is previous clockwise edge, etc.

In Fig. 11, for Face 2, we start with the edge e_9, identify the face f_2, as a clockwise adjacency, and corresponding next clockwise(ncw) edge as e_6 (Table 1). Following around the loop, we next identify e_1 and e_5 and eventually close the loop at the vertex v_5 by noting e_9 again.

The B-Rep model of a solid object can also be represented via vertex, edge, face, or even loop information using graph theory, where the nodes identify the individual elements and the branches define connectivity. (Some graphs are called "directed" graphs, since they identify adjacency direction.)

TABLE 1 Partial Winged-Edge Data Structure

Face	First edge	Edge	fcw	ncw	fccw	nccw
f_2	e_9	e_9	f_2	e_6	f_6	e_{12}
.	.	e_6	f_3	e_{10}	f_2	e_1
.	.	e_1	f_1	e_2	f_2	e_5
.	.	e_5	f_2	e_9	f_5	e_4
.	.	.	.	.	.	.
.	.	.	.	.	.	.
.	.	.	.	.	.	.
.	.	.	.	.	.	.

Information contained in a graph can be represented in a matrix form in order for algorithmic manipulation by CAD systems. Such an adjacency matrix is given here for the polyhedron's surfaces shown in Fig. 11, where 1 indicates adjacency:

Face	1	2	3	4	5	6
1	0	1	1	1	1	0
2	1	0	1	0	1	1
3	1	1	0	1	0	1
4	1	0	1	0	1	1
5	1	1	0	1	0	1
6	0	1	1	1	1	0

Adjacency matrices (and graphs) are commonly used in feature-based design for feature identification (extraction), as will be discussed in Section 4.4.

4.3.4 Model Conversions

Both solid-modeling methods discussed above, and others that were not detailed herein, have their virtues, which commercial CAD software designers take advantage of. CSG models are quite concise and have the advantage of being (relatively) easily convertible to B-Rep models, which in turn are useful for graphical outputs.

Solid modelers that allow user input, and subsequent data storage, in both CSG and B-Rep structures, are referred to as "hybrid modelers." Users of such a commercial CAD system, through a proprietary GUI, could model a part either through the CSG or the B-Rep modelers. In both cases, the part model is, subsequently, stored as a B-Rep data structure. However, segments of the solid model that are built through CSG will also have a CSG history tree for future CSG-based modifications, but not vice versa. Parametric modifications can be carried on both CSG and B-Rep built models, but parts of the model that were originally built via B-Rep cannot be modified using a CSG modellers (Fig. 12).

Both I-DEAS and Pro-Engineer CAD software packages allow designers to generate solid models using the CSG and B-Rep principles: first, a part's topological information can be "sketched" in two-dimensional space and subsequently "extruded" along three-dimensional curves to create simple primitives ("features"); several "features" can then be combined to create "parent–child" relationships. Both softwares keep track of the history

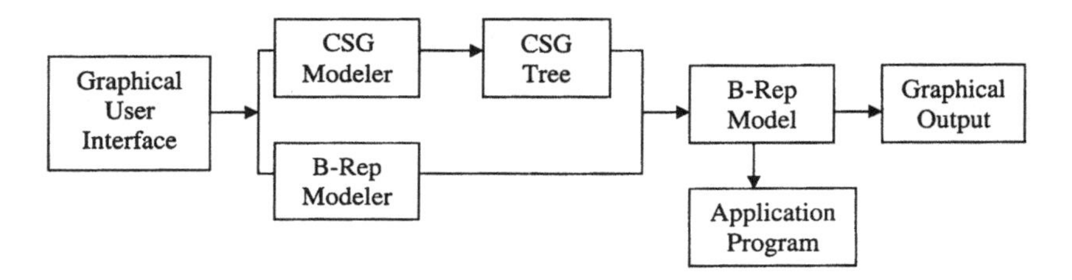

FIGURE 12 Architecture of a hybrid solid modeler.

of the Boolean operations and allow users to go back in history to modify the geometries of individual primitives.

4.4 FEATURE-BASED DESIGN

From a manufacturing engineering point of view, features can be seen as specific geometric shapes on a part that can be associated with certain fabrication processes. Thus it has been long advocated that if these features were highlighted during the modeling phase of a product's design process, in the subsequent production-planning phases, engineers could take advantage of this information in accessing historical data regarding the production of these features. Naturally, the engineers would have to be provided with material, tolerancing, and other pertinent data to complement the identified geometric (feature) information in reaching production decisions. In this chapter, as a continuation of the topic of geometric modeling, our emphasis will be on geometric (form) features and their utilization during the product-design process. That is, we will discuss the topic commonly referred to as design by features.

Features have been commonly classified by J. J. Shah and others as form, material, precision, and technological features. Form features identify geometric elements on the main body of a part (holes, slots, ribs, bosses, etc.) (Fig. 13). Material features capture material-composition and heat-treatment information. Precision features refer to tolerancing data. Technological features represent information related to the product's expected performance parameters.

The objective of design by features, as mentioned above, is twofold: (1) To increase the efficiency of the designer during the geometric-modeling phase, and (2) to provide a bridge (mapping) to engineering-analysis and process-planning phases of product development. The former can be achieved by providing designers with a library of features (not "primitives," as previously discussed in the context of CSG), from which they can pick

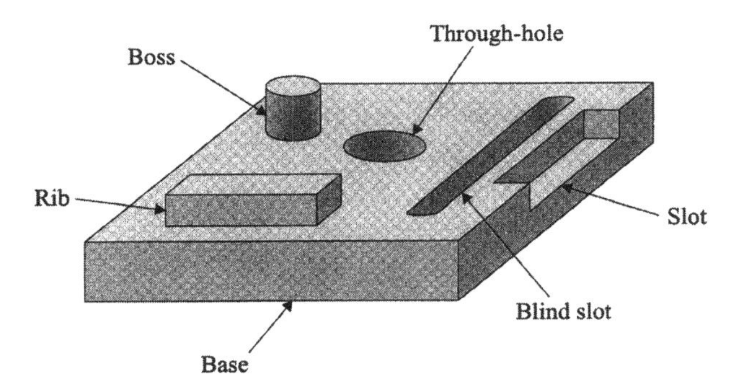

FIGURE 13 Common form features.

and place on the main configuration (body) of a part, or allowing them to extract (identical or similar) features from previous solid models of parts without an extensive feature library.

CAD research on design by features can be traced back to the work of several individuals at the University of Cambridge (A. R. Grayer, K. Kyprianou, and others) in the mid-1970s. Since then, there have been many noteworthy works that advanced the state of the art in feature-based design theory (by J. Shah, M. R. Henderson, R. Gadh, M. Mäntylä, and many others). Current numerical CAD packages have benefited from these works and do offer (limited) design-by-features capabilities. However, research in the field is still going on, the emphasis being on automatic recognition and identification of features from parts' solid models (primarily B-Rep models).

4.4.1 Design by Features

In feature-based design, parts' solid models are configured through a sequence of form-feature attachments (subtractions and additions) to the primary (base stock) representations of the parts, which can be as simple as a rectangular box (Fig. 14). These features could be chosen from a library of predefined (and sometimes application dependent, for example casting/molding/etc.) features or could be extracted from the solid models of earlier designs. The latter issue will be discussed in greater detail in subsection 4.4.2.

As is the case with many commercial CAD systems, form features can be individually modeled by the user explicitly using a B-Rep modeler (yielding unambiguous topological relationship information) or implicitly using a CSG modeler (yielding a tree representation of corresponding primitives and Boolean operators). Any attempt to generate a universal set of features must cope with the problem of database management—

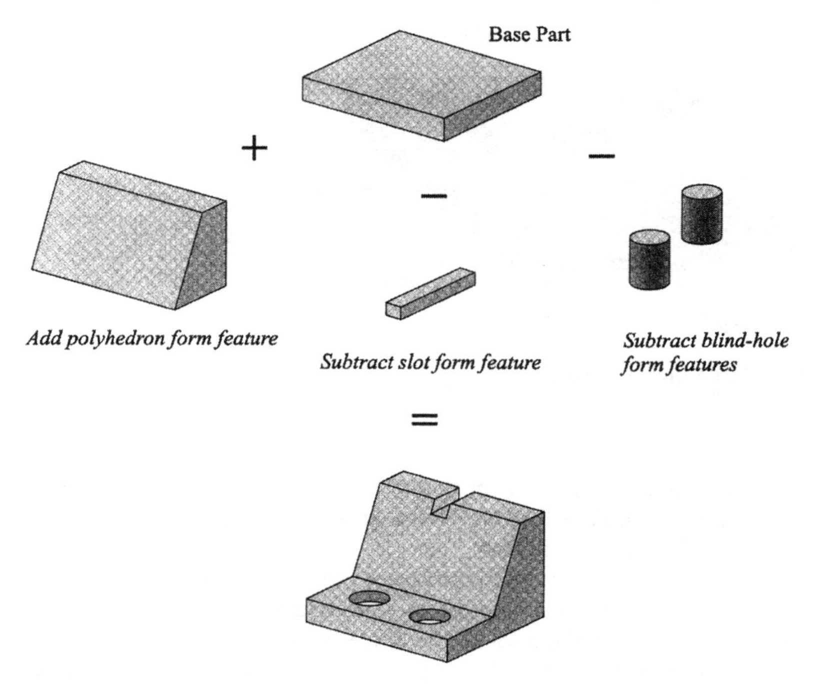

FIGURE 14 Design by features example.

storage and retrieval of form features, whose numbers may become unmanageable. A potential tool in dealing with such a difficulty would be the utilization of a logical classification and coding system for the form-feature geometries, such as a GT-based system (Chap. 3). In working with such a feature-based design system, the user would require the CAD system to search through the database of previous designs, identify similar features, and extract them for use in the modeling of the part at hand.

4.4.2 Feature Recognition

Automatic feature recognition normally refers to the examination of parts' solid models for the identification of features that have been predefined. The primary objective is not feature extraction per se but identification of the existence of a specific feature for the extraction of, for example, pertinent manufacturing information,. There have been numerous techniques proposed in the literature for the subsequent phase of feature extraction and use in solid modeling. However, one may question the need for the extraction of the geometric information of an already known entity. Thus recently there have been research efforts in developing extraction methods that would examine parts' solid models for the existence of geometric features, which have not been predefined, and extract them. Such features could then be classified and coded for possible future use in a GT-based CAD system. These features would continue to be part of the overall solid model of the part but be extractable in the future based on a user-initiated search for the most similar feature in the database via a GT code.

The two most important feature-recognition categories to date are (1) graph matching and (2) volume decomposition. Graph matching, normally, refers to topological matching in terms of the connectivity of faces that define form features within a B-Rep solid model. S. Joshi and T. C. Chang's work (based on the original work of Kyprianou) is most noteworthy in this subfield. Their work advocates the use of an attributed adjacency graph (AAG) for the definition of form features, where nodes represent faces and arcs represent adjacency with an assigned value of zero for convex and one for concave relationship. Using such a method a graph representation of a part's solid model is partitioned with respect to its features (Fig. 15). One must not, however, underestimate the computational effort required in trying to identify and match subgraphs for the recognition of form features.

The volume decomposition approach to feature extraction was developed by T. C. Woo in the early 1980s and later modified by numerous researchers, most notably by Y. S. Kim. In this approach, features are defined as volumes and decomposed from the part's solid model by subtraction (of primitives), yielding a tree structure, the nodes of which indicate Boolean operators (as in CSG), and the extracted features are the lowest leaves of the tree. Predefined features are then compared and matched to these volumes (features).

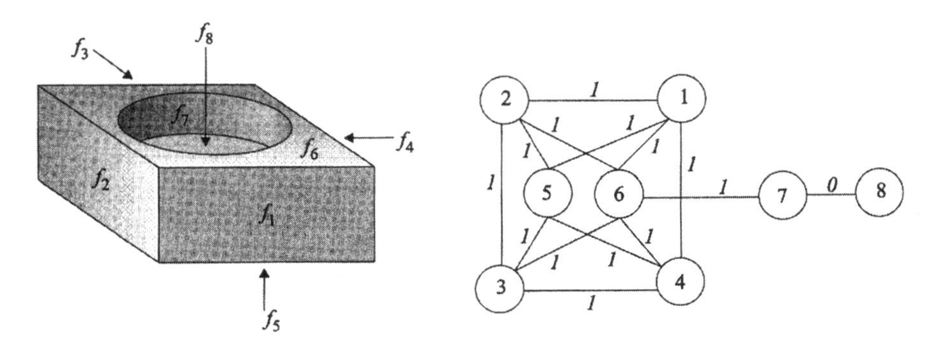

FIGURE 15 Face connectivity.

4.5 PRODUCT-DATA EXCHANGE

Despite intensive standardization efforts in the computing industry in the past two decades, almost all CAD hardware and software packages in commercial use today employ proprietary data-manipulation and data-storage formats. Thus, although large manufacturing companies can enforce the utilization of identical CAD systems within their enterprises, even they would face an uphill battle in data transfer between these systems, and other engineering analysis (CAE) manufacturing planning (CAM) software systems they employ. The problem gets quite complex due to variety of CAD/CAE/CAM systems used by the many companies that comprise the supply chain of different products.

4.5.1 IGES

The problem of exchanging design information between dissimilar systems has been under investigation since the late 1970s, even before the widespread commercial use of the Internet and the Web. The initial efforts concentrated on the exchange of graphics information between different CAD systems, which yielded the first version of IGES (Initial Graphics Exchange Specification), which was made available in 1980. IGES 1.0 was designed as a neutral format primarily for the exchange of mechanical part drawings (graphics).

This version of the IGES specification was based on Boeing's Database Standard Format (DBSF), which was influenced by the CAD systems then in use at Boeing, Computervision's CADDS 3 and Gerber IDS. Both relied on simpler geometric elements and their drafting packages included only basic text and dimensioning abilities. This version of IGES neither relied on a formal definition language nor did it require conformance to the specification.

IGES 2.0 followed the initial version and became available in 1983, with subsequent releases of IGES 3.0 in 1986, IGES 4.0 in 1988, and IGES 5.0 in 1990. These versions considerably extended the scope of the first version to include solid geometry (CSG and B-Rep) and finite-element modeling exchange capability. When additions to the specification list were considered, however, any entity had to exist in at least three major CAD systems before it would be considered for inclusion in IGES. IGES was targeted for the lowest common denominator, excluding many innovative unique features. The IGES standard, currently in its sixth revision, has been expanded to include most concepts used in major CAD systems. Although IGES has been intended as a neutral format, not tied to any particular CAD system, it still represents the entities of some CAD systems better than it does others'.

Exchanging data using IGES requires two modules, one on each CAD system: an IGES preprocessor on the first CAD system that would read the data file to be translated and produce an external file formatted in accordance with the IGES specification, and an IGES postprocessor on the second CAD system that would read the transferred file and translate it to the recipient's data format.

Many commercially available IGES processors today only support IGES Version 3, a few support IGES Version 4, and only a very few support IGES Version 5 and above (Version 5.3 being the latest). Support is generally best for elementary geometric entities, not so good for more complex geometric entities like B-splines and even annotations and dimensions, and almost nonexistent for concepts like features and assemblies. Thus it is advised that companies that rely on IGES rigorously test the capabilities of their processors to discover what does not transfer well. One can then either stop using the entities that do not work, or modify the output IGES file (manually or automatically) to work better with the second CAD system.

Currently, the IGES Specification is overseen by the IGES/PDES Organization (IPO). The IPO has been officially recognized by the U.S.A.'s National Institute for Standards and Technology (NIST) as the official organization responsible for the content of the IGES Specification. The IPO is also responsible for the U.S.A.'s input to the content of the PDES (Product Data Exchange using STEP) standard.

4.5.2 STEP

In mid 1980s, in response to foreseen serious deficiencies with IGES, the European Commission and U.S.A.'s NIST encouraged and funded projects for the development of a more comprehensive data-exchange specification. The primary result was the birth of PDES (Product-model Data Exchange Standard). Thus now the acronym PDES commonly refers to Product Data Exchange using STEP (STandard for the Exchange of Product model data). STEP, a derivative of PDES, was first proposed in 1984, resubmitted for approval both in 1988 in Tokyo and in 1989 in Frankfurt, but only achieved international standard status in 1994.

STEP - ISO 10303, provides a neutral computer-interpretable representation of product data intended to be used throughout the life cycle of a product, independent of any particular CAD/CAM system. As indicated above, its evolution and development took place under the auspices of the International Organization for Standardization (ISO) Technical Committee 184, Subcommittee 4. However, from the very beginning, it has been agreed that STEP needed to be developed in parts and offered as a replacement to

IGES, incrementally, as its parts reached maturity. STEP is currently organized as a series of parts that fall into one of the following categories: description methods, integrated resources, application protocols (APs), abstract test suites, implementation methods, and conformance testing.

APs define the information needed for a particular application and how this information is to be exchanged. These protocols draw on information encapsulated within the integrated resource models. STEP uses a formal specification language, EXPRESS, to specify precisely and consistently the product information to be represented. Some STEP APs that have achieved International Standard (IS) status are these:

AP201 Explicit Drafting
AP203 Configuration Controlled 3D Designs of Mechanical Parts and Assemblies
AP207 Sheet Metal Die Planning and Design
AP209 Composite and Metallic Structural Analysis and Related Design
AP210 Electronic Assembly, Interconnection and Exchange
AP213 Numerical Control Process Plans for Machined Parts
AP214 Core Data for Automotive Mechanical Design Processes
AP219 Manage Dimensional Inspection of Solid Parts or Assemblies
AP220 Process Planning, Manufacturing, Assembly of Layered Electrical Products
AP223 Exchange of Design and Manufacturing Product Information for Cast Parts
AP224 Mechanical Product Definition for Process Planning Using Machining Features
AP233 Systems Engineering Data Representation
AP235 Materials Information for the Design and Verification of Products

(Note that only AP201 and AP203 were part of the initial release of STEP in 1994. AP202 did not achieve ISO status until 1996, and APs 207 and 224 were published in 1999.)

AP203: Configuration-Controlled 3D Designs of Mechanical Parts and Assemblies

AP203 encompasses the following:

Product definition data and configuration control data pertaining to the design phase.
Five types of shape representations of a part that include wireframe and surface without topology, wireframe geometry with topology,

manifold surfaces with topology, faceted boundary representation, and boundary representation. (It excludes the use of constructive solid geometry for the representation of objects.)

Identification of other specifications for design, process, surface finish, and materials.

Data that identify the supplier of either the product or the design.

AP203 allows users to exchange geometry, topology, and configuration management data of a part or the whole product assembly. Although the parametric and layer information are not included, the solid-to-solid translation capability eliminates most of the modifications currently required when using alternative translation methods, such as IGES. AP203, being implemented by most CAD vendors today, is by far the most widely used application protocol.

AP214: Core Data for Automotive Mechanical Design Processes

AP214 encompasses the following:

Process plan information to manage the relationships among parts and the tools used to manufacture them

Product definition data and configuration control data pertaining to the design phase

Identification of standard parts, which have been classified according to national or industrial standards, and of library parts

Data that identify the supplier of a product and any related contract information

Any of eight types of representation of the shape of a part or tool: 2D-wireframe representation, 3D wireframe representation, geometrically bounded surface representation, topologically bounded surface representation, faceted boundary representation, boundary representation, compound-shape representation, and constructive-solid-geometry representation

Representation of portions of the shape of a part or a tool by form features

The simulation data for the description of kinematic structures and configurations of discrete tasks

Surface conditions and tolerance data

Although AP214's primary focus is the automotive industry, it includes many manufacturing-engineering processes common to other industries (for example, the aerospace industry). The capability of AP214 can be seen as a superset of AP203: it further includes the capability to exchange CSG-model, color, and layer information.

The recently released AP224, *Mechanical product definition for process plans using machining features*, may provide the necessary environment for the integration of part design with the process planning and production scheduling systems of an enterprise through the use of feature-based design. Although AP224 is presently available commercially on the ProEngineer platform via a third-party supplier, most other STEP-related products are limited to several conformance classes of AP203 and AP214. However, it is important to remember that STEP presents a powerful and robust technology beyond that currently implemented. STEP is still evolving, and it is now at a point when a significant number of APs will be reaching international standard status in the first few years of 2000.

Although, as in other formats, STEP-based data exchange is achieved through pre- and postprocessors, there exists an important difference between STEP and other data-exchange standards (IGES, Autodesk's DXF, and others): the alternatives normally deal only with particular application areas or products. STEP, on the other hand, is intended to store all data in an integrated form for a product throughout its life cycle without regard to discipline or application area. Data integration ensures that the information describing product design, manufacturing, and life cycle support is defined only once, thus eliminating redundancy and associated problems caused by maintaining redundant information.

Some industrial users of STEP are listed here:

Lockheed–Martin and some of its suppliers have collaborated on the design of the F-16 and F-22 jet fighter aircraft, while using STEP for exchange of technical data.

Bristol Aerospace has used STEP in the design of aircraft structures to customer requirements by allowing the sharing of three-dimensional solid modeling data.

Boeing Airplane and Pratt and Whitney, Rolls–Royce and GE Aircraft Engines have used STEP to help verify the form and fit of the parts that integrate the engines and the aircraft in the 777 and 767-400 planes.

General Motors has extensively used STEP to transfer the designs of vehicle models between its various divisions as well as its first-tier suppliers (Delphi Automotive Systems, Delco Electronics, and others).

REVIEW QUESTIONS

1. What is the primary purpose of geometric modeling in the context of computer-aided engineering (CAE)?

2. What is the primary purpose of solid modeling in the context of geometric modeling?
3. Why should designers be aware of various curve-fitting or surface-fitting techniques? What is the difference between a least-squares fit and a spline fit? Choose two products that have "free" surfaces (e.g., car body, dental implant) and recommend a least-squares fit or a spline fit. Justify your answers.
4. Why would an engineer decide to use Bézier (curve or surface) fits versus other approximations?
5. Although there exists a number of solid modeling techniques, users of various computer-aided design (CAD) packages may not be even aware of the variety and model objects in intuitive ways as opposed to using formal techniques. Review your knowledge of an existing CAD package and discuss the ways the software allows you to model three-dimensional objects while comparing those to the following formal techniques: primitive instancing and sweeping, constructive solid geometry, and boundary representation.
6. Describe the possible different object design features. Define the process design by features.
7. Why is feature recognition an important capability that would allow the widespread use of design by features?
8. Data-exchange protocols/standards/specifications allow users of different commercial CAD/CAE packages to work in a more coherent manner on the concurrent design of complex products in the virtual domain. Compare the two most common data-exchange specifications, IGES and STEP, respectively.

DISCUSSION QUESTIONS

1. Computer-aided design (CAD) of engineering products has been practiced as long as there have been computers in existence (since the early 1950s). The term CAD, however, has often been misused as solely "the employment of commercial engineering modeling software." These packages, which have been in existence only since the late 1970s, when compared to earlier ones, provide users with powerful graphical user interfaces (GUI) for the modeling of parts/products and graphical display of engineering analysis results. In the above context, discuss the role of current CAD packages in the design (synthesis and analysis) of engineering products.
2. Geometric modeling is a design task that can be efficiently implemented using a commercial CAD package. Discuss the importance of geometric modeling for the synthesis and analysis stages, or even potentially for

virtual reality modeling cases, of design. Address issues such as surface modeling versus solid modeling.

3. Part designs can be classified and coded using a group-technology (GT) technique, primarily to allow access to (past) similar designs. Would several different GT-based classification and coding systems be needed in a company for different objectives? That is, one system for design, one system for manufacturing planning, and yet another for cost engineering?
4. The use of design features has long been considered as improving the overall synthesis and analysis stages of products owing to the potential of encapsulating additional nongeometric data, such as process plans, in the definition of such features. Discuss feature-based design, where the user, through some recognition/extraction process, can access and retrieve individual similar or identical features on earlier product designs and utilize them for the design of the product at hand. Furthermore, compare such a design approach to a primitives-based design, where the user generates or accesses a limited-size database of geometric primitives for the design of a new product.
5. The majority of commercial CAD packages store geometric and nongeometric data using proprietary techniques. Discuss the importance of standardization of the data management process for CAD packages, including the use of data exchange interfaces, in the new economic reality of "distributed design" within large and complex supply chains.
6. In the near future, although the majority of engineering products will be modeled in the virtual (computer) space, representing the starting point of the design and manufacturing process, some products will still be crafted manually by artisans and/or industrial designers. Discuss the computer-aided modeling and analysis of such products, whose features are not originally defined by exact mathematical relationships. Consider and use some specific product examples in your discussion.
7. Computers and other information management technologies have been commonly accepted as facilitators for the integration of various design activities. Define/discuss "integrated design" in the modern manufacturing enterprise and address the role of computers in this respect. Discuss the role of suppliers in such design activities.

BIBLIOGRAPHY

Bedworth, David D., Wolfe, Philip M., Henderson, Mark R. (1991). *Computer-Integrated Design and Manufacturing*. New York: McGraw-Hill.

Bézier, Pierre (1977). Essai de Définition Numérique des Courbes et des Surfaces

Expérimentales. *Thèse de Doctorat èt Sciences*. Paris, France: Université Pierre et Marie Curie.
Bézier, Pierre (1986). *Courbes et Surfaces, Mathématiques et CAO* (Vol. 4,. Paris: Hermès.
Bloor, M. S., Owen, J. (June 1991). CAD/CAM product data exchange: the next step. *J. of CAD* 23(4):237–243.
Boungart, B. Geometric Modeling for Computer Vision. Ph.D. Diss., Stanford University, Stanford, CA, 1974.
Chang, Tien-Chien, Wysk, Richard A., Wang, Hsu-Pin (1998). *Computer-Aided Manufacturing*. Upper Saddle River, NJ: Prentice Hall.
Gadh, R., and Prinz, F.B. (1992). Recognition of geometric forms using the differential depth filter. *J. of CAD* 24(11):583–589.
Gilles, Susan. Introducing STEP—The Foundation for Product Data Exchange in the Aerospace and Defense Sectors. National Research Council Canada, Technical Report C2-447/1999, 1999.
Grayer, A. R. A Computer Link Between Design and Manufacture. Ph.D. Diss., University of Cambridge, Cambridge, England, September 1976.
Gu, P., Chan, K. (1995). Product modelling using STEP. *J. of CAD* 27(3):163–179.
Han, J.-H., Pratt, M., and Regli, W. C. (December 1999). Manufacturing feature recognition from solid models: a status report. *IEEE Trans. on Robotics and Automation* 16(6):782–796.
Hoffmann, Christoph M. (1989). *Geometric and Solid Modeling: An Introduction*. San Mateo, CA: Morgan Kaufmann.
Hoffmann, Christoph M., Rossignac, Jaroslaw R. (March 1996). Road map to solid modeling. *IEEE Transactions on Visualization and Computer Graphics* 2(1):3–10.
Joshi, S., and Chang, T. C. (1988). Graph-based heuristics for recognition of machined features from a 3-D solid model. *J. of CAD* 20(2):58–66.
Kemmerer, Sharon J. (July 1999). STEP—The Grand Experience. *NIST-U.S.A.*
Kim, Y. S. (1991). Form Feature recognition by convex decomposition. *ASME Int. Conf. on Computers in Engineering, Santa Clara, CA* (pp. 61–69).
Kyprianou, L. K., Shape Classification in Computer Aided Design. Ph.D. thesis, Christ's College, University of Cambridge, Cambridge, England, July1980.
Laurent, P.-J., Sablonnière, P. (2001). Pierre Bézier: an engineer and a mathematician. *J. of Computer Aided Geometric Design* 18:609–617.
Manjula, B., Waldron, Kenneth J., eds. (1996). *Mechanical Design: Theory and Methodology*. New York: Springer-Verlag.
Mäntylä, Martti (1988). *An Introduction to Solid Modeling*. Rockville, MD: Computer Science Press.
Mortenson, Michael E. (1997). *Geometric Modeling*. New York: John Wiley.
Murray, Dave (1998). *Inside SolidWorks*. Santa Fe, NM: OnWord Press.
National Research Council of Canada - IMTI - What is STEP? http://strategis.gc.ca, 2001.
Qamhiyah, Abir Ziyad. Form-Feature Extraction and Coding for Design by Features. Ph.D. Diss., Department of Mechanical Engineering, University of Toronto, Toronto, Canada. (1996.)

Qamhiyah, A. Z., Venter, R. D., and Benhabib, B. (Nov. 1996). Geometric reasoning for the extraction of form features of objects with planar surfaces. *J. of CAD* 28(11):887–903.

Salomons, O. W., van Houten, F. J. A. M., and Kals, H. J. J. (1993). Review of research in feature-based design. *SME Journal of Manufacturing Systems* 12(2):113–132.

SCRA. STEP Application Handbook. U.S.A. Defense Logistics Agency, 1 June 2000, Team SCRA - RAMP Product Data—http://ramp.scra.org/pdt_summary.html.

Shah, J. J. (June 1991). Assessment of features technology. *J. of CAD* Vol. 23(No. 4):331–343.

Shah, Jami J., Martti Mäntylä, and Nau, Dana S. eds. (1994). *Advances in Feature Based Manufacturing*. New York: Elsevier.

Sutherland, Ivan Edward (1980). *Sketchpad: A Man-Machine Graphical Communication System*. New York: Garland.

Tropak, Ihor. An Assessment of Feature Based Design and Manufacturing. M.Eng. thesis, Department of Mechanical Engineering, University of Toronto, Toronto, Canada. (1995).

Woo, T. C. (March 1982). Feature extraction by volume decomposition. *Conf. on CAD/CAM Technology in Mechanical Engineering, Cambridge, MA* (pp. 76–94).

Wozny, M. J., McLaughlin H. W., Encarnao, J. L. eds. (1988). *Geometric Modeling for CAD Applications: Selected and Expanded Papers*. New York: Elsevier.

Zeid, Ibrahim (1991). *CAD/CAM Theory and Practice*. New York: McGraw-Hill.

5

Computer-Aided Engineering Analysis and Prototyping

Engineering design starts with identifying customer requirements and developing the most promising conceptual product architecture to satisfy the need at hand (Chap. 2). This stage is often followed with a finer decision making process on issues such as product modularity as well as initial parametric design of the product, including its subassemblies and parts (Chaps. 3 and 4). The concluding phase of design is engineering analysis and prototyping facilitated through the use of computing software tools. Engineering students spend the majority of their time during their undergraduate education in preparation for carrying engineering analysis tasks for this phase of design, for example, ranging from mechanical stress analysis to heat transfer and fluid flow analyses in the mechanical engineering field. Students are taught many analytical tools for solving closed-form engineering analysis problems as well as numerical techniques for solving problems that lack closed-form solution models. They are, however, often reminded that the analysis of most engineering products requires approximate solutions and furthermore frequently need physical prototyping and testing under real operating conditions owing to our inability to model analytically all physical phenomena.

The objective of engineering analysis and prototyping can therefore be noted as the optimization of the design at hand. The objective function of the optimization problem would be maximizing performance and/or minimizing cost. The constraints would be those set by the customer and translated into engineering specifications and/or by the manufacturing processes to be employed. These would, normally, be set as inequalities, such as a minimum life expectancy or a maximum acceptable mechanical stress. The variables of the optimization problem are the geometric parameters of the product (dimensions, tolerances, etc.) as well as material properties. As discussed in Chap. 3, a careful design-of-experiments process must be followed, regardless whether the analysis and prototyping process is to be carried out via numerical simulation or physical testing, in order to determine a minimal set of optimization variables. The last step in setting the analysis stage of design is selection of an algorithmic search technique that would logically vary the values of the variables in search of their optimal values. The search technique to be chosen would be either of a combinatoric nature for discrete variables or one that deals with continuous variables.

In this chapter, we will review the most common engineering analysis tool used in the mechanical engineering field, finite-element modeling and analysis, and we will subsequently discuss several optimization techniques. However, as a preamble to both topics, we will first discuss below *prototyping* in general and clarify the terminology commonly used in the mechanical engineering literature in regard to this topic.

5.1 PROTOTYPING

A prototype of a product is expected to exhibit the identical (or very close to) properties of the product when tested (operated) under identical physical conditions. Prototypes can, however, be required to exhibit identical behavior only for a limited set of product features according to the analysis objectives at hand. For example, analysis of airflow around an airplane wing requires only an approximate shell structure of the wing. Thus one can define the prototyping process as a time-phased process in which the need for prototyping can range from "see and feel" at the conceptual design stage to physical testing of all components at the last alpha (or even beta) stage of fabrication prior to the final production and unrestricted sale of the product.

5.1.1 Virtual Prototyping

Virtual (analytical) prototyping refers to the computer-aided engineering (CAE) analysis and optimization of a product carried out completely within a computer (i.e., in virtual space). This process would naturally rely on the existence of suitable software that can help the designer to model the part (via solid modeling, Chap. 4) as well as to simulate a variety of physical phenomena that the part will be subjected to (commonly, via finite-element analysis, Sec. 5.2 below). In the past two decades, significant progress has been reported in the area of numerical modeling and simulation of physical phenomena, which however require extensive computing resources: computational fluid dynamics (CFD) is one of the fields that rely on such modeling and simulation tools.

The two primary advantages of virtual prototyping are significant engineering cost savings (as well reduced time to market) and ability to carry out distributed design. The latter advantage refers to a company's ability to carry out design in multiple locations, where design data is shared over the company's (and their suppliers') intranets. The design of the Boeing 777 airplane, in virtual space, has been the most visible and talked about virtual prototyping process.

Boeing 777

The Boeing company is the world's largest manufacturer of commercial jetliners and military aircraft. Total company revenues for 1999 were $58 billion. Boeing has employees in more than 60 countries and together with its subsidiaries they employ more than 189,000 people. Boeing's main commercial product line includes the 717, 737, 747, 757, 767, and 777 families of jetliners, of which there exist more than 11,000 planes in service worldwide. The Boeing fighter/attack aircraft products and programs include the F/A-18E/F Super Hornet, F/A-18 Hornet, F-15 Eagle, F-22 Raptor, and AV-8B Harrier. Other military airplanes include the C-17 Globemaster III, T-45 Goshawk, and 767 AWACS.

The Boeing 777 jetliner has been recognized as the first airplane to be 100% digitally designed and preassembled in a computer. Its virtual design eliminated the need for a costly three-stage full-scale mock-up development process that normally spans from the use of plywood and foam to handmade full-scale airplane structures of almost identical materials to the proposed final product.

The 777 program, during the period of 1989 to 1995, established and utilized 238 design/build teams (each having 10 to 20 people) to develop each element of the plane's frame (main body and wings), which includes more than 100,00 unique parts (excluding the engines). The engines have almost 50,000 parts each and are manufactured by GE, Rolls-Royce, or Pratt and Whitney and installed on the 777 according to specific customer demand.

Under this revolutionary product design team approach, Boeing designers and manufacturing and tooling engineers, working concurrently with Boeing's suppliers and customers, created all the airplane's parts and systems. Several thousands of workstations around the world were linked to eight IBM mainframe computers. The CATIA (computer-aided three-dimensional interactive application) and ELFINI (finite element analysis system), both developed by Dassault Systems of France, and EPIC (electronic preassembly integration on CATIA) were used for geometric modeling and computer-aided engineering analysis.

As a side note, it is worth mentioning that the 777's flight deck and the passenger cabin received the Industrial Designers Society of America Design Excellence Award. This was the first time any airplane was recognized by the society.

5.1.2 Virtual Reality for Virtual Prototyping

Virtual reality (VR) could be used as part of the virtual-prototyping process, in order to evaluate human–machine interfaces, for example, ease of operability of a device. The primary challenge in employing VR is to provide the user with a realistic visual sensation of the environment, normally achieved via head-mounted displays capable of generating stereoscopic images. The secondary challenge is to manipulate the environment through input devices, such as three-dimensional mice (also known as spaceballs) and intelligent gloves for simulating a one-way haptic interface (Fig. 1). However, no VR system can be fully useful if it cannot provide the user of the "virtual product" with haptic feedback—for example, a user must feel the effort required in opening a car door or lifting and placing luggage into a car's trunk.

The beginning of VR can be traced to I. Sutherland's work in the late 1960s on head-mounted display (Sutherland is also the designer and developer of the first known CAD system, Sketchpad, discussed in Chap. 4). However, VR significantly developed only more than a decade later with the introduction of high-definition graphic display hardware and surface-modeling software, as well as a variety of commercial interface devices (especially those developed for the entertainment industry) and flight-

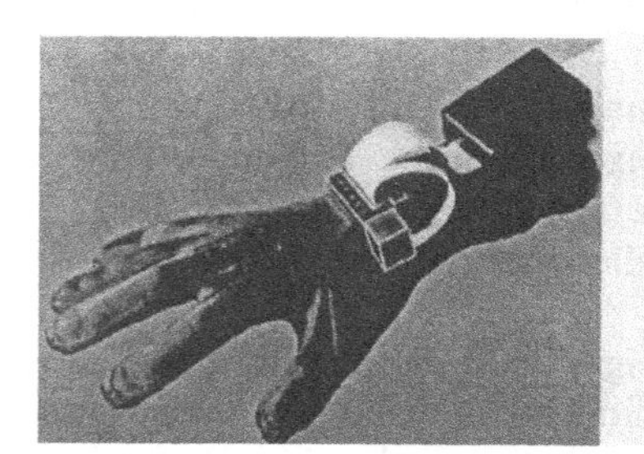
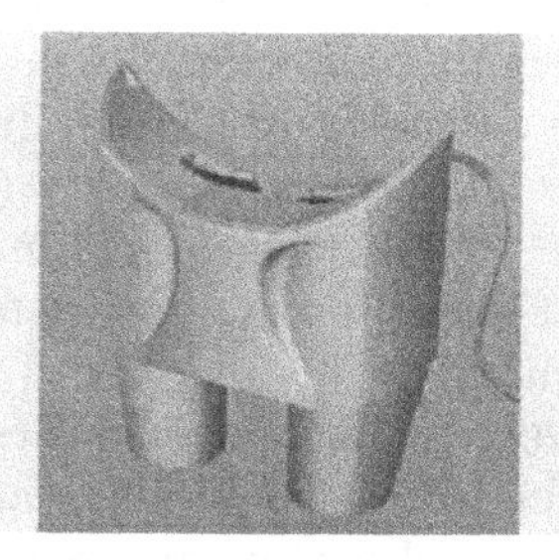
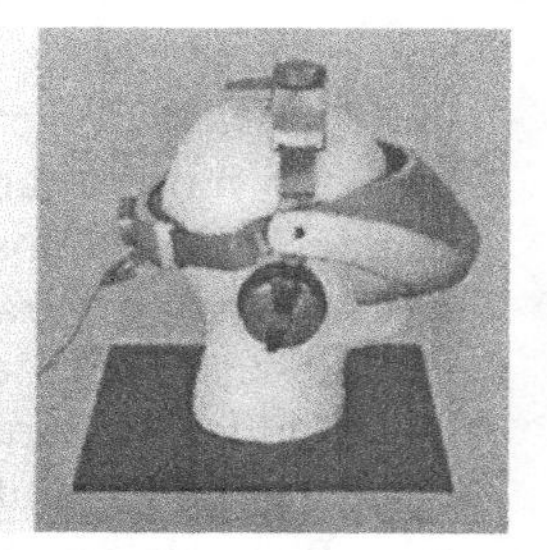

FIGURE 1 VR input/output devices. (Images courtesy of www.5DT.com.)

simulation applications. Naturally, not all CAD software packages provide easy interface to VR environments: CATIA with its SIMPLIFY module is one the few that not only can simplify geometric models for real-time manipulation but also can increase the quality of surface representations. VR users need to develop (nontrivial) interface programs for accessing CAD data stored by most other commercial packages, such as ADAMS/Car by Volvo, Renault, BMW, and Audi.

The automotive industry is the most common user of virtual reality in the design of commercial vehicles. Companies such as Chrysler, Ford, and Volkswagen utilize the CAD models of their vehicles to provide engineers with an immersive VR environment, for example, means of visualizing different dashboard configurations for visibility and reachability. Some have also experimented with VR to evaluate assembly (of door locks, window regulators, etc.) as well as disassembly (of tail lights, etc.) for maintainability. However, in almost all cases, users have been provided with only visual feedback and no force feedback. In numerous instances, integrated sensors have helped these users in detecting their head and hand movements and adjust the display of the virtual environment accordingly. It has been claimed that these users could evaluate the goodness of assembly plans, the suitability of tolerances, and the potential collisions with the environment.

5.1.3 Physical Prototyping

Despite intensive CAE and VR efforts and successes, as noted above, problems do arise both in the exact modeling of a product and in its (virtual) analysis process. It is thus common, and in most cases mandatory owing to governmental regulations, to manufacture physical product prototypes and test them under over-stressed or accelerated conditions (to mimic long-term usage or unusual circumstances). Such physical prototyping, however, should be restricted to the functional testing of the final optimized product or the fine-tuning of design parameters. It would be costly to use physical prototypes during the parameter-optimization phase, especially if tests require the destruction of the product under duress.

In response to lengthy physical-prototyping processes, since the late 1980s, numerous technologies have been developed and commercialized for "rapid prototyping" (RP). The common objective of these techniques has been the fabrication of physical prototypes, directly from their geometric solid models, in a time-optimal manner i.e., faster than existing conventional manufacturing techniques (Fig. 2). In most cases, however, prototypes fabricated using these material-additive and layered techniques can only exhibit a very limited number of a product's features, primarily because of

FIGURE 2 Layered manufactured parts.

material restrictions. A very successful use of RP technologies had been the generation of part models for the fabrication of sand-casting and investment-casting dies. Current research on RP concentrates on the development of new fabrication techniques that would yield functional prototypes with increased numbers of physical characteristics identical with (or very similar to) those of the real product itself. (Several RP technologies will be detailed in Chap. 9.)

5.2 FINITE-ELEMENT MODELING AND ANALYSIS

The finite-element method provides engineers with an approximate behavior of a physical phenomenon in the absence of a closed-form analytical model. The quality of the approximation can be substantially increased by spending high levels of computational effort (CPU time and memory). In this method, a continuum or an object geometry is represented as a collection of (finite) elements that are connected to each other at nodal points (nodes). Variations within each element are approximated by simple functions to analyze variables, such as displacement, temperature, velocity. Once the individual variable values are determined for all the nodes, they are assembled by the approximating functions throughout the field of interest.

Although approximate mathematical solutions to complex problems have been utilized for a long time (several centuries), the finite-element method (as it is known today) dates only back several decades—it can be traced to the earlier works of R. Courant in the 1940s and the later works of other aerospace scientists in the early 1950s. The first attempts at using the finite-element method were for the analysis of aircraft structures. In the past several decades, however, the method has been used in numerous engineering disciplines to solve many complex problems:

Mechanical engineering: Stress analysis of components (including composite materials); fracture and crack propagation; vibration analysis (including natural frequency and stability of components and linkages); steady-state and transient heat flow and temperature distributions in solids and liquids; and steady-state and transient fluid flow and velocity and pressure distributions in Newtonian and non-Newtonian (viscous) fluids.

Aerospace engineering: Stress analysis of aircraft and space vehicles (including wings, fuselage, and fins); vibration analysis; and aerodynamic (flow) analysis.

Electrical engineering: Electromagnetic (field) analysis of currents in electrical and electromechanical systems.

Biomedical engineering: Stress analysis of replacement bones, hips and teeth; fluid-flow analysis in blood vessels; and impact analysis on skull and other bones.

The finite-element modeling and analysis for the above-mentioned and other problems is a sequential procedure comprising the following primary steps:

1. *Discretization of the problem*: The object geometry or the field of interest is subdivided into a finite number of elements—the number, type, and size of the elements are closely related to the required level of approximation and should take into account existing symmetries and loading and boundary conditions.
2. *Selection of the approximating (interpolation) function*: The distribution of the unknown variable through each element is

approximated using an interpolation function—normally chosen in a polynomial form. The accuracy of the analysis can be improved by choosing higher-order (polynomial) representations, though at the expense of computational effort.

3. *Derivation of the basic element equations*: Based on the physical phenomenon examined (e.g., stress analysis), the equations that describe the behavior of the elements are derived (e.g., stiffness matrices and load vectors).
4. *Calculation of the system equations*: Individual element equations are assembled into an overall system model, and the boundary conditions are incorporated into this model.
5. *Solution of the system equations*: The system model is solved for the variable values at individual nodes (e.g., displacement).

In most cases, it is expected that an object model considered for finite-element analysis (FEA) would be developed in a CAD environment and imported using a preprocessor in the FEA software package (for example, one that interprets an IGES file). Similarly, the results of the FEA would be displayed to the user through a postprocessor in the FEA or CAD system.

In the following subsections, the above five-step process will be presented in greater detail. Mechanical stress, fluid flow, and heat transfer analysis problems will also be briefly addressed.

5.2.1 Discretization

The first step in FEA is the discretization of the domain (region of interest) into a finite number of elements according to the approximation level required. Over the years, numerous automatic mesh generators have been developed in order to facilitate the task of discretization, which is normally carried out manually by FEA specialists. If the domain to be examined is symmetrical, the complexity of the computations can be significantly reduced, for example, by considering the problem only in 2-D or even analyzing only a half or a quarter of the solid model (Fig. 3).

The shapes, sizes, and numbers of elements, as well as the location of the nodes, dictate the complexity of the finite-element model and greatly impact on the level of a solution's accuracy. Elements can be one-, two-, or three-dimensional (line, area, volume) (Fig. 4). The choice of the element type naturally depends on the domain to be analyzed: truss structures utilize line elements, two dimensional heat-transfer problems utilize area elements, and solid (nonsymmetrical) objects require volume elements. For area and

2D representation of a 3D problem.

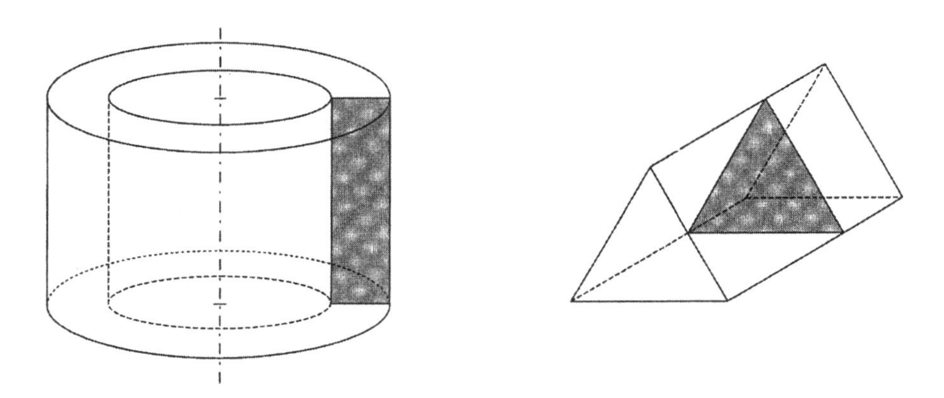

Analyzing a ¼ or a ½ of a problem.

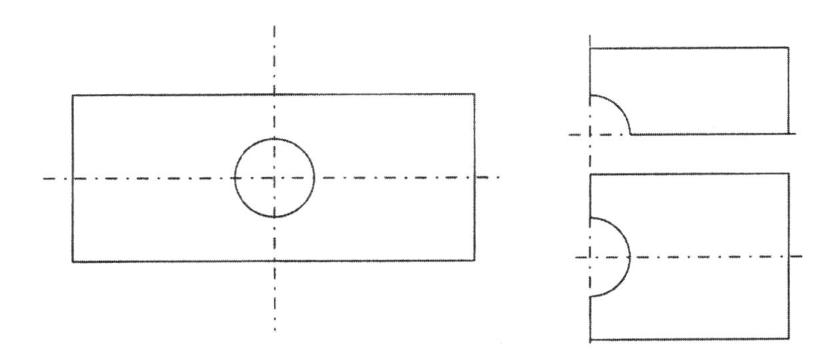

FIGURE 3 Reduction in finite element representation.

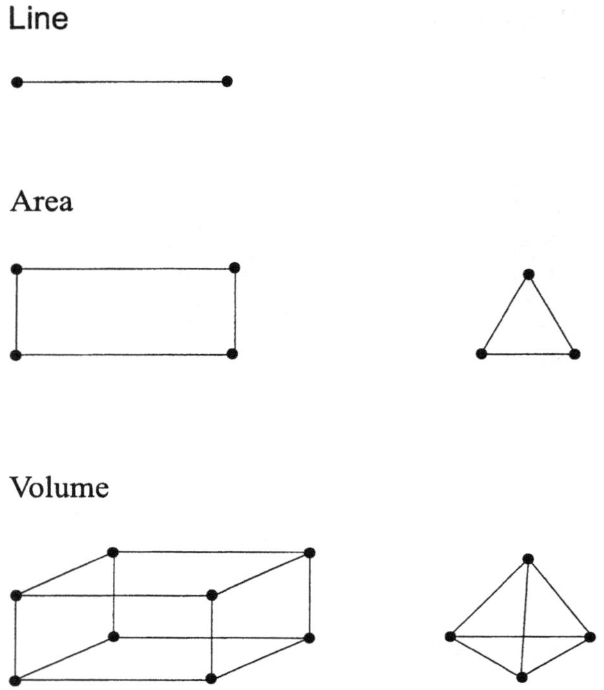

FIGURE 4 Basic element shapes.

volume elements the boundary edges do not need to be linear. They can be curves (Fig. 5—isoparametric representation).

The size of the elements influences the accuracy of FEA—the smaller the size, the larger the number, the more accurate the solution will be, at the expense of computational effort. One can, however, choose different element sizes at different subregions of interest within the object (domain) (Fig. 6), i.e., a finer mesh, where a rapid change in the value of the variable is expected. It is also recommended that nodes be carefully placed, especially at discontinuity points and loading locations.

5.2.2 Interpolation

Finite-element modeling and analysis requires piecewise solution of the problem (for each element) through the use of an adopted interpolation function representing the behavior of the variable within each element. Polynomial approximation is the most commonly used method for this

Area

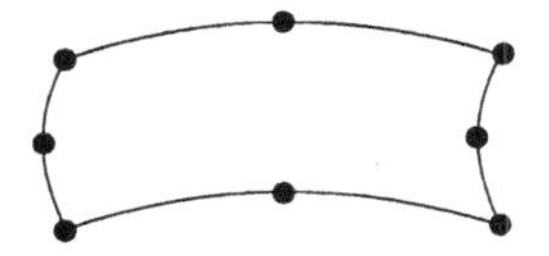

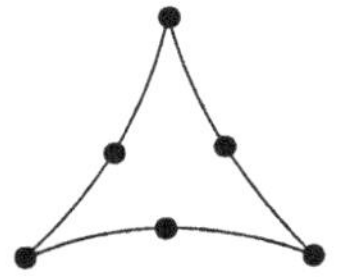

Volume

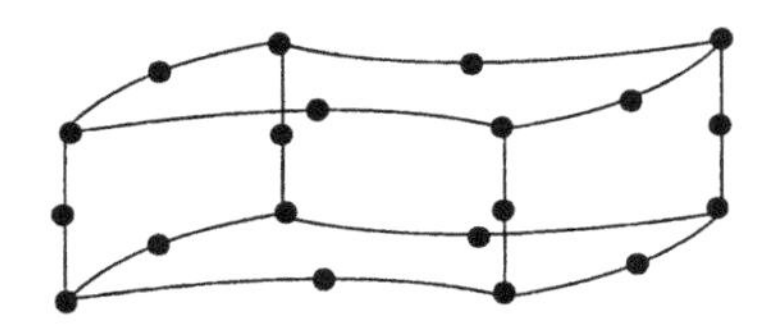

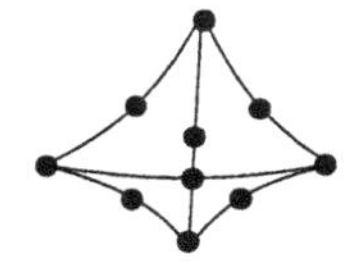

FIGURE 5 Curved elements.

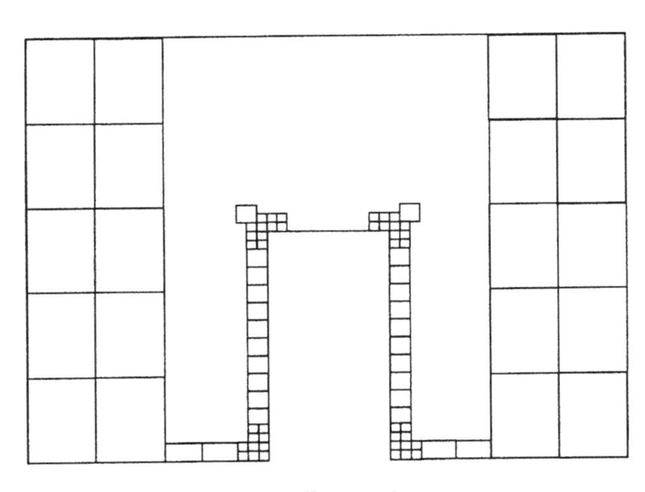

FIGURE 6 Elements of different size.

purpose. Let us, for example, consider a triangular (area) element, where the variable value can be expressed as a function of the Cartesian coordinates using different-order polynomial functions (Fig. 7): linear,

$$\phi(x, y) = \alpha_1 + \alpha_2 x + \alpha_3 y \tag{5.1}$$

and quadratic,

$$\phi(x, y) = \alpha_1 + \alpha_2 x + \alpha_3 y + \alpha_4 x^2 + \alpha_5 y^2 + \alpha_6 xy \tag{5.2}$$

One would expect that as the element size decreases and the polynomial order increases, the solution would converge to the true solution at the limit. However, one should not attempt to achieve unreasonable accuracies that would not be needed by the designers/and engineers, who would normally interpret the results of the FEA and use them as part of their overall design parameter optimization process (satisfying a set of constraints and/or maximizing/minimizing an objective function). It is thus common to find simplex (first-order) or complex (second-order) elements in most FEA solutions in the manufacturing industry, and not higher orders.

For the two-dimensional simplex element given in Fig. 7 and defined by Eq. (5.1), the variable's nodal values (e.g., $i = 1, j = 2, k = 3$) are defined as

$$\begin{aligned} \phi_i &= \alpha_1 + \alpha_2 x_i + \alpha_3 y_i \\ \phi_j &= \alpha_1 + \alpha_2 x_j + \alpha_3 y_j \\ \phi_k &= \alpha_1 + \alpha_2 x_k + \alpha_3 y_k \end{aligned} \tag{5.3}$$

where (α_1, α_2, and α_3) are the coefficients of the first-order polynomial. These coefficients can be solved for, using the above system of equations

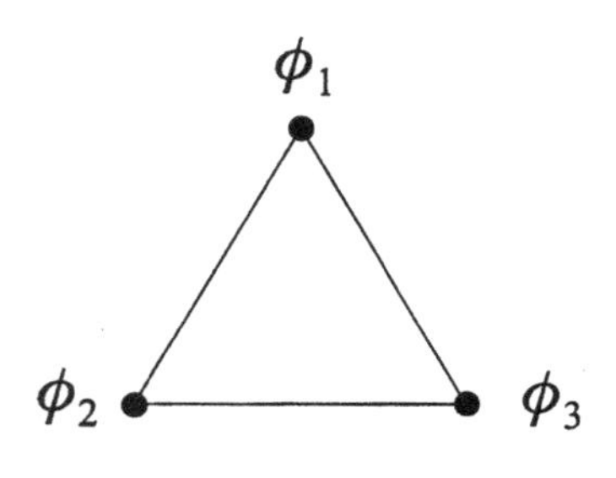

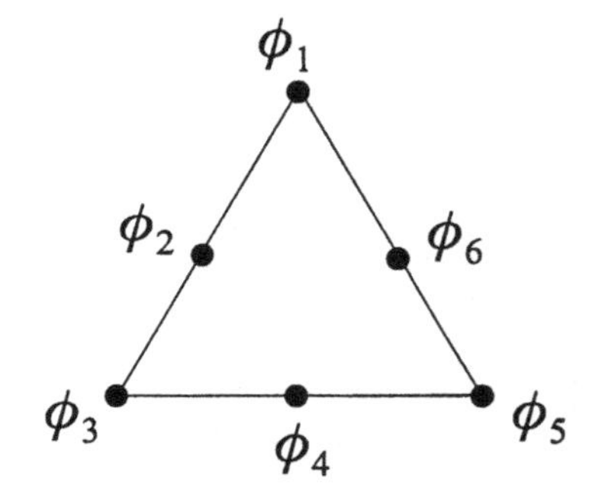

FIGURE 7 Two-dimensional element.

(i.e., three equations and three unknowns), in terms of the nodal coordinates and the function values at these nodes. Equation (5.1) can thus be rewritten as a function of the above nodal values as

$$\begin{aligned}\phi(x,y) &= N_i\phi_i + N_j\phi_j + N_k\phi_k, \\ &= [N]\{\phi\}\end{aligned} \tag{5.4}$$

where the elements of $[N]$, (N_i, N_j, and N_k), are functions of the (x, y) coordinate values of the three nodes,

$$\begin{aligned}N_i &= \frac{1}{2A}(a_i + b_i x + c_i y) \\ N_j &= \frac{1}{2A}(a_j + b_j x + c_j y) \\ N_k &= \frac{1}{2A}(a_k + b_k x + c_k y)\end{aligned} \tag{5.5}$$

$$A = \frac{1}{2}(x_i y_j + x_j y_k + x_k y_i - x_i y_k - x_j y_i - x_k y_j) \tag{5.6}$$

and

$$\begin{aligned}a_i &= x_j y_k - x_k y_j \qquad & a_j &= x_k y_i - x_i y_k \qquad & a_k &= x_i y_j - x_j y_i \\ b_i &= y_j - y_k & b_j &= y_k - y_i & b_k &= y_i - y_j \\ c_i &= x_k - x_j & c_j &= x_i - x_k & c_k &= x_j - x_i\end{aligned} \tag{5.7}$$

The value of $\phi(x, y)$ at any point (x, y) is assumed to be scalar in Eq. (5.4) (e.g., temperature). However, in most engineering problems, the variable at a node would be vectorial in nature (e.g., displacement along x and y). Thus the interpolation polynomial must also be defined accordingly in multidimensional space. For the simplex element above, let us assume that the variable ϕ will have two components u and v, along the x and y directions, respectively (Fig. 8). Then, based on Eq. (5.4),

$$\begin{aligned}u(x,y) &= N_i\phi_{2i-1} + N_j\phi_{2j-1} + N_k\phi_{2k-1} \\ v(x,y) &= N_i\phi_{2i} + N_j\phi_{2j} + N_k\phi_{2k}\end{aligned} \tag{5.8}$$

where N_i, N_j, and N_k are defined by Eq. (5.5), and the nodal values are defined as $u_i = \phi_{2i-1}$, $v_i = \phi_{2i}$, etc.

5.2.3 Element Equations and Their Assembly

Derivation of the element equations depends on the application at hand and can be carried out using a number of different methods. Since (mechanical) stress analysis is the most common (mechanical) engineering analysis

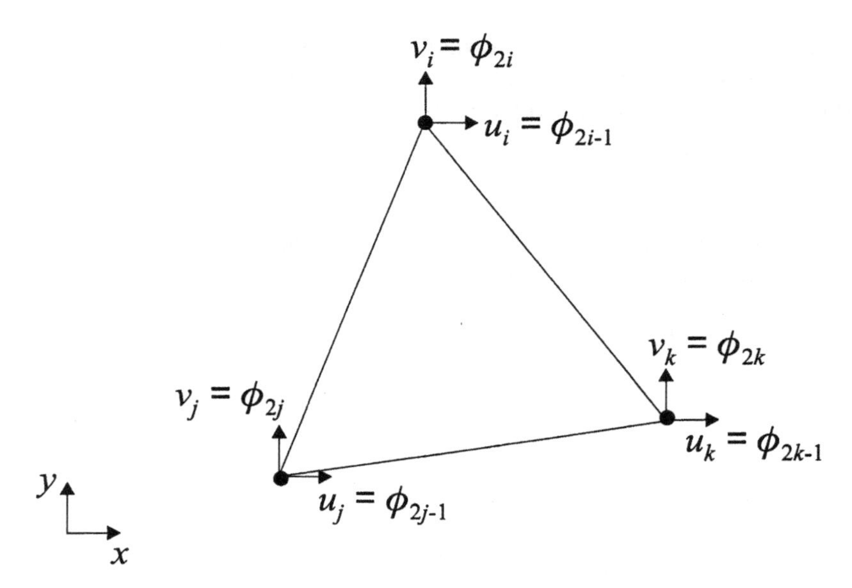

FIGURE 8 Two-dimensional simplex element.

problem, it will be utilized here as an example case study for the derivation of element equations. Other analysis problems will also be addressed in Sec. 5.2.5.

The three common modeling approaches used for elasticity analysis (i.e., stress analysis in the elastic domain) using finite elements are

The Direct Approach: Direct physical reasoning is utilized to derive the relationships for the variables considered. (This method is normally restricted to simple one-dimensional representations).

The Variational Approach: Calculus of variations is utilized for solving problems formulated in variational forms. It leads to approximate solutions of problems that cannot be formulated using the direct approach.

The Weighted Residual Approach: The governing differential equations of the problem are utilized for the derivation of the element's equations. (This method could be useful for problems such as fluid flow and mass transport, where we could readily have the governing differential equations and boundary conditions.)

The Variational Approach for Stress Analysis

Let us consider a two-dimensional stress–strain relationship:

$$\{\varepsilon\} = \begin{Bmatrix} \varepsilon_{xx} \\ \varepsilon_{yy} \\ \varepsilon_{xy} \end{Bmatrix} = [C]\{\sigma\} + \{\varepsilon_0\} = [C]\begin{Bmatrix} \sigma_{xx} \\ \sigma_{yy} \\ \sigma_{xy} \end{Bmatrix} + \begin{Bmatrix} \varepsilon_{xx0} \\ \varepsilon_{yy0} \\ \varepsilon_{xy0} \end{Bmatrix} \tag{5.9}$$

where $[C]$ is a matrix of elastic coefficients,

$$[C] = \frac{1}{E}\begin{bmatrix} 1 & -\nu & 0 \\ -\nu & 1 & 0 \\ 0 & 0 & 2(1+\nu) \end{bmatrix} \tag{5.10}$$

and $\{\varepsilon_0\}$ is the vector of initial strains. E is Young's modulus, and ν is the Poisson ratio. Equation (5.9) can also be written as

$$\{\sigma\} = [D]\{\varepsilon\} - [D]\{\varepsilon_0\} \tag{5.11}$$

where, for plane strain,

$$[D] = \frac{E}{(1+\nu)(1-2\nu)}\begin{bmatrix} 1-\nu & \nu & 0 \\ \nu & 1-\nu & 0 \\ 0 & 0 & \frac{1}{2}(1-\nu) \end{bmatrix} \tag{5.12}$$

The strain–displacement relationships are correspondingly defined as

$$\varepsilon_{xx} = \frac{\partial u}{\partial x} \quad \varepsilon_{yy} = \frac{\partial v}{\partial y} \quad \varepsilon_{xy} = \frac{\partial u}{\partial x} + \frac{\partial v}{\partial y} \tag{5.13}$$

where u and v are displacements along the (x, y) directions, respectively, each of which are functions of the coordinates (x, y).

Referring to finite-element displacement equations of a simplex, Eq. (5.8),

$$\begin{aligned} u(x,y) &= N_i u_i + N_j u_j + N_k u_k \\ v(x,y) &= N_i v_i + N_j v_j + N_k v_k \end{aligned} \tag{5.14}$$

or in the alternate notation for the nodal displacements, as in Eq. (5.8), Fig. 8,

$$\begin{Bmatrix} u \\ v \end{Bmatrix} = \begin{bmatrix} N_i & 0 & N_j & 0 & N_k & 0 \\ 0 & N_i & 0 & N_j & 0 & N_k \end{bmatrix} \begin{Bmatrix} u_{2i-1} \\ u_{2i} \\ u_{2j-1} \\ u_{2j} \\ u_{2k-1} \\ u_{2k} \end{Bmatrix} = [N]\{U\} \tag{5.15}$$

Using Eqs. (5.5), (5.13), and (5.15),

$$\begin{Bmatrix} \varepsilon_{xx} \\ \varepsilon_{yy} \\ \varepsilon_{xy} \end{Bmatrix} = \frac{1}{2A} \begin{bmatrix} b_i & 0 & b_j & 0 & b_k & 0 \\ 0 & c_i & 0 & c_j & 0 & c_k \\ c_i & b_i & c_j & b_j & c_k & b_k \end{bmatrix} \{U\} = [B]\{U\} \tag{5.16}$$

The stiffness matrix for the (two-dimensional) simplex element is then defined by

$$[k] = \int_V [B]^T[D][B]dV = [B]^T[D][B] \int_V dV \tag{5.17}$$

where the volumetric integral in the above equation can be replaced with (tA). t is the constant thickness of the element and A is the cross-sectional area.

Similarly, the element load vector due to initial strains, $\{P_i\}$, is defined as

$$\{P_i\} = \int_V [B]^T[D]\{\varepsilon_0\}dV = [B]^T[D]\{\varepsilon_0\}tA \tag{5.18}$$

and the element load vector due to body forces, $\{P_b\}$, is defined as

$$\{P_b\} = \int_V [N]^T \begin{Bmatrix} F_x \\ F_y \end{Bmatrix} dV = \frac{tA}{3} \begin{Bmatrix} F_x \\ F_y \\ F_x \\ F_y \\ F_x \\ F_y \end{Bmatrix} \tag{5.19}$$

where the vector $\{F_x \; F_y\}^T$ is the body-force vector per unit volume.

Equations (5.17) (5.18) to (5.19) and the concentrated forces vector, $\{P_c\}$, can be combined to complete the derivation of the element equations (excluding pressures applied on the element) summed over the entire domain (all the elements, $e = 1$ to E).

$$[K]\{U\} = \{P\} \tag{5.20}$$

where

$$\{P\} = \sum_{e=1}^{E} (\{P_i\} + \{P_b\})^e + \{P_c\} \tag{5.21}$$

and

$$[K] = \sum_{e=1}^{E} [k]^e \tag{5.22}$$

As shown above, the assembly of element equations, Eq. (5.20), is the combination of the element stiffness matrices into one global stiffness matrix, summing all the force vector components into one global force vector. The compatibility requirement must be met during this assembly process, that is, the values of the nodal parameters are the same for nodes that are shared by multiple elements. If the element matrices and vectors were calculated in local coordinates, it would be necessary to transfer them to a global (world) coordinate system. (Naturally, in a computer-aided analysis environment all above-mentioned transactions would be carried out automatically by the appropriate software module.) One must, finally, add the boundary conditions (geometric/essential and free/natural) onto the system's (assembled) model.

5.2.4 Solution

The finite-element method is a numerical technique providing an approximate solution to the continuous problem that has been discretized. The solution process can be carried out utilizing different techniques that solve the equilibrium equations of the assembled system. Direct methods yield exact solutions after a finite number of operations. However, one must be aware of potential round-off and truncation errors when using such methods. Iterative methods, on the other hand, are normally robust to round-off errors and lead to better approximations after every iteration (when the process converges). Common solution methods include

The Gaussian-Elimination "Direct" Method, which is based on the triangularization of the system of equations (the coefficient matrices) and the calculation of the variable values by back-substitution.

The Choleski Method, which is a direct method for solving a linear system by decomposing the (normally symmetric) positive definite FEA matrices into lower and upper triangular matrices and calculation of the variable values by back-substitution.

The Gauss–Seidel Method, which is an iterative method primarily targeted for large systems, in which the system of equations is solved one equation at a time to determine a better approximation of the variable at hand based on the latest values of all other variables.

For solving eigenvalue problems, FEA solution methods include the power, Rayleigh–Ritz, Jacobi, Givens, and Householder techniques; while for propagation problems, solutions include the Runge–Kutta, Adams–Moulton, and Hamming methods.

5.2.5 Fluid Flow and Heat Transfer Problems

In heat transfer problems, determination of temperature distribution within a conducting body is paramount to our understanding of heat dissipation and potential development of significant thermal stresses. The basic governing equation for heat transfer problems is

Heat inflow during dt

= (Heat outflow + Change in internal body energy) during dt

Both heat conduction and heat convection phenomena can be modeled and analyzed using a finite-element method. As in the (mechanical) stress analysis case, the first task at hand is the selection of the element type and division of the domain of interest into E elements. The next task is the choice of a temperature (variation) function within each element and to express it as a function of Cartesian coordinates and time. Next, the element conduction (or convection) matrix and equations can be developed using the variational approach. The last step in the formulation of the FEA problem is the assembly of the element equations and the incorporation of the boundary conditions to yield

$$[K]\{T\} = \{P\} \tag{5.23}$$

where $[K]$ is the overall conduction (or convection) matrix, $\{T\}$ is the nodal temperature vector, and $\{P\}$ is the heat-source vector.

In fluid mechanics, FEA has been widely applied in the past two decades to laminar as well as turbulent flows of Newtonian fluids (whose viscosity is not a function of velocity). Recently, however, FEA has been also applied to non-Newtonian fluids, especially by users of polymers. FEA for fluid and heat flows are similar—the process starts with the meshing of the domain; choice of a potential function and derivation of the element equations follows this step; the element equations are, then, assembled to yield

$$[K]\{\phi\} = \{P\} \tag{5.24}$$

where $\{\phi\}$ is the nodal velocity potential vector and $\{P\}$ is the input potential vector; the definition of the stiffness matrix $[K]$ is the same as in the cases of stress analysis and heat transfer analysis equations. Equation (5.24) can be solved, using any one of the methods mentioned in Sec. 5.2.4 for determining fluid velocity.

5.2.6 Commercial FEA Software

Commercial finite-element modeling and analysis packages can be categorized into comprehensive packages that provide FEA for several engineering fields, such as ANSYS, ALGOR, and MISC/NASTRAN, physical-phenomenon-specific packages that provide FEA for specific physical problems, such as FLUENT for computational fluid dynamics (CFD) problems and application-specific packages that specialize in unique engineering problems, such as MOLDFOW for injection-molding-related problems. All these CAE packages have been developed over the years to run on microcomputers (such as SUN) and lately on personal computers (mainly Windows-based platforms) as their CPU speeds become faster and RAM storage capability increases.

Although most FEA packages have evolved over the years in terms of the friendliness of their graphical user interfaces (GUIs) for domain modeling, it would be advisable to utilize the original CAD solid models of objects as our starting point and not to attempt to redefine these models within a

FEA package. At the opposite end of the spectrum, CAD packages have also significantly evolved in terms of their engineering analysis capabilities—SDRC (I-DEAS), for example, allows designers to run FEA on solid models for mechanical stress and heat transfer analyses. However, for complex problems (complex geometry, layered materials, two-phase flows, etc.), it would be advisable to utilize specialized FEA packages.

As discussed earlier, effective mesh generation is the precursor to any accurate FEA analysis. This step can be carried out on a CAD workstation. The outcome (domain model) can be transferred to a commercial FEA package using an available data-exchange standard (IGES or STEP) and prepared for analysis by being processed through a preprocessor. At this stage, the user is expected to add onto the geometric model the necessary boundary conditions (including loads) as well as material properties. Preprocessors are expected to verify the finite-element model by checking for distorted elements and modeling errors.

Once the solution of the problem has been obtained, a postprocessor can be run to examine the results (preferably graphically) via the GUI of a CAD system that would allow us to manipulate the output effectively—view it from different angles, cross-section it, etc. It is important to remember that the outcome of the FEA analysis is primarily a metric to be fed into an optimization algorithm that would search for the best design parameters.

5.2.7 An Example—Computer-Aided Injection Molding Analysis

Injection molding is a common plastics-processing technique used for the manufacture of containers, toys, electronic packaging, and automotive products. As simple as the process may be thought at first glance (i.e., filling of mold cavities with liquid polymer by injection at high speeds and pressures), the design of the mold is quite complex owing to the concurrent existence of several physical phenomena: flow of non-Newtonian fluid, heat transfer, and thermal stresses. A good mold design can significantly benefit from the usage of a FEA-based computer software package for the analysis of all the mentioned physical phenomena. Some of the design issues are discussed below, prior to a discussion of available commercial packages in the analysis of mold filling, cooling, and warpage issues (Fig. 9):

Cavities: Although the number of cavities on a mold base may be treated as a purely economic issue, their locations and arrangement affect injection pressure and clamping force. Furthermore, as discussed earlier in Chap. 3, part features (such as draft angles, sharp edges, the geometry of ribs) affect the flow of the molten material, the cooling time of the part, and its warpage.

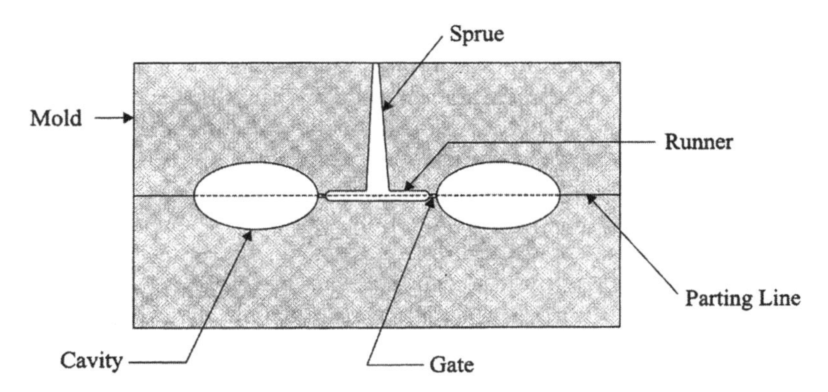

FIGURE 9 Mold-filling elements.

Gates: The type, geometry, location, and number of gates affect flow patterns during filling.

Sprue and runners: A mold-filling objective is minimization of the distance traveled by the molten material before it reaches the cavities. Other conflicting considerations include prolonged cycle times owing to excessive sizes of the sprue and runners and creation of undesirable flow patterns owing to insufficient diameters of the runners, etc.

Cooling: Effective cooling provides short cycle times and prevents defects such as warpage, poor surface quality, or even burn marks.

The injection molding process starts with the filling of the cavities with molten (normally thermoplastic) polymer and some additional melt to compensate for shrinkage. The fluid flow during the filling process is predominantly of the shear-flow type that is driven by pressure to overcome the melt's resistance to flow. Naturally, fluid temperature is an important factor, as the mobility of the polymer chains increases with increased temperature. Using FEA, the flow of fluid through the runner/gate/cavity assembly can be analyzed as a function of time, using a solid model of the overall system generated (and automatically meshed) on a CAD system. The two leading commercial FEA packages that can be used for this purpose are MOLDFLOW (Australia) and CMOLD (U.S.A.). Both packages can carry out automatic mesh generation and simulate mold filling.

During the mold filling analysis process, one can also examine the heat transfer characteristics of the mold configuration at hand (i.e., a mold design with specific locations and geometries for the sprue/runners/gates/cavities), concurrently with the fluid flow analysis mentioned above. Heat loss occurs through the circulating coolant (in the cooling channels) as well as through the mold surroundings. A considerable amount of cycle time is spent on cooling the molded parts. Thus one must examine temperature distributions during the filling process as well as during the postfilling period for different mold configurations and filling parameters. However, one must realize that mold cooling is a complicated problem and nonuniform mold cooling results in undesirable part warpage (during ejection) due to nonuniform residual stresses. Another important factor in part warpage is, of course, variations in shrinkage (due to flow orientation, differential pressure, etc.) Both the commercial FEA packages mentioned above provide users with corresponding modules for thermal analysis and warpage determination capabilities—(MF/COOL and MF/WARP by MOLDFLOW, and C-COOL and C-WARP by CMOLD).

Over the past two decades, many researchers have developed optimal mold design techniques that utilize the above-mentioned (and other) finite-element-based mold flow analysis tools, in order to relieve dependence on expert opinions and other heuristics. It should be mentioned here that most mold makers still heavily depend on human judgment rather than utilizing analytical methods in optimizing mold designs.

5.3 OPTIMIZATION

Engineering design is an iterative process, in which the outcome of the analysis phase is fed back to the synthesis phase for the determination of optimal design parameter values. That is, the parametric design stage is carried out under the auspices of a search algorithm whose objective is to optimize (through CAE analysis) an objective function (e.g., performance, cost, weight) by varying the product design parameters at hand. Most optimization problems encountered in engineering design are of the constrained type. An optimal solution ("best" parameter values) is selected among all feasible designs, subject to limits imposed on the variable design parameters. The variables are, normally, of a continuous type—i.e., they can be assigned any one of the infinite possible values. For example, the thickness of a vessel is a geometric (dimensional) continuous variable and can be assigned any (floating-point) value within a given range (t_{min} to t_{max}), while attempting to optimize a desired objective function.

As discussed above, a typical optimization problem aims at maximizing/minimizing an overall objective function Z, which is a function of a number of variables, x_i, $i = 1$ to n subject to (j) equality and (k) inequality constraints placed on the variables, whose optimal values we are trying to determine:

$$\min Z = Z(x_1, x_2, \ldots, x_n) \tag{5.25}$$

subject to $\phi_j(x_1, \ldots, x_n) = 0$ and $\psi_k(x_1, \ldots, x_n) \leq \psi_k{}^{\max}$

In most engineering design cases, the design team must decide what to optimize (i.e., what to choose as an objective function) and formulate the other desired specifications as equality and inequality constraints. However, in numerous cases, the team may be faced with a situation in which multiple objectives (sometimes in conflict with each other) must be optimized. Two common solutions to this problem are (1) to prioritize the objective functions and formulate a multilevel (nested) optimization problem, and (2) to combine the functions into a single weighted sum (overall) objective function. In the former case, a priority could be to reduce the number of fasteners used, for example, followed by determining the optimal geometrical parameters for each fastener. Thus one could achieve a required attachment strength by increasing the number of fasteners or by increasing their dimensions. At any iteration, for a given number of fasteners considered by the outer level of a two-loop search, the inner loop would select the (best) parameter values that would maximize fastening strength. Once determined, the search would return to the outer loop and check whether the number of fasteners could be further reduced. Otherwise, the optimal solution is considered to be reached.

For the latter multiobjective function case, an example task could be to attempt to maximize component life while minimizing the manufacturing cost:

$$\min Z = w_1\left(\frac{1}{L_n(x_1, \ldots, x_n)}\right) + w_2(C_n(x_1, \ldots, x_n)) \tag{5.26}$$

where L_n is the estimated (normalized) product life, C_n is the estimated (normalized) product cost, both functions of the variables x_1 to x_n, and w_1 to w_n are weighting coefficients. The choice of the weighting coefficients is application dependent.

In the above optimization problems, whether a single- or a multi-objective formulation, one must carefully examine the variables as well. Although in most design cases the variables would be of the continuous type, as mentioned in the above example, they could also be of a discrete or integer type. An objective function could have both types of variables or only one type. Solution techniques proposed in the literature, some of which are to be discussed herein, would be sensitive to the types of the variables.

Other factors that strongly affect the choice of a solution (search) method would include the expected behavior of the objective function—whether it has one or multiple extrema (single-mode, multimode functions); the order of the function (linear versus nonlinear) and whether its derivative can be calculated; and lastly the restrictions on the search domain—whether the problem is constrained or not constrained.

5.3.1 Overview of Optimization Techniques

Optimization procedures are widely applied in engineering, spanning from design to planning and to control. In this section, although we will overview a number of existing optimization solution techniques, our focus will be on those that are most useful in the engineering design cycle of synthesis → analysis → synthesis. Furthermore, among the most pertinent techniques, only a few will be detailed—it is expected that users of optimization will have to review carefully the complete existing spectrum on available search techniques.

It is important to acknowledge here that the field of numerical optimization reached recognition only after the 1940s and has been widely researched concurrently with the significant developments in computing hardware and software. The pioneers in the field (during the 1950s to the early 1980s) were W. C. Davidon, M. J. Powell, R. Fletcher, P. E. Gill, L. A. Wolsey, and G. L. Nemhauser, to mention a few. They and others classified optimization methods broadly into two main categories: continuous versus integer and combinatoric. In this section, our focus will be on the first category; the latter category deals with "process" problems, such as sequencing and network-flow analysis, in the context of planning for manufacturing.

5.3.2 Single-Variable Functions—Numerical Methods

Let us consider a simple case: a product's characteristic is a function of one design variable, $Z(x)$. Let us further assume that $Z(x)$ is a continuous function and can only be evaluated through a numerical simulation, such as FEA, and that derivatives of the function cannot be obtained. Based on experience (or preliminary investigation), we also know that $Z(x)$ is a single-mode function (one extremum). The problem at hand is to determine the optimal x value that would a minimize the objective function in the range $[a, b]$ for x:

$$\min Z = Z(x)$$

subject to $a - x \leq 0$ and $x - b \geq 0$.

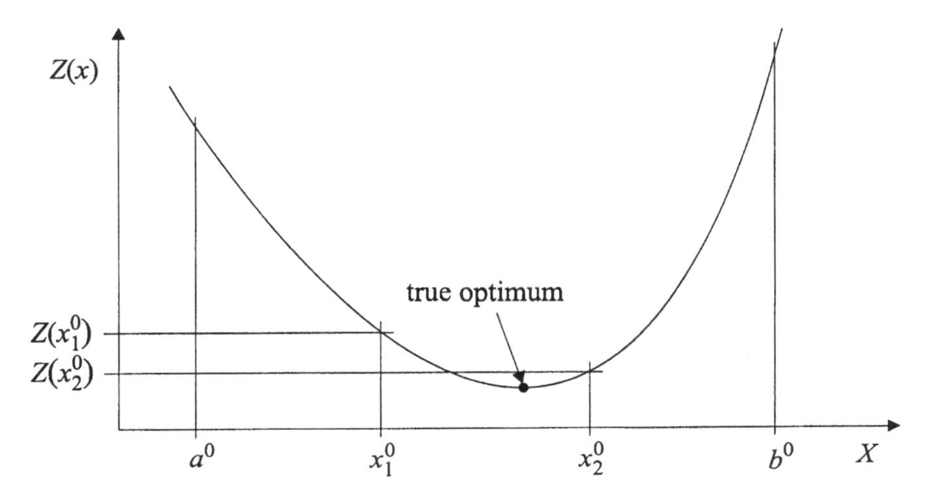

FIGURE 10 Golden section search—an example. First iteration.

The most popular numerical technique that can be used for the solution of the above optimization problem is known as the golden section search technique. It successively divides the available search range, specified as $[a^i, b^i]$ at every iteration, into two sections proportioned approximately as (0.319 and 0.681) and discards the one that does not contain the minimum. The number 0.681 has been discovered as the most efficient way for internal division by numerous mathematicians (whose derivation can be found in optimization books, such as the one by J. Kowalik and M. R. Osborne). The golden section search starts by choosing two x values, x_1^0 and x_2^0, which divide the interval $[a^0, b^0]$ into three thirds (Fig. 10), and proceeds to the evaluation of the function at these points, $Z(x_1^0)$ and $Z(x_2^0)$ (for example, through FEA), respectively.

The golden section iterative process compares the two function values, evaluated at x_1^i and x_2^i in Step i, and narrows the search domain accordingly:

(1) If $Z(x_1^i) > Z(x_2^i)$

$$a^{i+1} = x_1^i, b^{i+1} = b^i$$

$$x_1^{i+1} = x_2^i \quad \text{and} \quad x_2^{i+1} = b^{i+1} - 0.319(b^{i+1} - a^{i+1}) \tag{5.27}$$

(2) If $Z(x_2^i) > Z(x_1^i)$

$$a^{i+1} = a^i, \; b^{i+1} = x_2^i$$

$$x_1^{i+1} = a^{i+1} + 0.319\ (b^{i+1} - a^{i+1}) \qquad \text{and} \qquad x_2^{i+1} = x_1^i \tag{5.28}$$

The above search is normally terminated based on the size of the latest interval as a percentage of the initial interval,

$$\frac{(a^{i+1} - b^{i+1})}{(a^0 - b^0)} \leq \varepsilon \tag{5.29}$$

where ε is denoted as the convergence threshold.

A competing search method is the Fibonacci search technique that utilizes a number set named after the mathematician Leonardo of Pisa (also known as Fibonacci) who lived from 1180 to 1225. The Fibonacci numbers are defined as follows:

$$F_0 = F_1 = 1, \qquad F_i = F_{i-1} + F_{i-2} \qquad \text{for } i > 1 \tag{5.30}$$

The search divides the search domain of length $L = a - b$ into three sections by a proportion defined by

$$\Delta_i = L_{i-1} \frac{F_{i-2}}{F_i} \tag{5.31}$$

Either of the two outlying sections is eliminated based on the function values at x_1^i and x_2^i as in the golden section search method.

Although the Fibonacci method has been shown to have a slight advantage over the golden section search technique, the former requires an advanced knowledge of the size of the Fibonacci set (based on the desired ε). However, neither can cope with functions that may have multiple extrema. In such cases, one may have to search the entire domain, starting at one end and proceeding to the other at fixed increments in order to determine all the extrema and choose the variable value corresponding to the global extremum (Fig. 11). Over the years, numerous supplementary algorithms have been proposed in order to accelerate such brute force searches based on the availability of additional function values, normally obtained using a random search. Such supplementary algorithms allow the user to increase the size of the increments when it is suspected that the search is a distance away from an extremum (gobal or local).

Once can, naturally, argue the benefit of using any search technique in determining the minimum of a one-dimensional function at a time when it appears that we have "infinite" computing power, as opposed to using an exhaustive (brute force) method, when we test many, many x values. The counter-argument to the use of a brute force method would be that although the function may have only one variable, the function evaluations using, for example, FEA can consume enormous amounts of time if the search is

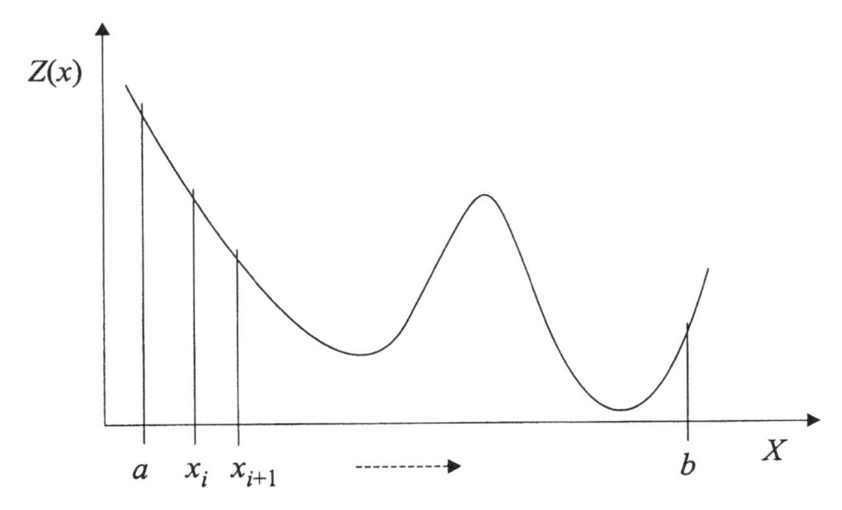

Figure 11 Multimode functions.

carried out in an ad hoc or random manner. The computation time problem would rapidly worsen for multivariable functions.

5.3.3 Multivariable Functions—Numerical Methods

Let us consider a product characteristic that is a function of multiple variables, $Z(x_1, x_2, \ldots, x_n)$. Let us further assume that $Z(x)$ is a continuous function and can only be evaluated through a numerical simulation (e.g., FEA), and that derivatives of the function cannot be obtained. The function is known to be single mode (the case of the existence of multiple extrema will be discussed at the end of this subsection), and there exist no restrictions on the variables. This optimization problem is called multivariable, single-mode, and unconstrained. However, despite all these simplifications, the "curse of dimensionality" increases the difficulty of solving the problem (compared to a single-variable function) hyperexponentially.

Although many solution techniques have been proposed over the years for the above problem, there does not exist a clear measure of efficiency in their comparison. Thus engineers are recommended to test several methods for their specific problem in regard to efficiency of convergence and choose the most suitable one. Most recommended search techniques vary the values of the variables simultaneously (in contrast to one at a time) and select the next point of evaluation based on past functional data. The general steps of a sequential search technique can be noted as follows:

1. Select one (or several) feasible point(s)—a point is a vertex of all variables, $\{x\}_0$ or $\{x\}_1$, $\{x\}_2$, ... $\{x\}_k$. (If the function has multiple extrema, these points should be widely separated.)
2. Evaluate the objective function at the initial points(s).
3. Based on the search technique utilized, choose the next feasible point.
4. Evaluate the objective function at this new point.
5. Compare the newest function value with earlier values and return to Step (3) if the search has not yet converged to the optimal solution, $\{x\}_{opt}$.

The specific search method reviewed in this section is the simplex method developed by J. A. Nelder and R. Mead, which lends itself to be adopted for constrained problems. The method has been often referred to as the flexible polyhedron search technique. As the name implies, the search utilizes a polyhedron in the hyperspace of the (multiple) variables. The simplex starts with four feasible vertices* labeled as follows:

$\mathbf{x}_h$ is the (multivariable) vertex that corresponds to $f(\mathbf{x}_h) = \max_i f(\mathbf{x}_i)$, i.e., the highest function value, for i vertices considered.
$\mathbf{x}_l$ is the vertex that corresponds to $f(\mathbf{x}_l) = \min_i f(\mathbf{x}_i)$, i.e., the lowest function value, for i vertices considered.
$\mathbf{x}_s$ is the vertex that corresponds to $f(\mathbf{x}_s) = \max_i f(\mathbf{x}_i), i \neq h$, i.e., the second-highest function value, for i vertices considered.
$\mathbf{x}_o$ is the centroid vertex of all $\mathbf{x}_i$, $i \neq h$.

$$\mathbf{x}_o = \frac{1}{k} \sum_{\substack{i=1 \\ i \neq h}}^{k+1} \mathbf{x}_i \tag{5.32}$$

As mentioned above, we will consider the simplex at hand for determining the next "point" (vertex) in our quest for the optimal variable values $\mathbf{x}_{opt}$. Once the initial simplex is constructed,

1. We first try a "reflection" operation to determine the next point as

$$\mathbf{x}_r = (1 + \alpha)\mathbf{x}_o - \alpha \mathbf{x}_h \tag{5.33}$$

where $\alpha > 0$ is a user chosen reflection coefficient.

---

* Each vertex is the set of all the variables $(x_1, \ldots, x_n)$ for the multivariable function considered. For clarity, bold lettering, $\mathbf{x}$, is used for $\{x\}$ in the description of the algorithm.

2. If $f(\mathbf{x}_s) > f(\mathbf{x}_r) > f(\mathbf{x}_l)$, we set $\mathbf{x}_h = \mathbf{x}_r$ and return to Step 1;
If $f(\mathbf{x}_r) < f(\mathbf{x}_l)$, we may expect the discovery of an even better point in the direction of $\mathbf{x}_r - \mathbf{x}_o$ and thus proceed to Step 3;
If $f(\mathbf{x}_h) > f(\mathbf{x}_r) > f(\mathbf{x}_s)$, we set $\mathbf{x}_h = \mathbf{x}_l$ and carry out a "contraction" by proceeding to Step 4.
If $f(\mathbf{x}_r) > f(\mathbf{x}_h)$, we contract without replacement, Step 5.
3. We "expand" as,

$$\mathbf{x}_e = \gamma \mathbf{x}_r + (1 - \gamma)\mathbf{x}_o \tag{5.34}$$

where $\gamma > 1$ is a user chosen expansion coefficient.

4. If $f(\mathbf{x}_l) > f(\mathbf{x}_e)$, we set $\mathbf{x}_h = \mathbf{x}_e$ and we return to Step 1. Otherwise, we set $\mathbf{x}_h = \mathbf{x}_r$ and we return to Step 1.

5. We "contract" as,

$$\mathbf{x}_c = \beta \mathbf{x}_h + (1 - \beta)\mathbf{x}_o \tag{5.35}$$

where $0 < \beta < 1$ is a user chosen contraction coefficient.

6. If $f(\mathbf{x}_h) > f(\mathbf{x}_c)$, we set $\mathbf{x}_h = \mathbf{x}_c$ and we return to Step 1. Otherwise, the simplex is "shrunk" as in Step 7.

7. We "shrink" the simplex as,

$$\mathbf{x}_i = \frac{1}{2}(\mathbf{x}_i + \mathbf{x}_l) \tag{5.36}$$

where $i = h, l,$ and s, and return to Step 1.

In the above algorithm, after each new function evaluation, the convergence criterion, as given below, must be checked:

$$\left\{ \frac{1}{k} \sum_{i=1}^{k+1} (f(\mathbf{x}_i) - f(\mathbf{x}_o))^2 \right\}^{1/2} < \varepsilon \tag{5.37}$$

where ε is the convergence threshold.

Nelder and Mead propose to have the initial polyhedron (simplex) be configured as widely spread as possible over the domain, in order not to be stalled at a local optimum. However, as with all other multivariable search techniques, there exists no convergence guarantee to the global optimum, even when the procedure is run several times starting with a different initial simplex.

The next problem to consider is having a constrained optimization problem, where the variables are restricted by numerous inequality constraints. A technique specifically designed for the Nelder–Mead method is the flexible tolerance method. A primary characteristic of this method is that if, during a reflection process, the new point lies within the infeasible region, it is not immediately dismissed as it would be with other constrained methods that build firewalls around the feasible domains and use penalty functions. (A penalty function would artificially increase the function value of the new infeasible point and thus severely discourage a search in the direction considered.) The flexible tolerance method, however, uses a more tolerant firewall approach: the firewall is initially placed further than its true location and moved back progressively as the search continues. That is, infeasible solutions are tolerated at lesser degrees as the search continues.

There are many other techniques that can be utilized for multivariable functions, especially if function derivatives can be evaluated (or approximated). The most effective methods are the Davidon–Fletcher–Powell method and the Rosenbrock method. As mentioned before, it is the application at hand that determines which method to use and how and when to stop it (i.e., declare convergence).

5.3.4 Integer and Combinatorial Optimization in Brief

Throughout this section, our primary focus has been on optimization problems in which the variables were allowed to be assigned any value in the continuous domain. In some design (and many manufacturing planning) instances, however, the variables can only be assigned an integer value (e.g., number of fasteners). One may attempt to use solutions developed for continuous optimization in these cases as well, and round the values to the nearest integer, though the solutions may no longer be optimal or even feasible. Thus it is strongly recommended that engineers should use suitable techniques for such problems—labeled as integer programming and combinatorics.

Gomory's Algorithm

This method solves an integer (linear programming) problem by using a constraint during the linear programming solution of the problem, which only allows integer solutions. The constraint mentioned is often called as the "Gomory's cutting plane."

Branch and Bound Method

Also known as the truncated enumeration method, this technique can be utilized to solve any bounded integer optimization problem. In such a method, the total set of solutions is systematically subdivided into smaller subsets; as the search progresses, some are discarded without examining any solutions within the subsets.

Dynamic Programming

This technique can be used to solve optimization problems with continuous as well as discrete (integer) variables. A problem of n variables is converted into n one-variable problems, which are then solved sequentially, yielding an optimal solution for the overall problem. Not all problems, however, lend themselves to being formulated as multiple subproblems. The three basic conditions for such a formulation can be summarized as follows:

The overall problem is divisible into n subproblems (stages), each of which (preferably) involves only one variable.

Relationships can be established between each pair of successive subproblems ($i-1$ and i).

The solution of the ith stage (subproblem) depends only on the solution of the $(i-1)^{th}$ stage.

Simulated Annealing

This method is a stochastic (probabilistic) method with a characteristic of occasionally allowing the choice of a sequence (in a combinatoric sequencing problem) that does not improve on the current minimum value of the objective function in order to prevent stalling at a local optimum. As in the flexible tolerance method for constrained (continuous) nonlinear problems, however, this acceptance of inferior sequences is reduced as the number of iterations increases.

Genetic Algorithms

Such algorithms attempt to mimic the natural selection process, where, in a population, the fittest is allowed to survive and reproduce. The optimization starts with a random generation of a population of genomes (different sequences of integer values for each of the variables) and their fitness evaluation (i.e., objective function value). "Offsprings" (new solutions) are generated by pairing two existing genomes that have the highest probability of yielding a better solution. (Different genetic algorithms (GAs) use different methods for generating offsprings.) Thus GAs can be classified as stochastic-type methods suitable for the sequencing of integer variables.

In conclusion to this subsection, an engineer once again should not assume that, since integer programming problems deal with finite spaces of solutions, one could use a brute force approach of simply testing every combination. The two primary counter-arguments are (1) the curse of dimensionality and (2) time spent in evaluating each solution.

REVIEW QUESTIONS

1. Define *engineering analysis*. How would computer hardware and software increase the efficiency of engineering analysis (i.e., computer-aided engineering analysis – CAE)?
2. What is a prototype of an engineering product?
3. How does the virtual prototyping process differ from physical prototyping? Discuss advantages/disadvantages of each.
4. How could CAE facilitate the prototyping process of engineering products?
5. What is the role of virtual reality in the prototyping process of engineering products?
6. Why would an engineer use finite element modeling and analysis (FEM/A) to obtain an approximate numerical solution?
7. Describe briefly the sequential process of FEM/A.
8. Would you recommend the use of FEM/A in manufacturing environments by engineers who do not have strong backgrounds in the field but have access to commercial software packages with FEM/A capability? Discuss both sides of the argument.
9. What is the role of optimization in the design process? In a typical engineering optimization process, what would the objective function, variables, and constraints represent, respectively?
10. What is the role of computers in an optimization process? Why should a design engineer attempt to choose the most efficient search method for the optimization of the variable values, as opposed to testing every possible combination? Consider problems with continuous and discrete variables.
11. In constrained optimization, strict firewalls could lengthen the search time significantly or may even prevent convergence. Discuss the use of strict firewalls versus tolerant, dynamic firewalls.

DISCUSSION QUESTIONS

1. Computers have often been considered as machines that automatically process information, versus (automated) production machines that process materials. Explain.
2. Discuss the role of computers in the different stages of the (iterative) design process: concept development, synthesis, and analysis.
3. In the mid part of the 20th century, design, planning, and control of manufacturing processes were argued to be activities that had more to do with the application of experience/knowledge gathered by experts in

contrast to the use of any mathematical models or systematic analysis techniques. As a contrary argument, discuss the basic physical phenomena (e.g., non-Newtonian fluid mechanics and heat transfer) that govern common manufacturing processes (e.g., casting of metals and molding of plastics) and their use in the design, planning, and control of these processes, Furthermore, discuss the role of computer-aided engineering (CAE) in these analyses.

4. Geometric modeling is a design task that can be efficiently implemented using a commercial CAD package. Discuss the importance of geometric modeling for the engineering analysis stage of design. Address issues such as surface modeling versus solid modeling.
5. A prototype of a product must meet some or all of its engineering specifications. These specifications can be geometric or functional. In this context, a large number of material-additive techniques were developed during the period 1980 to 2000 and classified as rapid prototyping (RP) processes. Justify through examples (production techniques and product features) why such layered manufacturing techniques may indeed be used to yield prototypes.
6. Discuss the concept of progressively increasing *cost of changes* to a product as it moves from the design stage to full production and distribution. How could virtual prototyping minimize potential future changes to a product, especially for those products that have very short development cycles?
7. Finite element modeling and analysis (FEM/A) methods have been developed to cope with the engineering analysis of complex product geometries and/or physical phenomena. Although these techniques would yield near-true results, the efficiency and correctness of software code may only be verified for simple scenarios. In this context, discuss the use of FEM/A during the (iterative) design process (i.e., synthesis–analysis) for the determination of optimal design parameter values. Furthermore, discuss the potential need for verifying the FEA results with physical tests.
8. Many conflicting interests affect the design of a product (i.e., its geometry, materials, configuration, etc.) as well as its fabrication/assembly techniques: profit, quality, environmental concerns, etc. As an engineer, when analyzing potential solutions, you may address the design process as a typical multivariable, multiobjective, constrained optimization problem. Discuss the utilization of such a mathematical approach to the determination of an optimal design versus utilizing an ad hoc approach.
9. Process planning in assembly (in its limited definition) refers to the optimal selection of an assembly sequence of the components. For

example, solving the traveling salesperson problem in the population of electronics boards. Although there are a number of search techniques for the solution of such problems, they could all benefit from the existence of a good initial (guess) solution. Discuss the role of group technology (GT) on the identification of such initial (guesses) sequences of assembly.

10. Design of experiments (DOE) is a statistical approach that can be used in the design of physical or simulation-based experiments for the determination of optimal variable values. Such factorial based experiments help engineers in the narrowing of the field of search to those parameters that have the greatest impact on the performance of the product as well as limiting the combinatoric number of variations of these variables. Discuss the role of DOE in the engineering analysis of a product.
11. Analysis of a production process via computer-aided modeling and simulation can lead to an optimal process plan with significant savings in production time and cost. Discuss the issue of time and resources spent on obtaining an optimal plan and the actual (absolute) savings obtained due to this optimization: for example, spending several hours in planning to reduce production time from 2 minutes to 1 minute. Present your analysis as a comparison of one-of-a-kind production versus mass production.
12. The use of design features has long been considered as improving the overall synthesis and analysis stages of products due to the potential of encapsulating additional nongeometric data, such as process plans, in the definition of such features. Discuss feature-based design, in which the user, through some recognition/extraction process, can access and retrieve individual similar or identical features on earlier product designs and utilize them for the design and engineering analysis of the product at hand.
13. Computer-aided engineering (CAE) analysis has been practiced as long as there have been computers in existence (since the early 1950s). Current commercial CAD software packages, which have been in existence only since the late 1970s, complement (older and new) CAE software with powerful graphical user interfaces (GUI) for the modeling of parts/products and graphical display of analysis results. In this context, discuss the complementary and potentially future competitive use of CAE and CAD packages in the design (synthesis and analysis) of engineering products.
14. In a majority of companies that do not employ engineers, and in some that do, decision making for design and/or production tasks relies on existing in-house expertise (human experts and/or written knowledge).

Discuss the diminishing role of experts in decision making as we improve our ability to model and analyze systems/phenomena and reach optimal solutions based on analytical/heuristic search techniques running on ever more powerful computers. Would such a trend lead to the disappearance of experience/knowledge-based "experts" and create a dependence on professionals (engineers/scientists with different backgrounds) who can develop and run these automated decision making systems?

15. The period 1980–2000 has witnessed the dismantling of the vertical integration of many large manufacturers and rapid movement towards supply-chain relationships. Discuss the impact of recent technological and management developments on this movement: short product lives, concurrent engineering carried out in the virtual domain (i.e., distributed design), minimization of in-process inventories, etc.

BIBLIOGRAPHY

Argyris, J. H., Kelsey, S. (Oct. 1954 to May 1955). Energy theorems and structural analysis. *Aircraft Engineering* Vols. 26 and 27.

Benzley, S. E., Merkley, K., Blacker, T. D., Schoof, L. (May 1995). Pre- and post-processing for the finite element method. *Journal of Finite Elements in Analysis and Design* 19(4):243–260.

Bernhardt, Ernest., ed. (1983). *CAE: Computer Aided Engineering for Injection Molding*. Munich, Germany: Hanser.

Beveridge, Gordon S. G., Schechter, Robert S. (1970). *Optimization: Theory and Practice*. New York: McGraw-Hill.

Boeing Co. Boeing News Release: Boeing 777 Digital Design Process Earns Technology Award. http://www.boeing.com/news/releases/1995/news.release. 950614-a. html, and http://www.boeing.com/commercial/777family/cdfacts.html

Cooper, Leon N., Steinberg, David (1970). *Introduction to Methods of Optimization*. Philadelphia: Saunders.

Courant, R. (1943). Variational methods for the solution of problems of equilibrium and vibrations. *Bulletin of the American Mathematical Society* 49:1–23.

Dai, Fan (1998). *Virtual Reality for Industrial Applications*. New York: Springer-Verlag.

Dieter, George Ellwood (2000). *Engineering Design: A Materials and Processing Approach*. New York: McGraw-Hill.

Edney, R. C. (1991). *Computer Aided Engineering for Mechanical Engineers*. New York: Prentice Hall.

Fagan, M. J. (1992). *Finite Element Analysis: Theory and Practice*. New York: John Wiley.

Fletcher, Roger., ed. (1969). *Optimization: 1968 Symposium of the Institute of Mathematics and Its Applications*. London, England: London Academic Press.

Fletcher, Roger (1987). *Practical Methods of Optimization.* New York: John Wiley.

Gen, Mitsuo (2000). *Genetic Algorithms and Engineering Optimization.* New York: John Wiley.

Gomes de Sá, A., Zachmann, G. (June 1999). Virtual reality as a tool for verification of assembly and maintenance processes. *J. of Computers and Graphics* Vol. 23(3):389–403.

Gomoroy, R.E. (1958). Outline of an algorithm for integer solutions to linear programs. *Bull. Amer. Math. Soc.* 64:275–278.

Huang, Hou-Cheng, Usmani, Asif S. (1994). *Finite Element Analysis for Heat Transfer: Theory and Software.* New York: Springer-Verlag.

Irani, R. K., Kim, B. H., Dixon, J. R. (Feb. 1995). Towards automated design of the feed system on injection molds BY integrating CAE, iterative redesign and features. *ASME Trans., Journal of Engineering for Industry* 117(1): 72–77.

Jacobs, Paul F. (1992). *Rapid Prototyping and Manufacturing: Fundamentals of Stereolithography.* Dearborn, MI: Society of Manufacturing Engineers.

Kobayashi, Shiro, Oh, Soo-Ik, Altan, Taylan (1989). *Metal Forming and the Finite-Element Method.* New York: Oxford University Press.

Kowalik, J., Osborne, M. R (1968). *Methods for Unconstrained Optimization Problems.* New York: Elsevier.

Lee, K., Gadh, R. (Oct. 1998). Destructive disassembly to support virtual prototyping. *IIE Transactions* 30(10):959–972.

Lepi, Steven M. (1998). *Practical Guide to Finite Elements: A Solid Mechanics Approach.* New York: Marcel Dekker.

Logan, Daryl L. (1997). *A First Course in The Finite Element Method Using ALGOR.* Boston: PWS.

Machover, Carl., ed. (1996). *The CAD/CAM Handbook.* New York: McGraw-Hill.

Nedler, J. A., Mead, R. (1965). A simplex method for function minimization. *Computer Journal, (No. 7):308.*

Nee, A. Y. C., Ong, S. K., Wang, Y. G. (1999). *Computer Applications in Near Net-Shape Operations.* New York: Springer-Verlag.

Nemhauser, George L., Wolsey, Laurence A. (1988). *Integer and Combinatorial Optimization.* New York: John Wiley.

Onalir, B., Kaftanoglu, B., Balkan, T. (June 1997). Application of computer-aided injection molding simulation for thermoplastic materials. CIRP, International Conference on Design and Production of Dies and Molds, Istanbul, Turkey, pp. 57–64.

Paviani, D. A., Himmelblau, D. M. (1969). Constrained nonlinear optimization by heuristic programming. *J. of Operations Research* 17:872–882.

Pironneau, Olivier (1989). *The Finite Element Methods for Fluids.* New York: John Wiley.

Sabbagh, Karl (1996). *21st Century Jet: The Making and Marketing of the Boeing 777.* New York: Scribner.

SAE (1998). *Virtual Prototyping: Computer Modeling, and Computer Simulation.* Warrendale, PA: Society of Automotive Engineers.

Schmitz, B. (May 1998). Automotive design races ahead. *J. of Computer-Aided Engineering* 17(5):38–46.

Siddique, Z., Rosen, D. W. (Dec. 1997). A virtual prototyping approach to product disassembly reasoning. *J. of Computer-Aided Design* 29(12): 847–860.
Singiresu, S. Rao (1999). *The Finite Element Method in Engineering*. Boston: Butterworth Heinemann.
Sutherland, I. (Dec. 1968). A head-mounted three dimensional display. *AFIPS, Proc. of Fall Joint Conference on Computers* 33:757–764 (San Francisco, CA).
Turner, M. J., Clough, R. W., Martin, H. C., Topp, L. J. (1957). Stiffness and deflection analysis of complex structures. *J. of Aeronautical Sciences* 23: 805–842.
Ulrich, K. T., Eppinger, S. D. (2000). *Product Design and Development*. Boston: McGraw-Hill.
Vose, Michael D. (1999). *The Simple Genetic Algorithm: Foundations and Theory*. Cambridge, MA: MIT Press.
Wang, K. K., Himasekhar, K., Chiang, H. H., Jong, W. R., Wang, V. W. Integrated CAE of injection molding using a three-layer approach. Soc. of Plastics Engineers, Proceedings of ANTIC, Vol. 37., Brookfield, CT, pp. 267–273.
Warwick, Kevin, Gray, John, Roberts, David (1993). *Virtual Reality in Engineering*. London: Institution of Electrical Engineers.
Wilson, John R. (1996). *Virtual Reality for Industrial Application: Opportunities and Limitations*. Nottingham, UK: Nottingham University Press.
Yevko, V., Park, C. B., Zak, G., Coyle, T. W., Benhabib, B. (Dec. 1998). Cladding formation in laser-beam fusion of metal powder. *J. of Rapid Prototyping* 4(4):168–184.
Zeid, Ibrahim (1991). *CAD/CAM Theory and Practice*. New York: McGraw-Hill.
Zou, Q., Horney, J. R. (1994). Minimize injection molded plastic part warpage by applying warpage analysis in mold design. Soc. of Plastics Engineers, Proceedings of ANTEC, Brookfield, CT, pp. 1104–1110.

Part II

Discrete-Parts Manufacturing

Manufacturing, in its broadest form, refers to "the design, fabrication (production), and, when needed, assembly of a product." In its narrower form, however, the term has been frequently used to refer to the actual physical creation of the product. In this latter context, the manufacturing of a product based on its design specifications is carried out in a discrete-parts mode (e.g., car engines) or a continuous-production mode (e.g., powder-form ceramic). In this part of the book, our focus is on the manufacturing (i.e., fabrication and assembly) of discrete parts. Continuous-production processes used in some metal, chemical, petroleum, and pharmaceutical industries will not be addressed herein.

In Chap. 6, three distinct fusion-based production processes are described for the net-shape fabrication of three primary engineering materials: casting for metals, powder processing for ceramics and high-melting-point metals and their alloys (e.g., cermets), and molding for plastics. In Chap. 7, several forming processes, such as forging and sheet forming, are discussed as net-shape fabrication techniques alternative to casting and powder processing of metals. One must note, however, that it is the manufacturing engineer's task to evaluate and choose the optimal fabrication process among all alternatives based on the specifications of the product at hand.

In Chap. 8, several traditional material-removal techniques, such as turning, milling, and grinding, collectively termed as "machining," are described. These techniques can yield parts that are dimensionally more accurate than those achievable by net-shape-fabrication methods. In practice, for mass-production cases, it is common to fabricate rough-shaped "blank" parts using casting or forming prior to their machining.

In Chap. 9, the emphasis is on nontraditional fabrication methods, such as electrical-discharge machining, lithography, and laser cutting, for part geometries and materials that are difficult to fabricate using traditional machining and/or forming techniques. Rapid layered fabrication of prototypes is also addressed in this chapter. A common constraint to all nontraditional (material-removal or material-additive) techniques is their restriction to one-of-a-kind or small-batch production.

In Chap. 10, several joining methods, such as mechanical fastening, adhesive bonding, welding, brazing, and soldering, are described as part of an overall discussion on product assembly. Automatic population of electronic boards and automatic assembly of small mechanical parts are also described in this chapter as exemplary applications of assembly.

In Chap. 11, workholding (fixturing) principles are discussed for the accurate and secure holding of workpieces in manufacturing. Numerous fixed-configuration (i.e., dedicated) jig and fixture examples are discussed for machining and assembly. Furthermore, several modular and reconfigurable systems are highlighted for flexible manufacturing.

In Chap. 12, common material-handling technologies, such as powered trucks, automated guided vehicles, and conveyors, targeted for the transportation of unit goods between manufacturing workcells, are described. The role of industrial robots in the movement of workpieces and tools within a workcell is also discussed in this chapter. The assembly of automobiles is addressed as an exemplary application area.

6

Metal Casting, Powder Processing, and Plastics Molding

This chapter presents net shape fabrication processes for three primary classes of engineering materials: casting for metals, powder processing for ceramics and (high-melting-temperature) metal alloys, and molding for plastics.

6.1 METAL CASTING

Casting is a term normally reserved for the net shape formation of a metal object by pouring (or forcing) molten (metal) material into a mold (or a die) and allowing it to solidify. The molten metal takes the shape of the cavity as it solidifies. Cast objects may be worked on further through other metal-forming or machining processes in order to obtain more intricate shapes, better mechanical properties, as well as higher tolerances. Over its history, casting has also been referred to as a founding process carried out at foundries.

6.1.1 Brief History of Casting

Casting of metals can be traced back in history several thousand years. Except in several isolated cases, however, these activities were restricted to the processing of soft metals with low melting temperatures (e.g., silver and gold used for coins or jewellery). An isolated case of using iron in casting has been traced to China, which is claimed to be possible owing to the high phosphorus content of the ore, which allowed melting at lower temperatures.

Casting of iron on the European continent has been traced back to the period A.D. 1200–1300, the time of the first mechanized production of metal objects, in contrast to earlier manual forming of metals. During the period A.D. 1400–1600, the primary customers of these castings were the European armies, in their quest of improving on the previously forged cannons and cannon balls. However, owing to their enormous weight, the large cannon had to be poured at their expected scene of operation.

The first two commercial foundries in North America are claimed to be the Braintree and Hammersmith ironworks of New England in early 1600s. Most of their castings were manufactured by solidifying molten metal in trenches on the foundry floor (for future forging) or poured into loam- or sand-based molds. Wood-based patterns were commonly used in the shaping of the cavities.

Despite the existence of numerous foundries in America, one of the world's most famous castings, the Liberty Bell (originally called the Province Bell) was manufactured in London, England, in 1775, owing to a local scarcity of bronze in the U.S.A. The bell, which cracked in 1835, has been examined and classified as a "poor casting" (being gassy and of poor surface finish). Cannon and bells were followed by the use of castings in the making of stoves and steam-engine parts. Next came the extensive use of castings by the American railroad companies and the Canadian Pacific Railroad. Their locomotives widely utilized cast-iron-based wheel centers, cylinders and brakes, among many other parts. Although the railroad continues to use castings, since the turn of the 20th century, the primary user of cast parts has been the automotive industry.

6.1.2 Casting Materials

The most common casting material is iron. The widely used generic term cast iron refers to the family of alloys comprising different proportions of alloying material for iron—carbon and silicon, primarily, as well as manganese, sulphur, and phosphorus:

Gray cast iron: The chemical composition of gray cast iron contains 2.5–4% carbon, 1–3% silicon, and 0.4–1% manganese. Due to its casting characteristics and cost, it is the most commonly used material (by weight). Its fluidity makes it a desirable material for the casting of thin and intricate features. Gray cast iron also has a lower shrinkage rate, and it is easier to machine. A typical application is its use in the manufacture of engine blocks. Gray cast iron can be further alloyed with chromium, molybdenum, nickel, copper, or even titanium for increased mechanical properties—strength, resistance to wear, corrosion, abrasion, etc.

Ductile cast iron: The chemical composition of ductile cast iron (also known as nodular or spheriodal graphite cast iron) contains 3–4% carbon, 1.8–2.8% silicon, and 0.15–0.9% manganese. First introduced in the late 1940s, this material can also be cast into thin sections (though not as well as gray cast iron). It is superior in machinability to gray cast iron at equivalent hardness. Its corrosion and wear resistance is superior to steel and equivalent to gray cast iron. Typical uses of ductile cast iron include gears, crankshafts, and cams.

Malleable iron: The chemical composition of malleable iron contains 2–3.3% carbon, 0.6–1.2% silicon, and 0.25–0.65% manganese. It can normally be obtained by heat-treating white iron castings. The high strength of malleable iron combined with its ductility makes it suitable for applications such as camshaft brackets, differential carriers, and numerous housings. One must note that malleable iron must be hardened in order to increase its relatively low wear resistance.

Other typical casting materials include

Aluminum and magnesium alloys: Aluminum is a difficult material to cast and needs to be alloyed with other metals, such as copper, magnesium, and zinc, as well as with silicon (up to 12–14%). In general, such alloys provide good fluidity, low shrinkage, and good resistance to cracking. The mechanical properties obtainable for aluminum alloys depend on the content of the alloying elements as well as on heat-treatment processes. Magnesium is also a difficult material to cast in its pure form and is normally alloyed with aluminum, zinc, and zirconium. Such alloys can have excellent corrosion resistance and moderate strengths.

Copper-based alloys: Copper may be alloyed with many different elements, including tin, lead, zinc, and nickel to yield, among others, a common engineering alloy known as bronze (80–90% copper, 5–20% tin, and less then 1–2% of lead, zinc, phosphorous, nickel, and iron).

Steel castings: These castings have isotropic uniformity of properties, regardless of direction of loading, when compared to cast iron. However, the strength and ductility of steel becomes a problem for the casting process, for example, causing high shrinkage rates. Low-carbon steel castings ($<$ 0.2% carbon) can be found in numerous automotive applications, whereas high-carbon cast steels (0.5% carbon) are used for tool and die making.

6.1.3 Sand Casting

Numerous advantages make casting a preferred manufacturing process over other metal fabrication processes. Intricate and complex geometry parts can be cast as single pieces, avoiding or minimizing subsequent forming and/or machining operations and occasionally even assembly operations; parts can be cast for mass production as well as for batch sizes of only several units and extremely large and heavy parts (thousands of kilograms) may be cast (as the only economically viable process of fabrication).

Among the numerous available techniques, sand casting is the most common casting process for ferrous metals (especially for large size objects such as automotive engine blocks). In sand casting, patterns are used for the preparation of the cavities, and cores are placed in the mold thereafter for obtaining necessary internal details. Due to the mostly mass production nature of the utilization of sand casting, the mold-making process and subsequent filling of the cavities is highly mechanized (usually in flow-line environments).

Pattern Making

Pattern making is the first step in the construction of a mold, with the exception of die-casting molds. Historically, mold cavities have been generated by building the mold, in an iterative manner, around a given pattern made of wear-resistance metal (for repeated use), plastics (for limited use), or wax (for one-time use). These patterns have been either manually prepared (i.e., cut or carved) by industrial designers or machined by numerous material removal techniques (Chap. 8) based on the object's CAD data. (The latest technology used in pattern making is layered manufacturing—one such commercially available rapid prototyping technology is *stereolithography,* commonly used for the fabrication of thermoset plastic parts—Chap. 9).

During pattern making, one can also include the gating system, through which the molten metal flows into the cavities, as part of the pattern (Fig. 1). Furthermore, patterns can be manufactured in two halves (called the "cope" and the "drag" patterns, or halves, of the mold), as opposed to a single-piece pattern, for the individual production of the two halves of the mold.

Although a pattern is used to produce the mold cavity, neither the pattern nor the cavity are dimensionally identical to the casting we intend to manufacture. Patterns must allow for shrinkage during solidification, for possible subsequent machining (namely, removal of some material to achieve better surface accuracy and finish), for distortion in large plates or thin-walled objects, and for ease of removal from the mold prior to casting.

Pattern making is followed by core making. Cores are patterns that are placed into the mold cavities and remain there during the casting process in order to yield the interior details of objects cast (Fig. 1). Naturally, they should be easily removable from the casting after the cooling period. In sand casting, cores are manufactured of sand aggregates.

One can realise that, for die casting applications, the pattern exists only in the virtual domain—i.e., as a CAD solid model. In such cases, the

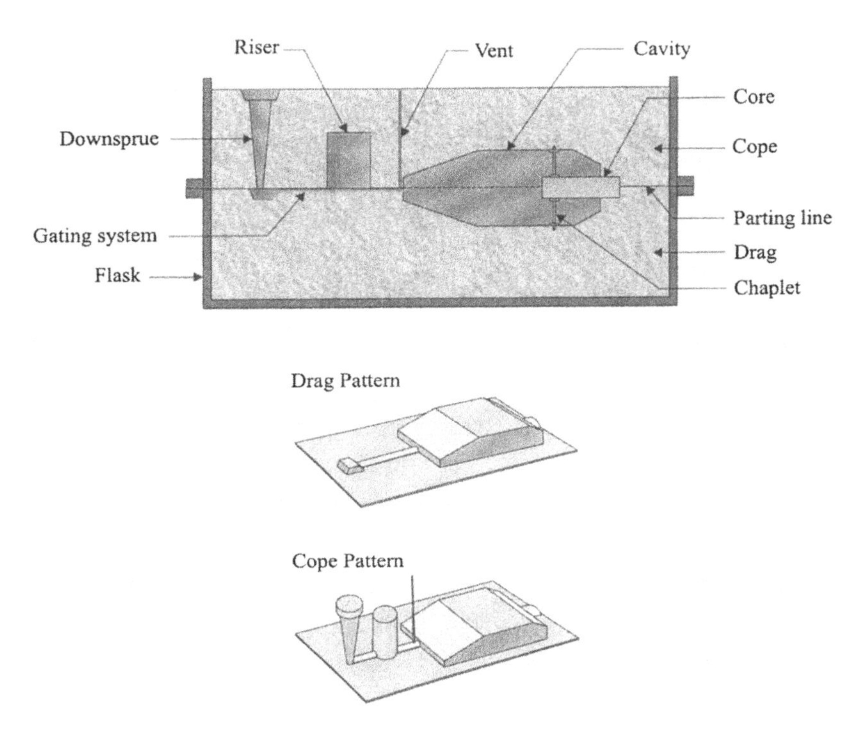

FIGURE 1 Sand mold.

mold is designed in the computer and its manufacturing operations are also planned in the same CAD domain.

Mold Making

As mentioned above, the sand casting mold is normally made of two halves—the cope and the drag. The sand used in making the mold is a carefully proportioned mixture of sand grains, clay, organic stretches, and a collection of synthetic binders. The basic steps of making a sand mold with two half patterns are as follows (Fig. 2):

1. The (half) pattern is placed inside the walls of the cope half of the mold.
2. The cope is filled with sand, which is subsequently rammed for maximum tightness around the pattern as well as around the gating system.
3. The pattern is removed.

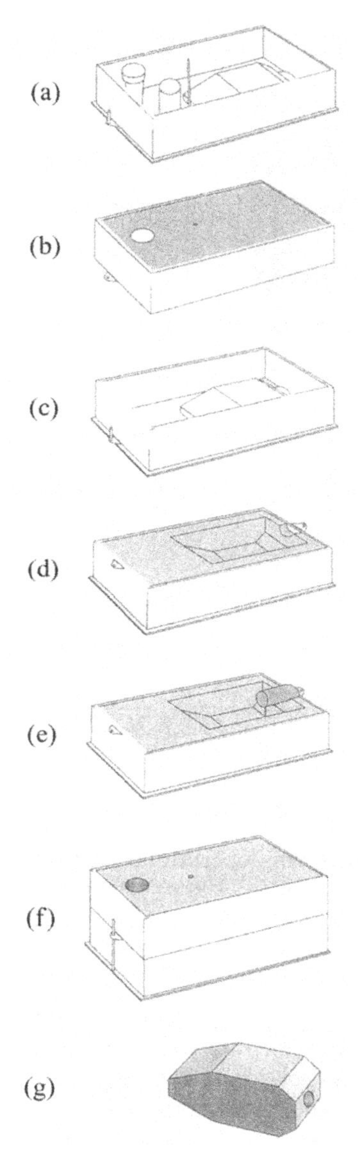

FIGURE 2 Mold-making and sand-casting process. (a) Cope pattern: ready to be filled with sand. (b) Cope filled with sand; pattern removed. (c) Drag pattern; ready to be filled with sand. (d) Drag filled with sand; pattern removed from drag. (e) Core placed inside drag. (f) Cope and drag assembled; molten metal poured into mold. (g) Metal cools and solidifies; casting removed from mold. Machining employed to remove the gating system; final product.

4. The second (half) pattern is placed inside the walls of the drag half of the mold.
5. The drag is filled with sand, which is subsequently rammed for maximum tightness around the pattern.
6. The pattern is removed and cores are placed if necessary.
7. The two mold halves are clamped together for subsequent filling of the cavities with molten metal.
8. The mold is opened after the cooling of the part and the surrounding sand (including the cores) are shaken out (through forced vibration or shot blasting).

Most sand cast parts would need subsequent machining operations for improved dimensional tolerances and better surface quality, which would normally be in the range of 0.015 to 0.125 in (app. 0.4 to 4 mm) for tolerance and 250 to 2000 μin (app. 6 to 50 μm) for surface roughness (R_a) (Chap. 16). However, one must note that sand casting can yield a high rate of production—hundreds of parts per hour.

6.1.4 Investment Casting

The investment casting process is also known as the lost wax process because of the expendable pattern (usually made of wax) used in forming the cavities. Although more costly than other casting processes, investment casting can yield parts with intricate geometries and excellent surface quality (15 to 150 μin, or approximately 1 to 6 μm). The term investment refers to the refractory mold that surrounds the wax pattern.

The basic steps of investment casting (mold making and casting) are as follows (Fig. 3):

1. An accurate metal die is manufactured and used for the large-scale production of wax patterns and gating systems.
2. The patterns are assembled into a multipart tree form and dipped into a slurry of a refractory coating material (silica, water and other binding agents). The tree is continuously lifted out and rotated to produce uniform coating and drainage of excessive slurry.
3. The tree is sprinkled with silica sand and allowed to dry.
4. The tree is invested in a mold with a slurry and allowed to harden (several hours to a day).
5. The mold is placed in an oven and the wax is melted off the investment casting mold (up to a day).
6. Molten metal is poured into the cavities while the mold is still at a high temperature.
7. The shells are broken and the castings cleaned.

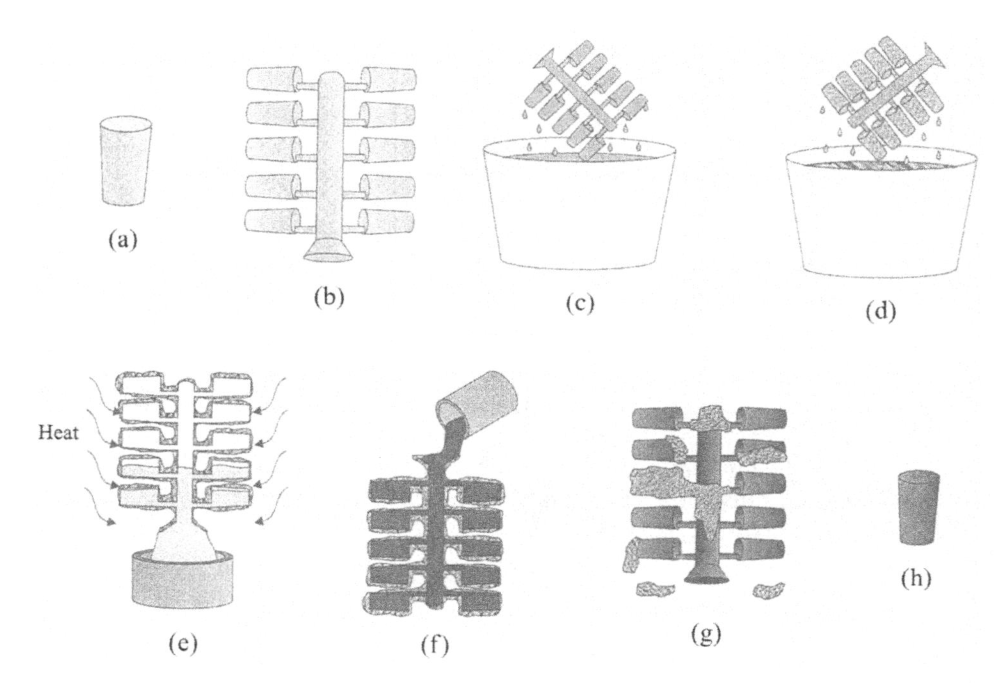

FIGURE 3 Investment casting. (a) Wax pattern. (b) Patterns attached to wax sprue. (c) Patterns and sprue coated in slurry. (d) Patterns and sprue coated in stucco. (e) Pattern melt-out. (f) Molten metal poured into mold; solidification. (g) Mold broken away from casting; finishing part removed from sprue. (h) Finished part.

Robots have been commonly used in the automation of the mold making process for investment casting: manufacture of wax patterns, assembly of trees, shell buildup, dewaxing, firing, casting, and cleaning.

6.1.5 Die Casting

Molds for multiuse must be made of comparably durable material (for example, tool-grade steel) and utilized for long runs in order to be economically viable. During the casting process, such molds would be sprayed (with silica-type fluid) prior to pouring of the molten metal, primarily to reduce wear. Molds are also be equipped with cooling systems in order to reduce cycle times, as well as to control the mechanical properties of the die cast part.

In the above context, die casting is a permanent mold process, where the molten metal is forced into the mold under high pressure, as opposed to pouring it in (under gravitational force). Die casting offers low cost, excellent dimensional tolerances and surface finish, and mass production capability (with low cycle times).

Die casting fabrication processes can be traced back to the mid-1880s, when it was used for the automatic production of metal letters. The development of the automotive industry in the early 1900s, however, is accepted as the turning point for die casting that first started with the production of bearings. Today, many automotive parts (door handles, radiator grills, cylinder heads, etc.) are manufactured through die casting (at rates of several thousands per hour). Most such parts are made of zinc alloys, aluminum alloys, or magnesium alloys.

As in other cases, a die casting mold comprises two halves. In this case one of the halves is fixed and the other is moving (the "ejector" half). After solidification, the casting remains in the moving half when the mold is opened. It is then ejected by (mechanically or hydraulically activated) pins. In order to prevent excessive friction with the fixed half and ease of ejection from the moving half, the part should have appropriate draft angles. Internal or external fins can be achieved by utilizing loose or moving die cores in the fixed half of the die. (Average wall thicknesses of die cast parts range from 1.0 to 2.5 mm for different alloys.)

There exist two primary die casting processes, whose names are derivatives of the machine configuration, more specifically, the locations of the molten metal storage units (Fig. 4): in the hot chamber machine, the molten metal storage unit is submerged in a large vat of molten material and supplies the die casting machine with an appropriate amount of molten metal on demand; on the other hand, for the cold chamber machine, a specific amount of molten metal is poured into the (cold) injection chamber that is an integral part of the die casting machine. Subsequently, this material is forced into the die under high pressure (typically, up to 150 MPa, or 20 ksi).

High-pressure cold chamber machines were originally supplied (ladled) manually by transferring molten metal from a holding furnace. However, since the 1970s, this process has been automated using mechanical ladles or machines that utilize pneumatic (vacuum) dispensers or electromagnetic pumps. Other automation applications in die casting have included the automatic lubrication of the die cavities by utilizing fixed or moving spray heads, as well as the use of robotic manipulators (ASEA, GM Fanuc and others) in the removal of parts from the dies (extraction), such as gasoline engines found in lawn mowers, snowmobiles, and garden tractors, and automotive fuel injection components.

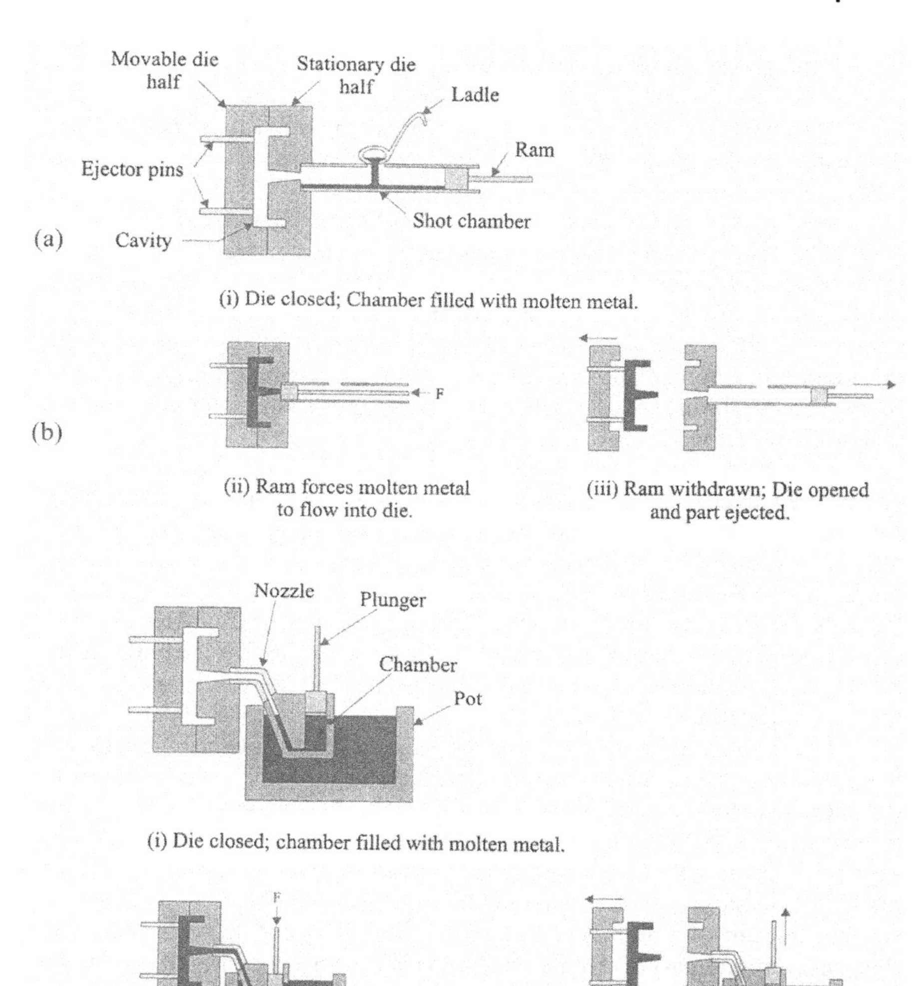

FIGURE 4 Die casting. (a) Cold chamber; (b) Hot chamber casting.

6.1.6 Design for Casting

The mechanical properties of a casting are of paramount concern to the user. Thus engineers must carefully design their parts and molds concurrently for optimizing a casting's performance. For example, parts can be designed to favor directional solidification for maximum strength and minimum chance of defects—columnar growth of dendrites would create weaknesses at sharp corners and must be avoided through the use of fillets. Furthermore, some metals are more susceptible to shrinkage during cooling and certain harmful shrinkage cavities—"hot spots." Such problems are more apparent at junctions, especially owing to changing wall thicknesses: they could be alleviated by utilizing small nonfunctional holes that would not affect the overall strength of the part (Fig. 5).

Some other casting-design guidelines are

Adjacent thin and thick sections cause porosity when cooling. Thus fillets and tapering should be used for projections, and when

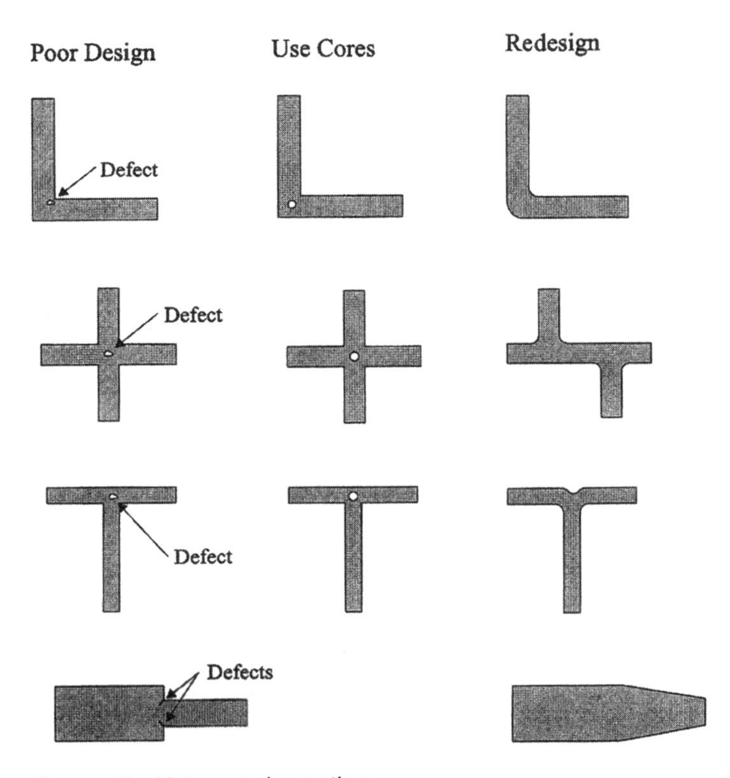

FIGURE 5 Hot spots in castings.

necessary local chilling should be employed as an additional measure.

It is generally more economical to drill out holes rather than using cores (especially for smaller holes).

Parting lines should be as straight as possible in order to prevent increased mold costs.

Casting threads (especially external) is more economical than machining.

Raised letters on parts (i.e., depressed shapes in the cavity) are cheaper to manufacture.

6.2 POWDER PROCESSING

Powder metallurgy, sintering, and powder processing have been synonymously used to describe the formation of discrete parts in mold/die cavities by compacting a mass of particles ($<$ 150 microns) under pressure. This net shape fabrication process is normally reserved for mass production of materials whose melting point makes them unsuitable for fusion techniques, such as casting. Here, the term powder processing will be utilized (versus the other two common terms) since we will discuss materials that are metal as well as nonmetal, and since sintering is only one of the primary steps in powder processing.

The basic steps of powder processing are powder production, compacting of powder, and sintering. The last phase involves heating the "preform" part to a temperature below its melting point, when the powder particles lose their individual characteristics through an interdiffusion process and give the part its own overall physical and mechanical properties. Sintering lowers the surface energy of the particles by reducing their (surface) areas through interparticle bonding.

6.2.1 Brief History of Powder Processing

The powder processing of ceramic pottery and platinum jewelry can be traced back several thousands years. With the introduction of forging and casting, powder processing took a pause until the early 1900s, except for occasional revival attempts along the way. The first commercially viable process in the early 1900s was the manufacture of tungsten wires used in electric (incandescent) bulbs. The production of tungsten carbide (with cobalt) followed in the 1920s. The next significant development was the fabrication of porous, self-lubricating bronze (90% copper and 10% tin powder) bearings (impregnated with oil) in the late 1920s.

The second half of the 20th century saw an explosive spread in the use of powder-processed modern materials, including a variety of cemented carbides, artificial diamonds, and cermets (ceramic alloys of metals). Today, such powder-processed components are used by many industries: aerospace (turbine blades), automotive (gears, bushings, connecting rods), and household (sprinklers, electrical components, pottery). Recent developments in efficient production techniques (such as powder injection molding and plasma spraying) promise a successful future for powder processing of light and complex geometry parts with excellent mechanical properties.

6.2.2 Powder Processing Materials

Materials for powder processed products are many, and new alloys are proposed yearly. In this chapter, only a representative subset will be discussed with the emphasis being on hard particles with high-melting temperatures.

Metals

Metal powders commonly used today for powder processing include iron and steel, aluminum alloys, titanium and tungsten alloys, and cemented carbides. There are numerous techniques for the production of metal powder:

Mechanical means can be effectively used to reduce the size of metal particles: Milling and grinding of (solid-state) metals rely on the fracture of the larger particles.

Melt atomization of metals can be classified as liquid or gas atomization. The former utilizes a liquid (normally, water) jet stream, which is fed with the molten metal, for the formation of droplets of metal (that has a low affinity to oxygen). Gas atomization is similar to liquid atomization, but it uses gases such as nitrogen, argon, or helium for melt disintegration.

Chemical reduction can also be used for the fabrication of metal powders from their (commonly) original solid state (for example, through the use of hydrogen).

Iron and steel are the most commonly (by weight) powder processed materials. Steels and alloyed steels are utilized for the production of bearings and gears in automotive vehicles, of connecting rods in internal combustion engines, and even of cutting tools and dies (high-speed steels, HSS). Powder processed steel parts can have homogenous distribution of (high-content) carbides with excellent isotropic properties for increased lifetime—a characteristic that cannot be easily obtained through casting or forming.

Although a preferred manufacturing technique for titanium alloy products is through melting, complex-geometry parts can be produced via powder processing. Tungsten products, on the other hand, are exclusively fabricated through powder processing owing to tungsten's high melting point (>3400 °C).

Cemented carbides (also known as hard metals), first developed in Germany in the 1920s, combine at least one hard compound and a binder metal—for example, tungsten carbide particles in a cobalt matrix. The hard metal provides the parts with high hardness and wear resistance, while the binder matrix provides them with mechanical and thermal shock resistance (toughness). The most common use for such carbides are cutting tools for the machining industry (and even for the mining industry).

Cermets

Cermet is a compound word indicating that the composition of the material contains at least one ceramic and one metallic component. Such materials have been fabricated since the mid-1900s. (The component with the highest volume fraction is considered to be the matrix.) Cermets are very suitable for high-temperature environments (e.g., metal-cutting tools, brake linings, and clutch facings). Metal-bonded diamond grinding wheels can be used to grind refractory materials, such as granite, fused alumina, and cemented carbides.

Ceramic powders can be produced through chemical reactions (solid–solid, solid–gas, and liquid–liquid). Some secondary mechanical processes (e.g., milling) can also be used for powder-size reduction.

6.2.3 Compacting

Bulk powder can be (automatically) transformed into ("green") preforms of desired geometry and density through compacting prior to their sintering. The first step in this process is effective mixing of the multimaterial powder. At this stage, lubricant, in the form of fine powder, is also added to the mixture (for reduced friction) if the powder is going to be formed in a closed die.

Most compacting operations, with the exception of processes such as slip casting and spray forming, are carried out under pressure: die compacting, isostatic compacting, powder rolling, extrusion of powder, and powder injection molding (PIM). Pressure-assisted compacting can be further categorized into cold (at ambient temperature) and hot (material-dependent enhanced-temperature) compactions.

Bulk powders are compressible materials—as the pressure is increased, the fraction of voids in the powder rapidly diminishes and the particles deform under (first elastic and then) plastic mechanisms (Fig. 6). The denser the preform is, the better are its mechanical properties and the less dimensional variation during sintering.

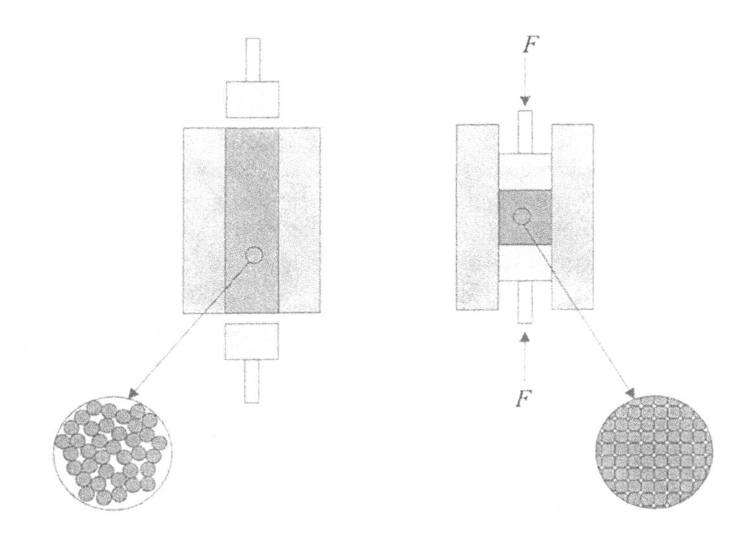

FIGURE 6 Compacting of powder.

Cold Compacting

Cold compacting (pressing), axial (rigid die) or isostatic (flexible die), is the most commonly utilized powder compacting method (Fig. 7). It requires only small amounts (and sometimes no amount) of lubricant or binder additions. In axial rigid die pressing, the powder is compacted by axially loading punches (one or several depending on the cross-sectional variations

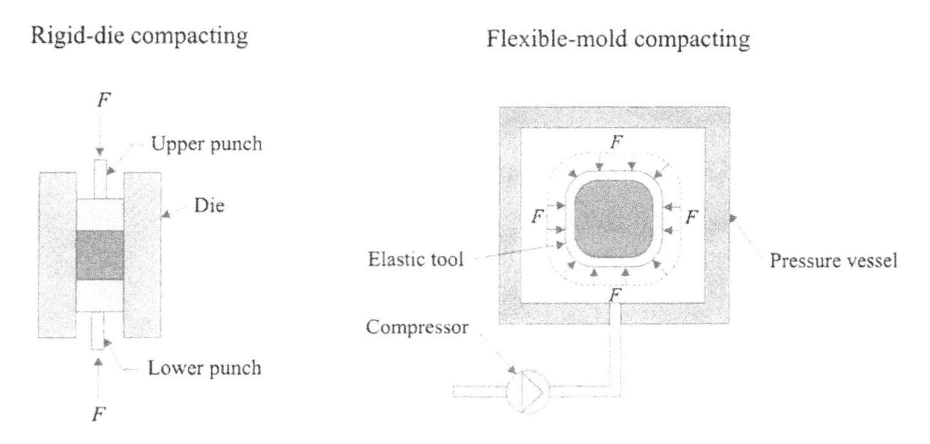

FIGURE 7 Rigid-die versus flexible-mold compacting.

of the part geometry), which are operated through mechanical or hydraulic presses. In isostatic compaction, a uniform pressure is applied to all the external surfaces of a powder body sealed in a flexible (elastomeric) envelope/mold. Incompressible liquids are normally utilized for exerting the required pressure. Although hydrostatic pressure would yield excellent uniformity in density, dimensional accuracy of the (green) preform is considerably less than it would be if manufactured in a rigid die.

Roll compacting can be utilized to fabricate (green) strips (or sheets) of powderprocessed (thin-walled) products. The powder can be fed into the rollers in vertical, inclined, or horizontal configurations (Fig. 8). Owing to the continuous nature of this process, however, the green product is usually fed (immediately) into a furnace on a rolling conveyor configuration. Frequently, the sintered product must be rolled again in order to reduce porosity.

Hot Compacting

The main hot compacting techniques are the axial and isostatic pressing processes and hot extrusion. Heating of the material in axial presses is achieved through direct heating of the powder or through heat transfer from the (heated) tool. In isostatic pressing, heating can be achieved by placing heating elements in the liquid enveloping the flexible mold.

Hot compacting of metals should be reserved for a select set of materials whose mechanical properties can indeed be improved during a heat-induced and pressurized compacting process. The process is expensive and difficult to operate and maintain. However, complex-shape products, when produced through such a technique, may be worth the effort—for example, jet-engine turbine disks fabricated from nickel-base superalloy powders. Temperatures in hot compacting can be as high as 1050–1100°C for beryllium and 1400°C for cemented carbides, or even higher (up to 2500°C for other materials).

Injection molding of powders, although occasionally considered as a hot compacting technique because of the elevated temperature of the plastic binding material (150° to 200°C), should be treated as a cold compacting

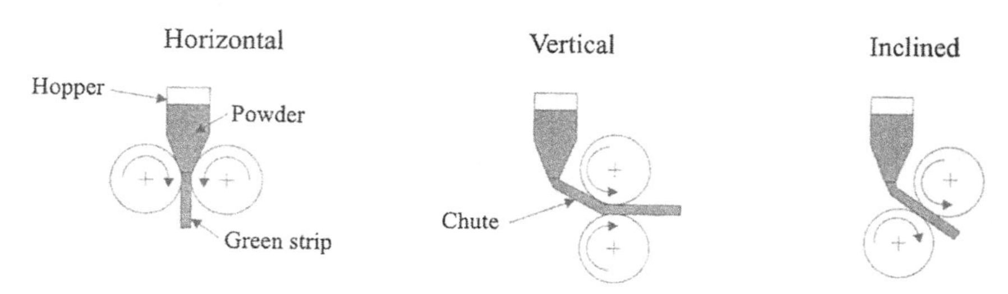

Figure 8 Roll compacting.

technique. The formation of green parts through this technique will be discussed following the presentation of the injection molding technique for polymers in Sec. 6.3 below.

6.2.4 Sintering

Sintering, the last stage in powder processing, is the thermal bonding of particles into a coherent, primarily solid structure. The mechanical properties of the original green compacted part are significantly improved through the elimination of the pores and the increase in density. However, it should be noted that the former phenomenon occurs at the expense of shrinkage and undesired dimensional changes. Thus maximum densities should be obtained at the presintering compaction phase.

Most sintering processes are carried out in pressureless environments and involve a partial liquid phase of the matrix component for multi-component materials. The presence of liquid (even for very short periods of time) improves the mass-transport rates and creates capillary pull. The application of heat can occur in batch or conveyor-type furnaces. Batch furnaces are easier to utilize, since the heating–cooling cycle is only dependent on the time a batch of parts spends in the furnace. In a continuous sintering furnace, the speed of the conveyor has to be carefully controlled, where parts are either placed on trays or directly on a metal screen belt. Sintered parts can be unloaded from furnaces using industrial robots.

Single-Phase Sintering

Sintering forms solid bonds between the particles, reducing the surface energy through grain growth and elimination of pores. Individual grain boundaries normally disappear by the end of the sintering process, and what remains behind is a solid cross section with distributed pores (Fig. 9). As is further shown in Fig. 10, individual grain boundaries are assumed to disappear through a neck growth process, in which two particles coalesce into a single larger particle.

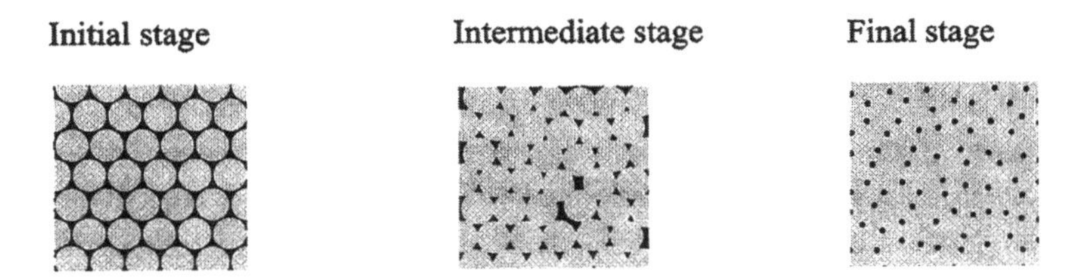

FIGURE 9 Sintering as a function of time.

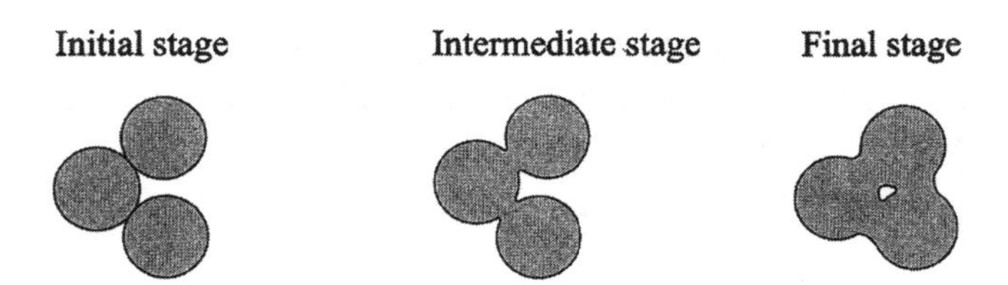

FIGURE 10 Neck growth in sintering.

Two mass-transport mechanisms contribute to grain bonding: surface transport and bulk transport. The former yields neck growth at lower temperatures without a change in particle spacing. Although bulk transport also contributes to neck growth, mass densification is the primary characteristic of this mechanism, which is achieved through volume diffusion, plastic flow, and viscous flow at high temperatures.

Sintering of multicomponent powder mixtures is normally carried out in the presence of a liquid phase of one of these components, as discussed below. However, sintering of mixtures can also be carried out in a single-phase (sintering) environment. In this case, neck growing predominantly occurs for the component with the lower melting temperature. Even if sintering times were prolonged for better mechanical properties, the individual rates of diffusion of the different powders would result in higher percentages of pores than those in single-component preforms.

Liquid-Phase Sintering

The presence of a liquid phase significantly increases the rate of sintering. Thus this process is commonly used in industry for both metal and ceramic alloys (e.g., cemented carbide cutting tools). Substantially full densities can be obtained through good wetting of the liquid on the solid particles, thus eliminating porosity. In this multistage process, the powder's temperature is first raised until the melting of one of the components. During this stage, solid-state sintering is already initiated. Subsequently, in the presence of the liquid phase, densification occurs through rearrangements (due to capillary forces), solution reprecipitation (i.e., grain growth), and final solid-state sintering.

6.2.5 Design for Powder Processing

As with casting, parts produced by powder processing are considered net shape and require few additional finishing processes. Due to processing requirements, especially the necessary high pressure for compacting, powder processed parts should not be too large. Thin geometrical details should also be avoided for ease of powder flow. The overall part geometry should be as simple and as uniform as possible. High length-to-diameter ratios (> 3) should be avoided. Sharp corners and edges weaken the part's mechanical properties. Undercuts and side holes, as well as threads, interfere with the ejection of the parts and should be machined after the parts have been sintered.

6.3 PLASTICS PROCESSING

Plastics have been one of the most controversial material groups of the 20th century. Despite their wide use in a large number of household and industrial products, they have been seen as a serious threat to the world's environment. However, as will be noted in this section, a large percentage of plastics are recyclable with minimum effort (in terms of having lower melting temperatures when compared to those of metals). Thermoplastic polymers constitute 85% of plastics in use—they can be recycled many times by simply repeating the heating and cooling cycle. Thermoset polymers, on the other hand, constitute the remaining 15% of plastics in use today and cannot be recycled.

Their resistance to corrosive degradation combined with their light weight have made plastics very suitable for use in industries such as construction, automotive, aerospace, and household products. They can be manufactured in continuous form (e.g., extrusion) or discrete form (e.g., injection molding). In the past several decades, plastics have also been reinforced with glass and carbon fibers to increase significantly their mechanical properties (strength and rigidity) to complement their excellent electrical and chemical properties.

6.3.1 Brief History of Plastics

The production of plastic products in modern times can be traced to the 1860s in England, where small moldings made of cellulose nitrate were made by A. Parkes. The 1930s witnessed the development of nylon and polyethylene by W. Carothers (working for DuPont de Nemours & Co.) and by ICI, England, respectively. The first uses of these products were in self-lubricating bearings and wire insulation. Today, plastics are used in the production of bottles, drums, toys, pipe fittings, wires, aircraft structures, and a variety of automotive parts.

Polymers and polymer composites have been used in automotive applications since the 1930s and 1950s, respectively. Today, approximately 8% (by weight) of material used in a North American automotive vehicle is plastics based. This percentage is also approximately 8% for European vehicles and 6% for Asian vehicles. Generally, 30% of plastics are used in the exterior of the car, 40% in the interior, 10% under the hood, and 20%

other (including structural components). Some examples include engine intake manifolds, instrument panels, side doors and door handles, fuel tanks, and fuel lines. It is expected that by 2015, the plastics content in a vehicle could rise to 12 to 15% (by weight). However, there are two competing factors that could affect this predicted composition: legislative recycling initiatives could keep the percentages at current levels or even force reductions, while legislative fuel-economy initiatives could force manufacturers to increase the usage of plastics up to 15 to 20% (by weight) in order to reduce the overall vehicle weight.

6.3.2 Engineering Plastics

Plastics refer to the family of polymers (organic materials), which are made of repeated collection of monomers produced through polymerization. The word polymer derives from the Greek words of *poly*, meaning many, and *meros*, meaning part. (Polyethylene, for example, comprises chains of ethylene, CH_2, monomers, as many as 10^6 of them per molecule.)

Polymers are classified based on their structures: linear chains, linear-branched chains and cross linked. The first two are called thermoplastic polymers; they can be solidified or softened (molten state) reversibly by changing their temperature. Cross-linked thermoset polymers, on the other hand, have their networks set after solidification and cannot be remelted, but only burned.

Thermoplastics

The four major low-cost, high-volume thermoplastic polymers are polyethylene, polypropylene, polystyrene, and polyvinyl chloride.

Polyethylene (PE) is a polymer comprising ethylene monomers. It has excellent chemical resistance to acids, bases, and salts. It is also easy to process (mostly through injection molding or extrusion), free from odor and toxicity, and reasonably clear when in thin film form. Major product lines of PE include bottles, toys, food containers, bags, conduits and wires, and shrink wraps.

Polypropylene (PP) is a fast growing low-cost polymer. Its heat resistance, stiffness, and chemical resistance is superior to those of PE. PP films can also be glass clear and be very suitable for food packaging when in coated (biaxially oriented) grade. Major product lines for PP include medical containers, luggage, washing-machine parts, and various auto parts (e.g., battery cases, accelerator pedals, door frames).

Polyvinyl chloride (PVC) is a polymer comprising vinyl and chloride monomers. It is always utilized with fillers and/or plasticizers (nonvolatile solvents), or even with pigments, lubricants, and extenders (e.g., parafins and oil extracts). PVC is the most versatile polymer; it can be rigid or flexible (when plasticized), it is resistant to alkalis and dilute mineral acids, and it can be a good electrical insulator. Major product lines of PVC include kitchen upholstery, bathroom curtains, floor tiles, blood bags, and pipes and fittings.

Polystyrene (PS) is a polymer comprising styrene monomers. It is the lowest-cost thermoplastic. Its major characteristic are rigidity, transparency, low water absorption, good electrical insulation, and ease of coloring. A significant limitation, however, is its brittleness—thus its rubber-modified grade of high-impact PS (containing up to 15% rubber). Its major product lines include mouldings for appliances, containers, disposable cutlery and dishes, lenses, footwear heels, and toys.

Thermosets

The four major thermoset polymers are polyester, epoxy, polyurethane, and phenolic. Although phenolics are historically the oldest thermosets, the largest thermoset family used today is the polyesters. Thermosetting polyesters are almost always combined with fillers, such as glass fibers, for yielding reinforced plastics with good mechanical properties. The automotive market is probably the largest consumer of such products. The high strength-to-weight ratio of polyester–glass laminates have led to their use also in aircraft parts manufacturing.

Composites

Composite plastics have two primary ingredients, the (thermoplastic or thermoset) polymer matrix and the reinforcement fibers/flakes/fillers/etc. The modulus and strength of the reinforced plastic is determined by the stiffness and the strength of the reinforcements and the bonding between them and the polymer matrix.

The most commonly used reinforcing material is glass fibers. They can be continuous fibers (woven into a laminated structure through filament winding) or (chopped) short fibers (mixed with the liquid polymer prior to being processed). E-glass (54% $Si0_2$) is the most widely used reinforcement: it has 76 GPa tensile modulus and 1.5 GPa tensile strength. Other reinforcing materials include carbon fibers, synthetic polymer fibers, and even silicon carbide fibers. DuPont's aramid polymer fiber (Kevlar 49) has found a niche market in aerospace and sports products, where superior performance is needed and cost is not a limiting factor. Kevlar's tensile modulus and strength are almost as twice those of E-glass fibers.

In the automotive industry, many companies (Ford, GM, Chrysler, Honda, etc.) have concentrated on the use of composite parts since the early 1980s, even in the primary vehicle structures, as a replacement for steel. The revolutionary car of the future could comprise 50% (by weight) aluminum and 50% composite plastics, thus achieving a 30 to 50% weight reduction in comparison to today's steel-based cars.

6.3.3 Thermoplastic Processes

The most widely used manufacturing processes for thermoplastic polymers are injection molding, extrusion, blow molding, rotational molding, and calendering. (Some of these can also be used for thermoset polymers, such as injection molding.) Extrusion, injection molding, and blow molding will be briefly reviewed here.

Extrusion

Although the focus of this book is on discrete parts manufacturing, the extrusion of plastics, which is a continuous process, is reviewed here because it is utilized in other plastics manufacturing processes to plasticize the polymer. The three primary elements of an extruder are the hopper, the barrel, which houses the screw, and the die (Fig. 11). Generally, the material (in granular form, pellets) is allowed to flow freely from the hopper into the throat of the extruder barrel (under gravity). As the screw turns in the heated barrel, a forward flow is generated. Frictional forces that develop within the barrel are the primary contributors to the melting (plasticizing) of the polymer. The molten material is fed into a die and exits the extruder (as it cools) assuming the cross-sectional shape of the die.

Besides pipes, tubes, and sheets, extruders can make hollow objects for blow molding (such as bottles) and provide injection molding machines with plasticized melt. In (noncontinuous) blow molding production, the resin flowing out of the extruder is fed into a mold and cut to dimension for yielding individual preforms (parisons), which are subsequently enlarged (and thinned in wall thickness) through blowing, as will be discussed below.

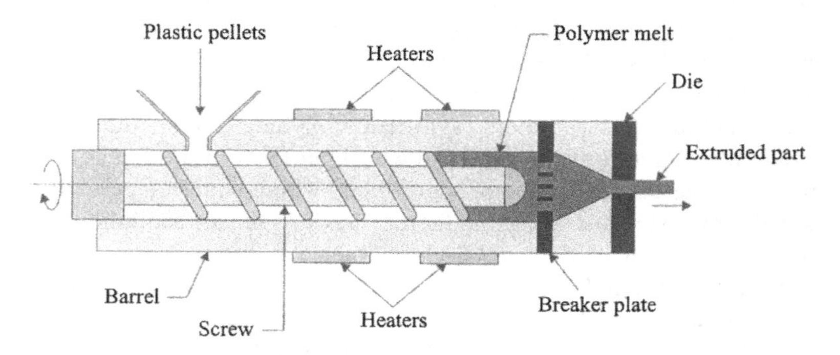

FIGURE 11 Plastics extrusion.

Blow Molding

Blow molding is primarily aimed at the production of thin-walled hollow plastic products. However, the process can be utilized for the fabrication of toys and even automotive parts. The basic steps of this process are: (1) the formation of a parison (a tube-like preform shape) in the molten state of the polymer, (2) sealing of one end of the parison and its inflation with blowing air injected from the other end—the parison then assumes the shape of the cavity of the mold, and (3) cooling and ejection from the mold (Fig. 12). The parison can be fabricated via a continuous or intermittent extrusion process linked to the blow molding machine. Parisons can also be injection molded in a cavity (of an injection mold) and then transferred to a second blowing mold.

Injection Molding

Injection molding is the most widely used process for thermoplastics in discrete parts manufacturing industries. The basic steps of injection molding are: (1) the transfer of resin (pure polymer or composite mixture) into a plasticizing chamber, (2) plasticizing of the resin and its transfer to the injection chamber (utilizing an extrusion screw or a cylinder), (3) pressurized

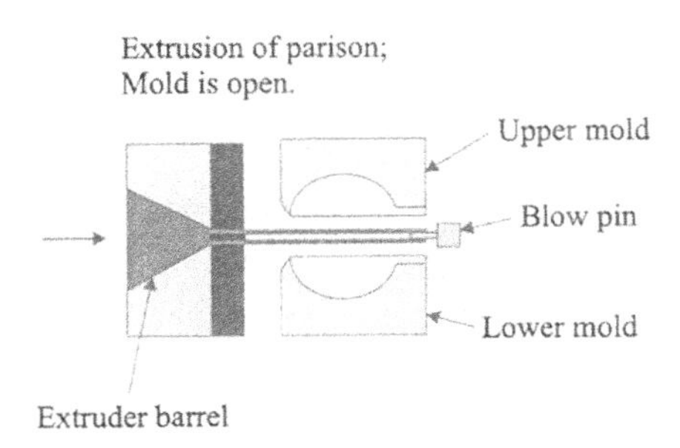

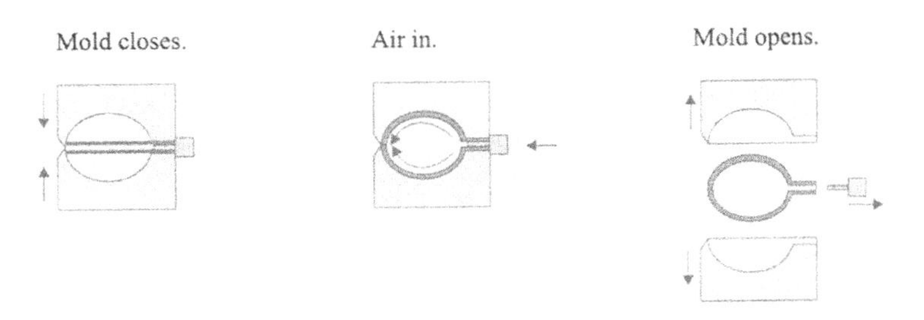

FIGURE 12 Blow molding.

injection of molten material into a closed mold (held tightly shut under great clamping forces), (4) solidification and cooling in the mold, and (5) ejection of parts from the cavities (Fig. 13).

Mold designs for injection molding are affected by many factors: the type of material to be molded and part geometry—affecting gating and

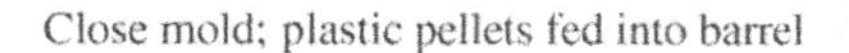

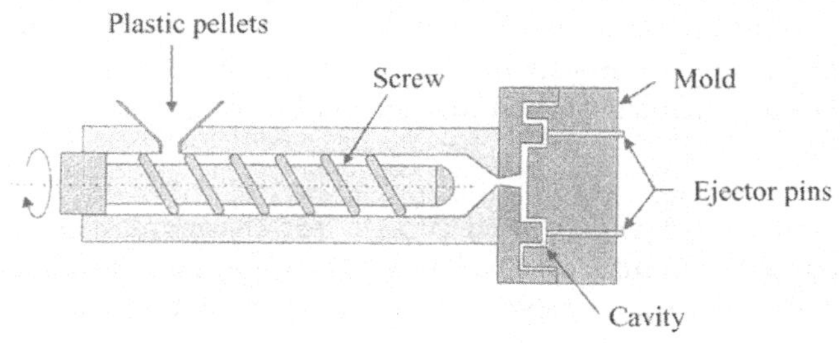

Plasticize plastic pellets.

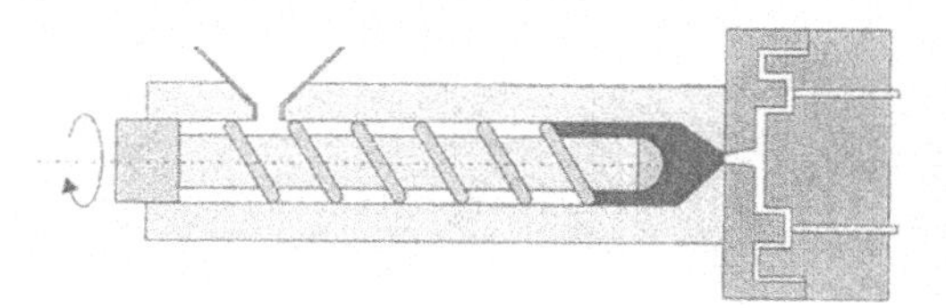

Inject molten plastic into mold; cool.

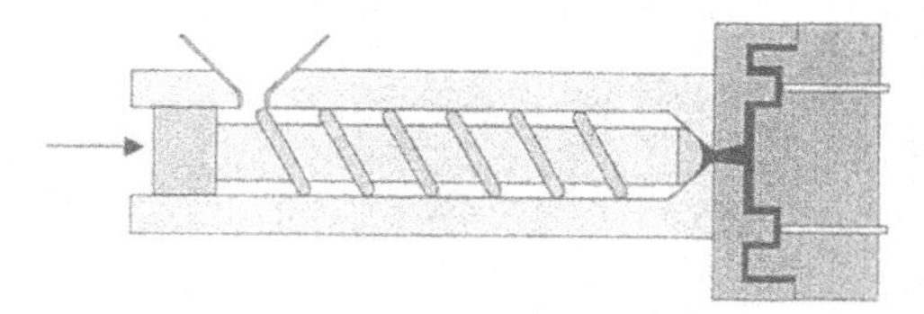

Eject part.

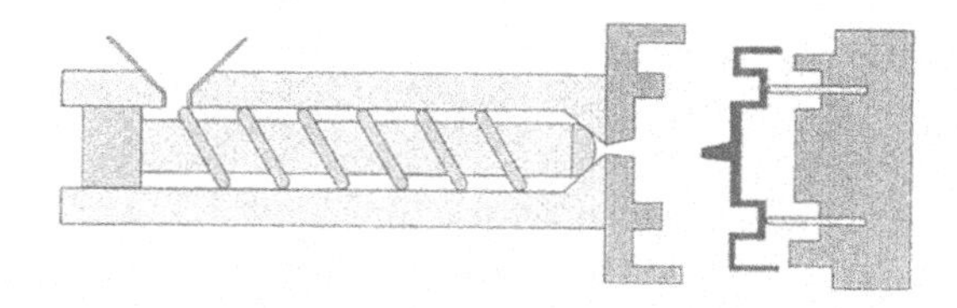

FIGURE 13 Injection molding.

ejection configurations, and production requirements—which dictate number of cavities, cooling rates, and mechanization. The mold may have two or three plates: one (or two) movable plates and one stationary plate. In two-plate molds, mold cavities are placed on the stationary plate and the ejectors on the moving plate (Fig. 14).

Injection molding can also be used for the fabrication of thermoset plastics. However, for such materials, the screw in the extrusion barrel must have a zero compression ratio—i.e., the depth of flight is uniform throughout the length of the screw. The materials themselves (most commonly phenolics) must also be modified for timely plasticization within the extruder. Normally, these thermoset materials are reinforced with short glass fibers, whose shrinkage characteristics must be considered during molding.

Recently, metal and ceramic powders have been also mixed with thermoplastic polymers for the fabrication of "compacts" (preforms) prior to a sintering phase in powder processing (Sec. 6.2). The polymer (commonly

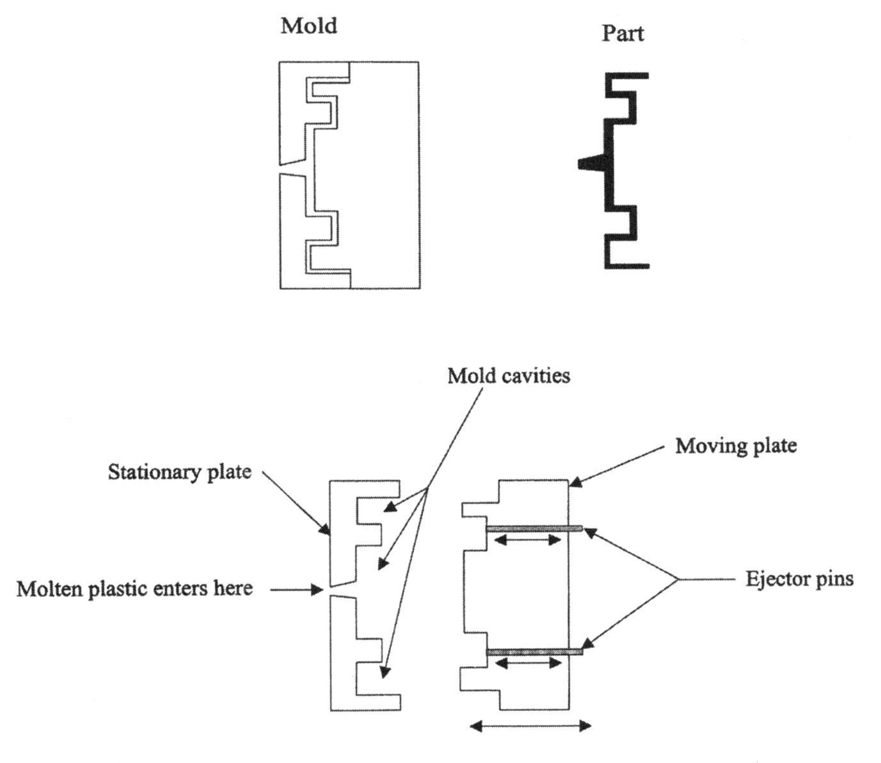

FIGURE 14 Injection mold design.

polyethylene) acts as a binder and is removed from the compact through thermal debinding or through the use of solvents. This injection molding process is commonly referred to as powder injection molding (PIM).

6.3.4 Thermoset Processes

Thermoset resins require higher temperatures (than do thermoplastics) to initiate polymerization (for the forming of a cross-linked matrix). Once processed, these plastics are temperature and chemical resistant, though they are also very brittle. Thus thermosets are seldom used without a reinforcement agent (such as fibers, glass, or synthetic polymers). As a result of these additives, processes for thermoset plastics must accommodate for the processing of two-phase mixtures (liquid matrix and solid additives). The three most widely used mold-based processes are compression molding, transfer molding, and, as mentioned above, injection molding. Other open-mold processes would include spray up, filament winding, and centrifugal casting.

Industrial robots are commonly used in spray up (or its derivative processes), where the robotic manipulator holds a spray gun that sprays catalyzed polyester resin mixed with chopped glass fibers onto a mold surface. Robots are also widely used in the removal of large parts from molds and in transferring them to other postcuring locations.

Compression Molding

Compression molding is the oldest method for the mass production of plastic products (thermoset as well as some polyethylene thermoplastics). This simple process includes two steps: (1) a controlled amount of resin (in pellet form) is placed into the cavity of a heated mold (150°C to 200°C) (2) the mold is subsequently closed under pressure and the resin is allowed to flow (to assume the shape of the cavity). Once ejected, parts can be transferred to a finishing area for the removal of flash (Fig. 15).

Compression molding can be used for high-reinforcement-content materials, with large surface areas and thicknesses, and it provides excellent uniformity in mechanical properties (isotropy). Also, since the polymer flows over short distances, concerns for large "frozen-in" stresses is reduced. Furthermore, the resin does not have to flow through a gating system (gate, sprue, and runners layout). However, the process is labor intensive (unless people are replaced by robots) and causes material waste (flash).

Transfer Molding

Transfer molding is a relatively new process developed (by L. E. Shaw) in response to the shortcomings of compression molding—especially for the production of parts with holes and recesses. The term "transfer" is in

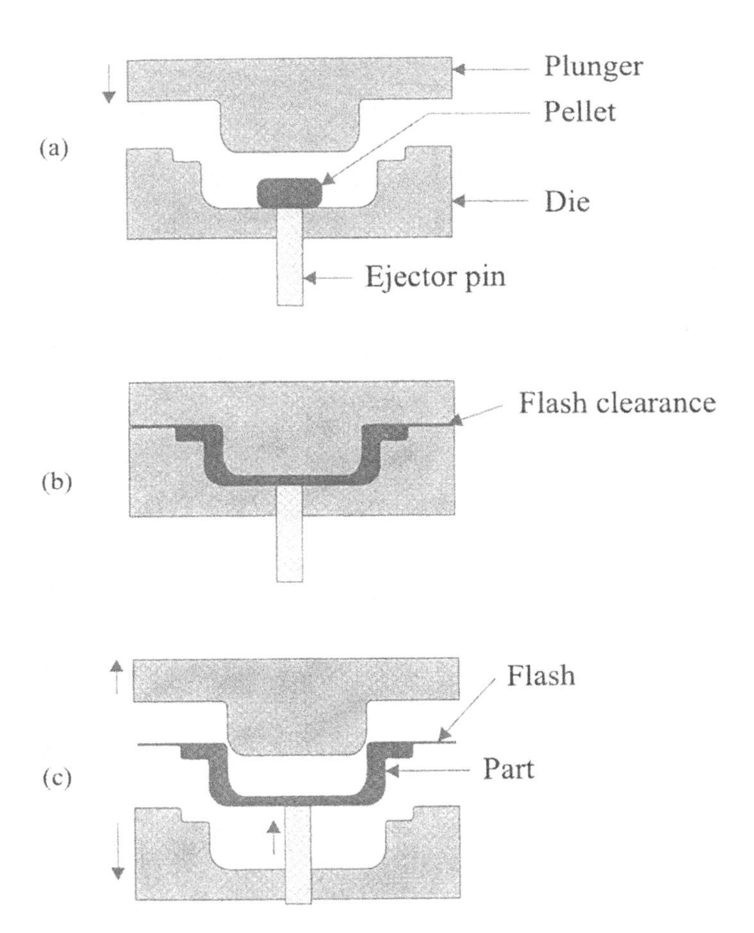

FIGURE 15 Compression molding. (a) Pallet loaded; mold closed. (b) Curing stage. (c) Mold opened; part ejected.

reference to the transfer of the molten resin, held in a middle plate of a three-plated mold, into the cavities, in the fixed plate of the mold, under pressure, through a gating system (Fig. 16). The process can be further automated by the utilization of an extruder screw that would provide the transfer molding press with controlled amounts of molten (plasticized) resin on demand.

In comparison to compression molding, transfer molding has the following advantages: good control of part thicknesses (owing to a totally-closed mold), production of intricate geometrical details, and better mechanical

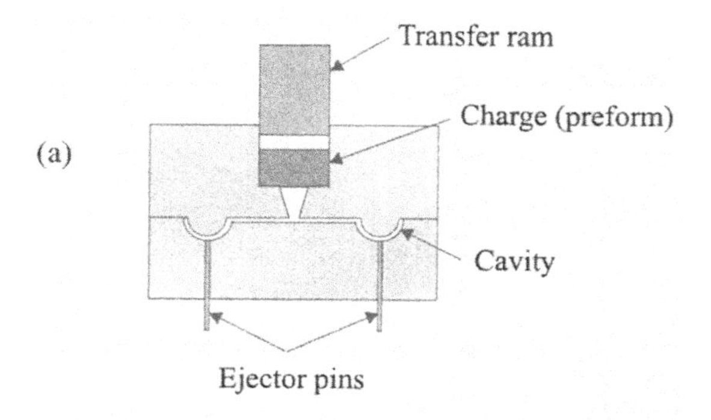

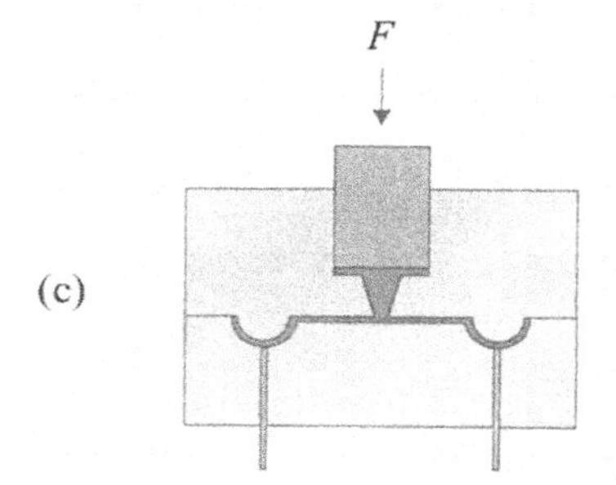

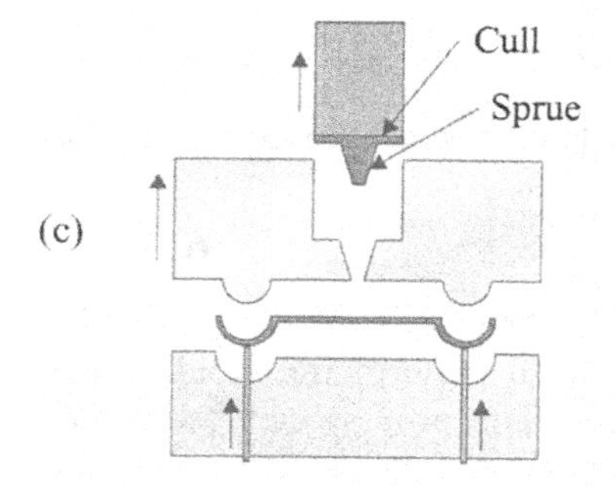

FIGURE 16 Transfer molding. (a) Charge loaded; (b) softened polymer pressed into cavity and cured; (c) part ejected.

properties (owing to less damage to fibers and furthermore their preferential alignment). However, material wastage in the gating system is greater here.

6.3.5 Design for Plastics Processing

The first step in designing a plastic product is the selection of an appropriate polymer (and reinforcing material, when needed). Factors in choosing a material for a specific application include mechanical properties, thermal and chemical properties (for example, resistance to ultraviolet sunlight), hazards (e.g., toxicity, flammability), appearance (e.g., transparency), and economics (including manufacturing costs). For the automotive parts industry, for example, engine parts must be resistant to automotive fluids and high temperatures. Similarly, body panels must be resistant to high paint-oven temperatures or, when not painted, they must have extra resistance to water absorbtion and UV light. Ski bindings, on the other hand, must be resilient to low temperatures and be very rigid.

From a manufacturing perspective, since most parts are fabricated in molds, part design strongly impacts on mold design and thus manufacturability. As discussed in Chaps. 3 and 5, the filling of the mold as well as the cooling of the part within the mold can be simulated using computer-aided engineering (CAE) analysis tools for better part design. Such analyses will remind designers to refrain from using sharp corners and/or sudden wall thickness changes that would disrupt the uniform flow of the resin in the cavities. Changes in wall thicknesses also result in additional shrinkage problems, such as stress concentrations, warpage, and even sink marks (Fig. 17). Sink marks predominantly occur opposite to ribs, flanges, and bosses, which are used for increasing stiffness and strength without adding weight to the part. Thus a rule of thumb is to have their thickness be 50 to 75% of the wall thickness they are reinforcing.

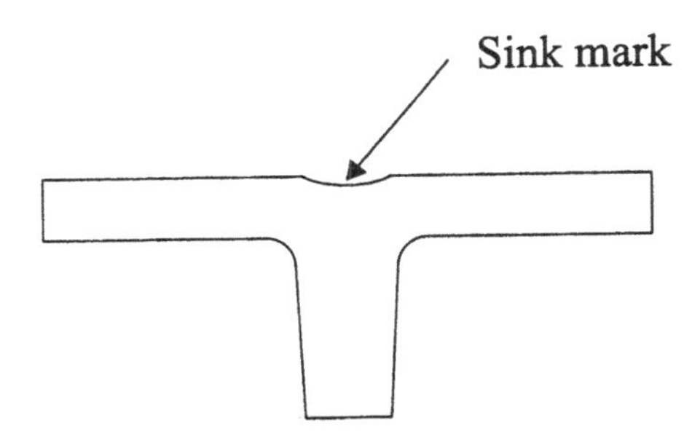

FIGURE 17 Sink mark.

Shrinkage problems are of major concern in the design of thermoset plastics. Mold design should accommodate for significant shrinkages in the curing of such materials. The problem is further complicated for composite parts, where fiber wetting as well as uniform fiber volume distribution are major concerns. A list of design guidelines for thermoset plastics and composites, which is also be applicable to thermoplastic parts, is given here:

Wall thicknesses must be kept as uniform as possible with gradual changes between sections through the use of fillets, tapers, etc.
Tapers should be used for ease of removal from the mold.
Side holes and/or undercuts should be avoided for low-cost molds.
Holes must not be placed too near to edges/faces to avoid fracture.
Fine screw threads should be avoided in composite part design, since even short fibers (less than 3 mm in length) would not be present at the threads.
Raised letters can be manufactured more easily (through engravings in the mold cavity).

REVIEW QUESTIONS

1. What is net shape (or near–net shape) fabrication? Identify several household products that are, or could have been, manufactured using a net shape process.
2. Why do casting/powder processing/plastics molding processes yield parts with mechanical properties better than those obtained with layered manufacturing (or lamination-type) processes?
3. What important property makes certain materials favorable for casting and not others?
4. Why is sand casting normally viewed as a process that lends itself easily to automation and mass production? Would you recommend sand casting for small-batch-size or one-of-a-kind production environments? Discuss both sides of the argument.
5. Why should one try to include the gating system into the pattern in sand casting? Furthermore, discuss the advantages/disadvantages of using halved patterns versus single-piece ones.
6. Discuss the investment casting process. Would you recommend investment casting for small-batch-size or one-of-a-kind production environments? Discuss both sides of the argument.
7. Define the two primary die casting processes and compare their uses.
8. Shrinkage is normally seen as a primary design concern in metal casting. How could one deal with this problem analytically?

9. What are the three basic steps of powder processing?
10. Why would one need to use powder processing instead of casting? If both processes were to be applicable, which one would you recommend?
11. What is a cermet?
12. Discuss the cold versus hot compacting of powder.
13. What is sintering? Why would one use a liquid-phase sintering process?
14. Thermoplastic polymers constitute 85% of plastics in use. They can be recycled many times by simply repeating the heating and cooling cycle. Thermoset polymers, on the other hand, constitute the remaining 15% of plastics in use today and cannot be recycled. Despite these facts, why do industries continue to manufacture thermoset-based products, including composites?
15. Describe the blow molding process.
16. Describe the injection molding process. What is powder injection molding (PIM)?
17. Discuss the fluid flow and cooling issues in plastics molding. You may refer to Chap. 5 for further information.

DISCUSSION QUESTIONS

1. Casting has commonly been used as a mass production technique. For example, in sand casting, highly accurate patterns and (when needed) mass produced cores are utilized for the production of thousands of identical parts. Review several of the common casting processes and discuss ways of using them profitably in high-variety production, for example by utilizing rapid prototyping techniques in the manufacture of patterns and cores.
2. Material removal techniques, as the name implies, are based on removing material from a given blank for the fabrication of the final geometry of a part. Compare material removal techniques to near–net shape production techniques, such as casting, powder processing, and forming, in the context of product geometry, material properties, and economics in mass production versus small-batch production environments.
3. Composite materials have been increasingly developed and used widely owing to their improved mechanical/electrical/chemical properties when compared to their base (matrix) material. For example, the use of glass, carbon, and Kevlar fibers in polymer base composites has significantly increased their employment in the automotive and sports products industries. Composite materials, however, may be in direct conflict with environmental and other concerns, which advocate that products should

be designed so that material mix is minimized or totally avoided for ease of manufacturing and/or recycling (including decomposition) purposes. Discuss the above issues in favor of continuing to use composite materials, otherwise propose alternatives.

4. In the mid part of the 20th century, design, planning, and control of manufacturing processes were argued to be activities that had more to do with the application of experience/knowledge gathered by experts in contrast to the use of any mathematical models or systematic analysis techniques. As a contrary argument, discuss the basic physical phenomena (e.g., non-Newtonian fluid mechanics and heat transfer) that govern common manufacturing processes (e.g., casting of metals and molding of plastics) and their use in the design, planning, and control of these processes. Furthermore discuss the role of computer-aided engineering (CAE) in these analyses.
5. Discuss potential postprocess defect identification schemata/technologies for parts that are manufactured using a casting or powder processing method. Furthermore, discuss possible sensing technologies that can be incorporated into different casting or powder-processing equipment for the on-line monitoring and control of the manufacturing process, while the parts are being formed.
6. Single-minute exchange of dies (SMED) is a manufacturing strategy developed for allowing mixed production (e.g., multimodel cars) within the same facility in small batches. The primary objective has always been to minimize the time spent on setting up a process while the machine is idle. This objective has been achieved (1) by converting as many on-line operations as possible to off-line ones (i.e., those that can be carried out while the machine is working on a different batch), and (2) by minimizing the time spent on on-line setup operations. Discuss the effectiveness of using SMED or equivalent strategies in the mass manufacturing of multimodel products, the mass manufacturing of customized products, and the manufacturing of small batch–size or one-of-a-kind products.

BIBLIOGRAPHY

Albertson, J. (Jan.–Feb. 1987). Some international trends in state-of-the-art die casting. *J. of Die Casting Engineering* 31(1):42–43.

Anon, S. (Jun.–Jul. 1985.). GM injection and emission systems. *Journal of Automotive Engineering* 10(3):38–39.

Barnett, S. (Mar. 1993). Automation in the investment casting industry: from wax room to finishing. *J. of Foundry International* 16(1):226–228.

Buckleitner, Eric., ed. (1995). *Plastics Mold Engineering Handbook*. New York: Chapman and Hall.
Bryce, Douglas M. (1996). *Plastic Injection Molding: Manufacturing Process Fundamentals*. Dearborn, MI: Society of Manufacturing Engineers.
Bryce, Douglas M. (1997). *Plastic Injection Molding: Material Selection and Product Design Fundamentals*. Dearborn, MI: Society of Manufacturing Engineers.
Bryce, Douglas M. (1998). *Plastic Injection Molding: Mold Design and Construction Fundamentals*. Dearborn, MI: Society of Manufacturing Engineers.
Chanda, Manas, Salil, Roy K. (1998). *Plastics Technology Handbook*. New York: Marcel Dekker.
Cook, Glenn J. (1961). *Engineered Castings: How to Use, Make, Design, and Buy Them*. New York: McGraw-Hill.
Doyle, Lawrence E., et al (1985). *Manufacturing Processes and Materials for Engineers*. Englewood Cliffs, NJ: Prentice-Hall.
German, Randall M. (1996). *Sintering Theory and Practice*. New York: John Wiley.
German, Randall, Bose, Animesh (1997). *Injection Molding of Metals and Ceramics*. Princeton, NJ: Metal Powder Industries Federation.
Groover, Mikell P. (1996). *Fundamentals of Modern Manufacturing: Materials, Processes, and Systems*. Upper Saddle River, NJ: Prentice Hall.
Heine, Richard W., Loper, Carl R., Jr., Rosenthal, Philip C. (1967). *Principles of Metal Casting*. New York: McGraw-Hill.
Hudak, G. (Apr. 1986). Robot automates diecasting cell. *American Machinist* 130(4):86–87.
Iwamoto, N., Tsuboi, H. (1985). Trend of computer-controlled die casting. *Society of Die Casting Engineers, Proceedings of the 13th International Die Casting Exposition and Congress, G-T85–035, Milwaukee, WI.*
Kalpakjian, Serope, Schmid, Steven R. (2000). *Manufacturing Engineering and Technology*. Englewood Cliffs, NJ: Prentice-Hall.
Lenel, Fritz V. (1980). *Powder Metallurgy: Principles and Applications*. Princeton, NJ: Metal Powder Industries Federation.
Maine, Elicia M. A. (1997). Future of Polymers in Automotive Applications. M.Sc. thesis, Department of Materials Science and Engineering, MIT, Cambridge, MA.
Mathew, J. (1985). Robotic applications in investment casting. *SME, Proceedings of Robots 9* 1:3.1–3.13 (Dearborn, MI).
McCrum, Norman G., Buckley, C. P., Bucknall, C. B. (1997). *Principles of Polymer Engineering*. New York: Oxford University Press.
Monk, J.F., ed. (1997). *Thermosetting Plastics: Moulding Materials and Processes*. London: Longman.
Rosato, Dominick V. (1998). *Extruding Plastics:7 A Practical Processing Handbook*. London: Chapman and Hall.
Sanders, Clyde Anton, Gould, Dudley C. (1976). *History Cast in Metal: The Founders of North America*. Des Plaines, IL: Cast Metals Institute, American Foundrymen's Society.
Schey, John A. (1987). *Introduction to Manufacturing Processes*. New York: McGraw-Hill.

Thümmler, Fritz, Oberacker, R. (1993). *An Introduction to Powder Metallurgy*. London: Institute of Materials.
Upadhyaya, G. S. (2000). *Sintered Metallic and Ceramic Materials: Preparation, Properties, and Applications*. Chichester, NY: John Wiley.
Upton, B. (1982–1983). *Pressure Diecasting—Volumes 1 and 2*. New York: Pergamon Press.

7

Metal Forming

Metal forming processes transform simple-geometry billets/blanks into complex-geometry products through the plastic deformation of the metal in open or closed dies. Due to the high costs of the dies, however, these processes are primarily reserved for mass production. Metals to be formed under (normally compressive) stress must be ductile and have low yield strength. These properties can be favorably induced, when necessary, by preheating the billets/blanks prior to their placement in the press. Furthermore, one should note that metal forming processes may take one or a few iterations (i.e., using one or multiple dies) in yielding near net shape desired geometries with no or little scrap.

Metal forming processes may be classified into two primary categories:

1. Massive forming processes (for bulk deformation), where parts undergo large plastic deformation.
2. Sheet-metal forming processes, where (thin-walled) sheets of metal undergo change in overall shape, but not much in their cross sections.

In this chapter, we will first briefly overview several common metal forming processes, but present detailed descriptions for only two of those that are targeted for discrete parts manufacturing (versus continuous production, such as for tubes and pipes): forging and sheet metal forming.

7.1 OVERVIEW OF METAL FORMING

7.1.1 Mechanical Behavior of Metals

Deformation of a solid body can be classified as elastic or plastic: when unloaded, an elastically deformed body always returns to its original shape regardless of history, rate, time, and path of loading; the plastic deformation of a body, on the other hand, depends on all these variables and is subjected to (permanent) loss of original shape when unloaded. Although the theory of elasticity is well established and yields accurate predictions of strain (due to mechanical stress), the theory of plasticity normally yields approximate solutions to plastic deformation problems.

The typical one-dimensional stress–strain curve shown in Fig. 1a for a tension test would normally be also applicable to the compression of ductile metals. As a load is applied on a metal part, it elongates in a linear proportion to the force until the stress level reaches the yield stress value, Y. At this critical point, when the load is released, the strain level of the part would be 0.2% or less. At any point before that, the part would completely recover its original shape. As the load is increased beyond the yield stress value, the part undergoes plastic deformation in a uniform-elongation phase until the stress level reaches the ultimate tensile strength value, UTS. At any point during this phase, if the load is removed, the part would recover the elastic strain portion of the deformation but permanently maintain the plastic elongation (or shortening in the case of compression) (Fig. 1b).

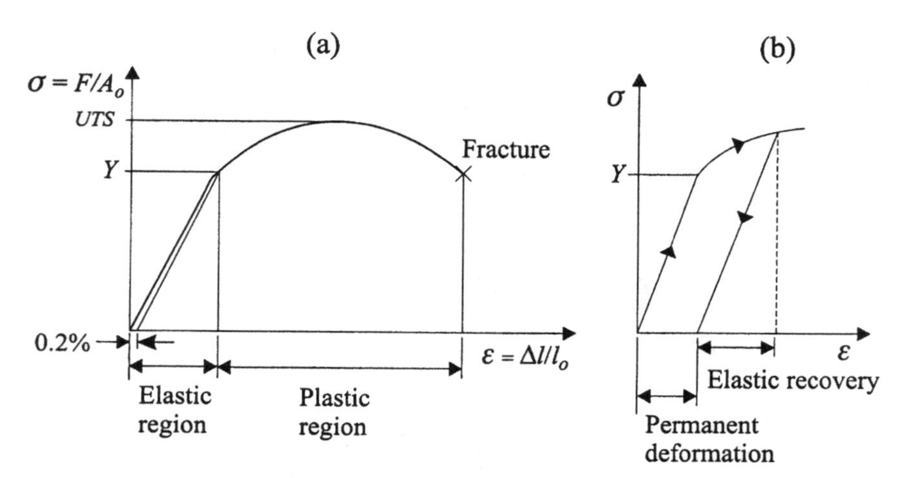

FIGURE 1 (a) Stress–strain curve for tension. (b) Loading-unloading cycle for plastic deformation: F, force; A_o, cross-sectional area; l_o, part's original length; Δl, incremental elongation.

Beyond the *UTS* stress level, the continuing application of load would lead to nonuniform elongation and eventual fracture of the part. In this context, *ductility* is the percentage of plastic deformation that the part undergoes before fracture.

As mentioned above, in metal forming the preference would be to process materials whose ductility is high (and that could be made even higher with increased temperature). Another important factor that we must take note of in metal forming is the rate of deformation (i.e., the amount of strain per unit time). It has been accepted that as the rate of deformation is increased, so would the necessary amount of stress to induce the required strain rate. As the temperature of the part is increased, however, one can obtain higher rates of deformation. Thus one can conclude that increasing temperature raises ductility, lowers yield stress, and thus shortens forming cycle times.

7.1.2 Common Metal-Forming Processes

Forming processes are broadly classified into massive forming and sheet metal processes. The former can be further divided into forging, rolling, extrusion, and drawing, while the latter include processes such as shearing/blanking, bending, and deep drawing. Some of these processes are briefly discussed below as preamble to a more detailed presentation of forging and sheet metal forming processes in Secs. 7.2 and 7.3, respectively. One must note, however, that most parts produced through metal forming could also be (geometrically) fabricated via casting or powder processing. It is the manufacturing engineer's responsibility to choose the most suitable fabrication method to satisfy the numerous constraints at hand, such as mechanical properties, dimensional requirements, and cost.

Forging

Forging is one of the oldest metal forming processes; it can be traced to early civilizations of Egypt, Greece, Persia, China, and Rome, when it was used in the making of weapons, jewellery, and coins. Forging, however, became a mainstream manufacturing process in the 18th century with the development of drop-hammer presses. Today, in closed-die forging, a part can be formed under compressive forces between the two halves of a die, normally in several steps, or in one step (with or without flash) (Fig. 2). The thin flash formed during closed-die forging cools quickly and acts as a barrier to further outward flow of the blank material, thus, forcing it to fill the cavity of the die.

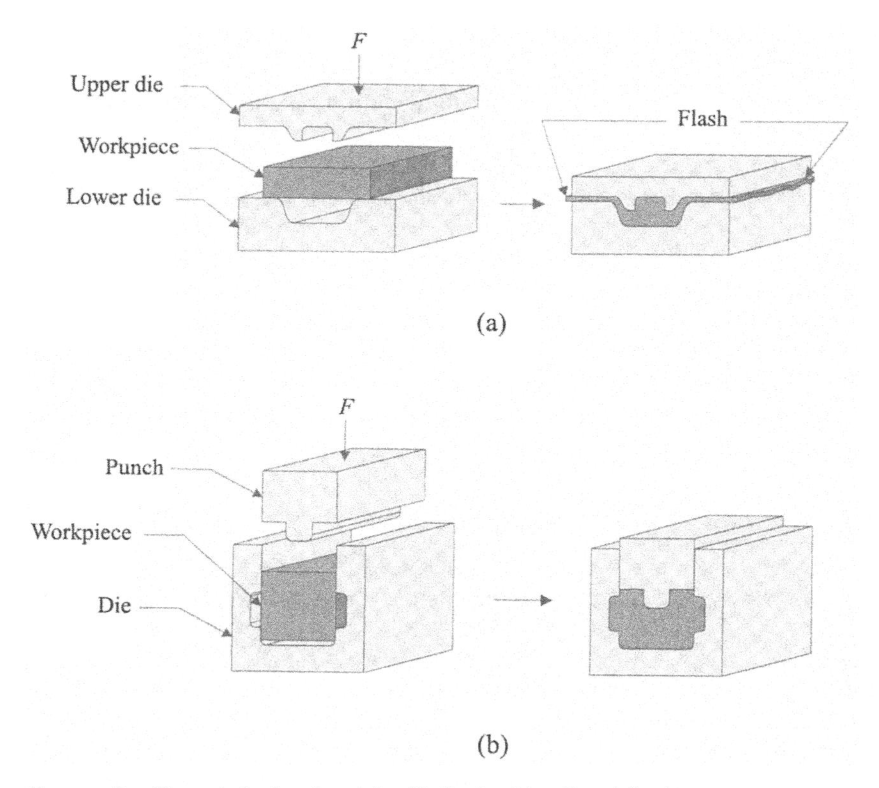

FIGURE 2 Closed die forging (a) with flash; (b) without flash.

Rolling

The rolling of metals can be traced to the 16th and 17th centuries in Europe—rolling of iron bars into sheets. Widespread rolling, however, was only initiated in the late 1700s and early 1800s for the production of railway rails. Today, rolling is considered to be mainly a continuous process targeted for sheet and tube rolling (Figs. 3a, 3b, respectively). Sheet rolling can be a hot or cold forming process for reducing the cross-sectional area of a sheet (or slabs and plates with higher thicknesses than sheets). The workpiece is forced through a pair of rolls repeatedly—each time reducing the thickness further. A rolling process can be utilized in shaping the cross section of a workpiece, such as I-beams or U-channels, or reducing the cross-sectional thickness and/or the diameter of a tube.

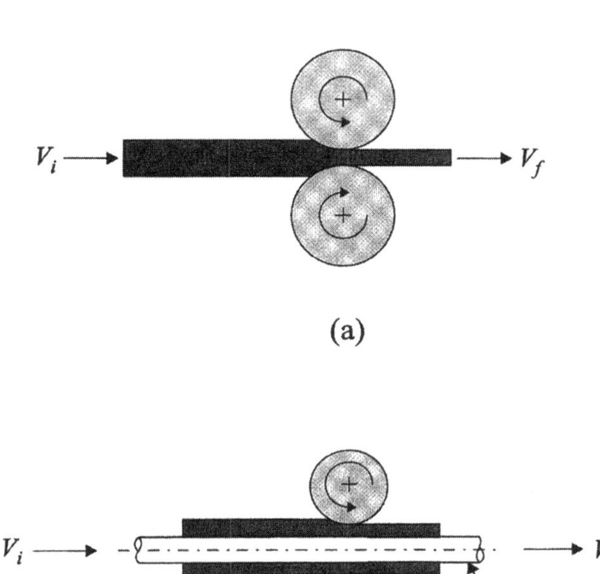

FIGURE 3 (a) Sheet rolling; (b) tube rolling.

Extrusion

The development and use of continuous extrusion can also be traced to Europe in the 1800s for the fabrication of pipes. Today, extrusion is utilized for the fabrication of simple as well complex cross-sectional solid or hollow products. It is based on forcing a heated billet through a die (Fig. 4). In direct extrusion, the product is extruded in the direction of the ram movement. In indirect extrusion, also known as backward or reverse extrusion, the (plastically) deformed product of hollow cross section flows in the opposite direction to the movement of the ram, (solid cross sections can also be obtained when utilizing a hollow ram).

Drawing

Drawing reduces the cross-sectional area of a rod, bar, tube, or wire by pulling the material (in a continuous manner) through a die (Fig. 5), in

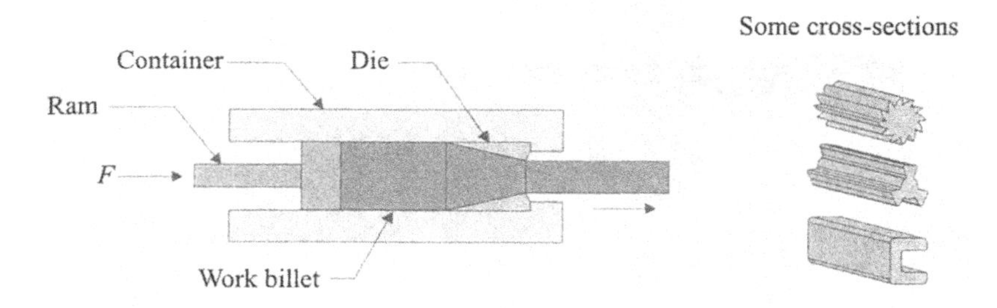

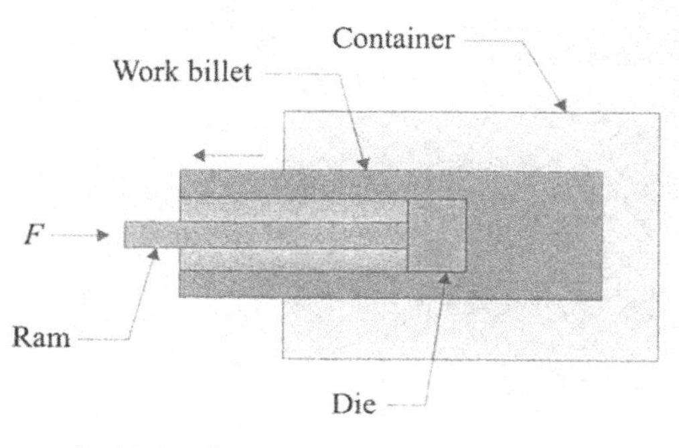

FIGURE 4 Extrusion.

contrast to the pushing action in extrusion. This process is normally a cold-working operation and can be carried out with a pair of undriven rolls instead of a die.

Sheet Metal Forming

Sheet metal forming refers to the forming or cutting/shearing of thin-walled sheets into discrete parts, including car body components and beverage cans. Little or no change in cross-sectional area is expected. In numerous cases, the amounts of elastic and plastic deformations are comparable, leaving the engineer to deal with "springback" effects. Commonly, sheet metal forming is performed on presses through the use of dies.

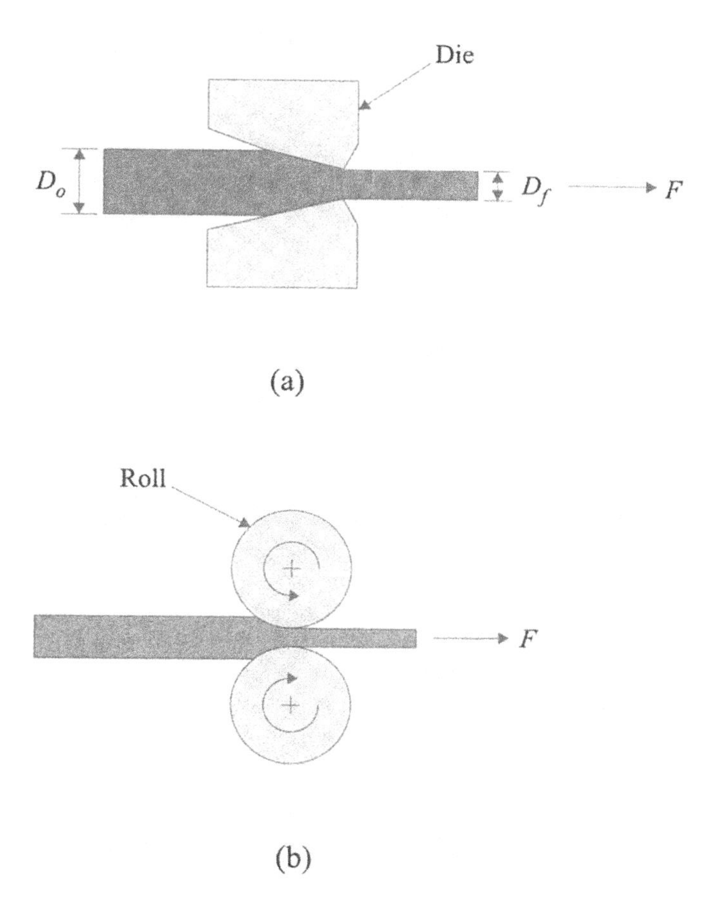

FIGURE 5 (a) Die drawing; (b) roll drawing.

7.1.3 Materials for Metal Forming

Formability of materials depends on the following factors: process temperature, rate of deformation, stress and strain history, and thermal/physical/mechanical properties of the material (including composition and microstructure). Ductile materials are ideal for forming. Brittle materials must be powder processed (Chap. 6). A representative list of materials suitable for metal forming processes is

Forging: Aluminum alloys, copper alloys, carbon and alloy steels, titanium alloys, tungsten alloys, stainless steel alloys, and nickel alloys.

Rolling: Aluminum alloys, copper alloys, carbon and alloy steels, titanium alloys, and nickel alloys.
Extrusion: Aluminum alloys, copper alloys, magnesium alloys, zinc alloys, lead alloys, titanium alloys, molybdenum alloys, and tungsten alloys.
Drawing: Aluminum alloys, copper alloys, alloy steels, stainless steels, cobalt alloys, chromium alloys, and titanium alloys.
Sheet metal forming: Low-carbon steels, aluminum alloys, titanium alloys, and copper alloys.

7.2 FORGING

Forging is a process in which metal billets are plastically deformed by compressive forces, normally within closed dies. Today, forging is the most common metal forming process for the fabrication of discrete solid (versus thin-walled) parts: connecting rods for the automotive industry, shafts for aircraft turbines, and gears for a variety of transportation equipment. Forged parts, small or large, although formed into net shape geometries, generally, require additional finishing operations for dimensional as well as mechanical properties improvements. Forging operations can be performed either cold or hot. Cold forging at room temperature requires greater forces than hot forging but yields much better dimensional accuracy and surface finish.

7.2.1 Forging Techniques

There are a large number of forging techniques, including open-die forging. Only four of these will be detailed below.

Closed Die Forging

In closed die forging, also known as impression-die forging, the billet acquires the shape of the cavity formed between the two halves of the die when closed under pressure (Fig. 2). The process is commonly carried out in several steps to reduce significantly the amount of force at each formation step and to minimize the possibility of defects as well as the amount of waste material (flash). The division of the overall objective into a smaller number of tasks is part geometry and material dependent. The design of the intermediate preform dies is a nontrivial task—it will be briefly addressed in Sec. 7.2.2.

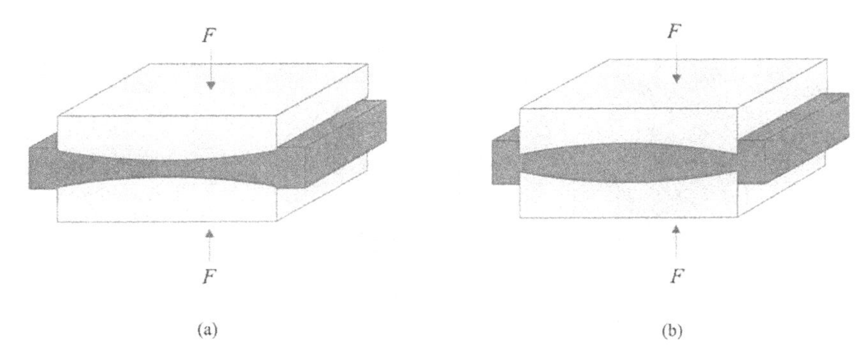

FIGURE 6 (a) Fullering; (b) edging.

The first task in closed die forging is the careful preparation of the billet/blank: it may be cut from an extruded bar or received directly from a casting process; subsequently, it is subjected to a preshaping process, normally through open die forging, when the material is distributed to different regions of the billet. Fullering distributes material away, while edging gathers it into an area/region of interest (Fig. 6). An important preparatory step in the forging process is lubrication through spraying (1) of the die walls with molybdenum disulfide or other lubricants for hot processes and (2) of the blank's surface with mineral oils for cold processes.

Built-in automation is widely utilized in closed die forging for the transfer of preforms from one cavity into another, commonly within the same die/press, as well as for the spraying of the die walls with lubricants. External industrial robotic manipulators have also been used in the placement of billets/blanks into induction furnaces for their rapid heating and their subsequent removal and placement into hot forging presses. Except in cases of flashless forging (Fig. 2b), these manipulators can also transport the parts into flash trimming and other finishing machines.

Extrusion Forging

Extrusion forging is normally a cold process and can be performed as forward or backward extrusion. In forward extrusion, a billet placed in a stationary die is forced forward through a die to form a hollow, thin-walled object, such as stepped or tapered diameter shafts used in bicycles (Fig. 7a).

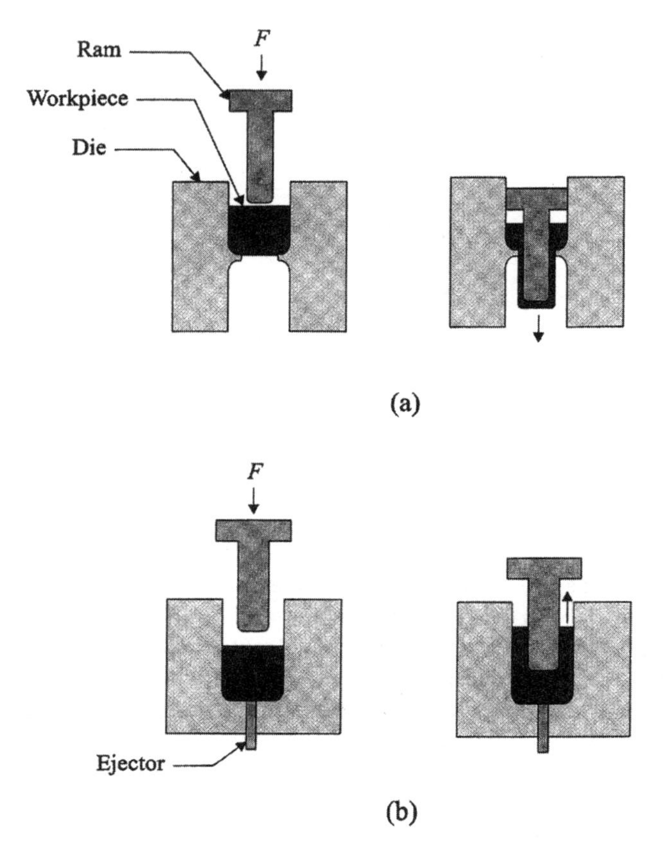

FIGURE 7 (a) Forward extrusion; (b) reverse extrusion forging.

In backward extrusion, also referred to as impact extrusion, a moving punch extrudes backward a billet placed in a (closed) cavity, also for the production of hollow, thin-walled objects (Fig. 7b).

Orbital Forging

In orbital forging, a metal blank is placed in the lower half of a die and deformed incrementally by the rotating upper half of the die. Synchronous to this rotation, the part can be raised upward by a piston that is part of the lower half of the die (Fig. 8). This process is also referred to as rotary forging and can be performed as a hot or cold operation. Bearing rings, bearing end covers, bevel gears, and various other disc-shaped and conical parts can be rotary forged.

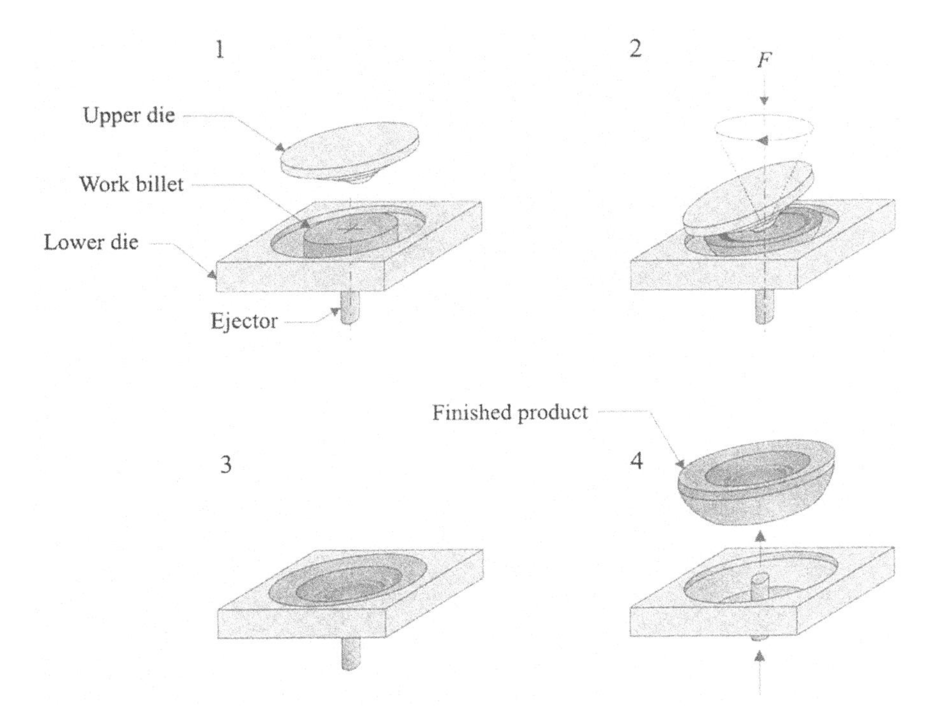

FIGURE 8 Orbital forging.

Roll Forging

Roll forging forms a metal blank into a desired shape by feeding it through a pair of rolls with shaped grooves (Fig. 9). The rolls are in operation for only a portion of their rotational cycle. This hot-forming process is termed forging although it does not employ a moving hammer/punch. It can be utilized for the production of long and thin parts, including tapered shafts, leaf springs, and, occasionally, drill bits (when the blank is also rotated with respect to the rolls as it advances between them). In a process similar to roll forging, alloyed steel gears can be manufactured by forming gear teeth on a hot blank fed between two toothed-die rolls (wheels).

7.2.2 Forgeability and Design for Forging

Forging produces parts of high strength-to-weight ratio, toughness, and resistance to fatigue failure. Metal flow within a die is affected by the resistance of the material to flow (i.e., forgeability), the friction and heat

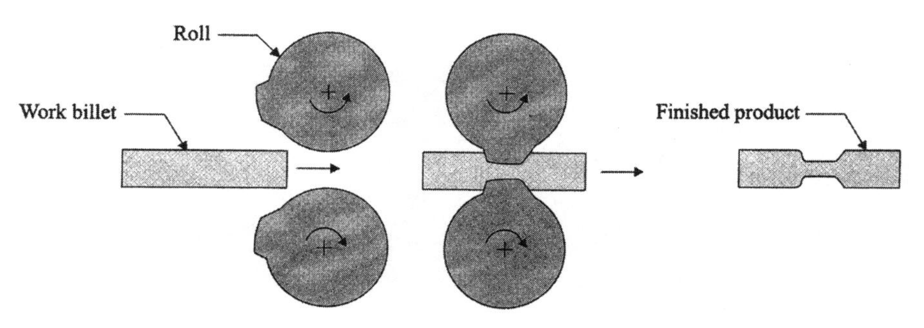

FIGURE 9 Roll forging.

transfer phenomena at the die/material interface, and the geometry of the part. Forgeability, in turn, is influenced by the metallurgical characteristics of the material and the actual process parameters, such as forming temperature and strain rates. Aluminum alloys are the least difficult to forge, normally at a temperature range of 400 to 550°C. Steels are more difficult to forge (at 1100 to 1250°C). Tungsten alloys are considered to be the most difficult materials to forge (at 1200 to 1300°C).

A forging process must ensure adequate flow of the material in the die cavity, thus preventing the occurrence of external and/or internal defects. As mentioned above, metal flow is affected by part geometry. Spherical and block like geometries are the easiest to forge in closed dies. Parts with long, thin sections or projections are more difficult to forge due to their high surface-area-to-volume ratios (i.e., increased friction during metal flow and severe temperature gradients during cooling). Wall thicknesses should be more than 1 mm for steel and more than 0.1 mm for aluminum. One must also make allowances for future machining operations and, most importantly, for material overflow.

As discussed above, complex part geometries require several preforming operations to achieve gradual metal flow. Thus the design of the intermediate die cavity geometries is one of the most important tasks in closed die forging. Although often referred to as art, the generation of the preform cavity geometries (i.e., process planning) would benefit from the use of computer-aided engineering (CAE) tools (such as finite element modeling) for metal flow analysis, as well as from the use of group technology (GT) tools for accessing past process plans developed for similar part geometries (Chaps. 3, 5).

One of the objectives of preforming is to minimize the material loss during forging—the flash. However, it is well established that forging loads increase as flash thickness decreases. Thus, one must optimally design for suitable flash loss while trying to minimize forging loads.

Other factors that affect closed die forging include

Draft angles: 2° to 4° draft angles could facilitate the removal of parts from die cavities when utilizing mechanical ejectors. These may have to be increased to 7 to 10° for manual removals.

Corner radii: Sharp corners must be avoided for increased ease of metal flow.

Parting line: The position of parting lines affects the ease with which billets can be placed in die cavities and the subsequent removal of the preforms and finished parts. It also impacts on the grain flow within the part, and thus on its mechanical properties.

7.2.3 Forging Machines

Presses and hammers are used in the forging of discrete parts. They are primarily chosen according to the part geometry and material as well as production rates. Hydraulic mechanical, and screw presses are used for both hot and cold forging, while hammers are mostly used in hot forging.

Hydraulic Presses

Hydraulic presses can be configured as vertical or horizontal machines and can operate at rates of up to 1.5 to 2.0 million parts per year. Although they operate at much lower speeds than do mechanical presses, the ram speed profile can be programmed to vary during the stroke cycle.

Mechanical Presses

Mechanical presses can also be configured as vertical or horizontal. The driver system (crank or eccentric) is based on a slider–crank mechanism (Fig. 10). Since the ram is fitted with substantial guides and since the press is a constant stroke machine, mechanical presses yield better dimensional accuracy than do hammers. Knuckle joint (mechanical) presses that can produce larger loads for short stroke lengths are often used for cold coining operations. The primary power sources for large mechanical presses are DC motors.

Screw Presses

Screw presses utilize a friction, gear transmission, electric or hydraulic drive to accelerate a flywheel–screw subassembly for a vertical stroke (Fig. 10). In the most common friction drive press, two driving disks (in continuous

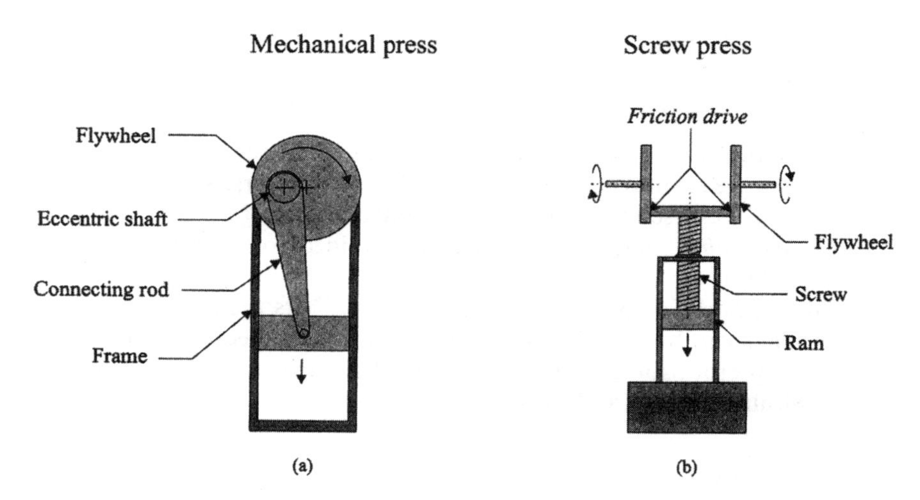

FIGURE 10 (a) Mechanical forging press; (b) screw press.

motion) are utilized to engage a flywheel through friction (one disk at a time, for upward and downward motion). The flywheel, in turn, accelerates the screw attached to it in a downward/upward motion, where maximum speed is achieved at the end of the stroke.

Hammers

A hammer press is a low-cost forging machine that transfers the potential energy of an elevated hammer (ram) into kinetic energy that is subsequently dissipated (mainly) by the plastic deformation of the part. The two most common configurations are the gravity-drop hammer and the power-drop hammer (Fig. 11). As the name implies, the former utilizes only gravitational acceleration to build up the forging energy. The latter type supplements this energy through the utilization of a complementary power source—most commonly hydraulic—for increased vertical acceleration.

The selection of a suitable forging machine for the task at hand is influenced by several factors: part material and geometry and desired rate of deformation (i.e., strain rate). Hydraulic presses can achieve a stroke speed of up to 0.3 m/s and apply a force of typically up to 500 MN in closed die forging. Mechanical presses can achieve a stroke speed of up to 1.5 m/s and apply a force of typically up to 100 MN. (A power-drop hammer, in contrast, can achieve a stroke speed of up to 9 m/s.) Presses are normally preferred for more ductile materials than those for hammers (e.g., aluminum versus steel).

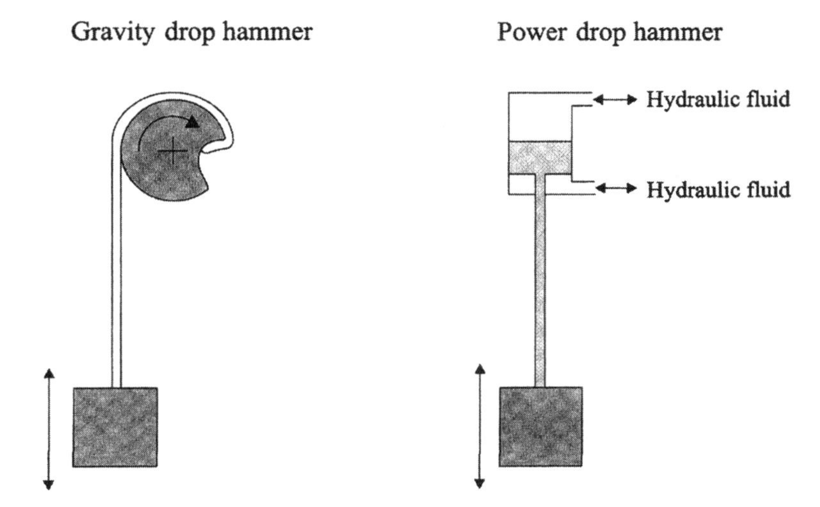

FIGURE 11 Hammers for forging.

7.3 SHEET METAL FORMING

In sheet metal forming, a sheet blank is deformed, normally, into a three-dimensional object—the deformation usually changes the shape of the part but not its cross-sectional thickness. Among the numerous sheet metal forming operations, only a selective few that are most pertinent to discrete manufacturing will be detailed in this section. They are deep drawing, blanking/stamping, and bending. Products manufactured through these processes include desks and cabinets, appliances, car bodies, aircraft fuselages, and a variety of cans.

7.3.1 Sheet Metal Forming Processes

Blanking

The terms blanking and stamping have been used interchangeably to describe the shearing of planar blanks out of a metal sheet, mostly for their subsequent forming into three-dimensional objects via other forming operations. Typically, the sheet metal is secured and a punch/die combination is utilized to shear a desired cross-sectional geometry. Although the outcome of shearing is a blank with not-so-smooth edges, if necessary a fine blanking operation, developed in the 1960s, can be utilized to obtain smooth and vertical edges, for products that will not be further plastically deformed.

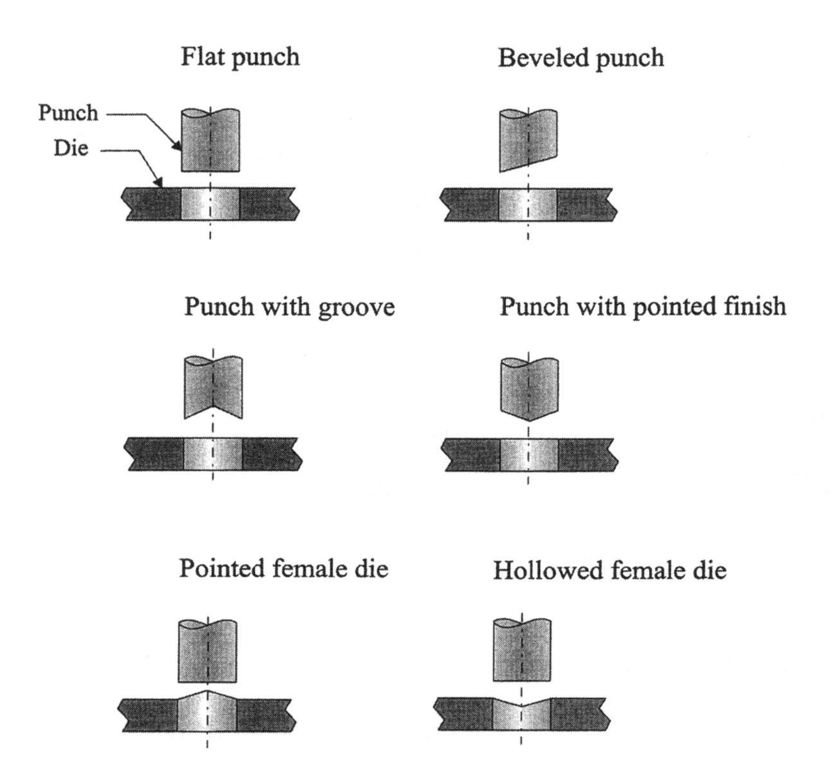

FIGURE 12 Blanking punch die configurations.

As shown in Fig. 12, the blanking force can be reduced if a beveled punch with an oblique shearing move or other nonflat punches are utilized. However, if the part is intended to be used with no further forming, the punch should be either flat or symmetical (with a groove, pointed end, or hollow face). One must also determine a suitable gap between the punch and the die (typically 2 to 8% of the sheet's thickness) for smooth fracture (shearing). In fine blanking, where the sheet is held in place by a pressure pad (with V-shaped projections that penetrate the sheet metal for a better grip), the gap is only about 1%. (Lower clearances are normally reserved for thin and ductile metals.)

Deep Drawing

Deep drawing is a metal forming process targeted for the production of thin-walled cup/can shape objects through a combined compression–tension operation. As shown in Fig. 13, a blank is forced into a die cavity by a punch and assumes the shape of the punch while being held by the blank

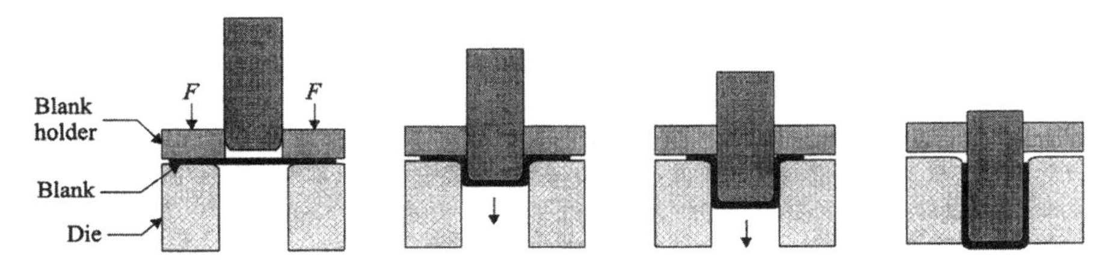

FIGURE 13 Deep drawing.

holder. The process normally maintains the thickness of the sheet metal and can be used for shallow or deep parts. In an alternative configuration, reverse drawing, the location of the punch can be reversed, leading to an upward motion (against gravity).

In certain applications, parts can be deep-drawn in several steps—redrawing. At each step, the cup becomes longer (deeper) and its diameter is reduced. However, if the wall thickness needs to be reduced as well, an ironing operation is implemented. In this process, as the part is redrawn, it is forced through an ironing ring (like an extra die) placed inside the cavity (Fig. 14). Ironing is the preferred operation for the fabrication of beverage cans.

In order to achieve production efficiency, it has been proposed that multiple dies can be vertically aligned in a tandem configuration, thus allowing greater reductions in wall thicknesses in a single stroke. However, due to misalignment problems and the necessary long stroke, an alternative arrangement was developed, a stepped die. In this single-die design, successive reductions can be achieved within a shorter stroke.

Bending

Bending is one of the simplest, yet widely used, metal forming operations. Bending of large metal sheet plates into auto body or appliance body parts

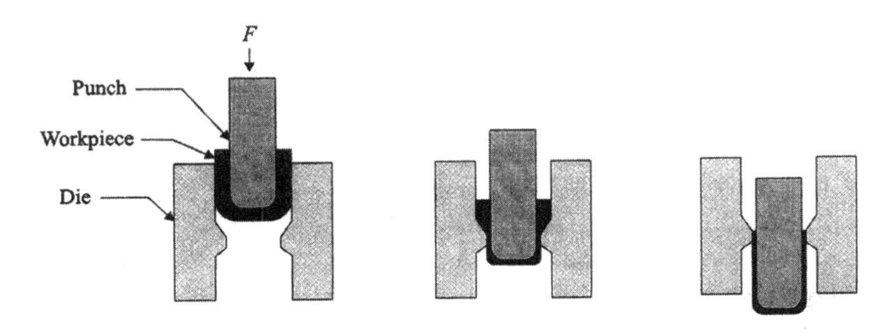

FIGURE 14 Ironing.

are achieved in mass production presses. The operation involves the forcing of a plate (or parts of it) into simple die cavities (or against a wall) by a punch. Since elastic and plastic deformations are typically of the same order of magnitude, the resultant springback effect must be compensated for by overbending the plate. An alternative to overbending would be to implement localized plastic deformations for increased resistance to the springback effect.

7.3.2 Formability and Design for Sheet Metal Forming

Although sheet metal forming operations may seem to be simple techniques, their analysis is complex owing to the possibility of the presence of several failure mechanisms. Despite the existence of empirically determined "formability" curves for different materials, users are always advised to utilize CAE tools for stress analysis. Formability has been formally defined as the ability of the sheet metal to undergo the desired plastic deformation without failure. In deep drawing, for example, failure or defects can occur owing to nonuniform thinning of the cup. Tearing can occur at the bottom of the cup or wrinkling can result at the flange of the cup. The former can be avoided by allowing strain hardening to occur at preferred rates. Wrinkling, on the other hand, can be controlled by applying suitable clamping forces.

As implicitly discussed above, the properties of the blank material influence formability in addition to the process parameters. Alloyed steels, copper alloys (including bronze), and some aluminum alloys are considered to have excellent formability characteristics because of their high strain hardening capabilities. Steels are commonly used in the automotive industry (body parts, bumpers, shock absorbers, exhaust systems, etc.) and the home appliance industry, copper alloys are used for a variety of small finished parts (ballpoint pen cartridges, zip fasteners, screws, etc.) and aluminum alloys are used in the automotive industry, the aircraft industry, and even in the shipbuilding industry. Some titanium alloys have also been sheet formed into parts for the aircraft and aerospace industries.

Layout Planning in Blanking

Optimal positioning of blanks on a strip or a plate can significantly reduce scrap and therefore result in cost savings (minimizing material to be recycled) (Fig. 15). The ultimate solution would naturally be having zero scrap (Fig. 16). One could also have different shapes mixed on a single strip, for better utilization. Overall, the problem is a classical mathematical optimization problem, where the variables are the position and orientation of the blanks on the strip, and the objective function to be minimized is the surface area of the leftover scrap. (An additional variable set could include

Poor configuration

Better configuration

FIGURE 15 Optimal part configuration on a strip.

the outer dimensions of the strip or the plate—for example, if a single row of circular blanks yields 40% waste, by increasing the number of rows to 6 and packing the circles, we could reduce the waste percentage to 25%.)

7.3.3 Dies and Presses for Sheet Metal Forming

As with forging, sheet metal forming may require several steps to obtain the exact shape of the product: the dies must be designed accordingly, and presses should be selected for optimal production. Also, as with forging, manufacturers may decide to combine several preforming cavities (or operations) into a single die (or a single forming station).

Large sheet metal parts (as those found in the automotive industry) are almost always manufactured using single-cavity dies installed in large presses and transferred (sequentially) from one station to another using conveyors or large robotic manipulators. For smaller parts, several single dies can be mounted on a common base plate at one press station, where parts are moved from one die to another (within the same station) automatically.

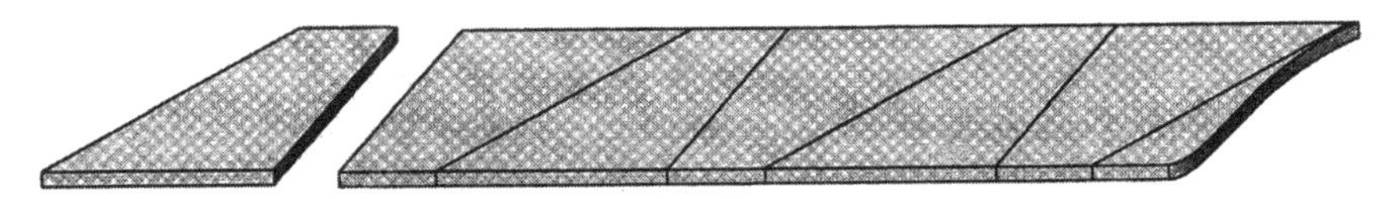

FIGURE 16 No-scrap production.

In stamping, "progressive" dies can be used to blank a part in several stages—i.e., each punch performs one of the many blanking operations needed on one part. The strip on which the part is mounted (or is part of) progresses forward after every blanking operation and finally removed (sheared off) from the strip after the last blanking operation.

Today, manufacturers can decide on choosing dedicated presses for their specific product at hand or choose universal presses that can do both stamping and forming (for small-size parts). As described in Sec. 7.2.3, these presses can be either mechanically or hydraulically driven. However, in the mass production of large parts, an engineer must also carefully design a transportation system for the movement of semifinished (preformed) parts from one station to another. These transportation systems can be of the continuous-line type or targeted for the transfer of small batches. In either case, robotic manipulators with magnetic grippers are very widely utilized throughout the sheet metal forming industry for the loading/unloading/transfer of parts. These manipulators could be of the stand-alone type or built in into the press (the dedicated type) and can handle parts weighing above 50 kg each.

As will be discussed below, in addition to the selection of the most suitable dies, presses, and transport devices, manufacturers must also pay special attention to die changing systems. A quick die exchange technique can significantly increase production efficiency.

7.4 QUICK DIE EXCHANGE

Tactical flexibility in manufacturing requires companies to respond to market demand fluctuations in a timely and profitable manner. A key requirement is to have operational flexibility on the factory floor, whereby production models and batch sizes of parts can be varied without disruptions. Group technology was discussed in Chap. 3 as a potential facilitator for the production of families of (similar) parts within (physical or virtual) workcells. Productivity gains can be achieved in such environments by having common setup tools and procedures, so that setup transformation from one part model to another does not require an excessive amount of time.

In this section, we will briefly review the topic of quick die exchange, which is at the heart of productivity improvement through the elimination of waste (i.e., activities that do not add value to the product). In this context, the single-minute exchange of dies (SMED) philosophy proposed by S. Shingo stands out as an excellent starting point. Shingo's SMED approach is a vital part of a comprehensive manufacturing strategy that he has advocated since the early 1950s: stockless production, the minimization of in-process inventories. SMED is a companion to just-in-time (JIT) manufacturing and defect-free production tactics in this quest. Many hundreds of applications of the SMED philosophy around the world have reduced setup times from several hours to a few minutes, especially in environments of metal forming, metal casting, and plastics molding.

The above-mentioned time savings, via rapid die exchanges, yield increased machine utilization in mixed production environments. This objective is achieved by distinguishing between on-line (internal) and off-line (external) setup activities and increasing efforts to reduce the former activities and thus minimizing the time the machine has to be down. Shingo reports that typically in a setup process on-line activities take up to 70% of the overall die exchange time. Two thirds of this time, in turn, is spent on final adjustments and trial runs. Shingo proposes a two-step approach to waste reduction:

1. Identification and separation of current on-line and off-line setup activities, whereby subsequently maximizing the latter by converting as many of the (current) on-line setup tasks as possible into off-line ones,
2. Reduction of time spent on all on-line and off-line setup activities, with the greater emphasis being on the on-line tasks

Effective a priori preparation of setup tools and their efficient transportation can significantly reduce time spent on on-line activities. For example, mechanization of die mounting through moving bolsters, roller conveyors, revolving die holders, or even through the employment of air cushions will save setup time. Additional operations that were previously carried out on-line, but now are classified as off-line, can also be efficiently carried out to minimize overall setup time. A typical example would be standardization of the functional elements of different dies—modification of die geometry for clamping height standardization, use of centering jigs, and so on.

Reduction of time spent on on-line set-up activities constitutes the primary objective of any productivity improvement attempt. No long list of generic guidelines for this objective exists, so tool and die designers must evaluate every application individually for savings through ingenuity and innovation. Shingo does highlight, however, three generic (common) guidelines:

The use of clamping techniques that miminize the time spent on securing the die in the press should be a priority. Examples include

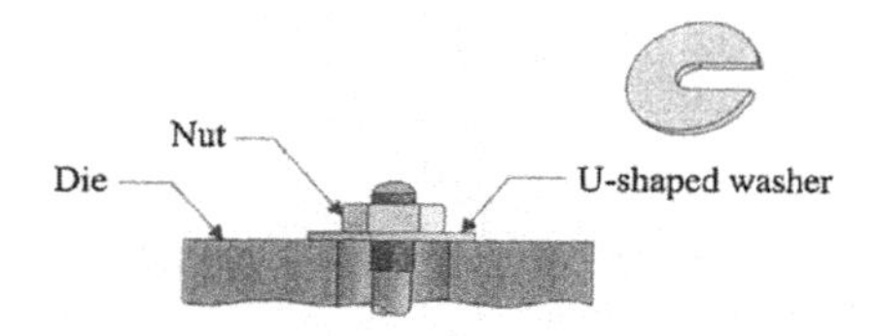

The U-shaped washer is fastened/removed by a single turn of the nut, whose diameter is smaller than that of the hole.

(a)

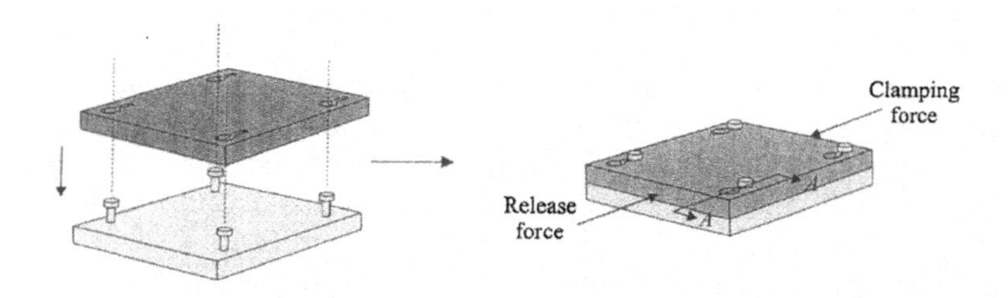

A-A Cross section

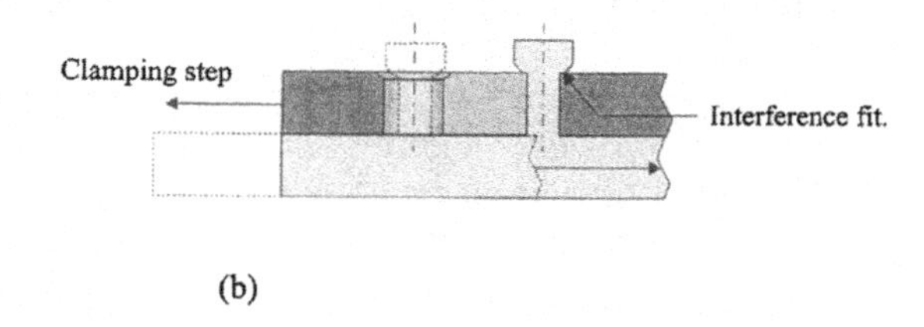

(b)

A computer mouse

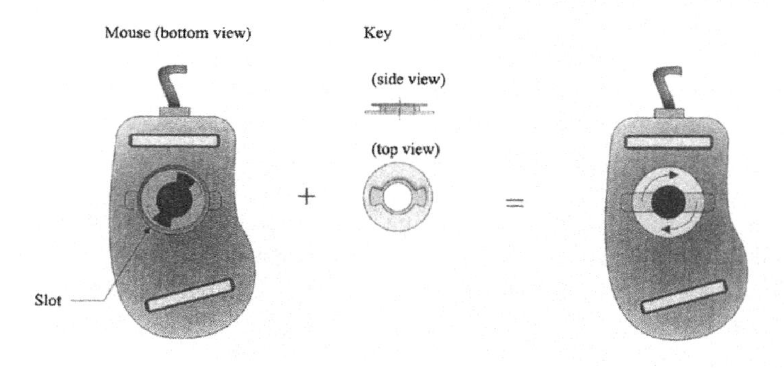

The key is inserted into the slot at the bottom of the mouse; it is then turned a fraction of a complete rotation to lock it into place.

(c)

FIGURE 17 One-turn clamping.

one-turn attachments (where several points of connection are achieved with the turn of one mechanism) (Fig. 17), cam and clamps, spring-loaded pins, etc.

The elimination of as many adjustments as possible through the use of guiding pins, locators, height gages, or even electronic indicators/sensors.

The concurrent (versus sequential) implementation of several on-line tasks. Safety issues should be a paramount concern, however, when utilizing multiple operators.

As a complementary point to the above discussion, the reader must be aware that the frequency of exchanging dies is a direct function of total (on- and off-line) time spent of the setup. Although a die can be removed from a press and replaced by a different one within minutes, this naturally does not imply that it can be remounted after a short while. Off-line die preparation dictates the length of that time. Thus manufacturing engineers must not neglect the issue of minimizing off-line setup times, even when they have reduced the on-line activities to several minutes and maximized machine up-time.

REVIEW QUESTIONS

1. Is metal forming an elastic or a plastic deformation process? Explain.
2. What is material ductility and how does it affect metal forming?
3. Which typical fabrication processes are utilized in the preparation of billets/blanks for forging?
4. Why is it preferable to have forging carried out in multiple steps?
5. Describe forward and backward extrusion forging, respectively.
6. Why are long and thin part sections difficult to forge?
7. Discuss the flash formation process in forging and define its advantages/disadvantages.
8. Discuss the selection of a suitable forging machine.
9. Describe the fine blanking process.
10. Describe the redrawing and ironing processes as well as the deep drawing process that uses multiple dies in tandem or a stepped die.
11. What is the springback effect in metal bending?
12. Discuss the optimal positioning of blanks on a strip for blanking operations.
13. What are progressive dies in forging as well as in blanking?
14. Discuss the topic quick die exchange. Differentiate between on-line (internal) and off-line (external) setup activities.

DISCUSSION QUESTIONS

1. Forging is normally a multistep process: the final shape of the part is achieved via multiple forming operations in a single die with multiple cavities using one forging machine. The parts in progress are moved forward from one cavity to the next after every cycle of the forging. Discuss methods/technologies that allow users optimally to (process) plan the manufacturing process. That is, minimize cost (or time) of manufacturing, subject to achieving the desired geometric and mechanical properties of the part. Similarly, discuss the load-balancing issue for multi-cavity (progressive) forging dies: that is, optimal balancing of the force and energy requirements for the plastic deformation processes in all the cavities, using computer-aided engineering analysis tools, so that the forging press is better configured.
2. Single-minute exchange of dies (SMED) is a manufacturing strategy developed for allowing the mixed production (e.g., multimodel cars) within the same facility in small batches. The primary objective has always been to minimize the time spent on setting up a process while the machine is idle. This objective has been achieved (1) by converting as many on-line operations as possible to off-line ones (i.e., those that can be carried out while the machine is working on a different batch), and (2) by minimizing the time spent on on-line setup operations. Discuss the effectiveness of using SMED or equivalent strategies in the mass manufacturing of multimodel products, the mass manufacturing of customized products, and the manufacturing of small batches or one-of-a-kind products.
3. During the 20th century, there have been statements and graphical illustrations implying that product variety and batch size remain in conflict in the context of profitable manufacturing. Discuss recent counterarguments that advocate profitable manufacturing of a high variety of products in a mass production environment. Furthermore, elaborate on an effective facility layout that can be used in such environments: job-shop, versus cellular, versus flow-line, versus a totally new approach.
4. Analysis of a production process via computer-aided modeling and simulation can lead to an optimal process plan with significant savings in production time and cost. Discuss the issue of time and resources spent on obtaining an optimal plan and the actual (absolute) savings obtained due to this optimization. For example, spending several hours in planning to reduce production time from 2 minutes to 1 minute. Present your analysis as a comparison of one-of-a-kind production versus mass production.

BIBLIOGRAPHY

Altan, Taylan, Oh, Soo-Ik, Gegel, Harold L. (1983). *Metal Forming: Fundamentals and Applications*. Metals Park, OH: American Society for Metals.

Avitzur, Betzalel (1983). *Handbook of Metal-Forming Processes*. New York: John Wiley.

Blazynski, T. Z. (1989). *Plasticity and Modern Metal-Forming Technology*. London: Elsevier Applied Science.

Carle, D., Blount, G. (1999). The suitability of aluminum as an alternative material for car bodies. *J. of Materials and Design* 20:267–272.

Doyle, Lawrence E., et al (1985). *Manufacturing Processes and Materials for Engineers*. Englewood Cliffs, NJ: Prentice-Hall.

Groover, Mikell P. (1996). *Fundamentals of Modern Manufacturing: Materials, Processes, and Systems*. Upper Saddle River, NJ: Prentice Hall.

Hayashi, H., Nakagawa, T. (Nov. 1994). Recent trends in sheet metals and their formability in manufacturing automotive panels. *J. of Materials Processing Technology* 46(3–4):455–487.

Hosford, William F., Caddell, Robert M. (1993). *Metal Forming: Mechanics and Metallurgy*. Englewood Cliffs, NJ: Prentice Hall.

D. Beadle, John (ed.). (1971). *Metal Forming*. Basingstoke, UK: Macmillan.

Kalpakjian, Serope, Schmid, Steven R. (2000). *Manufacturing Engineering and Technology*. Upper Saddle River, NJ: Prentice Hall.

Kobayashi, Shiro, Oh, Soo-Ik, Altan, Taylan (1989). *Metal Forming and the Finite-Element Method*. New York: Oxford University Press.

Marciniak, Z., Duncan, J. L. (1992). *The Mechanics of Sheet Metal Forming*. London: E. Arnold.

Pearce, Roger (1991). *Sheet Metal Forming*. New York: Hilger.

SAE (2000). *Sheet Metal Forming: Sing Tang 65th Anniversary Volume*. Warrendale, PA: Society of Automotive Engineers.

Schey, John A. (1987). *Introduction to Manufacturing Processes*. New York: McGraw-Hill.

Schuler, GmbH. (1998). *Metal Forming Handbook*. Berlin: Springer-Verlag.

Talbert, Samuel H., Avitzur, Betzalel (1996). *Elementary Mechanics of Plastic Flow in Metal Forming*. New York: John Wiley.

Wagener, H. W. (Dec. 1997). New developments in sheet metal forming: sheet materials, tools and machinery. *J. of Materials Processing Technology* 72(3):342–357.

Wagoner, R. H., Chenot, Jean-Loup (1997). *Fundamentals of Metal Forming*. New York: John Wiley.

8

Machining

Machining refers to cutting operations that are based on the removal of material from an originally rough-shaped workpiece, for example via casting or forging. Thus, in the literature, such operations have been often called metal cutting, material removal, and chip removal techniques. Herein the term machining is used as an all-encompassing term that includes the fabrication of metal as well as nonmetal parts.

Machining operations are considered to be the most versatile manufacturing techniques for the production of highly accurate part geometries. They can be utilized for the fabrication of one-of-a-kind products as well as for mass production. Recently, Tlusty estimated that the annual value of machining operations in the U.S.A. is above $160 billion based on the existence of almost 1.87 million machine tools.

Machining operations can be classified according to the geometry of the object's profile—rotational versus prismatic, as well as to the sizes of the object features. External and internal rotational object profiles can be achieved through turning and boring operations, respectively, carried out on lathes and/or boring machines. Prismatic profiles can be fabricated through milling operations carried out on a variety of milling machines. All these techniques would yield acceptable surface quality for the majority of the machined parts. However, for higher surface finish quality, there exist a variety of abrasive techniques, such as grinding, lapping, and polishing.

Parts with small dimensions that cannot be machined on conventional material removal machines have to be fabricated on nonconventional machines, such as electrochemical, electrical discharge and laser beam cutting machines (Chap. 9). As we move towards nanoscale technologies, other modern technologies, such as electron beam–based material removal techniques, are expected to be utilized in pertinent manufacturing fields.

Machine tools for material removal have probably been in existence for two millennia. Turning operations carried out on lathes operated by cords attached to flexible wood sticks, for turning workpieces back and forth, can be traced to the Middle Ages in Europe. However, as discussed in Chap. 1, the development of industrial machine tools for commercial metal cutting purposes took place first in England and then in the U.S.A. during the period 1750 to 1900. The primary motivation for the development of these early machine tools was to be able to fabricate parts with higher accuracies than those producible by casting and forging and also to machine better dies for use by these net shape techniques.

Innovations in the past century focused on the following primary issues: high-speed machining for cost reduction, harder tools for enlarging the class of materials that can be machined, better mathematical modeling of the mechanics of cutting for increased product quality (via reduced vibrations) and for longer tool life (via lower cutting forces), and automation. Except for the last issue, automation through numerical control (to be discussed in Part III of this book), all other issues will be addressed in this chapter.

In Sec. 8.1 below, several representative nonabrasive machining techniques will be reviewed and critical material removal rate variables such as cutting velocity and feed rate will be introduced. Economics of machining, which attempts to minimize costs, utilizes these variables in the derivation of the necessary optimization models. Thus in Secs. 8.2 and 8.3 of this chapter we will address the relationship of cutting tool wear to machining process parameters. We will conclude the chapter with a discussion of representative abrasive machining methods in Sec. 8.4.

8.1 NONABRASIVE MACHINING

Numerous conventional nonabrasive fabrication techniques have been utilized in the past two centuries for the machining of parts with highly complex geometries. These operations have been typically classified as single-point or multipoint machining: the latter type utilizes multipoint cutting tools (such as drills, reamers and milling cutters). These operations have also been referred to as continuous versus intermittent machining: in continuous cutting, the tool is in continuous contact with the workpiece until the end of the pass; in intermittent cutting, since the tool has multiple (discontinued) cutting points, every point remains in contact with the workpiece only for a part of the tool-holder's rotation, i.e., it cuts the workpiece intermittently, where overall continuity is achieved due to the existence of many cutting points.

8.1.1 Turning

In one-dimensional turning, a (single-point) cutting tool mounted on a carriage travels parallel to the axis of rotation of the workpiece, normally held by a chuck and a tailstock (for longer parts) (Figs. 1a, 1b). This feed motion of the tool reduces the radius of the rotational workpiece by an amount equal to the depth of the cut in a direction normal to the feed motion axis (in the same plane). In two-dimensional turning, the tool travels and cuts into the workpiece in the feed direction as well as in the perpendicular depth-of-cut direction, thus yielding workpiece profiles with a variable diameter (Fig. 1c). Both one-dimensional and two-dimensional turning operations can be carried out on manual or on automatically controlled lathes.

The major process variables in turning are the feed rate, f, the cutting velocity, V, and the depth of cut, a. The feed rate of turning is equal to the travel rate of the tool in the feed direction, normally defined in the units of mm/rev (or inches/rev)—i.e., distance traveled by the tool per each revolution of the spindle/workpiece. The cutting velocity of turning refers to the linear velocity of the workpiece at the point of contact with the tool:

$$V = \pi N \left(\frac{d_1 + d_2}{2} \right) \tag{8.1}$$

where N is the spindle's (i.e., workpiece's) rotational speed, defined in the units of revolutions per minute (rpm), d_1 and d_2 refer to the initial and post-cutting diameters of the workpiece, respectively, defined in the units of meters or feet, together, yielding the units of m/min (or ft/min) for V. (For example, we could machine a stainless steel workpiece with a TiN-coated cutting tool at up to $f = 0.75$ mm/rev and $V = 200$ m/min for $a = 0.5$ mm.)

Turning of a workpiece is normally carried in several passes: in the first pass (or several initial passes), the objective is removal of material at increased rates (achieved by selecting a high feed rate) at the expense of surface finish quality; and in the last fine-turning pass the objective is meeting dimensional integrity and surface quality requirements using a reduced feed rate for the same cutting velocity, so that for each rotation of the spindle, the distance that the tool travels in the feed direction is considerably shortened, thus providing maximum continuity on the workpiece's surface.

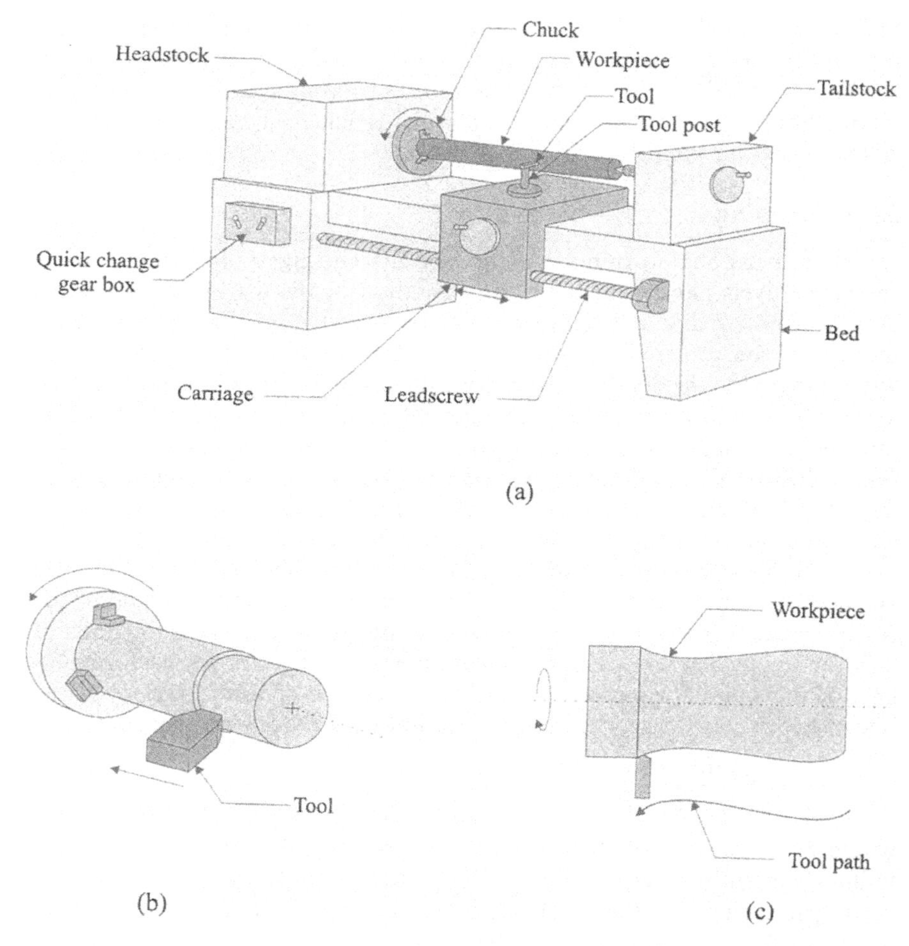

FIGURE 1 (a) An engine lathe; (b) one-dimensional turning; (c) two-dimensional turning.

As mentioned above, turning operations are carried out on lathes (Fig. 1a). The workpiece is held in a three- or four-jawed chuck that is normally manually tightened or power actuated. (For small-diameter cylindrical parts, one may choose to utilize collets, instead of multijawed chucks, for increased tightness.) Most lathes are of the bench type (mounted on a frame with cabinets for tool storage) and are commonly referred to as engine lathes—a term given to the first lathes, which were operated using belts attached to external engines. Today, lathes have built-in electric motors and provide spindle speeds up to 10,000 rpm (for smaller workpiece diameters). Most lathes are capable of performing a number of different cutting operations on rotational workpieces beyond turning—drilling, boring, and thread cutting. Furthermore, some, like turret lathes, can (sequentially) carry out several operations on the same workpiece utilizing a multitool turret, often under the control of an on-board microprocessor based controller.

8.1.2 Boring

The boring operation is the internal turning of workpieces-namely, enlargement of a hole through material removal (Fig. 2). Thus issues discussed above for turning normally apply to boring as well. Boring can be carried out on a lathe if the workpiece size allows it. Otherwise, there exist horizontal and vertical boring machines especially designed for the fabrication of large-diameter internal holes with high accuracies (Fig. 3). Unlike in turning, however, moderate cutting speeds and feed rates are utilized (for small depths of cut) to achieve these accuracies.

The vertical boring machine is reserved for the machining of large workpieces (above 1 m in diameter, up to 5 to 10 m). Multiple tools can be mounted on the overhead tool holders that engage the workpiece fixtured on a turning worktable. On horizontal boring machines the option of having the tool or the workpiece rotate exists. In the former, the workpiece is fed into the cutting tool.

8.1.3 Drilling

Drilling is the most common (multipoint) cutting technique targeted for the production of small-diameter holes; but the complexity of its tool geometry makes it difficult to model mathematically. Normally a rotating tool

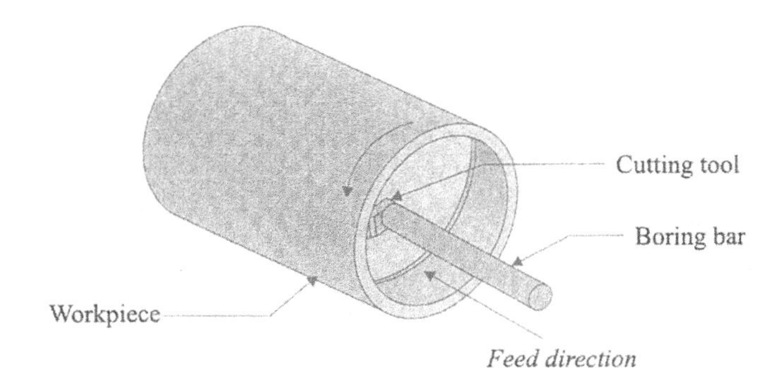

FIGURE 2 Boring on a lathe.

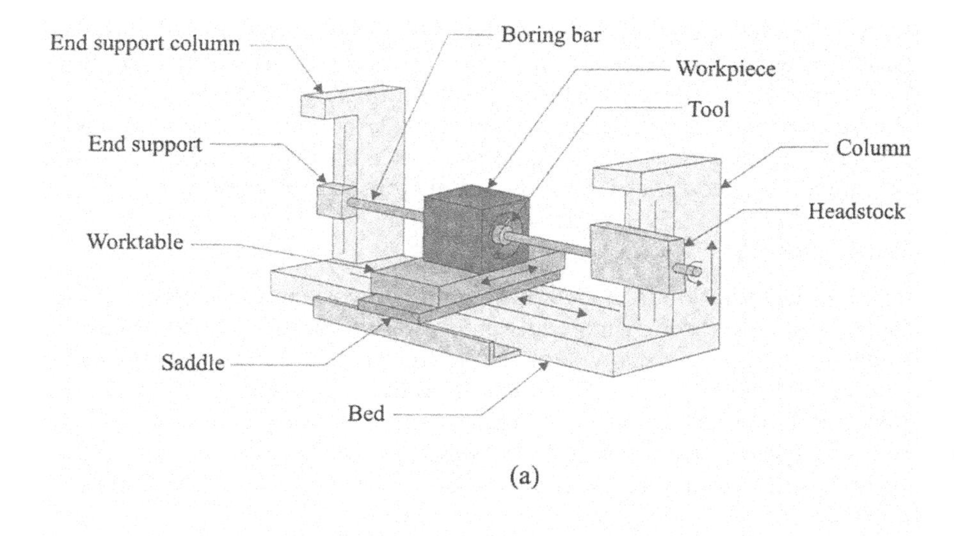

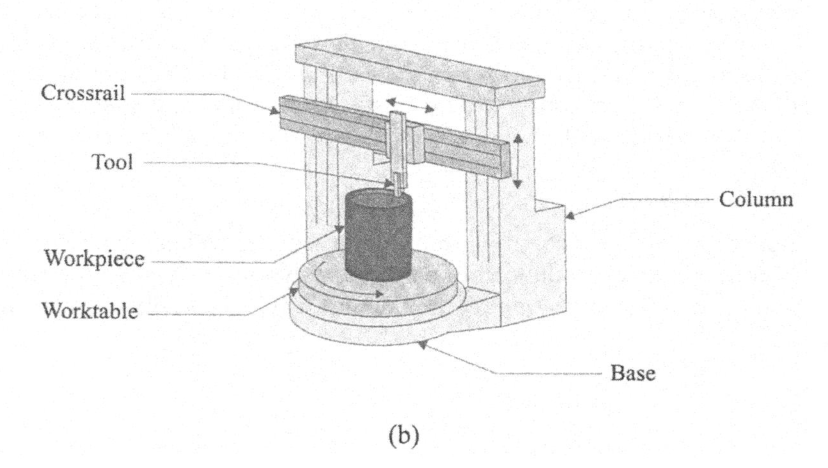

FIGURE 3 (a) Horizontal boring machine; (b) vertical boring machine.

mounted on a spindle is fed into a fixtured stationary workpiece (Fig. 4). Occasionally, drilling can be carried out on a lathe, where a rotating part is fed into a stationary drilling tool held in a turret or a tailstock.

The major process parameters in drilling are the diameter of the drill (or the hole to be machined), d, the feed rate, f, and the cutting velocity, V. The feed rate of drilling is equal to the travel rate of the tool into the

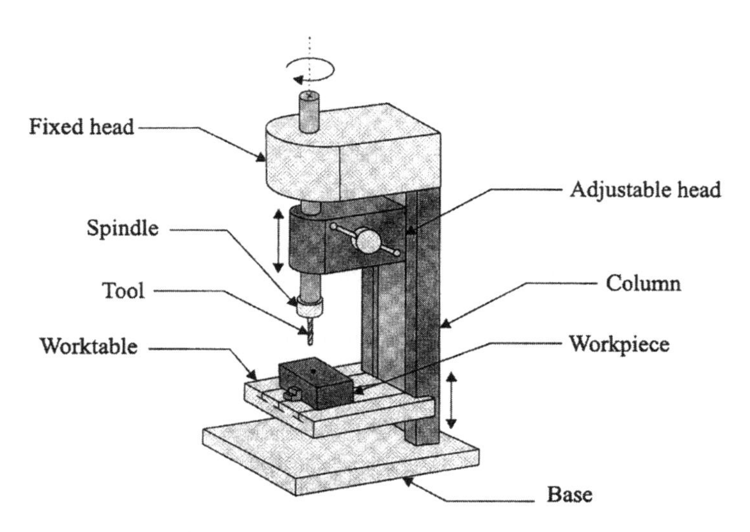

FIGURE 4 Drilling.

workpiece, defined in the units of mm/rev. The cutting velocity of drilling refers to the linear velocity of the tool at the point of contact with the workpiece:

$$V = \pi N d \tag{8.2}$$

where N is the spindle's (i.e., the tool's) rotational speed (rpm). The effective feed rate, f_e, and the depth of cut, a, for drilling are defined per tooth: For a 2-flute drill, $f_e = f/2$, and $a = d/2$, unless the drilling operation is an enlargement of a diameter from d_1 to d_2 —then, $a = (d_2 - d_1)/2$.

In drilling a new hole (in a solid workpiece), the drill with its chisel edge yields a very rough cut under the axial point contact of the tool with the workpiece. The surface of interest, however, is the side wall of the hole, which is burnished by the rubbing action of the "twisted" flutes of the drill and the material that escapes outward. If one considers this an unacceptable machining operation, a reaming tool can be used for improving significantly the side-wall's surface quality. A reaming tool is also a multipoint cutter, but due to its geometry with "straight" flutes, it acts like a boring tool and yields excellent surface finish. (The depth of cut during reaming is, generally, between 0.25 to 0.70 mm).

Drill presses can be utilized for drilling holes and their subsequent potential tapping (threading) or reaming. Most drill presses are of the bench type and may allow for drilling holes at angles different from the vertical. There also exist turret-type press drills with turrets that hold many cutting tools (drills, reamers, etc.) for a set of sequential operations on a workpiece fixtured on an X–Y table. Such drill presses would be utilized for fast and accurate relative positioning of the tool with respect to the fixed drilling location of the workpiece.

8.1.4 Milling

Milling is a material removal process for nonrotational objects: it uses a multipoint tool that rotates about a fixed axis while the prismatic workpiece is fed into the tool according to a prespecified travel path (Fig. 5). This intermittent cutting process is commonly classified as face (or end) milling versus peripheral (or plain) milling (Figs. 5a, 5b, respectively).

In two-dimensional end milling, the axis of rotation of the cutter remains orthogonal to the travel plane of the workpiece while a desired profile is machined. Once the planar cutting operation is completed, the workpiece can be elevated in the vertical direction (to the machining plane) for the next planar profiling operation. This stop-and-go milling operation is normally referred to as 2 and one half dimensional machining, since the workpiece only moves incrementally in the third direction (yielding a staircase effect for curved surfaces). Milling operations, however, can be carried out as up to five-axis, three-dimensional machining, where the position of the workpiece is continuously varied in all three orthogonal Cartesian axes while the tool's axis is rotated simultaneously with respect to two orthogonal axes (Fig. 6). Such coordinated and synchronized motions of the tool and the workpiece can yield highly accurate spherical surfaces (or any other three-dimensional surface) with improved surface finish.

The cutting process parameters in milling are similar to those in single-point turning: the thickness of the material removed is considered to be the

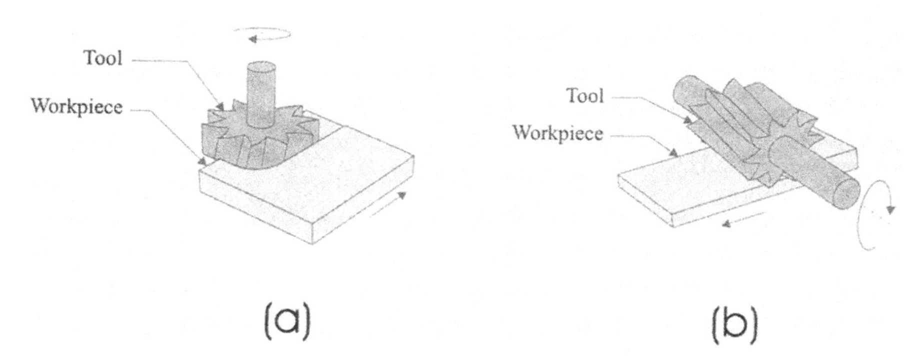

FIGURE 5 (a) Face/end milling; (b) peripheral/plain milling.

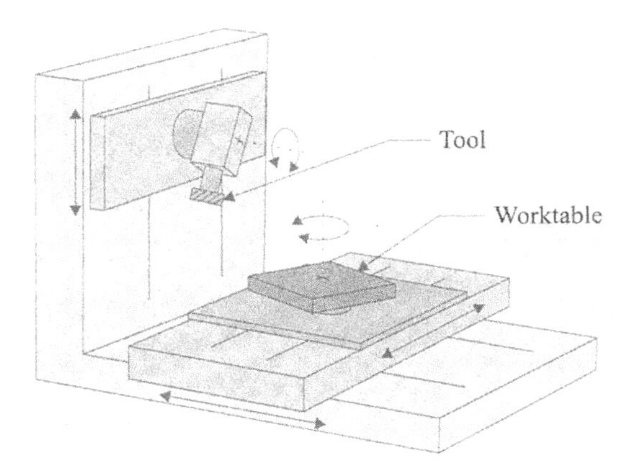

Figure 6 Five-axis milling.

depth of cut (the travel of the workpiece along the orthogonal direction to the planar motion of cutting), while the effective feed rate per tooth is given as the travel rate (feed rate) of the tool holder, f, along the desired profile divided by the number of teeth. As in drilling, the cutting velocity is the (relative) linear velocity of the tool tip as it engages the workpiece. If one assumes that the velocity of the tool is much greater the velocity of the workpiece (i.e., the feed rate) at the instant of engagement, then in a simplified form,

$$V = \pi N d \tag{8.3}$$

where N is the rotational speed of the fixed-axis tool (or spindle) in the units of rpm and d is the diameter of the tool in the units of mm (or inches). (For example, in milling a stainless steel workpiece with a coated TiN cutting tool, a feed rate of up to 0.4 mm/tooth can be achieved for a cutting velocity of up to 500 m/min.)

As in turning, the surface finish of a workpiece in milling is a function of the depth of cut (thickness layer) as well as the feed rate. Most applications would require numerous rough cuts carried out at high feed rates and large depths of cut, as would be allowed by the tool characteristics and the milling machine power, for a maximum material removal rate (i.e., minimum cost). A subsequent fine cut would be carried out at much lower values of both process variables (especially the effective feed rate, f_e).

The most common milling machine is the knee-and-column milling machine (Fig. 7), which can have a spindle configuration either for peripheral (horizontal) or for face (vertical) milling. The column provides necessary rigidity to the cutting tool and normally houses the electric motor that

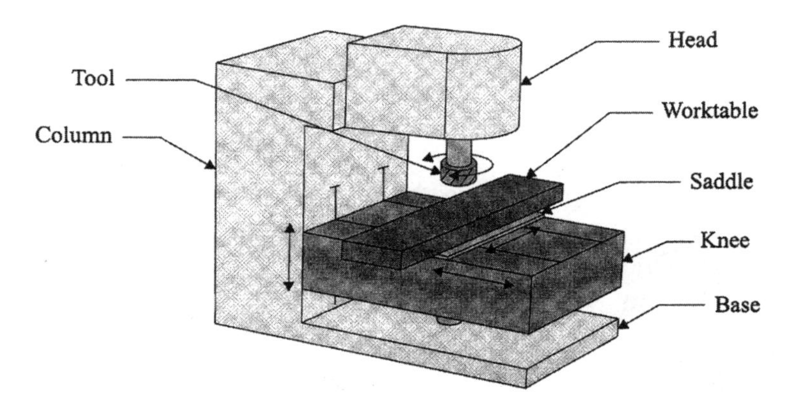

FIGURE 7 Knee-and-column milling machine.

drives it; the knee supports the worktable (and its actuators) capable of motion in three orthogonal directions. Other machining operations that can be carried out on milling machines include gear-machining, planing, and broaching (though frequently the latter operations are carried out on special purpose planing/shaping and broaching machines).

8.1.5 Design for Machining

A large number of part geometry and machining process parameters affect the quality and cost of parts manufactured through machining operations. The impact of process parameters on part quality and cost will be discussed in Sec. 8.2 in the context of tool wear and surface finish. In this section, our primary focus will be on part geometry requirements.

Machining, though very versatile, is an expensive technique for the fabrication of parts when compared to net shape techniques, some of which were presented in Chaps. 6 and 7. Thus machining should be used sparingly and in an optimized manner. From a part geometry point of view, engineers must choose the most suitable process/machine and be aware of the technical difficulties in the fabrication of intricate features. Furthermore, the initial stock part geometry should be selected for minimal material removal—i.e., with dimensions as close as possible to the final part dimensions. In regard to achievable tolerances, users should consult existing handbooks and specifications for the machines on the factory floor. For example, turning and boring yield different tolerance levels for different hole radii (e.g., ± 0.01 to ± 0.1 mm tolerance for radii of 20 to 1000 mm). Drilled holes' radial dimensional tolerances can be improved using reaming from ± 0.05–0.20 mm to ± 0.025–0.125 mm. As will be discussed later in this chapter, all these tolerance levels can of course be further improved using abrasive surface finishing techniques.

Setup times add considerable cost to the machining of parts. Thus parts should be designed to require minimum setup changes as well as cutting tool changes. Undercuts, for example, though feasible to manufacture, are significantly cost adding features. Furthermore, long and narrow parts, large and flat parts, and thin walled parts are not easily machinable.

Sharp corners, tapers, and major variations in profiles should be avoided in turning. Tool travel along the external part profile for turning and along the internal profile for boring should be as much as possible free of interruptions (which would require the tool to disengage and reengage somewhere along the profile, as opposed to an easy engagement at the start of the profile). Through holes are preferable to blind holes in both boring and drilling operations. (Through holes are easier to ream; for blind holes, the hole should be drilled deeper than necessary if reaming is to follow.) Furthermore, holes are easier to drill on flat surfaces perpendicular to the tool's motion axis.

In milling operations, external intersections of surfaces should be chamfered, if necessary, instead of honed with a small radius of curvature. For internal cavities (through or blind), one must be aware of the minimum radius of curvature of the vertical edges, which would be determined as a function of the depth of the cavity (the deeper the cavity, the longer the cutting tool necessary, the larger its diameter for rigidity).

Some design issues for rotational and prismatic object geometries are shown in Fig. 8.

8.2 MECHANICS OF CUTTING—SINGLE-POINT TOOLS

The objective of this section is to address two important factors that affect machining productivity in terms of cost and quality: tool wear and chatter. A number of process parameters and tool- and workpiece-material properties impact these factors, most notably, cutting forces and chip formation. The mechanics of cutting in terms of cutting forces will be addressed first in Sec. 8.2.1, which will be followed by a discussion on chip formation and control in Sec. 8.2.2.

8.2.1 Cutting Forces

The study of the mechanics of cutting can be traced to the mid-1880s in Europe, to the works of French scientists Cocquilhat and Joessel, who reported on cutting forces and their relationship to tool geometry. At the start of the 1900s, we note the work Reuleaux on chip formation and the

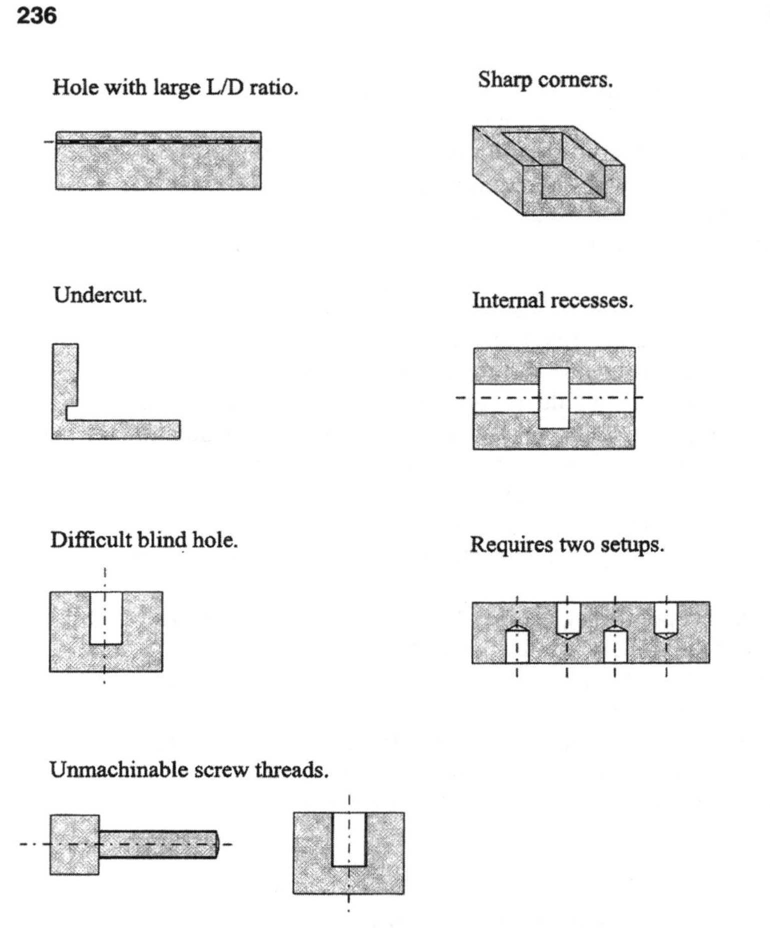

FIGURE 8 Design features to avoid in machining.

work of Taylor on tool life. In the more modern history of metal cutting (1940–1970), the primary contributors were Ernst (chip formation), Merchant (chip geometry), Kronenberg (tool geometry), Armarego (mechanics of cutting), and Tlusty (vibrations and chatter).

As addressed by the above and other researchers, it is commonly accepted that machining through material removal is a plastic flow process. Material (chip) removed from the workpiece at the tool contact point flows over the tool's surface and eventually breaks away. The mechanics of this process has been often studied through single-point cutting-edge models and extrapolated to multipoint cutting. These models have been classified as orthogonal cutting and oblique cutting operations (Fig. 9). The former is the

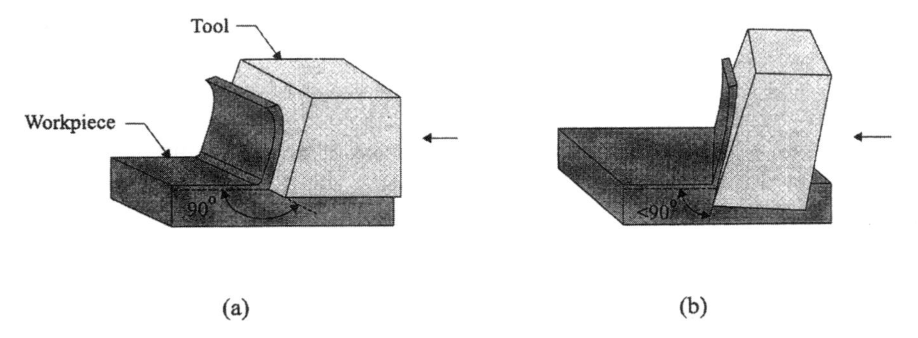

FIGURE 9 (a) Orthogonal cutting; (b) oblique cutting.

simpler case that assumes the cutting edge is orthogonal to the cutting velocity direction. The latter one assumes a nonzero inclination of the tool's cutting edge.

Figures 10a and 10b define the cutting geometrical parameters and the primary forces acting on the tool, respectively. The cutting force F_c is in the same direction as the cutting velocity, V, the thrust (feed) force, F_t, is in the direction the tool is forced into the material (feed direction):

$$F_c = \frac{F_s \cos(\beta - \alpha)}{\cos(\phi + \beta - \alpha)} \tag{8.4}$$

$$F_t = \frac{F_s \sin(\beta - \alpha)}{\cos(\phi + \beta - \alpha)} \tag{8.5}$$

where F_s is the shear force that causes the chip to deform at a ϕ angle to the direction of cutting, α is the tool rake angle, and β is the angle between the

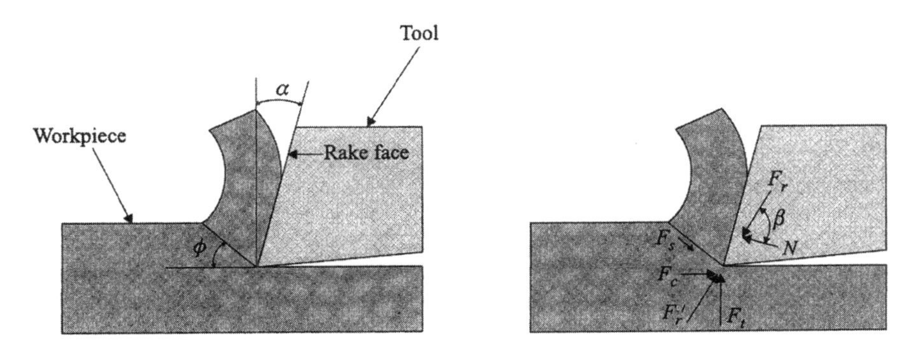

FIGURE 10 (a) Orthogonal cutting geometry; (b) forces.

resultant force F_r and the normal to the rake surface of the cutting tool, N. Merchant's equation states that $2\phi = 90° + \alpha - \beta$.

As noted above, the primary machining forces (F_c and F_t) are directly affected by the tool's geometry, as well as by the workpiece material properties, which in turn impact on the choice of power requirements and cutting tool materials. For example, decreasing the rake angle, α, increases machining forces, which in turn indirectly contributes to faster tool wear rates, and vice versa.

It has been common practice to measure machining forces through dynamometers. These sensors are normally placed under the tool carrier, in turning, and underneath the workpiece, or between the tool and the spindle, in milling. The primary objective of monitoring machining forces is to monitor indirectly the tool wear and adjust the cutting parameters such as feed rate, when necessary.

8.2.2 Chip Formation and Control

Chip formation is a plastic flow (shear) phenomenon initiated along the shear angle—the primary shear zone (Fig. 11). As the chip moves along the tool's surface, fracture contributes to its breakage. There are large number of different chip geometries that can be produced during machining based on the workpiece material, cutting conditions, and possible chip breakers employed.

The three classes of chips are

Continuous chips: The formation of continuous chips is most common to ductile materials (copper, aluminum, mild steel, etc.) machined at high cutting speeds and/or with high rake angles. Although such chip formation yields good surface finish, in practice, long chip strips would cause entanglement with the cutting tool and be detrimental to the safety of machine operators and even damage the surface of the workpiece. Continuous chips can be broken, however, at appropriate lengths, through the use of chip breakers, while maintaining good surface finish.

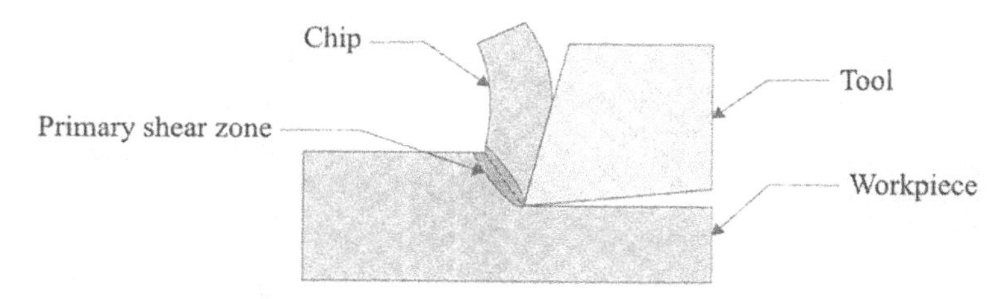

FIGURE 11 Shear zone in orthogonal cutting.

Continuous chips with built-up edges (BUE): At lower cutting speeds and lower rake angles, edges (of workpiece material) can build up between the tool and the under surface of the chip. Such BUE can weld to the chip as well as to the workpiece and get fractured away as machining progresses, while causing surface imperfections as well as contributing to rapid tool wear. They can be avoided using effective coolant liquids and increased rake angles.

Discontinuous chips: Material brittleness, low rake angles, and large depths of cut contribute to the formation of discontinuous chips (not necessarily separated but fractured). This intermittent phenomenon may cause the tool to vibrate (chatter), which will in turn adversely affect the quality of the object's surface and potentially cause increased tool wear.

Friction in Metal Cutting

In metal cutting, the area of contact between the tool and the chip experiences very high cutting forces and temperatures. This phenomenon is further complicated by the fact that the chip's under surface moving along the rake face of the tool is chemically clean. The combination of these factors leads to high values of the coefficient of friction and potentially to a stick-and-slip type of relative motion of the chip. That is, for a large portion of the contact length, the workpiece material adheres to the tool's rake face. This motion of the chip is characterized as plastic flow. An appropriate increase in the rake angle could significantly reduce the normal force on the rake face and subsequently reduce the coefficient of friction, thus indirectly prolonging the tool's life.

Chip Flow Control

Chip flow control refers to the breaking of continuous chips in a controlled manner. There are two common methods of achieving chip breakage: the obstruction type (commonly used in turning operations) and the groove type (incorporated into almost all modern tool inserts). In obstruction-type chip breakers, an obstruction of optimal height that is placed a short distance from the cutting edge interrupts the flow of the continuous chip and forces it to curl (Fig. 12a). In groove-type chip breaking, an optimal geometry groove is built into the tool (at the time of its fabrication) for curling the chip and causing it to break directly, or act as a guide to curl the chip and push it against an obstruction for its breakage (Fig. 12b). In drilling, chip breakage and improved removal can be achieved through suitable flute profile design (e.g., by incorporating chip breaking groove features into the flute or flank profile).

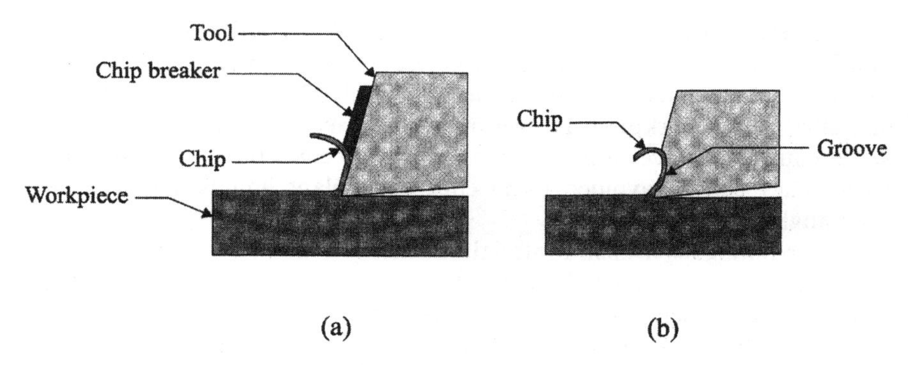

FIGURE 12 (a) Obstruction; (b) groove breaking.

8.2.3 Vibrations in Machining

Cutting tool and workpiece vibrations during machining can be quite detrimental to tool wear and negatively affect the dimensional accuracy and surface finish of the workpiece. Machine tool vibrations have been investigated by many academic and industrial researchers, most notably by Tlusty's group since 1963. The objective has been to model and predict vibrations through analytical and finite element modeling–based numerical works in order (1) to design the optimal machine tool configuration and (2) to determine (off-line) the best cutting process parameters (velocity, feed rate, and depth of cut). The overall conclusion has been that, no matter how rigid the machine tool structure is, cutting vibrations are inevitable. They should be managed through passive vibration isolation techniques or active (real-time) control methods.

Machine tools commonly experience three types of vibrations:

Free vibrations: These occur owing to an impulse force applied on the machine tool, such as a sudden reversal of the milling table's direction of motion. These vibrations are considered to be transient and rapidly decay. A suitable vibration isolation system would effectively damp free vibrations.

Forced vibrations: These occur owing to a (normally periodic) dynamic force applied on the machine tool, such as intermittent cutting forces in milling, periodic variations in depth of cut, and imbalances in the drive system of the machine tool. Their impact is the greatest when the existing frequency is near one of the natural frequencies of the machining operation, potentially leading to the very undesirable instability of the system. Forced vibrations can be managed most effectively by varying the cutting process parameters and the cutting geometry (i.e., adjusting the amplitude and direction of the cutting forces and therefore the amplitude and frequency of vibrations).

Self-excited vibrations: These vibrations, also known as "chatter," are primarily due to variations in the cutting conditions, while machining under specific (constant) process parameters. Chatter is undesirable and difficult to predict because of the combinatoric nature of machining parameters (e.g., workpiece material, tool material, cutting geometry, velocity, feed rate, and even cutting fluid used). Machining while chatter is present yields unacceptable surface finish and leads to rapid tool failure, including catastrophic cutting edge failure.

There are two main mechanisms to self-excited tool vibrations: mode coupling, which occurs when the tool vibrates in at least two directions in the plane of cutting (that includes the cutting velocity vector) with the same frequency and a phase shift normally due to nonoverlapping passes of cut; and regenerative instability, which is due to a relative periodic vibration between the tool and the workpiece, since surface waviness in successive cuts would never overlap (statistical impossibility) and thus cause variations in chip width (Fig. 13). This continuous regeneration of surface waviness causes periodic variations in cutting forces that lead to self-excited vibration. The most effective way of dealing with chatter is its (often experimental) prediction (i.e., of the possible instability) and adjustment of the cutting parameters for its avoidance.

Stability of machining dynamics can be achieved by coping with all sources of vibration—free, forced, and self-excited. As mentioned above, machine configuration optimization by utilizing vibration isolation mechanisms and high-quality components (e.g., radial bearings) will significantly contribute to this objective. Selecting optimum cutting conditions and tool geometries will also reduce vibrations (especially chatter). In milling, for example, one can use cutters that have variable tooth spacing to break the regular waviness of the surface, when chatter becomes a difficult problem.

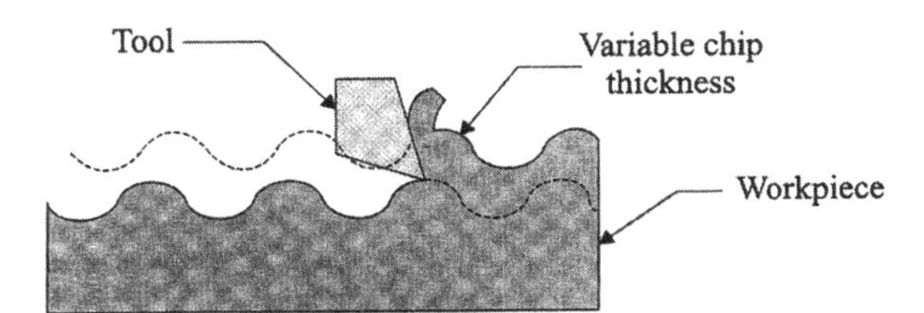

FIGURE 13 Regenerative chatter.

Also, honed or chamfered tools are less likely to cause chatter than those with sharp cutting edges.

8.2.4 Cutting Temperature and Fluids

Energy expended during the plastic deformation of the chip and overcoming frictional forces is (almost all) converted into thermal energy (i.e., heat). Significant research in this area has shown that the hottest zone in machining is where the chip goes through its stick-and-slip motion over the rake face of the tool (especially along the "stick" length of contact). It is thus no surprise that major tool wear occurs in this region of the rake face. Cutting temperatures can be over 1000°C in this region of the rake face and 600° to 700°C on the flank face near the cutting tool edge when cutting steel with carbide tools (Fig. 14).

Cutting temperatures increase with cutting velocity and feed rate. In intermittent cutting, as expected, the temperature profile is periodic as a function of time, thus subjecting the tool to thermal impact in addition to

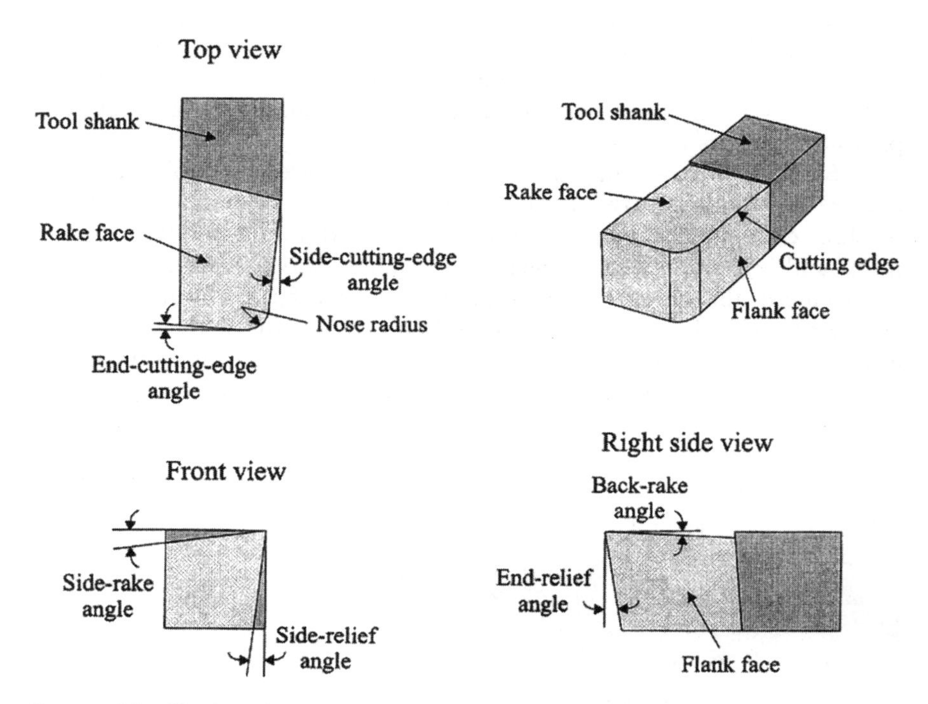

FIGURE 14 Single-point tool geometry; orthographic views.

mechanical impact. Thermal fatigue failure is a common problem, for example in tungsten carbide tools (especially at high cutting velocities and when coolants are utilized).

Cutting fluids can be used as effective coolants to reduce the temperature of the tool at the interface with the chip. Such fluids also provide a cooling effect to the workpiece to prevent thermal distortions, and, very importantly, they act as lubricators for friction reduction on the rake face. Cutting fluids can be water-based or oil-based. Water-based fluids may contain fatty soaps, sulphur, organic salt, and sulphides to provide better cooling action. Oil-based fluids may contain mineral oils, chloroform, phosphates, and polymeric ethers to provide better lubrication at the tool–chip interface. The application method of cutting fluids (flood versus mist) and the rate of application depend on cutting conditions (cutting velocity and feed rate).

8.3 TOOL WEAR AND SURFACE FINISH

The life of a cutting tool directly impacts machining costs. The longer the tool life is, the less the tooling cost, and furthermore the fewer the tool changes (i.e., lower setup and preparation costs). Despite a century of research in this area, most wear mechanisms reported in the literature should be treated as well-investigated conjectures, and not as proven theorems. Today, research in this area has shifted toward monitoring and predicting tool wear in real time, in order to prolong tool usage and also to prevent catastrophic failure of the tool. In many instances, breaking the tool's cutting edge (tip) can cause irreparable damage to the workpiece's surface. Thus it would be preferable to stop using the cutting tool well ahead of this point.

Although cutting tools have been classified as single-point and multipoint tools, many of the wear mechanisms for the former apply to the latter as well. This commonality is further strengthened by the widespread use of generic inserts (held in single- and multipoint tool holders) for turning, milling, and even for large-diameter drilling.

The tool regions of particular interest, from the wear point of view, are shown in Fig. 14. The rake face is the primary surface of contact between the chip and the cutting tool, and the flank of the tool is a region (especially at the cutting edge) where the tool comes into contact with the workpiece. The geometries of both rake and flank surfaces significantly affect cutting forces and surface quality. For example, cutting-edge strength would significantly diminish for back rake angles above 5°. In contrast, the cutting-edge strength would increase as the back-rake angle becomes negative and approaches an optimal value around $-5°$. The end-relief angle prevents contact between the end flank and the workpiece, though angles above 5° will start weakening the cutting edge.

8.3.1 Cutting-Tool Materials

The interface between a cutting tool and the workpiece can be characterized by high forces, severe friction, high temperature, and, in the case of intermittent cutting operations, high-frequency impact. Thus the ideal cutting tool should have both high hardness and high toughness—a rarity in the field. Ceramic tools, for example, have excellent hardness properties (even at temperatures of 500°C and above), but low toughness. High-speed steel (HSS) tools, on the other hand, have excellent toughness, but their hardness rapidly diminishes at high cutting temperatures (above 500°C).

HSS tools were developed at the turn of the 20th century for higher speed machining. They can have about 10% molybdenum or up to 18% tungsten as the alloying element. The former is tougher and thus widely used in drilling and end milling. HSS tools normally cut at velocities lower than 50 m/min. Carbide tools (also known as cemented or sintered carbides) were developed in the 1930s for hardness and cutting velocity characteristics that are better than those of HSS tools. Primary carbide inserts today are of tungsten or titanium type. They can cut steel at velocities of up to 200 m/min.

Carbide tools can be also coated for improved mechanical and thermal properties, as well as for lower friction and increased resistance to chemical reactions between tool and workpiece materials. They can cut steel at velocities of up to 400 m/min. Typical coating materials are titanium carbide on tungsten carbide, aluminum oxide (Al_2O_3)—ceramic, and even synthetic diamonds. Coating is achieved through either chemical or physical vapor deposition.

Ceramic tools (introduced in 1950s) and cubic boron nitride (cBN) tools (introduced in the 1960s) are typically used for machining conditions that require high hardness at elevated temperatures. (cBN is almost the hardest material in existence, being second only to diamond). These tools can cut steel at velocities of up to 700 to 800 m/min.

The cutting edges of most carbide inserts are either chamfered or honed for strengthening purposes. It has been conjectured that such an edge preparation could prevent premature chipping (microfracture) of the cutting edge in the first (accelerated-wear) period of tool wear and thus prolong tool life (especially in high-impact, intermittent machining operations). Typical chamfers would be 20° to 45° inclined and up to 0.2 mm wide, while the radius of curvature of a honed edge would be 0.025 to 0.180 mm (Fig. 15).

8.3.2 Tool Wear Mechanisms

Tool wear has probably been the one machining issue most researched since the beginnings of machine tool history. The objective of toolmakers has always been to develop harder and tougher cutting tools in order to machine engineering materials that have been continuously evolving. Although significant progress has been made in the development of wear resistant tools, these still rapidly wear out under severe machining conditions present at the tool–chip interface. After a century of research, we still do not have an accurate wear prediction model and frequently use the tool life formula developed by Taylor almost a century ago (Sec. 8.3.3). However, with increasing achievements in the area of "intelligent machining," it is anticipated that tool wear will be able to be monitored in real time, where (software-based) digital filters will be utilized for predictions into the near-future behavior of the tool.

Cutting temperature is accepted as one of the most important process parameters that affect tool wear: cutting tools have much lower mechanical hardness and toughness at elevated temperatures. As one would expect, the other factor is the thermal and mechanical periodic impact that a tool gets subjected to in intermittent cutting, such as milling.

The exact mechanisms of wear in machining is still an open topic for research. However, the following list contains some mechanisms of wear that have been accepted by many researchers:

Adhesion: This wear mechanism is based on the formation of welded junctions and their subsequent breakage, i.e., tool particles on the rake face being removed by the chips. built-up edge (BUE) has also been noted as an adhesive wear mechanism, potentially causing chipping of the tool edge.

Abrasion: Hard particles in the workpiece material (e.g., carbides) abrade and dislodge (micro)particles from the surface of the cutting tool. This wear mechanism (as is adhesion) is more prominent at higher

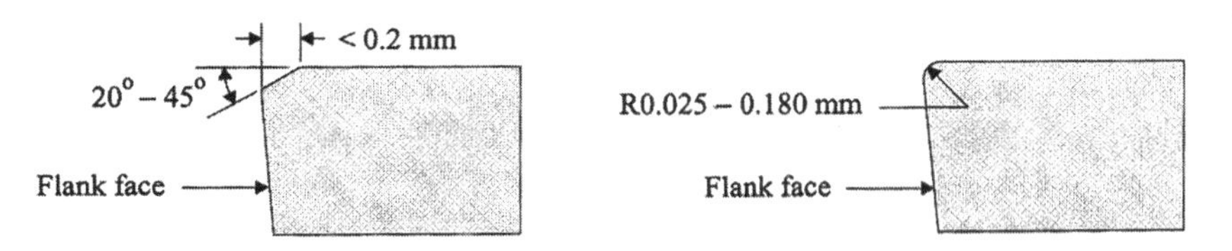

FIGURE 15 (a) Chamfered; (b) honed edge.

cutting velocities that yield increased cutting temperatures and weaken the tool's hardness.

Diffusion: This wear mechanism is present because of potential chemical affinity between the tool material and the workpiece material (e.g., cobalt, in tungsten carbide tools, diffusing into the steel workpiece chips).

Fatigue: Thermal and mechanical loading of the cutting tool results in microcracks that lead to chipping and, at worst, to catastrophic failure of the cutting edge (i.e., significant tool-edge breakage).

The progressive wear of cutting tools can be quantified by the following two metrics:

Crater wear: Also known as rake-face wear, crater wear corresponds to a formation of crater like shallow cavity on the rake face of the tool very near to the tool edge (Fig. 16a). All wear mechanisms discussed above contribute (at different degrees) to crater wear, typically measured by the depth of the crater. Although potentially advantageous at the beginning, lowering cutting forces owing to increased rake angles, this wear weaken the cutting edge and causes its failure through fracture.

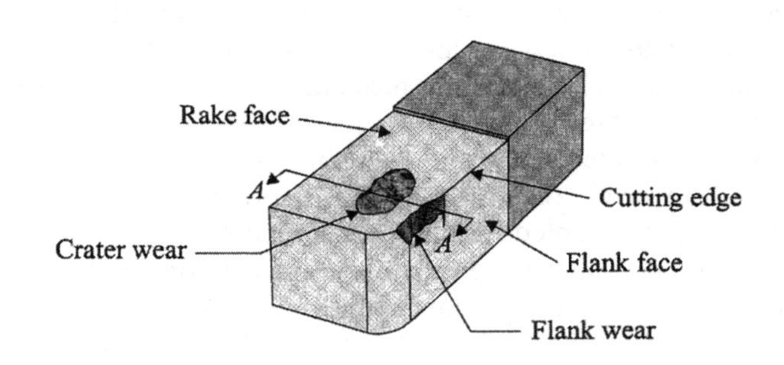

A-A Cross section

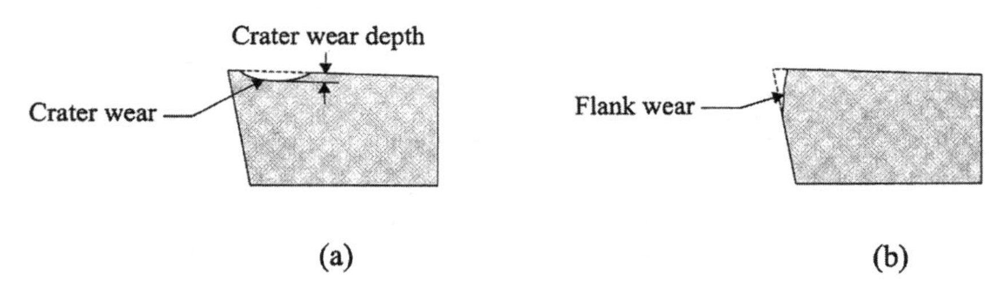

FIGURE 16 (a) Crater wear; and (b) flank wear.

Flank wear: In contrast to crater wear, which is most prominent in the machining of ductile materials, flank wear can be present under almost all cutting conditions. It refers to the wearing of the flank face, starting at the cutting edge and progressively developing downward and sideways (Fig. 16b). Flank wear results in the reduction of the cutting edge's sharpness, leading to higher cutting forces, eventually leading to tool fracture. It is primarily caused by abrasion.

8.3.3 Tool Life Equation

Over the past several decades, it has been established that tool flank wear (its width) can be expressed as a function of time utilizing a three-region (period) tool life curve (Fig. 17a). The first region refers to the exponen-

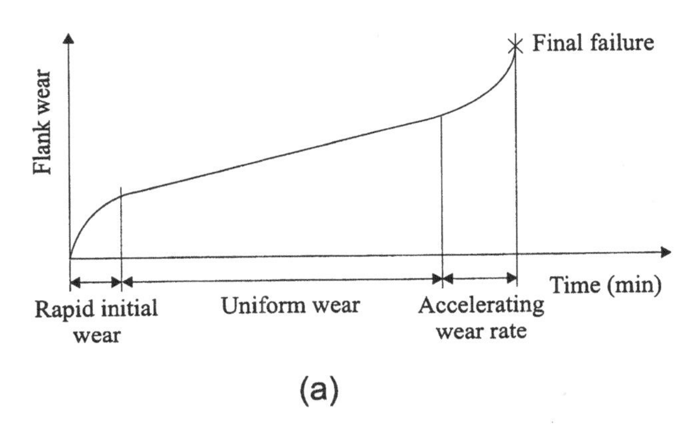

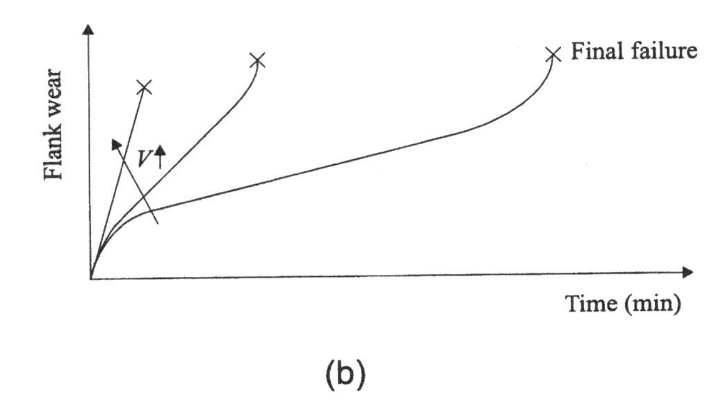

FIGURE 17 Flank-wear curve.

tial degradation of the tool edge area and could last up to 2 to 5 sec for carbide tools cutting steel. The second region can be approximated by a linear (uniform rate) relationship. This period can be assumed to represent the useful life of the tool. The third and last region corresponds to the final exponential degradation of the tool prior to its total failure. It is strongly advised to halt machining once the tool enters this period of its life.

As also shown in Fig. 17b, flank wear is a strong function of cutting velocity, V: the flank wear rate increases as the cutting velocity is increased. Based on much empirical data, Taylor proposed that the relationship between tool life and velocity can be expressed as a logarithmic function. In the logarithmic domain, tool life, T, is approximately a linear function of cutting velocity:

$$VT^n = C \tag{8.6}$$

where C is the cutting velocity (m/min) achievable for the tool–workpiece combination at hand that would correspond to one minute of tool life, n is the slope of the relationship in the logarithmic domain and primarily depends on the tool material (0.1 to 0.17 for HSS tools, 0.3 for titanium coated tungsten carbide tools, and up to 0.6 to 1.0 for ceramic tools).

Taylor's tool life formula has been modified over the years to include feed rate, f, and depth of cut, a:

$$VT^n f^{n_1} a^{n_2} = K \tag{8.7}$$

where K is a proportionality constant, and n, n_1, n_2 are tool material dependent (constant) exponents (typically, 0.5 to 0.8 for n_1 and 0.2 to 0.4 for n_2).

8.3.4 Workpiece Surface Finish

Surface finish (i.e., roughness) is an important dimensional requirement in machining and typically acts as a constraint on feed rates—the higher the feed rates, the worse the surface finish. The literature on machining categorizes factors that affect surface finish into two: those that affect the ideal finish of the workpiece surface and those that affect the natural finish. The former can be formulated (and accurately estimated) as a function of feed rate and tool geometry. Figure 18a shows the surface finish model for a (single-point) turning tool, while Fig. 18b shows the model for a face-milling tool. Although the topographic traces left on the workpiece are different for turning and milling (rotational versus planar motion), the profiles of finish are very similar.

The primary factors that contribute to natural surface finish are: the occurrence of BUE, chatter or other vibration mechanisms, inaccuracies in

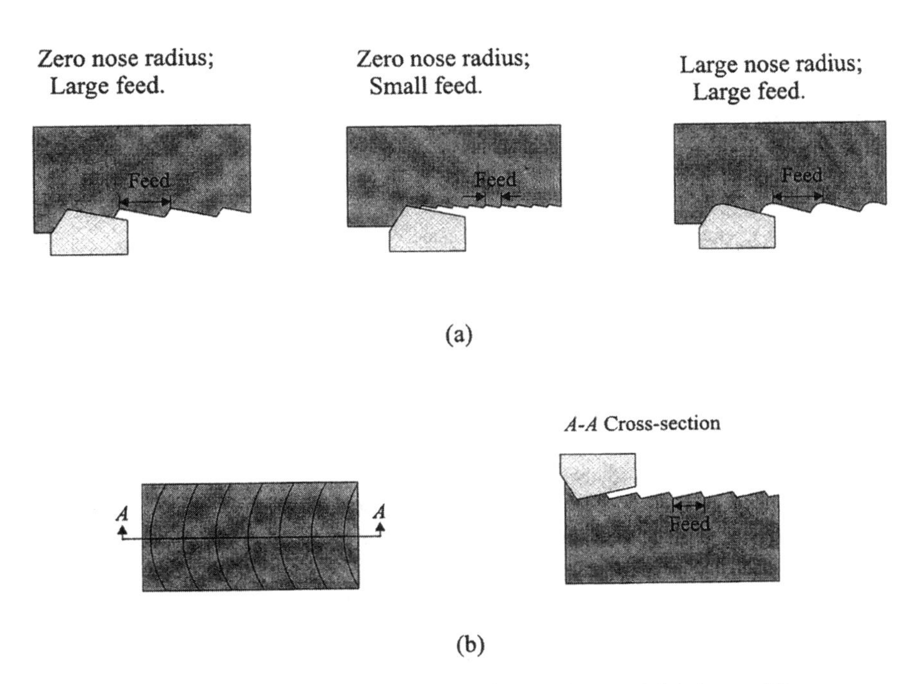

FIGURE 18 Ideal surface finish model for (a) turning and (b) face milling.

tool/workpiece motions, workpiece material inhomogeneuity, and tool wear. Surface roughness is almost impossible to model analytically because of these factors. Engineers would have to rely on past empirical data and additional run-time measurements to adjust the process parameters to reduce surface roughness to an acceptable level. The following guidelines can be used in this endeavor: increasing cutting velocity (to reduce BUE) and reducing feed rate—note that the former may cause chatter and the latter reduce productivity; increasing the tool nose radius and decreasing the cutting edge angle, as much as chatter would allow; and, in milling, tilting the spindle slightly in order to prevent contact between the tool and the already machined part of the workpiece behind the cut.

Surface integrity must also be considered as a measure of surface finish in machining. Residual tensile stresses are very common in machined surfaces due to severe temperature gradients that develop during metal cutting. Such stresses lead to microstructure damage (microcracks) and reduce the fatigue strength of the workpiece. Residual stresses can be reduced by utilizing a variety of surface treatment methods that yield high compressive residual stresses and a smooth surface and thus increased fatigue life. Shot peening, where the workpiece surface is bombarded with cast steel, glass, or ceramic balls of diameter up to 5 mm, is such a technique. Other similar techniques include laser peening and water jet peening.

8.4 ABRASIVE CUTTING

Abrasive cutting processes are primarily utilized as postmachining operations for improving surface quality in terms of reducing roughness. They may, however, add on further residual stresses and occasionally lead to surface burning at high cutting rates, especially in grinding. Thus care has to be exercised in the use of abrasive cutting tools even though they have been around for several millennia (dating back to the use of abrasive stones for the sharpening of hunting tools).

The most common abrasive cutting processes are grinding, honing, lapping, superfinishing, and polishing. All these processes use bonded hard, sharp, and friable abrasive grains for the removal of very thin layers of metals. Although grinding is the most versatile technique, it also yields the worst surface finish amongst the abrasive processes:

Grinding: This process utilizes an abrasive wheel for the internal and external machining of cylindrical as well as prismatic workpieces (Fig. 19a).

Honing: This process utilizes a set of abrasive "sticks" (stones) bonded on a mandrel for the (internal) machining of holes (bores) through a rotational motion that is in sync with a vertical reciprocating motion of the mandrel (Fig. 19b). Honing can yield a surface finish that is as twice as good (half the roughness) as one produced by grinding.

Lapping: This process utilizes a loosely bonded abrasive material (abrasive particles suspended in a viscous fluid) placed between the workpiece and a rotating lap tool (following an 8-shaped trajectory in three-dimensional space) (Fig. 19c). Lapping yields a surface finish of excellent quality—commonly utilized for optical lens, bearing surface, and gage surface machining (finishing).

Superfinishing: This process is similar to honing but differs in the high frequency of reciprocation of the tool (up to 1500 strokes per minute) and the shorter strokes (Fig. 19d). The result is a best achievable (mirrorlike) surface finish. Superfinishing can follow other abrasive processes for further refinement of the surface finish.

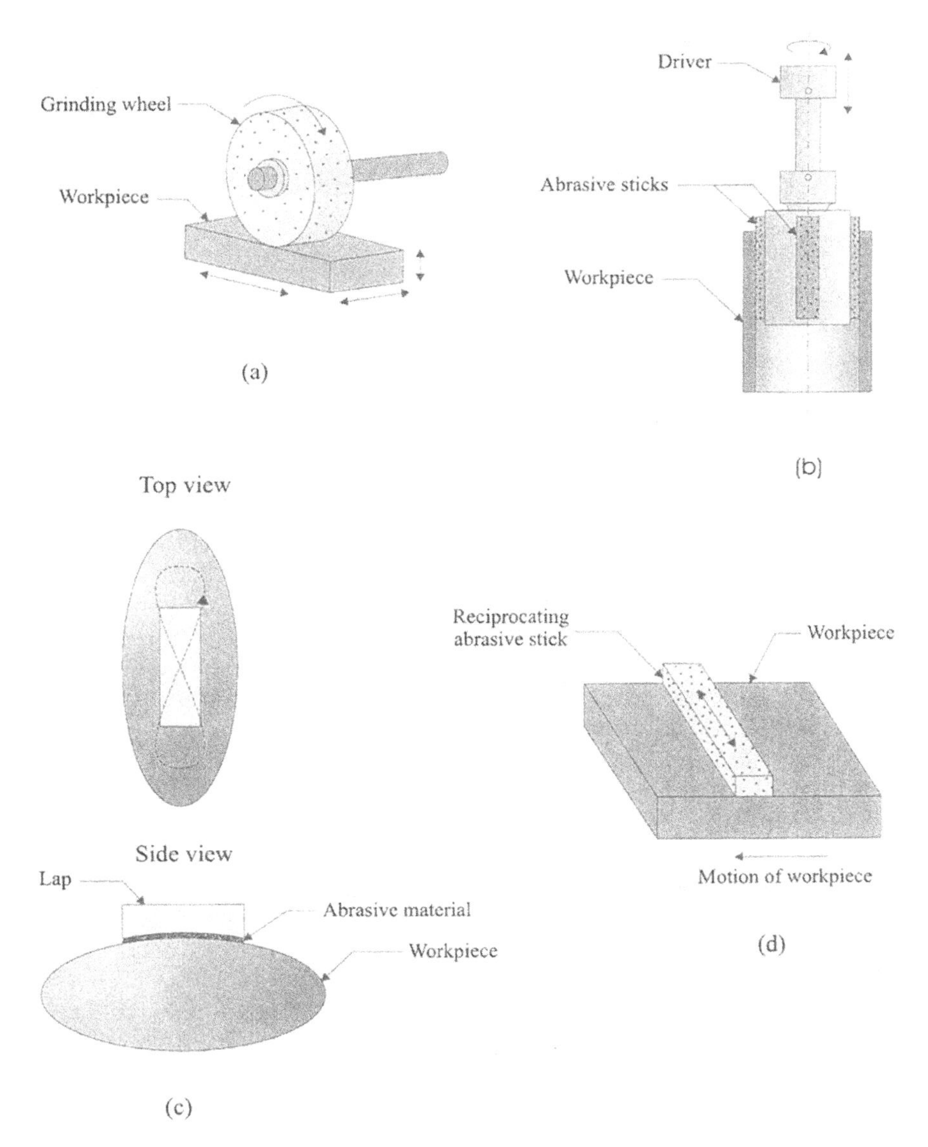

FIGURE 19 (a) Grinding; (b) honing; (c) lapping; (d) superfinishing.

Polishing: This process utilizes a high-speed polishing wheel/disc (made of leather, felt, or even paper): the abrasive grains are glued to the periphery of the wheel, for the removal of fine scratches or burrs.

In this section, only the grinding process will be detailed.

8.4.1 Grinding Operations

The grinding wheel has abrasive grains enveloped in a matrix of bonding material (Fig. 20a). These grains are of irregular shape and randomly dispersed within the matrix. Owing to this random dispersion, three mechanisms of interactions exist between the grains and the workpiece: cutting, plowing, and rubbing. Only cutting causes material removal, while plowing only causes deformation of the surface (Fig. 20b).

The three main types of grinding are

Surface grinding: This process is used for machining flat surfaces. The cutting process parameters, feed rate, cutting velocity, and depth of cut are defined as those for peripheral and face milling (Fig. 21). The table on which the workpiece is placed can translate or rotate in a planar motion with respect to a fixed-axis rotating grinding wheel. The table (i.e., the workpiece) is brought up by an increment equal to the depth of cut once the current pass has been completed (typically achieved by lowering the grinding wheel, as opposed to raising the table).

Cylindrical grinding: This process is used for the internal or external machining of rotational workpieces (Fig. 22). Normally, as in turning and boring, the grinding wheel (i.e., the cutting tool) translates with respect to a fixed-axis rotating workpiece in the feed direction, at a constant depth of cut for the pass at hand.

Centerless grinding: This process is also used for the internal or external machining of rotational workpieces (Fig. 23). In contrast to cylindrical grinding, the workpiece is not held by a chuck, but rotates freely between support rolls and a control wheel for internal grinding, or between a grinding wheel and a control wheel for external grinding. In the former configuration, the control wheel pushes the workpiece toward the internally placed grinding wheel.

The control wheel is tilted at an angle in order to feed the object forward (in the feed direction) (Fig. 23b). A continuous line of parts can be fed into the external grinding system, whereas parts are machined one at a time in internal grinding.

The primary advantage of this process over cylindrical grinding is reduction in setup time.

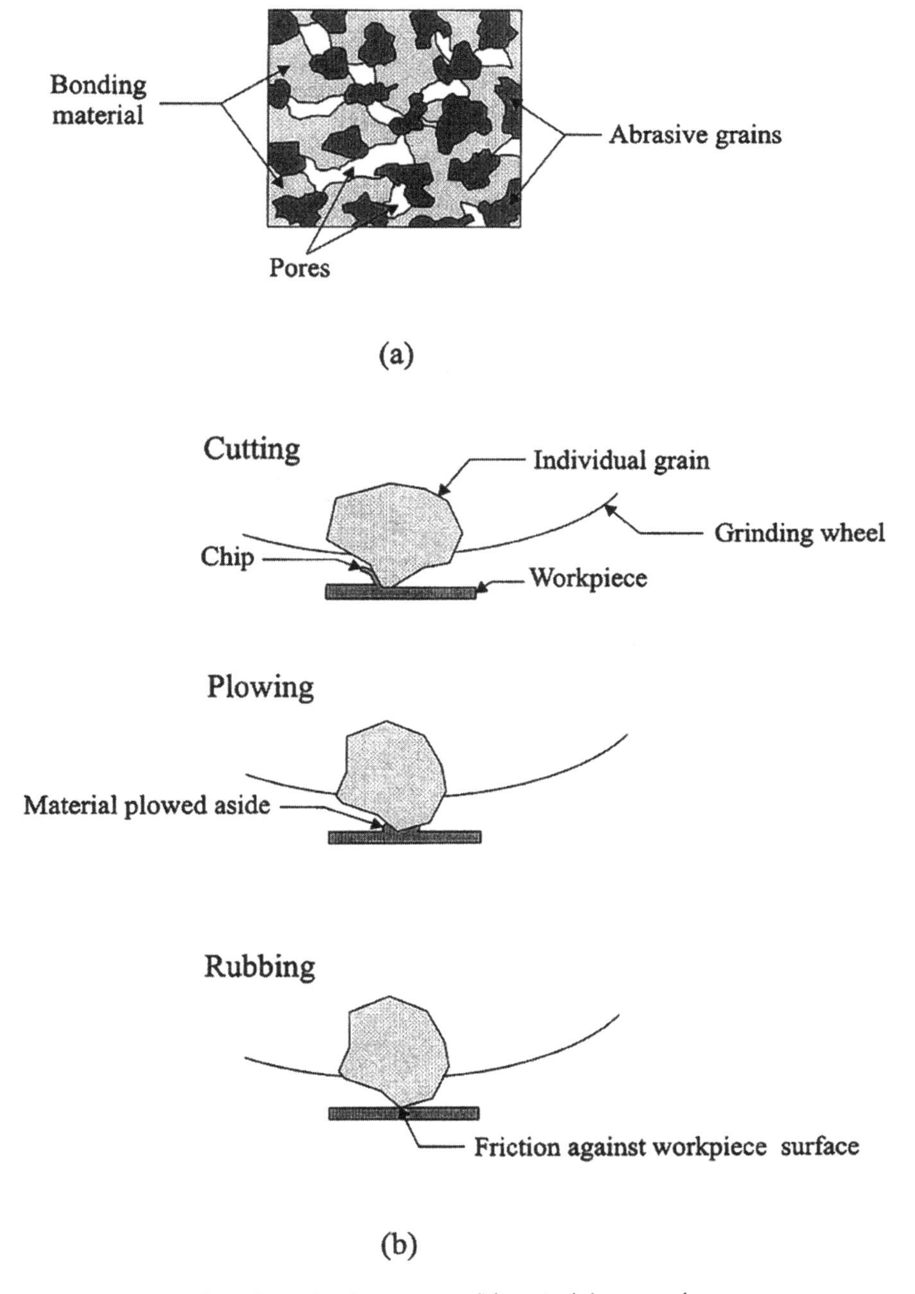

FIGURE 20 (a) Grinding wheel structure; (b) material removal.

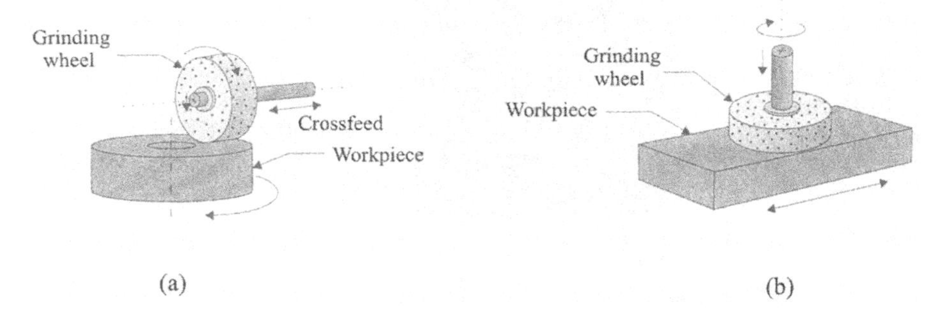

FIGURE 21 Surface grinding with (a) a horizontal spindle; (b) a vertical spindle.

8.4.2 Tool Materials and Tool Wear for Grinding

The grinding wheel has abrasive particles bonded together and formed into a desired shape for the specific grinding application. The abrasive particles are hard, brittle refractory materials that are classified according to their hardness, toughness, and friability (capacity to fracture and yield another cutting edge, in contrast to gradual wear into a dull shape). The hardest of abrasive materials, such as diamond and cBN, are often referred to as superabrasives.

Common abrasives used in grinding wheels include aluminium oxide (Al_20_3) and silicon carbide (SiC). The latter is harder and has much better friability, but it is not as tough as the former. Superabrasives include natural diamond (or graphite-based synthetic) and cBN. Superabrasives are two to four times harder than common abrasives. Synthetic diamonds are more friable than natural diamonds. cBN crystals need to be etched or coated for ease of bonding into a grinding wheel.

Common bonding materials (used as the matrix) in grinding wheel production include vitrified (a mixture of feldspar mineral and clay), silicate

FIGURE 22 Cylindrical (a) internal and (b) external grinding.

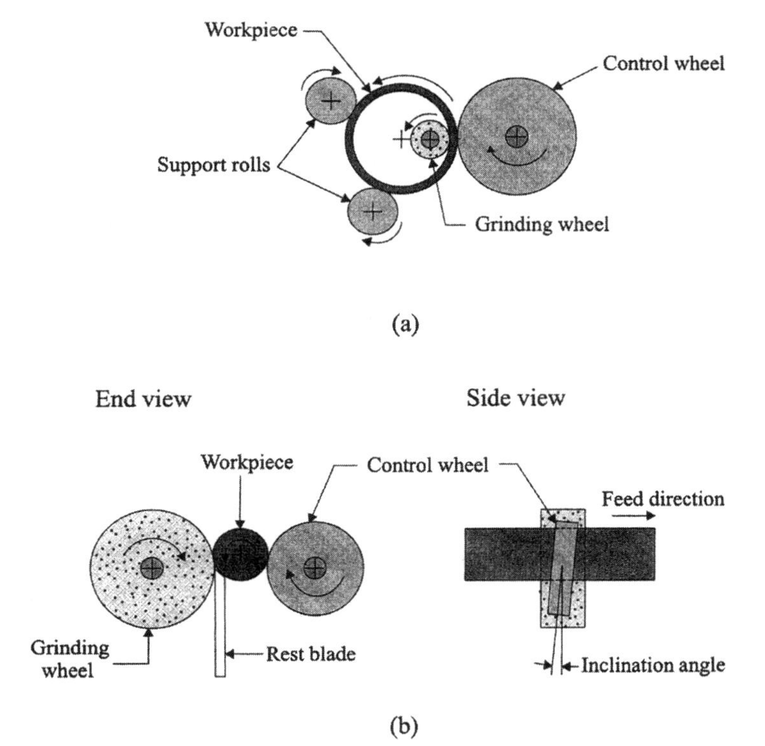

FIGURE 23 Centerless (a) internal and (b) external grinding.

(sodium silicate), shellac, resinoid (thermoset resin), rubber, and metallic (bronze, aluminum, etc.).

Grinding Tool Wear

Grinding occurs at elevated cutting temperatures (up to 1700°C), owing to the high negative rake angles of the particles when forming the chips, and high friction. Besides leading to rapid tool wear, high temperatures can also severely affect the dimensional and surface integrity (i.e., high residual tensile stresses) of the workpiece. As in other machining operations mentioned in this chapter, a variety of cutting fluids can be used in grinding for cooling and lubrication purposes.

The wear of the grinding wheel can be attributed to three common wear mechanisms: attrition wear (dulling of the individual grains), grain fracture wear (the breaking away of parts of the grain, yielding new sharp edges), and bond fracture wear (the dislodging of the grains from the wheel through the fracture of their bonds). The combination of these wear mechanisms yields a wear curve very similar to the three-region wear curve depicted in Fig. 17 for single-point cutting tools: For grinding tools, the Y-axis of the tool wear curve represents the volume of wheel wear, and the X-axis represents the volume of workpiece material removed.

REVIEW QUESTIONS

1. Machining is considered to be one of the most versatile fabrication processes for parts with complex geometries. What constraints (geometrical, material, batch size, etc.) would make such a manufacturing choice impractical?
2. List four primary issues researched in the past century for the innovation of machining processes.
3. Define continuous versus intermittent machining. Give examples of each.
4. Define depth of cut, feed rate, and cutting velocity in turning.
5. Since boring can be carried out on lathes, why would one use dedicated boring machines?
6. If holes fabricated via drilling were considered unacceptable due to poor dimensional or surface quality reasons, what would you recommend as a remedy? Explain.
7. Define depth of cut, feed rate, and cutting velocity in milling.
8. Define 2.5-, 3-, and 5-axis milling, respectively. Give some part geometry examples.
9. Would you recommend to fabricate (i.e., preshape) the blank to be machined to be as near as possible to the final desired geometry? Explain.
10. Why are through holes preferable to blind holes?
11. Why are holes easier to machine on flat surfaces perpendicular to the tool's motion axis?
12. Define the cutting and thrust forces in turning and milling. How would one measure forces in machining? What is the primary objective of monitoring machining forces?
13. How does chip formation affect surface quality? How can one control the chip formation process?
14. Define the need for cutting fluid use in machining.
15. Define chatter in machining. Explain the two mechanisms that cause chatter.
16. Define crater wear and flank wear. Explain the mechanisms that cause tool wear in machining.

17. Describe the primary machining (i.e., tool–workpiece interaction) mechanisms in grinding.
18. Why would one choose centerless grinding over cylindrical grinding?
19. Describe tool wear in grinding.

DISCUSSION QUESTIONS

1. Material removal techniques, as the name implies, are based on removing material from a given blank for the fabrication of the final geometry of a part. Compare material removal techniques to near–net shape production techniques, such as casting, powder processing, and forming, in the context of product geometry, material properties, and economics in mass-production versus small-batch production environments.
2. The mechanics of material removal operations (single-point and multipoint cutting) has been modeled extensively. Such models, when combined with heat transfer models, can help engineers predict chip formation, surface finish, tool wear, etc. Discuss the utilization of analytical (or heuristics-based) models in off-line process planning as well as in on-line adaptive control that would be based on the utilization of a variety of sensors for force, vibration, and temperature measurements.
3. Woodworking is a topic rarely addressed in manufacturing books since wood is not considered an engineering material. However, even when excluding the pulp-and-paper and construction industries, the large furniture industry is a testimony to the importance of woodworking. Discuss the issues of fabrication and assembly for wood-based products in comparison to metal-based products. Include in your discussion the problem of irregularities, defects, and other features of natural materials that the production engineer has to cope with.
4. When presented with a process planning problem for the machining of a nontrivial part, different (expert) machinists would formulate different process plans. Naturally, only one of these plans is (time or cost) optimal. Considering this and other issues, compare manual (operator-based) machining versus NC-based machining, as enterprises are moving toward integrated and computerized manufacturing. Formulate at least one scenario where manual machining would be favorable.
5. Process planning in machining (in its limited definition) refers to the optimal selection of cutting parameters: number of passes and tool paths for each pass, depths of cut, feed rates, cutting velocities, etc. It has been

often said that computer algorithms should be utilized in the search for the optimal parameter values. Although financially affordable for mass production environments, such (generative) programs may not be feasible for utilization in one-of-a-kind or small-production environments, where manufacturing times may be comparatively very short. Discuss the utilization of group technology (GT)–based process planners in such computation-time-limited production environments.

6. Several fabrication/assembly machines can be physically or virtually brought together to yield a manufacturing workcell for the production of a family of parts. Discuss the advantages of adopting a cellular manufacturing strategy in contrast to having a departmentalized strategy, i.e, having a turning department, a milling department, a grinding department, etc. Among others, an important issue to consider the transportation of parts (individually or in batches).
7. Machining centers increase the automation/flexibility levels of machine tools by allowing the automatic change of cutting tools via turrets or tool magazines and carry out a variety of material removal operations. Some machining centers also allow the off-line fixturing of workpieces onto standard pallets, which would minimize the on-line setup time (i.e., reduce the downtime of the machine). That is, while the machine is working on one part fixtured on Pallet 1, the next part can be fixtured on Pallet 2 and loaded onto the machine when it is has finished operating on the first part. Discuss the use of such universal machining centers versus the use of single-tool, single-pallet, unipurpose machine tools.

BIBLIOGRAPHY

Armarego, E. J. A., Brown, R. H. (1969). *The Machining of Metals*. Englewood Cliffs, NJ: Prentice-Hall.

Benhabib, B. (1982). Flank Wear of Carbide Tools in the First Period. M.Sc. thesis, Faculty of Mechanical Engineering, Technion, Haifa, Israel.

Ber, A., Friedman, M. Y. (1967). On the mechanism of flank wear in carbide tools. *CIRP Annals* 15:211–216.

Boothroyd, Geoffrey G., Knight, Winston A. (1989). *Fundamentals of Machining and Machine Tools*. New York: Marcel Dekker.

Childs, Thomas, et al (2000). *Metal Machining: Theory and Applications*. London: Arnold.

Cocquilhat, M. (1851). Expérience sur la résistance utile produites dans le forage. *Annales des Travaux Publics en Belgique* 10:199.

DeGarmo, E. Paul, Black, J. T., & Kohser, Ronald A. (1997). *Materials and Processes in Manufacturing*. Upper Saddle River, NJ: Prentice Hall.

DeVries, Warren R. (1992). *Analysis of Material Removal Processes*. New York: Springer-Verlag.
Doyle, Lawrence E., et al (1985). *Manufacturing Processes and Materials for Engineers*. Englewood Cliffs, NJ: Prentice-Hall.
Drozda, Thomas J., Charles, Wick (eds.) (1998). *Tool and Manufacturing Engineers Handbook*. Dearborn, MI: Society of Manufacturing Engineers.
Ernst, Hans, et al (1938). *Machining of Metals*. Cleveland, OH: American Society of Metals.
Ernst, H. (1938). Physics of metal cutting. *Machining of Metals*. Cleveland, OH: American Society of Metals.
Fenton, R. G., Oxley, P. L. B. (1969). Mechanics of orthogonal machining: allowing for the effects of strain rate and temperature on tool-chip friction. *Proceedings of the Institute of Mechanical Engineering* 178:417–438.
Fermer, Hugh (1995). *Machine Tools: A History 1540–1986*. Amberley, England: Amberley Museum.
Groover, Mikell P. (1996). *Fundamentals of Modern Manufacturing: Materials, Processes, and Systems*. Upper Saddle River, NJ: Prentice Hall.
Hine, Charles R. (1970). *Machine Tools and Processes for Engineers*. New York: McGraw-Hill.
Jahanmir, Said, Ramulu, M., & Koshy, Philip (1999). *Machining of Ceramics and Composites*. New York: Marcel Dekker.
Joessel, Philip (1864). Experiments on the most favorable form of tool in workshops from the point of view of economy of power. *Annuaire de la Société des Anciens Élèves des Écoles Impériales d'Arts et Métiers* 16.
Kalpakjian, Serope, Schmid, Steven R. (2000). *Manufacturing Engineering and Technology*. Upper Saddle River, NJ: Prentice Hall.
Kronenberg, M. (1943). Cutting angle relationship on metal cutting tools. *Mechanical Engineering* 65:901.
Kronenberg, M. (1966). *Machining Science and Applications*. Oxford: Pergamon Press.
Malkin, Stephen (1989). *Grinding Technology: Theory and Applications of Machining with Abrasives*. New York: John Wiley.
Merchant, M. E. (1944). Basic mechanics of the metal cutting process. *Journal of Applied Mechanics* 15:A-168.
Nee, John G. Ed. (1998). *Fundamentals of Tool Design*. Dearborn, MI: Society of Manufacturing Engineers.
Reuleaux, F. (1900). Über den Taylor Whiteschen Werkzeugstahl. *Verein zur Beförderung des Gewerbefleisses in Preussen, Sitzungsberichte* 79:179.
Schey, John A. (1987). *Introduction to Manufacturing Processes*. New York: McGraw-Hill.
Shaw, M. C. (1984). *Metal Cutting Principles*. Oxford: Oxford University Press.
Shaw, Milton Clayton (1996). *Principles of Abrasive Processing*. Oxford: Oxford University Press.
Stephenson, David A., Agapiou, John S. (1997). *Metal Cutting Theory and Practice*. New York: Marcel Dekker.

Taylor, F. W. (1907). On the art of cutting tools. *Transactions of the American Society of Mechanical Engineers* 28:31–350.
Tlusty, J. (2000). *Manufacturing Processes and Equipment*. Upper Saddle River, NJ: Prentice Hall.
Walsh, Ronald A. (1994). *Machining and Metalworking Handbook*. New York: McGraw-Hill.

9

Modern Manufacturing Techniques

In Chaps. 6 to 8 of this book, several primary manufacturing processes were presented for the fabrication of metal, plastic, and ceramic parts. The casting, molding, powder processing, metal forming, and conventional machining techniques described in these chapters dominated the manufacturing industry until the mid-1900s. Their total dominance, however, has been reduced with the introduction of numerous new commercial (nontraditional) manufacturing techniques since the 1950s, ranging from ultrasonic machining of metal dies to the nanoscale fabrication of optoelectronic components using a variety of lasers.

The first such processes were developed in response to the common drawbacks of traditional material removal techniques discussed in Chap. 8, for faster and more accurate machining of modern engineering materials. These nontraditional machining processes (introduced mainly in the late 1940s) were originally targeted for the production of complex geometry as well as microdetailed aerospace parts. Today the emphasis remains on reduced scale manufacturing (micro and nano level) with extensive use of lasers for noncontact, toolless fabrication of parts for all industries: household, automotive, aerospace, and electronics.

Modern manufacturing techniques have often been classified according to the principal type of energy utilized to remove or add material—mechanical, electrical, thermal, and chemical.

Mechanical processes: Ultrasonic machining and abrasive jet machining are the two primary (nontraditional) mechanical processes. Material is removed through erosion, where hard particles (in a liquid slurry) are forced into contact with the workpiece at very high speeds.

Electrochemical processes: Electrochemical machining is the primary representative of this group. It uses electrolysis to remove material from a conductive workpiece submerged in an electrolyte bath; particles depart from the anodic workpiece surface toward a cathodic tool and get swept away by the high-speed flowing electrolyte liquid.

Thermal processes: Electrical discharge machining, electron beam machining, and laser beam machining are the three primary thermal energy–based processes. Metal removal in electrical discharge machining is achieved through high-frequency sparks hitting the surface of a workpiece submerged in a dielectric liquid bath. In electron beam machining, a high-speed stream of electrons impinge on a very small focused spot on the surface of the workpiece and, as in electrical discharge machining, vaporize the material (this is preferably carried out in a vacuum chamber). Laser beam machining is utilized for the cutting of thick-walled parts as well as micromachining of very thin walled plates through fusion. Lasers are also commonly used in additive processes, lithography-based or sintering-based, for the solidification of liquids and powders. Naturally, the types of lasers used in these applications are quite varied.

Chemical processes: Chemical machining, also known as etching, refers to the removal of material from metal surfaces through purely chemical reactions. It can favorably be used in etching shallow depths (or holes) in metals such as aluminum, titanium, and copper, which are vulnerable to erosion by certain chemicals (most notably hydrochloric, nitric, and sulphuric acids). Due to difficulties in focusing on small areas, most chemical processes use chemical-resistant masks to protect surfaces from unwanted etching.

All above-mentioned modern material removal or material additive processes are characterized by the following common features: higher power consumption and lower material removal (or additive) rates than traditional fabrication processes, but yielding better surface finish and integrity (i.e., less residual stress and fewer microcracks). A large number of these processes also are capable of fabricating features with dimensions several orders of magnitude less than those obtainable by traditional processes.

In this chapter, we will first review several (nontraditional) processes that belong to the class of material removal techniques in two separate sections: nonlaser versus laser-based fabrication. Subsequently, we will discuss several modern material additive techniques commonly used in the rapid fabrication of layered physical prototypes.

9.1 NONLASER MACHINING

In this section, we will introduce the following nontraditional machining processes: ultrasonic machining, electrochemical machining, electrical discharge machining, and chemical machining. The first three methods utilize machining tools while the last process does not.

9.1.1 Ultrasonic Machining

Ultrasonic machining (USM) is an indirect abrasive process, in which hard, brittle particles contained in a slurry are accelerated toward the surface of the workpiece by a machining tool oscillating at a frequency up to 100 kHz. Through repeated abrasions (material removal), the tool machines a cavity of a cross section identical to its own (Fig. 1). The gap maintained between the tool and the workpiece is typically less than 100 μm.

The literature reports on a British patent issued in 1942 to L. Balamurth as the first design of a USM device. The period for the introduction of the first commercial machines was 1953–1954. Currently, modern USM machines can be used for the fabrication of complex cavity profiles through axial vibration and displacement (Fig. 2a), as well as two-dimensional profiles through a relative planar movement of the workpiece with respect to the machining tool (as in milling) (Fig. 2b).

USM is used primarily for the machining of brittle materials (dielectric or conductive): boron carbide, ceramics, germanium, glass, titanium carbides, ruby, and tool-grade steels. The machining tool must be highly wear resistant, as are low-carbon steels. The abrasives used in the slurry are the same as those used in most grinding wheels: boron carbide, silicon carbide, and aluminum oxide, or when affordable, diamond and cubic boron nitride. Abrasives (25–60 μm in diameter) are normally mixed with a water-based fluid (up to 40% by solid volume) to form the slurry, which may also act as a coolant, in addition to removing the chipped workpiece particles from the interface zone.

Various investigations have shown that higher material removal rates (up to 6 mm/min) can be achieved with (1) increased grain size (up to an optimal diameter) and concentration of abrasives in the slurry, and (2) increased amplitude and frequency of the oscillations of the tool. Increased material removal rates naturally result in increased tool wear rates. Furthermore, harder workpiece materials cause larger tool wear (tungsten

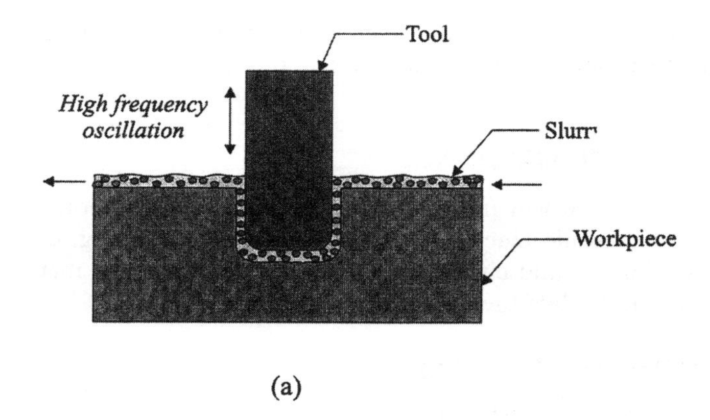

(a)

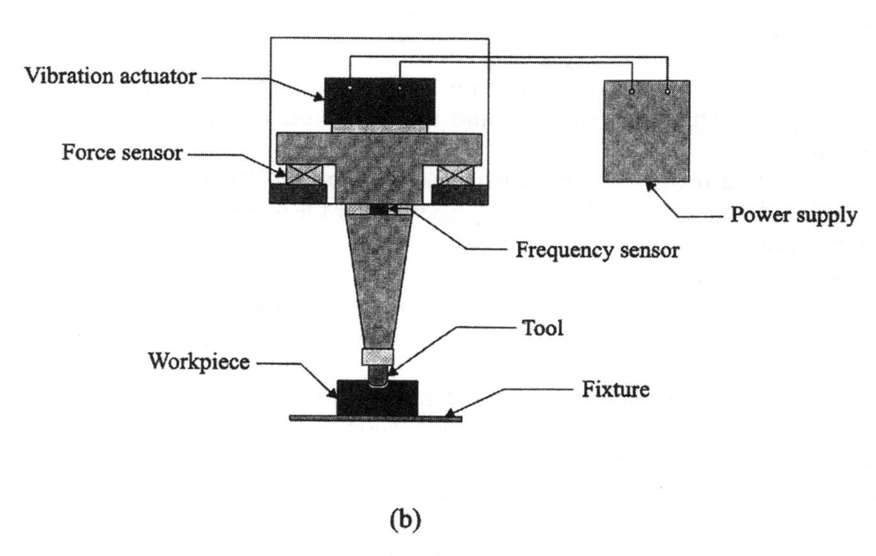

(b)

FIGURE 1 Ultrasonic machining (a) process and (b) device.

carbide versus glass). Surface finish in USM can be an order of magnitude better than that achievable through milling.

USM competes with traditional processes based on its strength of machining hard and brittle materials as well as on the workpiece geometry complexity. For example, via USM, we can fabricate holes (many at a time) of diameters as small as 0.1 mm. For such accurate holes, USM can be carried out in two steps, a rough cut and then a finer cut. Another typical

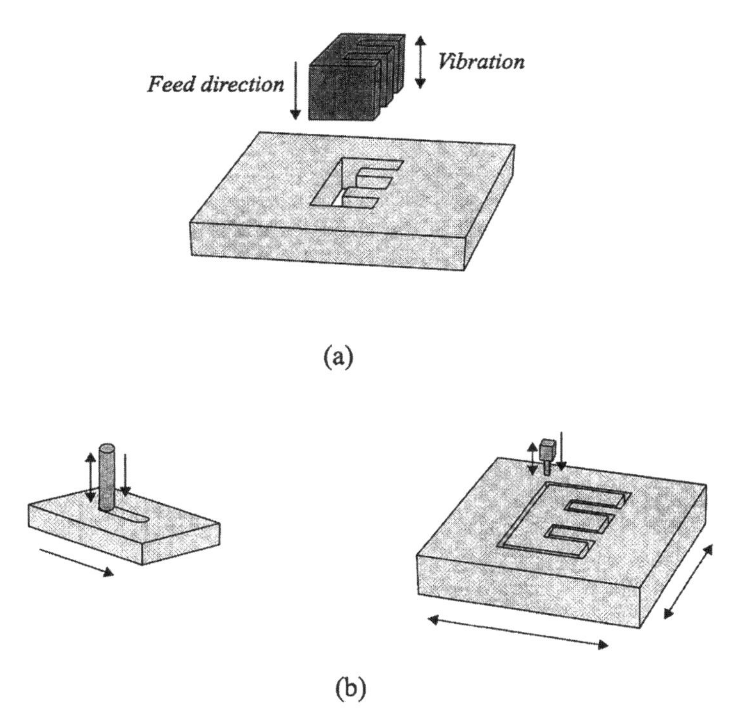

FIGURE 2 (a) Axial and (b) planar USM.

application of USM is the machining of dies of complex geometry to be used in metal forming.

Ultrasonic machine tools resemble small milling machines and drill presses in size and in operation. The major components of such machines are the vibration generator and the slurry storage and pumping unit (Fig. 1b). Those that provide planar motion for the workpiece have appropriate motion controllers as well. There also are some horizontal versions of ultrasonic machines.

9.1.2 Electrochemical Machining

Electrochemical machining (ECM) is a metal removal process based on the principle of reverse electroplating. Since Faraday's work in the early 1800s, it has been known that if two conductive materials are placed in a (conductive) electrolyte bath and energized with a direct current, particles travel from the surface of the anodic material toward the surface of the cathodic material. In ECM, the workpiece is made the anodic (positive) source and a machining tool is made the cathodic (negative) sink (Fig. 3). However, unlike in electroplating, a strong current of electrolyte fluid carries away the deplated material before it has a chance to reach the machining tool. The final shape of the workpiece is determined by the shape of the tool.

Although electroplating can be traced back to the discoveries of M. Faraday (1791–1867), application of ECM to metal removal was first reported in the British patent granted to W. Gussett in 1929. The commercialization of this process is credited to the U.S. company, Anocut

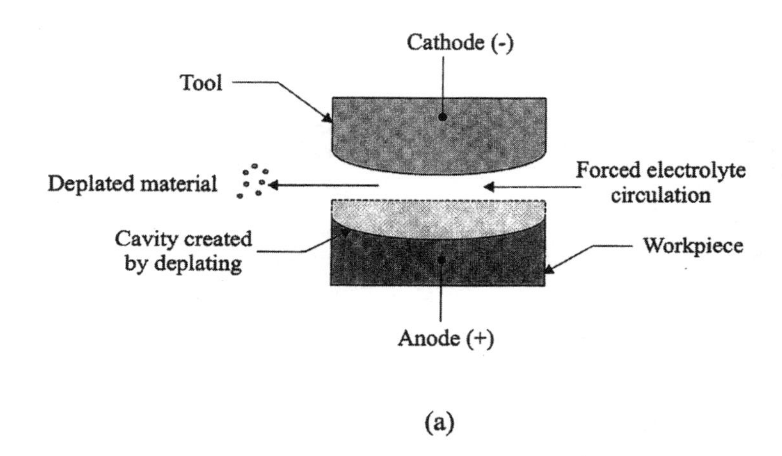

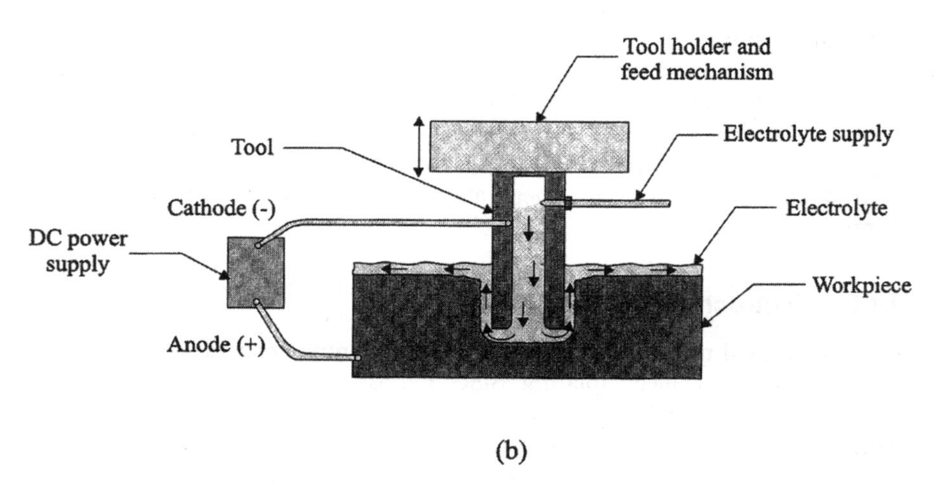

FIGURE 3 Electrochemical machining (a) process and (b) device.

Engineering, in the early part of the 1960s. Today ECM is one of the most widely utilized processes for the fabrication of complex geometry parts. Since ECM involves no mechanical process, but only an electrochemical one, the hardness of the workpiece is of no consequence.

All conductive materials are candidates for ECM, though it would be advantageous to use this costly process for the hardest materials with complex geometries. Also, since there exists no direct or indirect contact between the machining tool and the workpiece, the tool material could be copper, bronze, brass or steel, or any other material with resistance to chemical corrosion. It should be noted that, in certain applications, such as hole drilling, the side surfaces of the tool must be insulated to prevent undesirable removal of material from its surface (Fig. 4). Thus, in such cases, only the tip of the tool is utilized for deplating. The electrolyte must have an excellent conductivity and be nontoxic. The most commonly used electrolytes are sodium chloride and sodium nitrate.

The material removal rate in ECM (the highest of the nontraditional processes) is a direct function of the electrical power, the conductivity of the electrolyte, and the actual gap, maintained between the tool and the workpiece during the feed operation (a few mm/min). The larger the gap, the slower the removal rate will be, though short-circuiting is a danger when the tool and the workpiece come into contact

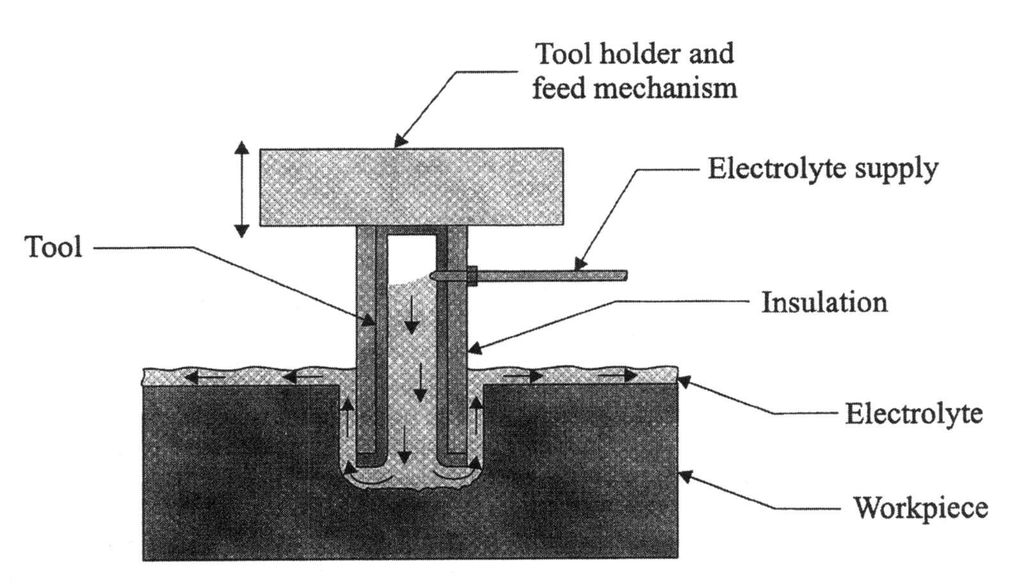

FIGURE 4 Insulation for ECM.

or are in very close proximity. Thus gap control is an important process parameter in ECM. The surface quality in ECM is worse than in ultrasonic machining but still much better than in milling.

ECM is a very versatile process that can be used for profiling and contouring, multiple hole drilling, broaching, deburring, sawing, and most importantly the fabrication of forging die cavities (die sinking) at rates of 10 times those achievable by electrical discharge machining. One must notice that owing to the nature of deplating, sharp corners are not machineable by ECM.

ECM machines exist in very large sizes, as well as in sizes of typical milling machines. They exist in horizontal and vertical configurations. ECM machines utilize 5 to 20 volts DC for deplating, though at current levels of up to 40,000 amps. Most modern ECM machines employ numerical control (NC)–based processors for the control of the workpiece motion with respect to the tool, as well as to regulate all other functions, such as the flow of the electrolyte.

9.1.3 Electrical Discharge Machining

Electrical discharge machining (EDM) is a metal removal process based on the principle of spark-assisted erosion. As in ECM, the workpiece and the shaped tool are energized with opposite polarity, 50 to 380 volts DC and up to 1,500 amps, in a bath of dielectric fluid. As the cutting tool (the electrode) is brought to the vicinity of the workpiece, electrical discharge, in the form of a spark, hits the surface of the workpiece and removes a very small amount of material. The frequency of discharge is controlled; it is typically between 10 and 500 kHz. This is a thermal process; the region of the spark reaches very high temperatures, above the melting point of the metal workpiece (Fig. 5).

The history of the modern EDM process can be traced to the independent work of two groups: B. R. Lazarenko and N. I. Lazarenko in Russia (in the former USSR) and H. L. Stark, H. V. Harding, and I. Beaver. Today EDM is one of the most widely used nontraditional metal cutting processes. It exists commercially in the form of EDM die sinking machines, wire cutting machines (EDWC) and grinding (EDG). In die sinking (Fig. 6a), a shaped electrode (cutting tool) is used to make complex geometry cavities or cutouts in metal workpieces. The workpiece can be of any hardness since there is no mechanical action–based cutting. As expected, the material removal rate is a direct function of the discharge energy and the melting temperature of the workpiece material. In wire-EDM (EDWC) (Fig. 6b), a small diameter (e.g., copper or tungsten) wire travels slowly along a prescribed contour and cuts the entire thickness of the

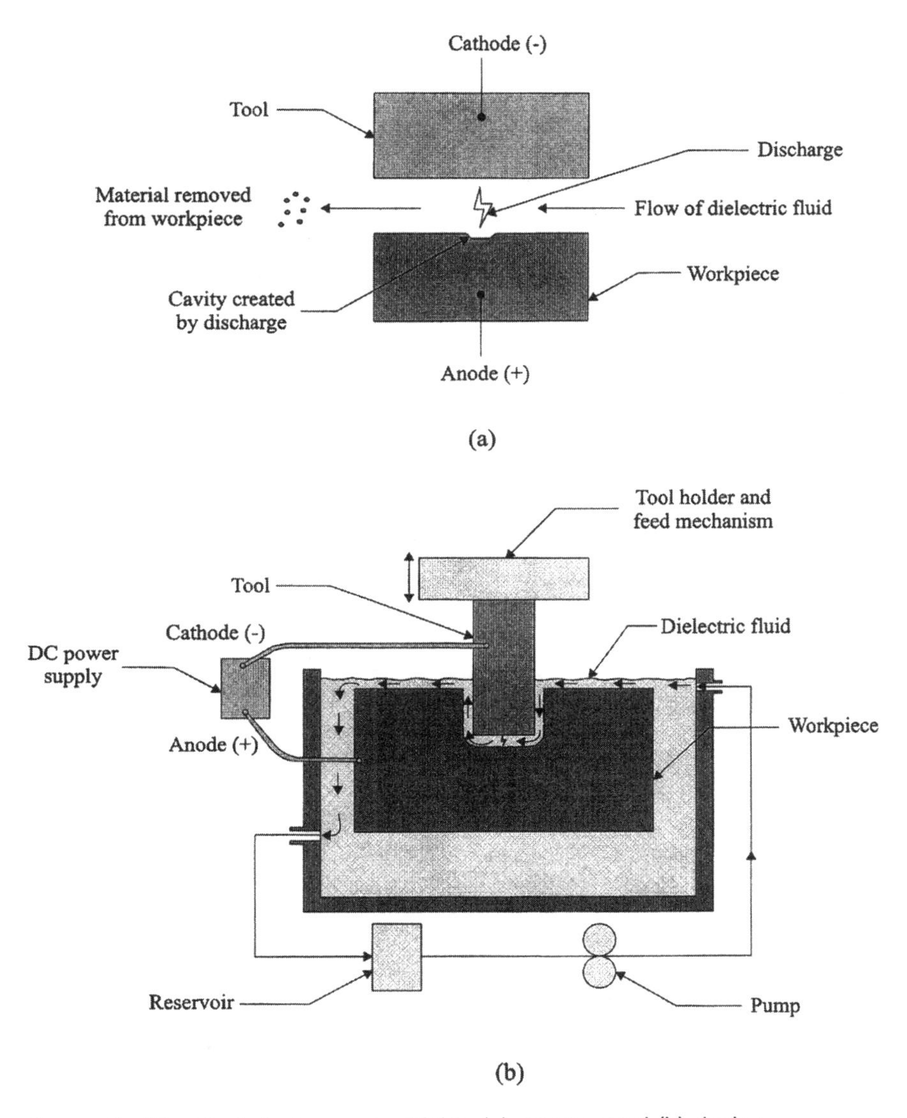

FIGURE 5 Electrical discharge machining (a) process and (b) device.

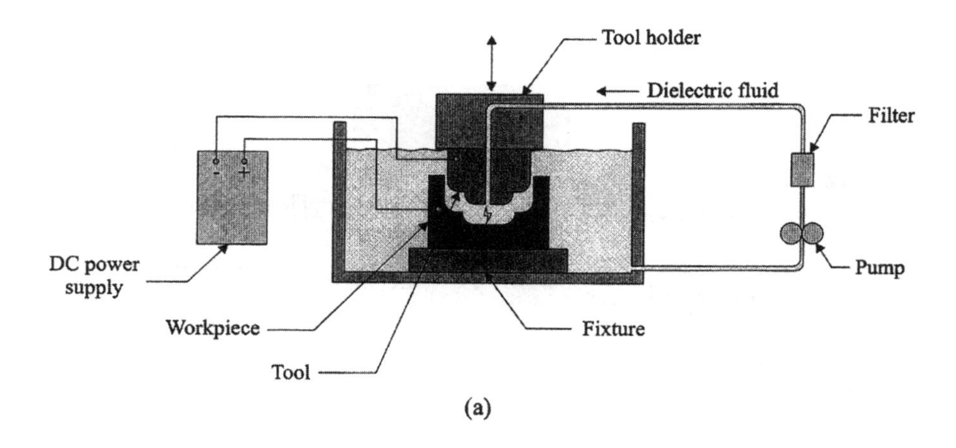

(a)

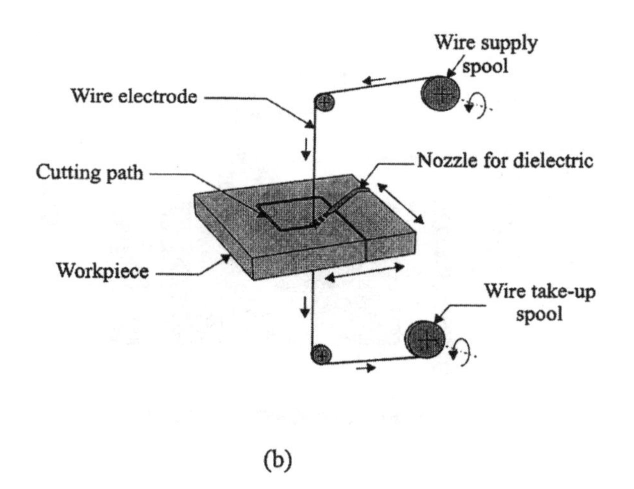

(b)

FIGURE 6 EDM (a) die sinking and (b) wire cutting.

workpiece (as in sawing) using the principle of spark erosion. This process can cut workpiece thicknesses of up to 300 mm with a wire of 0.16 to 0.3 mm. The lower the workpiece thickness, the faster is the feed rate.

A primary disadvantage of EDM is tool wear. Thus it is common to utilize several identical geometry cutting tools during the machining of one profile. These tools can be fabricated from the following materials using a variety of casting/powder processing/machining techniques: graphite, copper, brass, tungsten, steel, aluminum, molybdenum, nickel, etc. The principal tool wear mechanism is the same as the spark erosion mechanism that removes particles from the workpiece surface. The dielectric fluid (hydrocarbon oils, kerosene, and deionized water) is an insulator between the tool and the workpiece, a coolant, and a flushing medium for the removal of the chips.

Despite the above serious disadvantage, EDM can yield part geometries not achievable by other nontraditional processes, primarily because it can be configured into multiaxis cutting machine tools (small to very large). As in milling, the rotation of the cutting tool can be synchronized with a planar (X–Y) motion of the workpiece table to obtain a variety of profiles, including internal threads, teeth, etc. The surface finish achievable in EDM is comparable to ECM or slightly worse. Thus it can be utilized in the fabrication of both tools and dies (e.g., stamping/extrusion/molding dies) and individual parts.

9.1.4 Chemical Machining

The use of chemical etchants in the removal of material from metal parts' surfaces is commonly referred to as chemical machining (CHM). This process is based on the controlled removal of metal particles from a part's surface through targeted etching using acids and alkaline solutions.

The first generic step to all chemical machining processes is the creation of a mask on the surface of the workpiece that is resistant to the etchant used (the terms resist and maskant are interchangeably used to describe the thin film placed on the surface of the part). The next step is etching—material removal from the unprotected sections of the workpiece (Fig. 7).

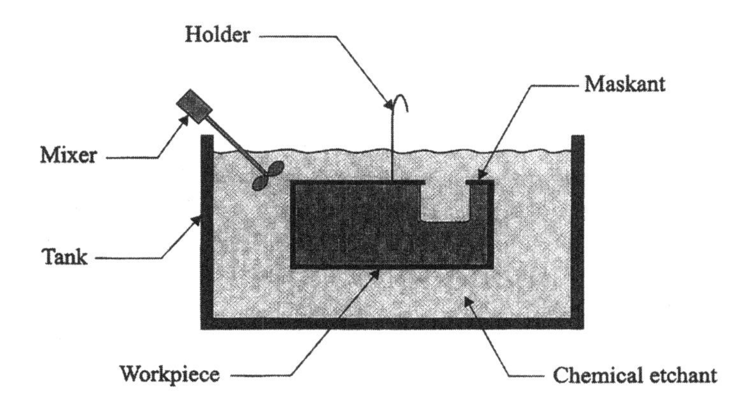

FIGURE 7 Chemical machining.

Chemicals have been used for many centuries in the engraving of decorative items and jewellery, as well as in printing and photography since the 17th century. The widespread use of etching in the manufacturing industry can be traced back only to the early 1940s, when airplane manufacturers started to use chemicals in removing material from airplane structural elements, primarily, for weight reduction. Today the electronics industry is probably the largest user of this technology in the fabrication of printed-circuit boards (PCBs) and integrated-circuit (IC) devices. Therefore in this section we will first review CHM for the fabrication of (relatively) large metal parts and then discuss photolithography for microscale manufacturing.

Chemical Milling and Blanking

In chemical milling, shallow cavities are etched on the surface of a metal workpiece—most commonly on large aerospace structures. In chemical blanking, through holes are blanked in thin plates by etching the unprotected locations (circular profiles) on the part from above and under.

Prior to the individual discussion of both above-mentioned processes, it would be beneficial to review the common set of issues:

Workpiece material: All metals are candidates for CHM. The most common ones include aluminum alloys, magnesium alloys, copper alloys, titanium alloys and steel alloys.

Maskants: Maskants and resists are commonly classified according to the technique utilized in their application and removal. Cut and peel maskants—very common in CHM milling—are applied via dipping or spray coating and removed by cutting (manually or by a laser) and peeling Photoresists—common in CHM blanking and electronics manufacturing—are applied via dipping, spray coating, or roll coating and removed via washing; screen resists are applied via screening (i.e., through a metal mesh placed on the part, which acts as a "negative") and thus there is no need to remove resists from areas to be etched. Naturally, although maskants come in a large variety, they must be utilized according to the material at hand and the etching method to be used: polymers/neoprene for aluminum alloys, polyethylene for nickel, neoprene for brass, and so on.

Etchants: The selection of an etchant depends on the workpiece material, maskant material, depth of etch, and surface finish required. Common etchants include sodium hydroxide (NaOH) for aluminum, sulphuric acid (H_2SO_4) for magnesium, and hydrofluoric acid (HF) for titanium. Material removal rates (i.e., etch rates) using these etchants can vary between 0.01 mm/min and 0.05 mm/min.

Chemical milling (Fig. 8) starts with the preparation of the workpiece surface: removal of residual stresses from the surface (e.g., through shot peening) and cleaning/degreasing. Maskant is applied next. Global

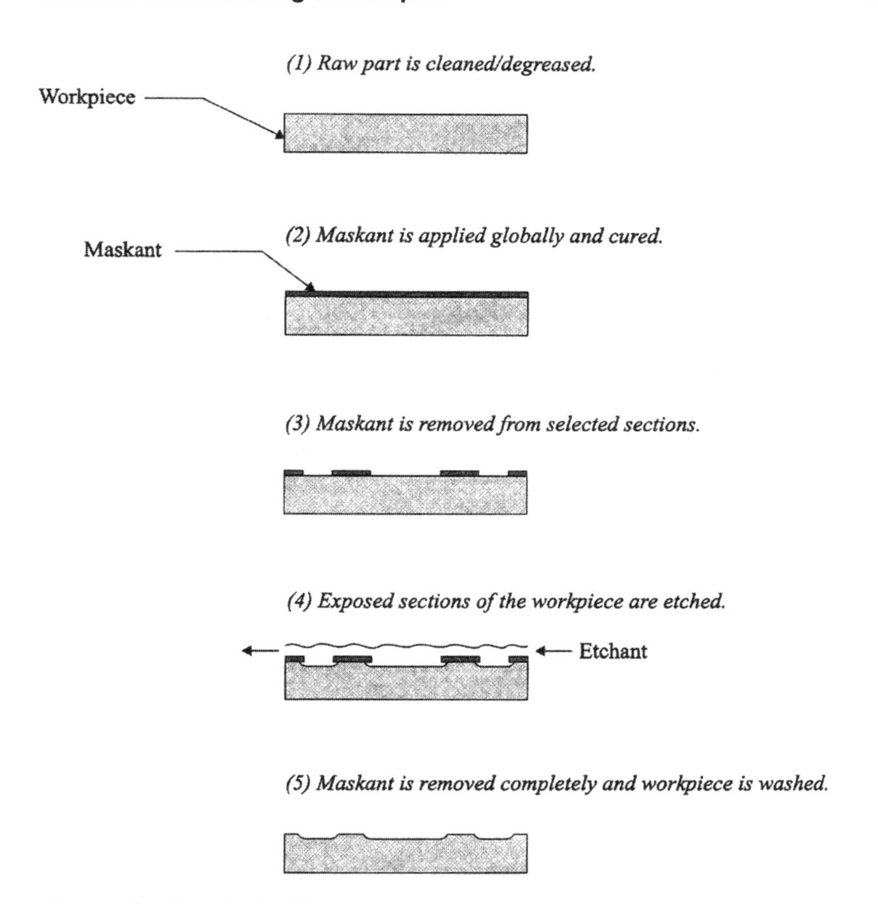

FIGURE 8 Chemical milling.

application and curing of the masking material is followed by the removal of necessary maskant sections (demasking). Exposed workpiece sections are etched using a flow of etchant. The last step is the removal of maskant from all unetched areas and the washing of the workpiece.

Chemical blanking follows the same process of chemical machining, except, in this case the etchant attacks the exposed metal surfaces of the thin workpiece (less than 0.75 mm) to fabricate simple through holes or complex cutout profiles (Fig. 9).

Chemical milling/blanking can be effectively utilized for the machining of airplane wing skins, helicopter vent screens, instrument panels, flat springs, artwork and so on. Although the equipment utilized is generally simple and easy to maintain, one must not underestimate the safety and

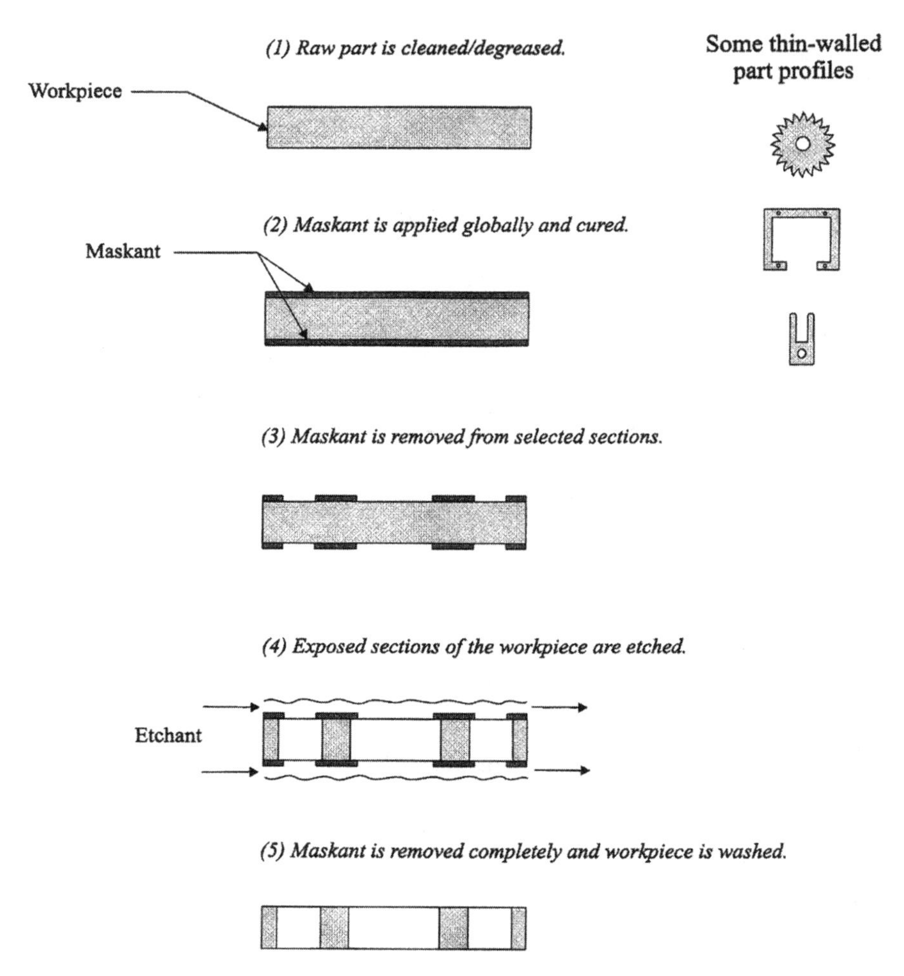

FIGURE 9 Chemical blanking.

environmental precautions necessary when dealing with highly toxic chemicals (maskants and etchants).

Microlithography

Lithography refers to transferring a pattern contained in a photomask into a photoresist polymer film through its curing and then utilizing this resist mask to replicate the desired pattern in an underlying thin conductor film. Microlithography refers to the lithographic process for the manufacturing of microscale patterns, normally in silicon-based wafers used in the fabrication of IC devices.

Although photochemical processes have been in existence for many decades, microlithography owes its start to the invention of the monolithic IC by J. Kilby and R. Noyce in 1960. Since that time, exponential increases in device densities on modern ICs necessitated corresponding innovative mass production techniques for microscale patterns. These techniques have included photolithography, x-ray lithography, electron beam lithography, and ion beam lithography.

The oldest *photolithographic technique* (prior to the 1970s) utilized a (chrome on glass) mask, pressed into contact with a photoresist-coated wafer, and flood exposure of the complete wafer with ultraviolet (UV) light for the curing of the (photopolymer) resist. A more robust technique, developed in 1973, projection lithography, uses optical imaging to reflect directly the photo mask onto the maskant/resist. This technology can yield features of 0.2 to 1.5 μm, only limited by the wavelength of the UV light source (200–450 nm).

X-ray lithography, in existence since the mid-1970s, can yield feature resolutions better than photolithography through the use of x-ray light sources in combination with suitable (polymer) resists. However, since no material is totally transparent to x-rays, mask fabrication is one of major disadvantages of this technique (Fig. 10). Furthermore, it is difficult to collimate or focus x-rays. Thus, owing to excellent resolution improvements in photolithography in the late 1990s, x-ray lithography may never become a commercial success.

Electron beam lithography has evolved from a basic technology utilized in scanning electron microscopy in the 1960s, to become a competing technique to photolithography in the early 1970s. Owing to their extremely short wavelengths (0.01 nm), electron irradiations can be utilized for high-resolution fabrication of IC devices (or the photomasks). However, such high resolutions (below 100 nm and frequently as low as a few nanometers) come with a very high price tag. Thus this technology is often called e-beam nanolithography, and it is primarily targeted for the fabrication of prototype ICs or nanoscale devices.

Ion beam lithography, researched in the late-1970s, offers the promise of better resolution than e-beam lithography, since ions scatter much less than electrons. However, this technology is dependent on the development of high-brightness energy sources, suitable lenses, and stable masks before it can become a commercially viable technique.

In lithography, as in other chemical machining techniques described in this section, the formation of a resist mask on the thin film conductor substrate is followed by an etching operation. The accurate transfer of the

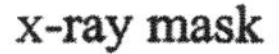

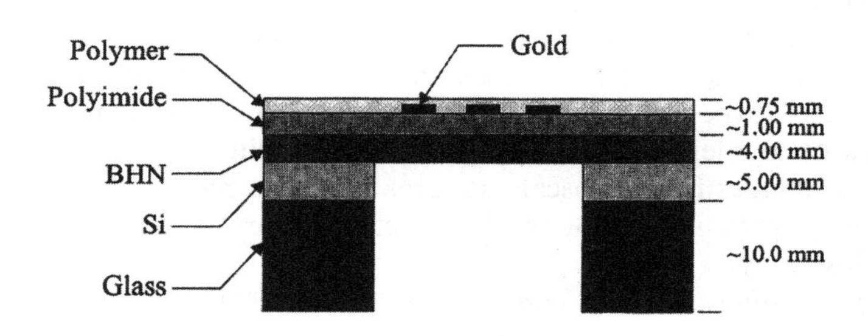

Optical photomask

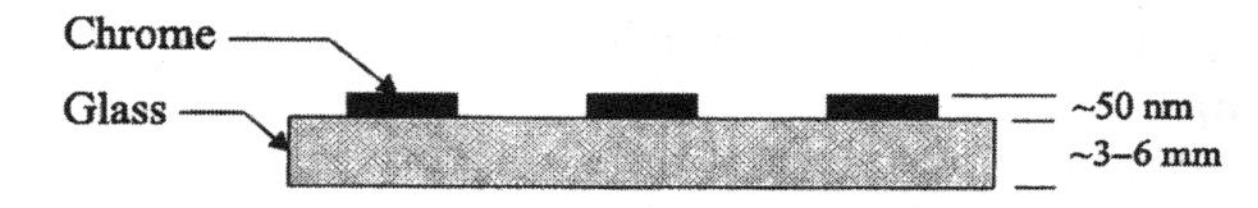

FIGURE 10 X-ray versus optolithography masks.

desired pattern onto the substrate (e.g., silicon, aluminum, silicon nitride) requires vertical side walls, smooth line edges, and no residues, accomplished at a rate that can be tolerated by the masking resist layer. The etching must be highly directional with no lateral etching (Fig. 11). This objective can be achieved via dry glow discharge, high-ion-density plasma etching. (A plasma is a partially ionized gas that includes electrons, ions, and a variety of neutral species—it can achieve metal removal rates of up to 1 μm per min.) Wet etching is undesirable, since it may cause the photoresist to lose adhesion and cause dimensional accuracy problems. Furthermore, dry etching lends itself to automation better than wet processes. Typically, chlorocarbon and fluorocarbon gases (e.g., CCl_4, CF_4) are used for etching metal films.

As a final step in microlithography, the resist layers are removed using O_2 plasmas; then there is a final cleaning process.

9.2 LASER BEAM MACHINING

Laser beam machining is a thermal material removal process that utilizes a high-energy coherent light beam to melt and vaporize particles on the surfaces of metallic and nonmetallic workpieces (Fig. 12). The term LASER is an acronym—light amplification by stimulated emission of radiation. As

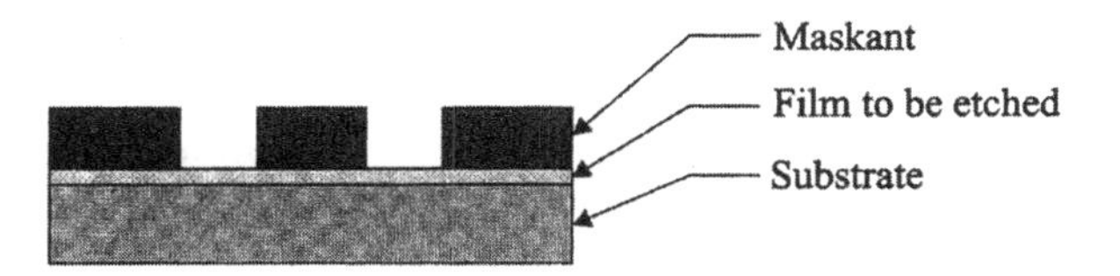

Plasma etching

FIGURE 11 Directional plasma etching.

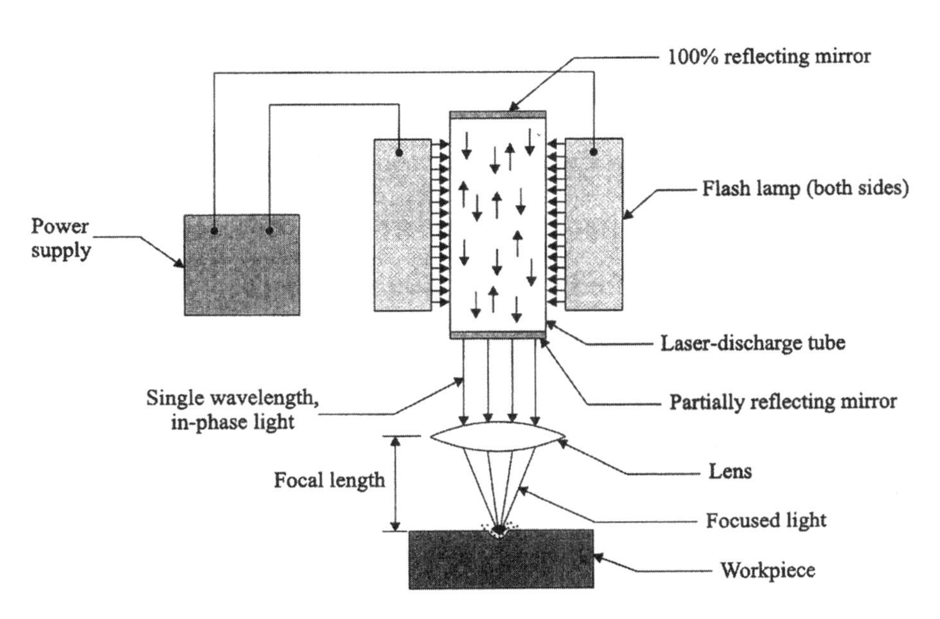

FIGURE 12 Laser beam machining.

the name implies, a laser converts electrical energy into a high-energy density beam through stimulation and amplification. Stimulation refers to the excitement of the electrons, which results in a stream of photons with identical wavelength, direction, and phase. Amplification refers to the further stimulation of the photons through an optical resonator to yield a coherent beam.

The study of light can be first traced back to Newton's work in the 1700s, who characterized it as a stream of particles, and later to Maxwell's work on electromagnetic theory. In the early 1900s, Einstein propounded the quantum concept of light, which lead to the theory of quantum mechanics in the early 1920s. The first device utilizing stimulated emission is attributed to J. P. Gordon, H. J. Zeiger, and C. H. Townes (1955). The first laser device is attributed to T. H. Mainman (1960). Most of today's modern laser devices were developed, subsequently, in the first half of 1960s, with the exception of the "excimer" laser developed in mid-1970s.

The three primary classes of lasers, classified based on the state of the lasing material, are gas, liquid, and solid. All lasers operate in one of the two temporal modes: continuous wave and pulsed. The three most commonly used lasers in manufacturing are

Nd:YAG: The neodymium-doped yttrium–aluminum–garnet ($Y_3Al_5O_{12}$) laser is a solid-state laser. Although very low in efficiency, its compact configuration, ease of maintenance, and ability to deliver light through a fiber-optic cable has helped it to be widely used (app. 25%) in the manufacturing sector. A Nd:YAG laser can provide up to 50 kW of power in pulsed mode and 1 kW in continuous wave mode.

CO_2: The carbon dioxide laser is a (molecular) gas laser that emits light in the infrared region. It provides the highest power for continuous wave mode operations (up to 25 kW versus 1 kW for Nd:YAG) and is the most commonly utilized laser source in manufacturing (though not with fiber optics).

Excimer: These short-wavelength gas lasers, though not as nearly as powerful as CO_2 or Nd:YAG lasers, can focus the light beam into very small spots. The term excimer is a shortened compound word for excited dimer, meaning two molecules (dimer) of the same (exci) molecular composition, such as H_2, O_2, N_2, and C_2. Common excimer lasers, however, have diatomic molecules of two different atoms, such as argon fluoride (ArF), and krypton chloride (KrCl). The laser vessel is normally prefilled with a mixture of gases, including argon, halogen, and helium.

9.2.1 Laser Beam Drilling

Like other solid-state lasers, Nd:YAG lasers are best suited for operation in a pulsed mode for maximum energy output. That is, energy is stored until a threshold value is reached and then rapidly discharged at frequencies of up to 100 kHz (but typically operated below 1 kHz for maximum pulse energy output). Owing to their preferred pulsed mode operation at high-energy outputs, Nd:YAG lasers are best suited for drilling operations (besides welding and soldering). They can, however, also be used for (continuous) contour cutting in continuous wave mode for low-power applications, whereas the CO_2 laser would be used for high-power applications.

In drilling, energy transferred into the workpiece melts the material at the point of contact, which subsequently changes into a plasma and leaves the region (Fig. 13). A gas jet (typically, oxygen) can further facilitate this phase transformation and the departure of material. (A pulsed mode laser can cause a microdetonation effect as repeated pulses hit the workpiece.)

Laser drilling can be utilized for all materials, though it should be targeted for hard materials and hole geometries that are difficult to achieve with other methods. For example, using a robot and an Nd:YAG laser, holes can be drilled at any inclination on a solid part without the need to orient the part—the robot end effector that carries the end point of a fiber-optic cable attached to the laser source can orient it with a five-degree-of-freedom mobility (x, y, z, ϕ, ψ) anywhere within its workspace (Fig. 14). Furthermore, laser drilling can also be used for superfast hole drilling, at rates above 100 holes per second, of diameters of as low as 0.02 mm, when using a fast-moving mirrors/optics arrangement Fig. 15b in Sec. 9.2.2 below. Large holes can be achieved by a "trepanning" approach, where the laser starts at the center of the hole and follows a spiral path that eventually cuts the circumference of the circle in a continuous wave mode.

Typical past examples of laser drilling have included small nickel–alloy cooling frames for land-based turbines, with almost 200 inclined holes; gas-turbine combustion liners for aircraft engines, with up to 30,000 inclined

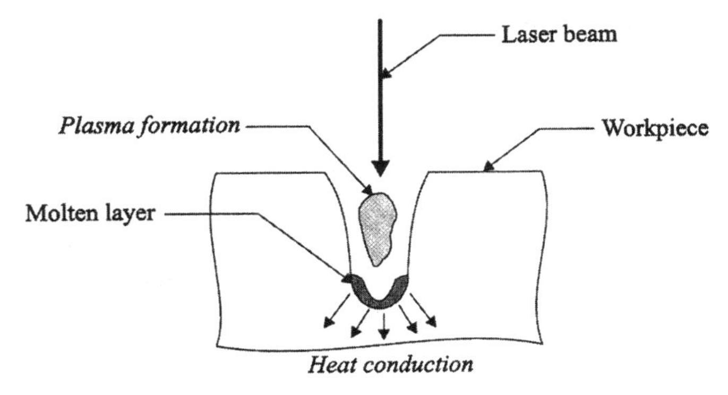

FIGURE 13 Interface region for laser drilling.

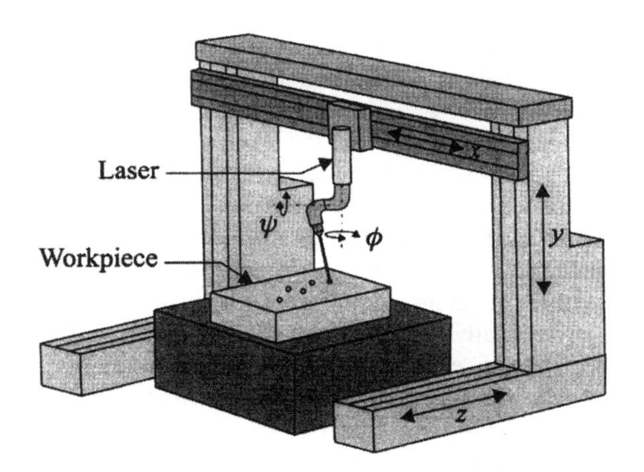

FIGURE 14 Five-degree-of-freedom laser drilling.

holes; ceramic distributor plates for fluidized bed heat exchangers; and plastic aerosol nozzles. Some more recent cases of laser drilling include bleeder holes for fuel-pump covers and lubrication holes in transmission hubs in the automotive industry, and fuel-injector caps and gas filters in the aerospace industry.

9.2.2 Laser Beam Cutting

The term laser cutting is equivalent to (continuous) contour cutting in milling: a laser spot reflected onto the surface of a workpiece travels along a prescribed trajectory and cuts into the material. Multiple lasers can work in a synchronized manner to cut complex geometries.

Continuous wave gas lasers are suitable for laser cutting. They provide high average power and yield high material removal rates and smooth cutting surfaces, in contrast to pulsed-mode lasers that create periodic surface roughnesses.

The CO_2 laser has dominated the laser cutting industry since the early 1970s. Since CO_2 lasers cannot be easily coupled to fiber-optic cables, CO_2-based cutting systems come in three basic configurations: moving laser, moving optics, and moving workpiece (Fig. 15). Moving laser systems cannot operate at large speeds and are normally restricted to flat-sheet cutting. Moving optics systems can provide cutting speeds in excess of 100 m/min (owing to fast moving mirrors) and can machine three-dimensional static or moving workpieces. Moving workpiece systems are equivalent to

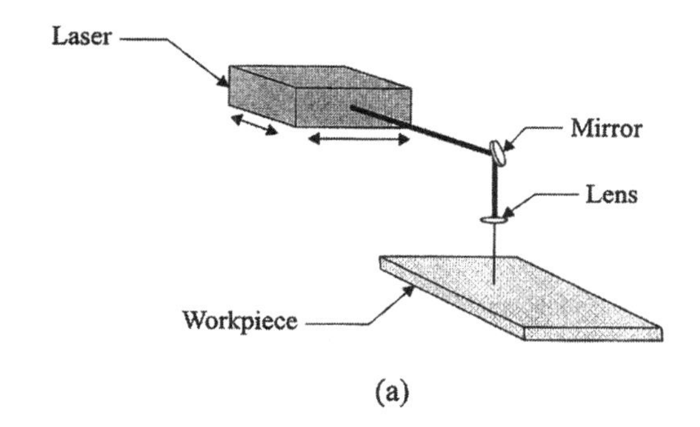

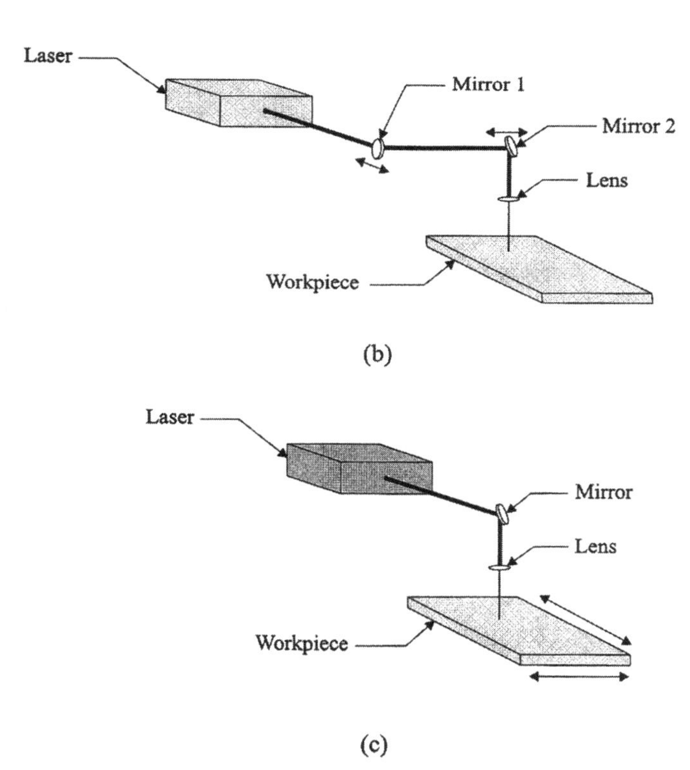

FIGURE 15 Continuous laser cutting: (a) moving laser; (b) moving optics; (c) moving workpiece.

traditional turning (rotational machining) and milling (prismatic machining) machine tools.

The material removal mechanisms in laser cutting are similar to those in laser drilling: in steady-state operation, the input energy is balanced primarily by the conduction energy that melts and vaporizes the material. As the light beam moves forward, a continuous molten (erosion) front forms because of high temperature gradients (Fig. 16). The kerf (narrow slot) left behind has parallel walls: for thin-walled metal workpieces the kerf width is typically less than 0.5 mm, so that there is very little material waste. In a large number of cases, fast-flowing gas (e.g., oxygen) streams are utilized to assist laser cutting: they remove material and keep the focusing lens clean and cool.

Although most metals can be cut by lasers, materials with high reflectivity (e.g., copper, tungsten) can pose a challenge and necessitate the application of an absorbent coating layer on the workpiece surface. Furthermore, for most metals, the effective cutting speed exponentially decreases with increasing depth of cut.

CO_2 and Nd:YAG lasers can also be utilized in the cutting of ceramics and plastics/composites. Overall, typical industrial applications of laser cutting include removing flash from turbine blades, cutting die boards, and profiling of complex geometry blanks.

Analysis of Laser Cutting

Over the past two decades a large number of studies have been reported in the literature on the analytical and numerical analysis of laser cutting. As mentioned above, laser machining is a thermal process, where radiant

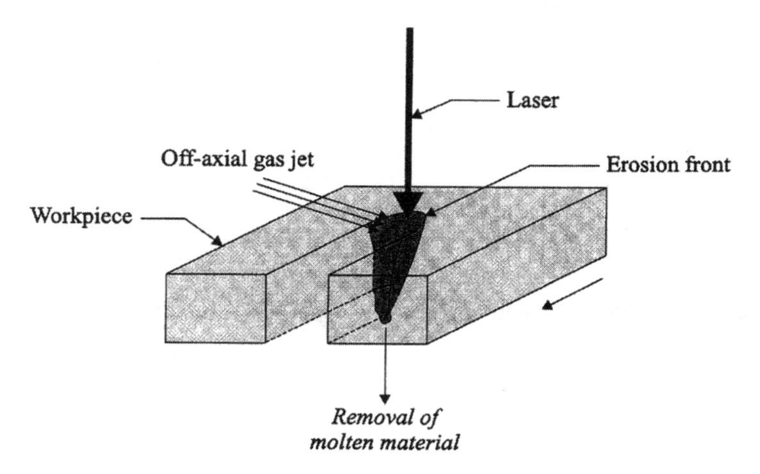

FIGURE 16 Material removal in laser cutting.

energy from a laser light source is utilized for material removal. As this radiant energy is absorbed by the workpiece material, the local temperature rises, leading to melting and vaporization (i.e., dissipation of heat). Although the contact region is a multiphase environment, solid, liquid, and gas, most studies have concentrated on the dissipation of heat in the solid workpiece via conduction and through the surroundings via convection (Fig. 17).

Numerical methods, commonly based on finite element analysis (FEA), are utilized to determine optimal cutting parameters for a given part material and material removal task: laser power, spot size of the laser light (as it hits the surface of the workpiece), depth of cut (for grooving tasks), and cutting speed. Naturally, a desired optimization objective would be the maximization of cutting speed—though surface quality (smooth cutting profile) and surface integrity (minimum residual stresses) can also be considered in selecting cutting parameters. One must realize that pulsed mode lasers can be modeled in the same way as continuous wave mode lasers, in which the input of energy is modeled as a function of time. In such cases, the pulsing frequency (when adjustable) becomes another cutting parameter whose value is optimized.

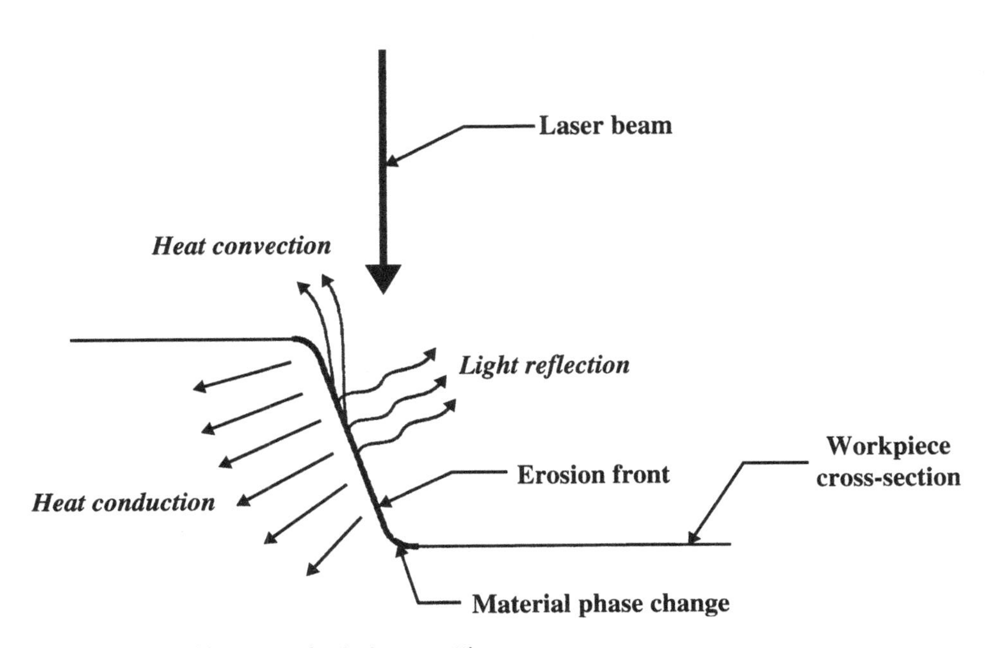

FIGURE 17 Heat transfer in laser cutting.

9.3 RAPID LAYERED MANUFACTURING

A number of material additive methods have been developed during the period of 1985 to 2000 for the rapid, layered fabrication of one-of-a-kind parts. Collectively called rapid prototyping (RP) manufacturing techniques, most such commercial systems can only fabricate parts (plastic, metal, or ceramic) for limited engineering analysis and testing purposes. On the other hand, a common primary advantage to these RP techniques is their utilization of computer-aided design (CAD) part models, where significantly shortened production times of fabrication can be obtained for complex geometry parts with almost no manual intervention during the build phase.

As will be noted later in this section, the underlying principles of today's RP techniques have been in existence and in commercial use for several decades, though they are now applied in a novel manner for the building of three-dimensional prototypes using readily available parts' CAD solid models. These principles include photolithography, sintering, and laser cutting.

Brief History

The key concept of layered manufacturing, also known as solid free-form fabrication (SFF), is decomposition of a three-dimensional CAD solid model into thin (virtual) cross-sectional layers, followed by physically forming the layers, using one of the material additive RP techniques, and stacking them up layer by layer. The creation of three-dimensional parts in such a layered fashion can be traced to the creation of the Egyptian pyramids more than 3,000 years ago. In more modern times, however, layered fabrication has been primarily used for topography and photosculpture.

The use of layered techniques in topography can be traced back to the early 1890s, when Blanther suggested a layered method for making a mold for the fabrication of topographical relief maps by stacking up layers of wax plates (cross sections cut to proper dimensions) and manually smoothing the curvatures in the third dimension. After creating the positive and negative halves of such a mold, a printed paper map is pressed between them to create a raised relief map. A similar lamination technique has also been long used by architects and urban planners for the creation of relief maps or structures by cutting contour lines on cardboard sheets and then stacking and pasting them to form three-dimensional representations.

Photosculpture refers to the creation of three-dimensional replicas of parts or human forms. The concept can be traced back to the work of F. Willème in the 1860s; he simultaneously used 24 cameras (equally spaced on the circumference of a circle) to photograph a figure and then create its replica by creating 24 individual parts of the figure and assembling them.

Although, as discussed in Section 9.1.4, photolithography has been successfully used for many decades in the creation of masks in chemical machining, H. Kodama of Japan is credited with the first reporting of a photopolymer-based, layered rapid prototyping system in 1981. This early attempt is very similar to today's commercial systems in using a UV light source for layer solidification and consecutive layer buildup by controlled submersion of the already built section of the part into a vat of liquid polymer. Within a decade of this report, the RP industry blossomed by the development and commercialization of numerous techniques. The first was the stereolithography apparatus (SLA) by 3D Systems, in around 1985; next came the selective laser sintering (SLS) system by DTM, the laminated object manufacturing (LOM) system by Helisys, and the fused deposition modeling (FDM) system by Stratasys, to mention a few.

In this section, our emphasis will be on the two most common RP techniques: stereolithography for the fabrication of plastic parts, and selective laser sintering for the fabrication of primarily metal parts. As a preamble to the detailed presentation of these two methods/systems, we will, however, first review CAD data preparation as a common task to all RP techniques.

9.3.1 CAD-Based Part Data Preparation

Layered manufacturing techniques rely on the input of a CAD system regarding the geometric information of every layer to be produced. Thus the first step in every RP process is the creation of the geometric solid model of the part to be utilized in the determination of layer data, also known as slice data. Once such a solid model of the part is available, the next step is process planning: we must determine the orientation of the part build (i.e., choose the vertical axis of build that will be normal to the planar layers), and then slice the part model into layers of desired thickness. The last step in CAD data creation is verification of information and translation into a format to be understood by the RP machine. These steps are individually addressed below.

Step 1: Solid Model of Part

The solid model of the part to be built by the RP method can be generated using any one of the available commercial CAD packages. As expected, the CAD data must represent a valid geometric model, whose boundary surfaces enclose a finite volume.

Step 2: Orientation of Build

Determining the orientation of the build is the most important task in process planning for layered manufacturing. Three criteria are normally considered during the selection of an optimal build direction (i.e., part orientation): surface quality and dimensional accuracy, mechanical strength of the finished part, and fabrication time:

Due to the staircase effect of layered building, the curved surfaces of a part must be positioned optimally with respect to the build direction in order to minimize deviations from the desired curvature (Fig. 18). (For example, a cylindrical part built by one circular cross section stacked on top of the other would have a smooth side surface, whereas the same part built by stacking up rectangular cross sections would not.)

Since layered (laminated) structures have transversely isotropic mechanical properties (i.e., isotropic in the plane of the layer, but different in the third orthogonal build direction), the build direction (i.e., the part orientation during the build) must be chosen according to the expected future loading conditions of the part. The primary mode of failure of layered parts is assumed to be delamination due to excessive transverse shear stresses: the building direction in laminated parts has the weakest mechanical properties owing to weak interlayer bonding. FEA can be used to determine the stresses that develop in such parts.

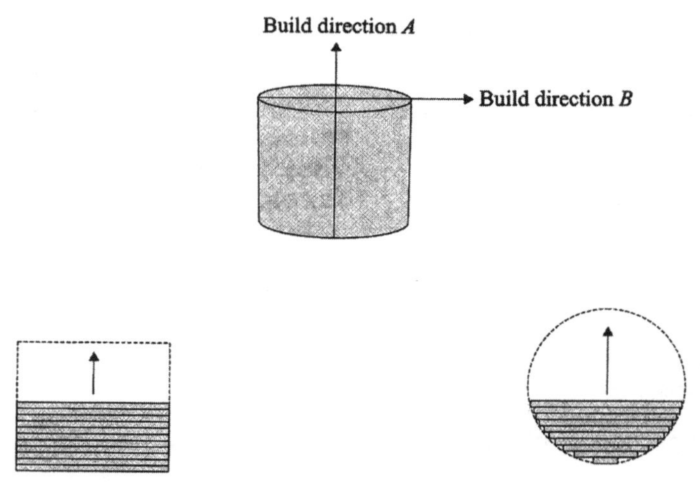

FIGURE 18 Part orientation.

When the time spent on preparing a layer for solidification is more than the build time spent on fabricating (solidifying) it, one would be advised to minimize the number of layers by choosing an appropriate build direction, and vice versa. (For example, it would be faster to build a rectangular part with a build direction that is normal to the surface with the greatest surface area.)

Frequently, however, process planners may have to choose the build direction (part orientation) subject to all of the above criteria and others simultaneously. In such cases, one could formulate the optimization problem along the guidelines provided in Sec. 5.3 of this book.

Step 3: Slice Data

The last step in data preparation for layered fabrication of parts is the slicing of the geometric solid model of the part in order to determine the outer boundaries of individual cross sections to be solidified and stacked on top of each other. The principle of slicing is to intersect parallel planes (orthogonal to the build direction) with the geometric model of the part and determine the intersection contours between these planes and the surfaces of the part (Fig. 19).

There are numerous techniques for slicing CAD models. The most common (yet the most inaccurate) method requires solid models (generated via B-Rep or CSG techniques) to be converted into tesselated representations. In a tesselated representation, all part surfaces (curved or planar) are approximated by planar triangular elements (surfaces) yielding the commonly known STL (stereolithography) file (Fig. 20). The sizes of the triangular elements (facets) can be automatically determined by commercial CAD package translators according to a tolerance value provided by the user (i.e., maximum allowable deviation of the planar facet from the actual curved surface—along the surface normal direc-

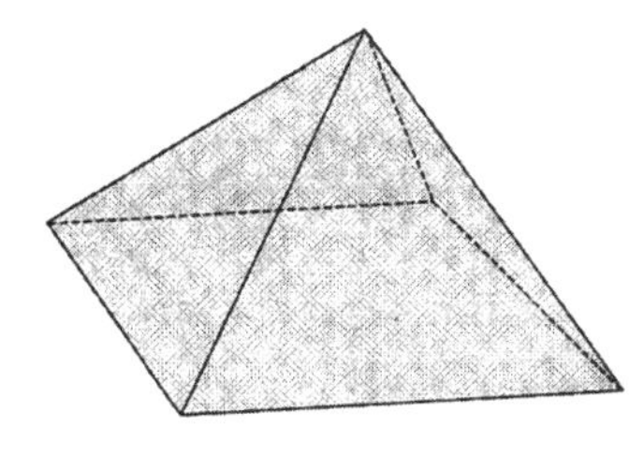

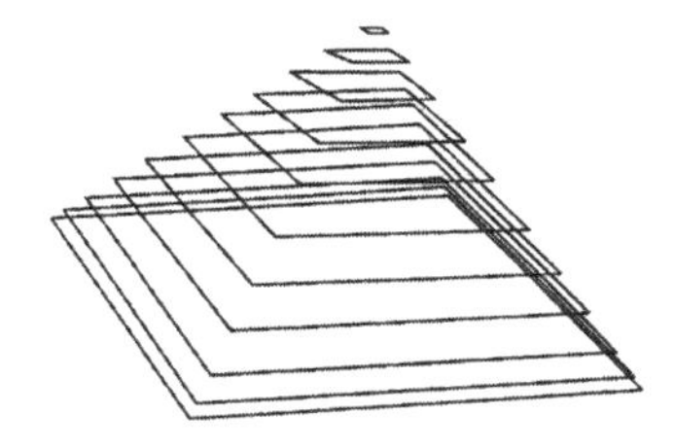

Figure 19 Slicing for RP.

FIGURE 20 Tesselated object model.

tion). High accuracy requirements naturally yield very large STL files (possibly with millions of facets), where the slicing process (i.e., the intersection of the planes with the tesselated part model) would accordingly take a long time and be prone to errors.

An alternative method to generating STL files through tesselation is direct slicing of the solid model of the part and the use of NURBS (nonuniform rational B-spline) curves for approximating the intersection contours.

The cutting planes used for the generation of slice data are normally placed at equal intervals, corresponding to the constant thickness of the layer to be solidified by the RP technique. However, one can place these planes at variable distances by employing a variable layer thickness method. Larger distances (i.e., thicker layers) can be used for those sections of the part with vertical side surfaces, as limited by the power of the energy source, and smaller distances (i.e., thinner layers) for those with curved side surfaces (Fig. 21).

Once the slice data has been obtained, the RP system would determine a raster plan (i.e., path of scanning), if it uses a laser beam–based solidification/cutting source. The objective is minimum time spent on solidifying/cutting the layer at hand while ensuring dimensional integrity. This subject is RP technique dependent and thus will be detailed further in the following sections when we address the different layered manufacturing techniques.

9.3.2 Layered Manufacturing Techniques

Layered manufacturing techniques have been classified over the past decade according to the energy source, material type, material phase,

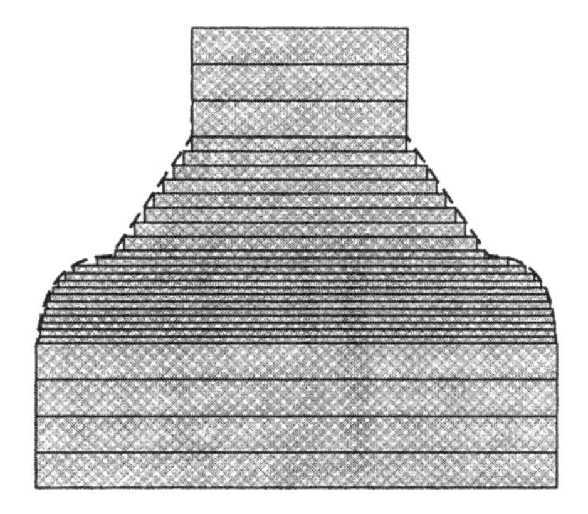

Figure 21 Variable layer thickness in RP.

and so on. Herein, RP techniques are grouped according to the principle of part fabrication that they utilize: lithography, powder processing, deposition, and cutting. Since these techniques are mostly targeted for the prototyping of parts of very different materials, we will not attempt to say that one is a better technique than another. The discussions will, though, attempt to include information on their specific applications in different manufacturing industries.

Lithography-Based Methods

Lithography-based layered manufacturing techniques rely on the selective solidification of a liquid monomer via UV light. The four major commercial systems are the stereolithography apparatus (SLA), U.S.A., the Stereos, Germany, the solid creation system (SCS), Japan, and the solid ground curing (SGC), Israel. The first three utilize a laser light source and are very similar in their fabrication of the layered polymer parts. The SGC system, on the other hand, is more loyal to the original photolithography systems developed several decades earlier: It uses a mask that blocks the flooded UV light from hitting undesirable regions on the liquid layer surface.

Laser lithography: Techniques that employ utilize laser lithography build parts on a platform attached to an elevator, whose vertical motion is accurately controlled. The platform is placed in a vat of photopolymer (typically acrylic- or epoxy-based): the part is built in a bottom-up approach using selective curing by a UV laser (Fig. 22).

The first step in laser lithography is the preparation of the liquid layer: The elevator is lowered into the vat (by a distance equal to the thickness of

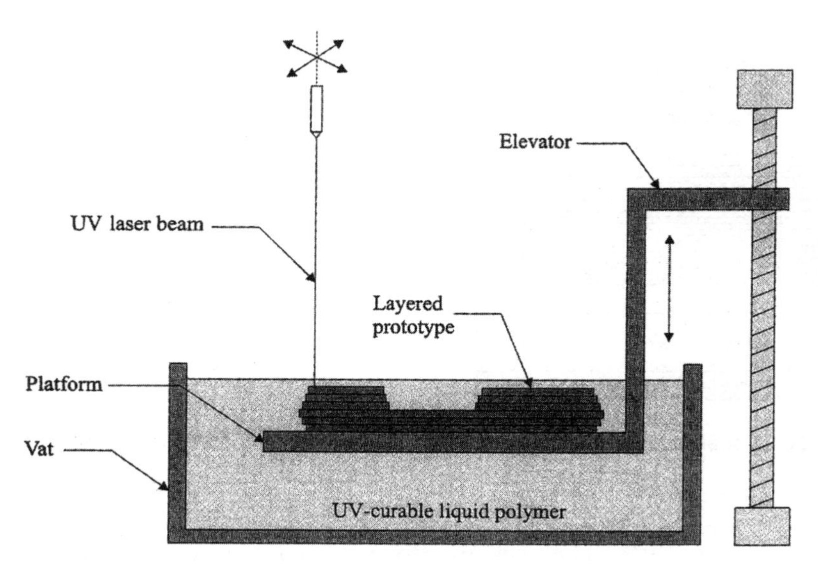

FIGURE 22 Laser lithography.

the layer to be solidified) and liquid polymer is spread (over the previously solidified part) using a coating/wiping mechanism. Next, a UV laser beam scans the surface (according to a given raster pattern) to cure the necessary regions selectively. Light beam delivery is normally achieved by using fast and very accurate servocontrolled galvanometer mirrors. The two-step process is repeated until all the layers of the part are built. The "green" part is then removed from the vat, washed, and placed in a postcuring oven for final solidification via flood-type UV light and heat.

Photolithography: Techniques that utilize photolithography for the layered manufacturing of three-dimensional parts differ from laser lithography systems, in principle, only in the method they use to solidify the photopolymer layers. Instead of laser-based selective pointwise curing, these techniques flood the liquid layer's surface with UV light (from a strong source) that passes through an erasable/replaceable mask. In the SGC system, for example, the mask is produced by charging a glass plate via an ionographic process and then developing the negative image (of the layer to be solidified) with an electrostatic toner, just as in photocopying processes (Fig. 23).

A feature unique to SGC is the removal of excess photopolymer from the vat once a layer has been solidified and its replacement with liquid wax which is allowed to solidify before the build of the next layer (Fig. 23). The advantage of this step is the formation of a solid fixturing structure around

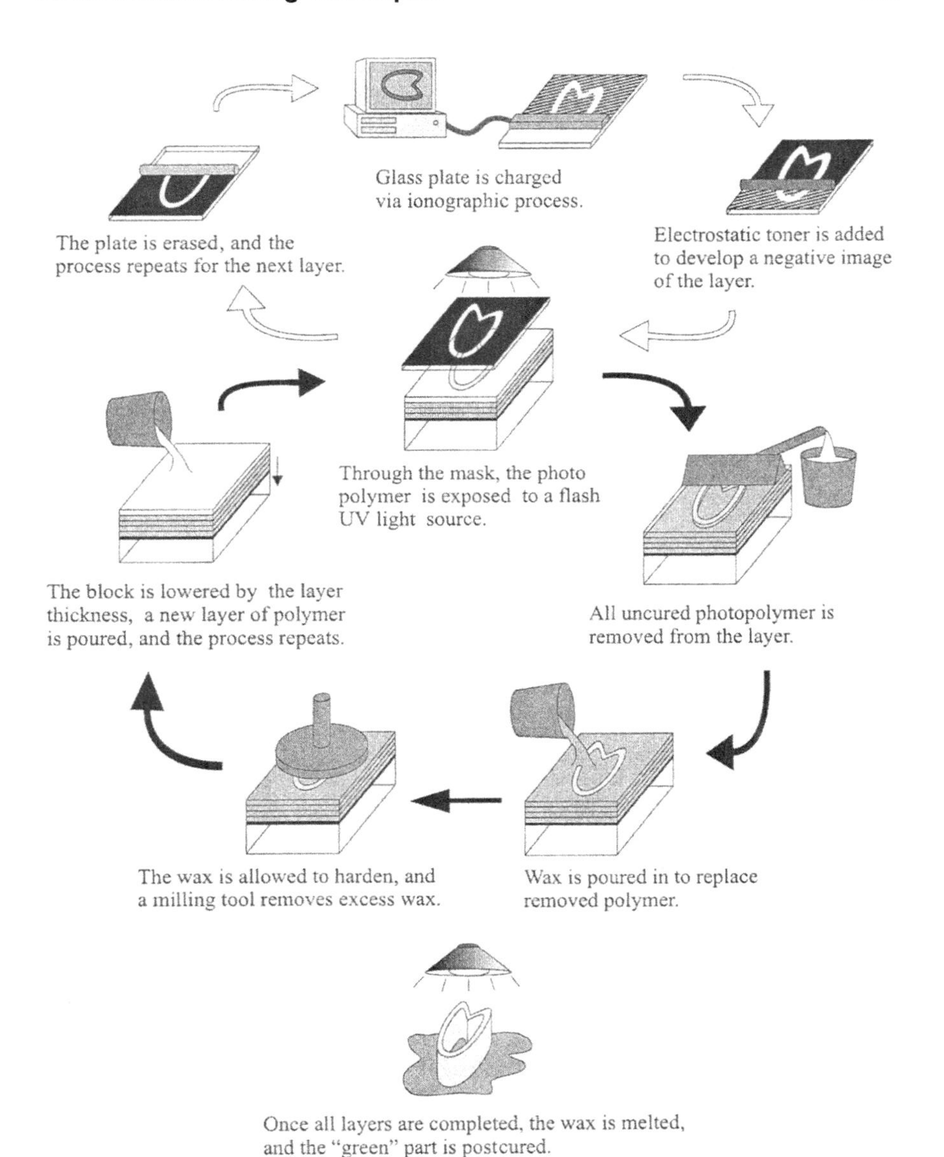

FIGURE 23 Photolithography for RP.

the part (built in a layered manner), which ensures the part's stability in a vat of liquid, as is not the case with other lithography-based commercial systems. In SGC, once the wax has solidified and thus a totally solid structure (that includes the polymer part) is formed in the vat, a milling tool is utilized to remove excess wax and prepare the system, after the lowering of the elevator by a layer thickness, for the deposition of the next photopolymer layer. Once the iterative process of building the entire part has been accomplished, the wax is melted and the green part is postcured.

As one would expect, the layer formation step in SGC systems is quite time consuming, as opposed to the rapid solidification of the photopolymer layer. Thus, as discussed earlier, for these systems it would be beneficial to select a build direction (part orientation) that minimizes the number of layers to be built.

Powder Processing Methods

Layered manufacturing techniques that rely on powder processing methods differ primarily in the way that they bind the particles (plastic, metal, or ceramic). The two major commercial systems are selective laser sintering (SLS), U.S.A., and three-dimensional printing (3DP), U.S.A. The former utilizes a CO_2 laser for sintering the particles in selective regions of a vat of powder, whereas the latter uses ink-jet printing technology in depositing a binding material to "glue" the particles together.

Laser sintering: In laser sintering–based systems, parts are built in a chamber that can be lowered in a controlled manner for the formation of thin powder layers (Fig. 24). Typically, the following steps are followed: (1) The powder bed is lowered by the desired layer thickness; (2) a roller is used to drag a controlled amount of powder from an external material source and spread (and compact) it in the build chamber; (3) a laser is used to scan selectively the necessary regions of the deposited layer in order to bind the particles into a desired cross-sectional geometry and to bind the new layer with the old one. The layer formation and solidification steps are repeated until the entire layered part is manufactured.

Binder sintering: The 3DP process, also known as the Direct Shell Production Casting (DSPC) process, is the best known commercial binder sintering technique. The process is similar to laser sintering of powder particles, with the exception of the binding agent: a fluid binder is injected/sprayed selectively onto the powder, instead of using a CO_2 laser heat source (Fig. 25). The binder droplets are deposited using a similar technique used by the scanning head of ink-jet printers. As in selective laser sintering, the surrounding extra unbound powder provides the part with a rigid support during the build and is removed (shaken away) from the part once the

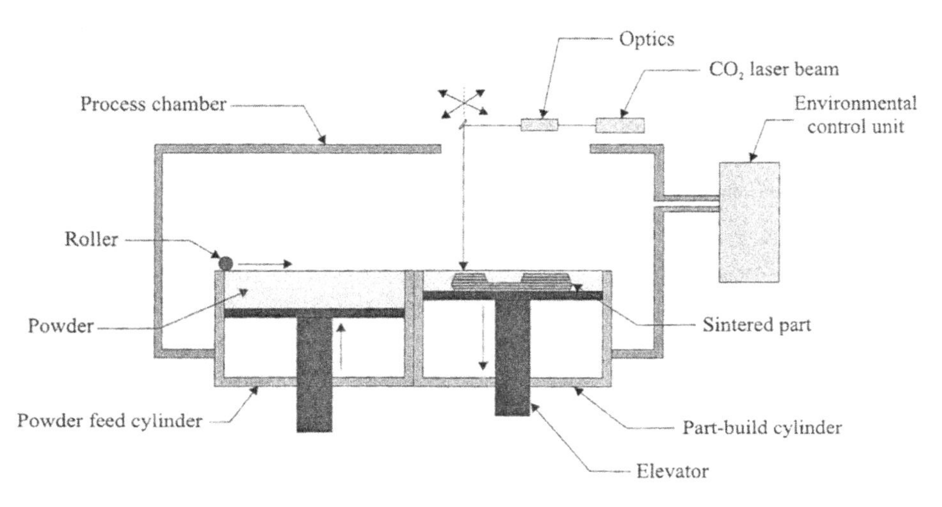

FIGURE 24 Laser sintering.

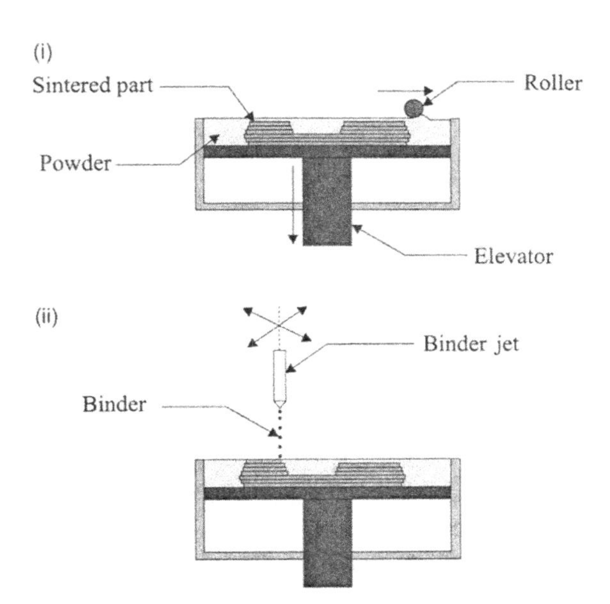

FIGURE 25 Binder sintering. (i) The elevator is lowered, and a roller is used to drag a controlled amount of powder from an external material source and spread it in the build chamber. (ii) A fluid binder is selectively sprayed onto the powder.

fabrication process is over. The green part is then treated in appropriate ovens for further densification. Typical powder materials used include stainless steel, tungsten, tungsten carbide, and ceramic alloys, which are bound with colloidal silica or polymeric binders.

Deposition Methods

Deposition-based RP techniques selectively deposit molten material for the direct building of layered parts. Although there have been many academic attempts in the past decade (the 1990s), only a very few commercial systems exist today: fused deposition modeling (FDM), U.S.A. model maker, U.S.A., and ballistic particle manufacturing (BPM), U.S.A. All three systems utilize (low-melting-point) thermoplastics. While the first method (FDM) uses a direct contact deposition (extrusion) of molten plastic, the other two use ink-jet type print heads for selective scanning of plastic droplets (similar to 3DP's binder deposition).

In FDM, a continuous filament of thermoplastic polymer (e.g., polyethylene and polypropylene) or investment casting wax is fed into a heated extruding head. The filament is raised about 1 °C above its melting point and directly deposited onto a previously built layer by the x–y scanning extruding head (Fig. 26). Once a layer is formed, the part (built on an elevator platform) is lowered by the thickness of the next layer. Solidification of molten material lasts about 0.1 sec, and the layered part needs no further processing.

Cutting Methods

Cutting-based methods, also known as lamination processes, use laminates of paper sheets or plastic or metal plates as the raw material for layer formation, a binding agent for gluing the layers together and a cutting mechanism (a laser or a mechanical cutter) for selectively cutting the contours of the layer's geometrical x–y profile. Current commercial systems include laminated object manufacturing (LOM), U.S.A., solid center (SC), Japan, and hot plot, Sweden.

In LOM, the process starts with the lowering of the platform by the layer thickness and the deposition of an adhesive binding agent on the previous layer (unless the paper has already been impregnated with heat-activated adhesive) (Fig. 27). A new layer (laminate) is rolled and pressed over the previous layer for good adhesion. Then a CO_2 laser is utilized to cut the contours of the layer selectively and crosshatch the remaining excess material for its easy removal once the part has been completely built. Postprocessing activities in LOM include sanding, polishing, painting, and sealing for moisture resistance (via urethane, epoxy, or silicon sprays).

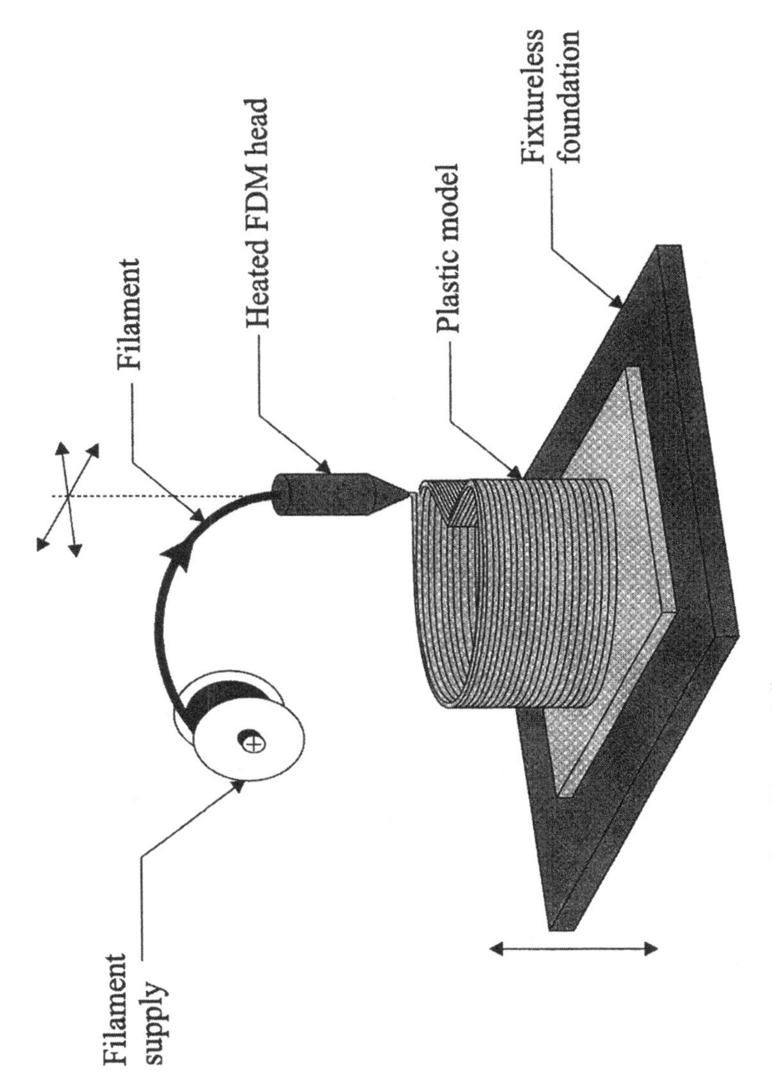

FIGURE 26 Fused deposition modeling.

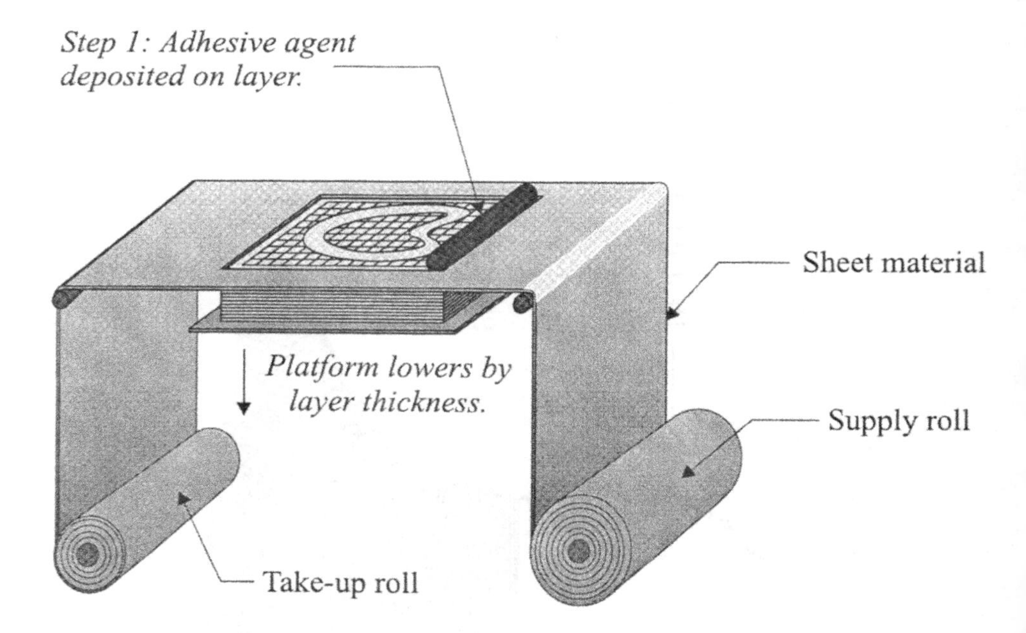

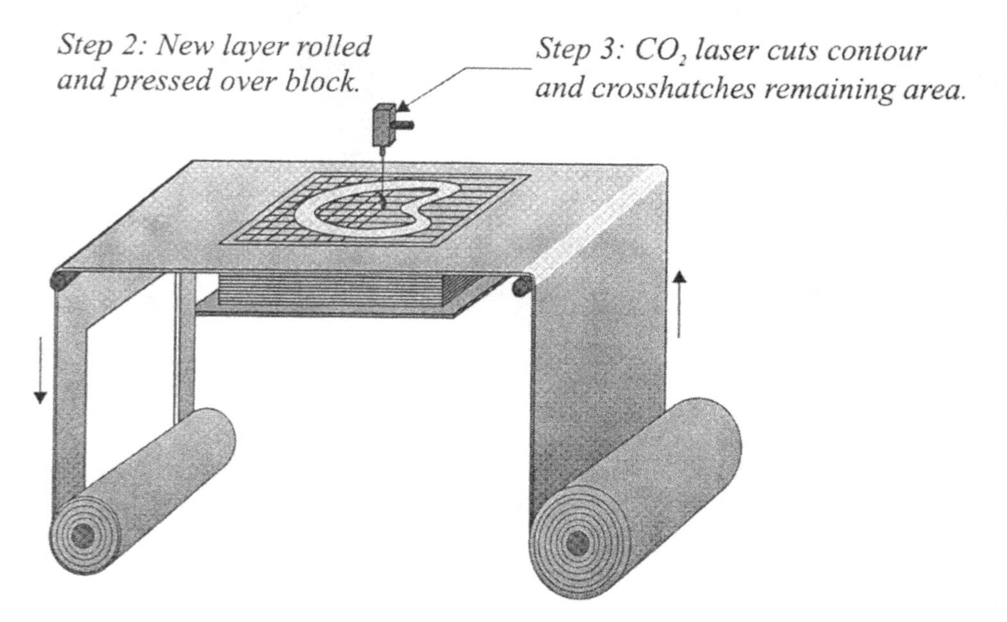

FIGURE 27 Laminated object manufacturing.

Some Industrial Examples

The industrial uses of prototypes built by the above-mentioned methods are still quite limited. However, it is expected that, with the movement of numerous metal-based layered manufacturing techniques from research laboratories to commercial enterprises, the uses of rapid prototyping in the manufacturing industry will substantially grow during 2000–2010.

Lithography: Polymer parts have been used as patterns in sand casting of wiper motor covers (Ford), engine blocks (Mercedes Benz), and so on. Polymer parts have also been used as prototypes of plastic toys, jewellery, spectacle frames, electrodes in electrical discharge machining, electric wire connectors, etc.

Powder processing: Metal parts have been used as prototypes for car engine cylinder heads (Porche), garden hedge trimmers, limited use mold cavities, automotive turbocharger housing units, etc. Ceramic parts have been used as shells for investment (lost-wax) casting, aircraft fuel control systems, etc.

Deposition: Plastic parts have been used as prototypes for ski bindings, child car seat chest clips, golf clubs, window/patio door elements, freezer light fixtures, etc.

Lamination: Laminated parts have been used as prototypes for automotive transaxle housings, crankshafts, intake manifolds, for a variety of toys, and even for footwear sole masters.

9.3.3 Stereolithography

The stereolithography (SL) process, also called initially three-dimensional printing, was developed in 1982 by C. W. Hull and commercialized in 1986 by 3D systems leading to the first stereolithography apparatus (SLA) Model 190. SLA-190 was capable of manufacturing layered photopolymer parts in a work volume of $190 \times 190 \times 250$ mm (app. $7.5 \times 7.5 \times 10$ in.) with layer thicknesses as low as 0.1 mm. A HeCd laser of 7.5 mW of power provided the system with its UV light source. The price of the unit was approximately \$70,000–\$100,000 US.

As introduced in Sec. 9.3.2, Fig. 23, the layer formation process on the original SLA machines comprised three basic steps: (1) deep dip—the elevator is lowered by a distance of several layer thicknesses to allow the viscous liquid in the vat to flow easily over the latest solidified layer; (2) elevate—the elevator is raised back up the deep dip distance minus the layer thickness (of the next layer); and (3) wipe—a recoater blade is used to level the liquid and sweep away excess liquid. In the newer SLA models, since the mid-1990s, a deposition-from-above recoater is utilized: as the wiping blade is drawn across the vat's surface, it releases the required amount of resin to cover one layer. This new technique eliminates the deep dip step and provides thinner and more uniform layers.

Selective solidification of the photopolymer layer using a UV laser light source requires an optimal hatching plan: curing of a solid cross section with maximum dimensional accuracy. This requires high fidelity to the contours specified, when tracing them with the laser light, as well as not to overhatch, which could cause unnecessary shrinkage. Since the release of their first SLA model, 3D Systems has developed a series of newer and better hatching styles, while companies such as Ciba-Geigy have developed better photopolymers (better mechanical properties, faster curing, less shrinking, etc.).

The laser light source cures the photopolymer by drawing lines (straight or curved) across the surface of the liquid layer. Since in practice, a laser spot would have a Gaussian distribution intensity (i.e., a bell curve with maximum intensity at its center), if held momentarily at one spot, it would cure a volume of a parabolic cone shape (Fig. 28a). The depth of cure (i.e., the height of the cone) is a direct function of the time the laser spot is held constant in the same position. (The stronger the energy source is, the faster the curing rate.) As the laser spot is translated along a trajectory, the small volumes overlap and yield a cured line of a semi-parabolic cylinder (Fig. 28b). The allowable speed of travel is a direct function of the energy source. (The stronger the energy source, the faster the spot can be scanned, while providing a sufficient depth of cure.) For better layer adhesion, it is advised to have cure depths larger than the layer thickness.

The accuracy of scanning is a direct function of the galvanometer, servocontrolled mirrors, and the optics configuration. The best circular shaped spot is achieved when the laser beam is in focus and orthogonal to the liquid surface. As the laser spot is translated around, the incidence angle changes and the focal distance increases. A possible solution to such spot size inconsistencies would be to place the mirror at a large distance. An alternative solution would be to use a flat field lens of variable focal length.

Table 1 provides comparative data for the various SLA machines developed over the past two decades. As is noticeable from this table, these systems have improved over the years to built parts at faster rates, with thinner layers (i.e., better dimensional accuracy) and with improved mechanical properties. In parallel to these developments, academic researchers have reported (1) on the use of SLA machines for building ceramic parts by using the monomer liquid as a matrix for the ceramic particles and then burning off the solidified polymer and (2) on the use of laser lithography for the rapid manufacturing of (glass) short fiber

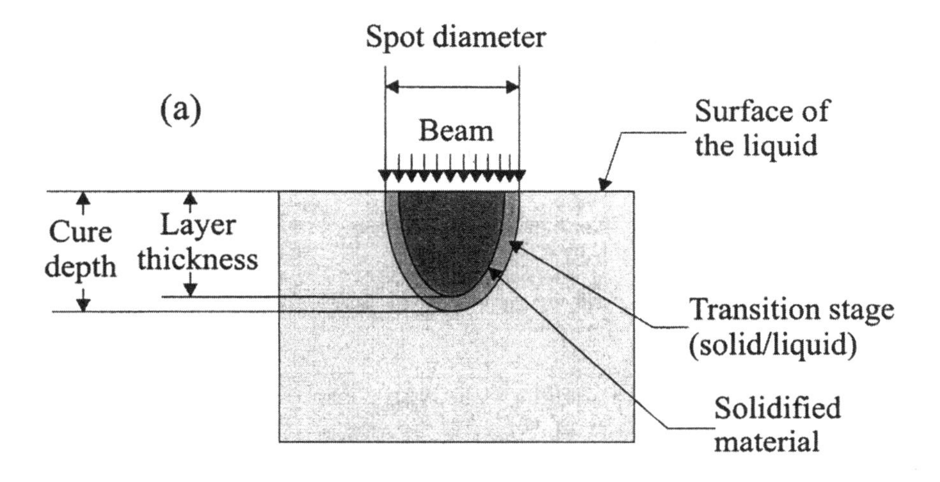

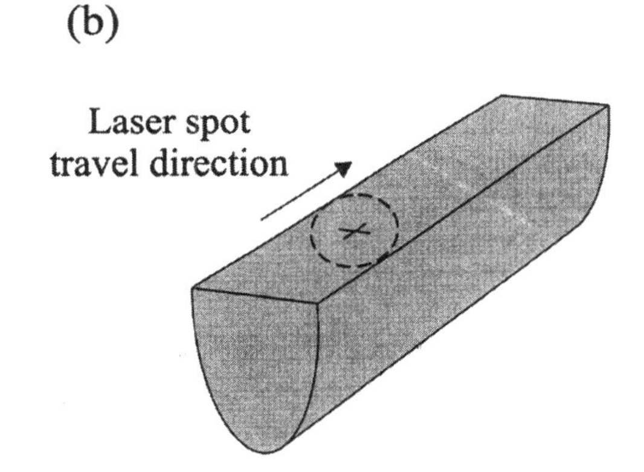

FIGURE 28 (a) Depth of cure; (b) line of cure.

reinforced (layered) polymer parts as direct functional prototypes for many automotive applications.

9.3.4 Selective Laser Sintering

The selective laser sintering (SLS) process was developed at the University of Texas at Austin in early 1980s by C. Deckard and commercialized by the DTM Corporation during the latter part of the 1980s. The first SLS machine, the Sinterstation 2000, was shipped in 1992. This machine used two cylindrical chambers: one for storing the raw powder

TABLE 1 SLA Specifications

Machine model	Workspace (mm×mm×mm)	Laser type (mW)	Maximum scan speed (mm/s)	Minimum layer thickness (mm)	Spot size (mm)
SLA-190	190×190×250	HeCd (7.5)	760	0.1	0.20–0.28
SLA-250	250×250×250	HeCd (6–24)	635–762	0.0625–0.15	0.06–0.28
SLA-3500	350×350×400	Nd:YVO_4 (160)	2540	0.05–0.1	0.20–0.30
SLA-5000	508×508×584	Nd:YVO_4 (216)	5000	0.05–0.1	0.20–0.30
SLA-7000	508×508×600	Nd:YVO_4 (800)	2540–9520	0.0254–0.127	0.28–0.84

material and another for building the part. For every layer, the first cylinder was raised to deposit sufficient powder in front of the powder leveling roller, to be transported to the second cylinder, which is lowered by the thickness of the layer to be solidified by the CO_2 laser (Fig. 24).

The selective sintering process occurs in the processing chamber, which is supplied with inert gas in order to prevent oxidation or explosion of fine metal powder particles. The temperature of the particles (in a region to be solidified) is raised locally to induce sintering but not melting. As in laser cutting, the continuously moving laser light transfers its radiant energy into the powder bed, where it propagates through conduction. The depth of useful heat transfer is a direct function of the laser's power and scanning speed of the spot. (The laser light is delivered from a stationary source through moving mirrors, as with a configuration used in SLA machines.)

A two-phase (liquid–solid) sintering is utilized in SLS systems for intralayer and interlayer bonding. As addressed in Chap. 6, better sintering is achieved by melting the low melting temperature component of a multi-component powder material (liquid phase) and keeping intact the solid particles of the other material. Wetting of the solid particles can be enhanced by utilizing various types of fluxes: metal chlorides and phosphates. Possible material combinations include metal particles coated by polymers, composite (binary) metal powders (one with a lower melting temperature than the other), cermets, and composite ceramic blends:

Metal/metal: Cu-Ni, Fe-Co, and W-Mo
Cermets: Al_2O_3-Fe, Al_2O_3-Ni, and WC-Co
Ceramic/ceramic: Al_2O_3-ZrO_2

Postprocessing of laser sintered parts first have to include the burning out of the polymer binder (for coated metal and ceramic particles), followed

TABLE 2 DTM Specifications

Machine model	Workspace (mm×mm×mm)	Laser (W)	Maximum scan speed (mm/s)	Minimum layer thickness (mm)	Spot size (mm)
2000	ϕ300×381(H)[a]	CO_2 (50)	914	0.076–0.51	0.40
2500	381×330×457	CO_2 (25 or 100)	7,500	0.076–0.51	0.45

[a] Cylindrical vat.

by a conventional sintering method for increased density (for coated and uncoated multicomponent materials). If desired, a secondary metal component (e.g., bronze) can be added to the sintering furnace and be allowed to melt and infiltrate the pores of the part fabricated in the SLS machine for achieving near 100% density.

Table 2 gives some of the specifications of the first- and second-generation DTM Sinterstation machines.

REVIEW QUESTIONS

1. Why have numerous nontraditional machining processes been developed since the mid-20th century? What advantages do they offer over traditional material removal techniques?
2. Describe the uultrasonic machining (USM) process. What type of material is USM normally targeted for?
3. What are the advantages of electrochemical machining (ECM) over USM?
4. Discuss the setting of an optimal gap between the tool and the workpiece in ECM.
5. How can the accuracy of ECM-based drilling of holes be improved?
6. Discuss the electrical discharge machining (EDM) in die sinking applications.
7. Compare ECM with EDM.
8. Describe chemical milling.
9. Why would one use chemical blanking over other traditional or nontraditional hole making techniques?
10. Define lithography-based manufacturing and compare the four available techniques: photolithography, x-ray lithography, electron beam lithography, and ion beam lithography.
11. What is dry etching? Why is it preferable to wet etching?

12. Describe the three primary classes of lasers.
13. Describe the laser beam drilling process and discuss why would it be preferable to other nontraditional hole making processes, such as chemical blanking.
14. Describe the laser beam cutting process and discuss why would it be preferable to other traditional continuous path machining processes such as milling.
15. Discuss the data preparation process common to all layered, rapid manufacturing techniques. In your discussion pay particular attention to the optimization of build parameters (e.g., layer thickness, build orientation).
16. Compare the two most common lithography-based rapid manufacturing techniques: laser lithography (e.g., SLA) versus photolithography (e.g., SGC).
17. Compare the two most common powder processing–based rapid manufacturing techniques: laser sintering (e.g., SLS) versus binder sintering (e.g., DSPC).
18. In newer SLA machines, a deposition-from-above recoater is utilized. Justify the use of such a technique.
19. Describe the postprocessing stage in SLS machines.

DISCUSSION QUESTIONS

1. Integrated circuit (IC)–based electronic component manufacturing processes can be argued to be the modified versions of fabrication and assembly techniques that have long been in existence (e.g., photolithography). These modifications have been mainly in the form of employing optical means for the fabrication of increasingly smaller components. Do you agree with this assessment? Discuss this specific issue and expand the argument to recent rapid layered manufacturing techniques. Have there recently been any truly new manufacturing techniques developed? If yes, give some examples.
2. Material removal techniques, as the name implies, are based on removing material from a given blank for the fabrication of the final geometry of a part. Compare material removal techniques to near net shape production techniques, such as rapid layered manufacturing, in the contexts of product geometry, material properties, and economics in mass production versus small batch production environments.
3. Discuss potential postprocess defect identification schemata/technologies for parts that are layer manufactured using a lithography or powder processing method. Furthermore, discuss possible sensing technologies that can be incorporated into different lithography or powder

processing equipment for the on-line monitoring and control of the manufacturing process, while the parts are being formed.

4. Process planning in (traditional and nontraditional) machining and even in rapid layered maufacturing (in its limited definition) refers to the optimal selection of cutting parameters: number of passes and tool paths for each pass, depths of cut, feed rates, cutting velocities, etc. It has been often advocated that computer algorithms be utilized in search of the optimal parameter values. Although financially affordable for mass production environments, such (generative) programs may not be feasible for utilization in one of a kind or small production environments, where manufacturing times may be comparatively short. Discuss the utilization of group technology (GT)–based process planners in such computational time limited production environments.
5. A prototype of a product must meet some or all of its engineering specifications. These specifications can be geometric or functional. In this context, a large number of material additive techniques were developed during the period 1980 to 2000 and classified as rapid prototyping (RP) processes. Justify through examples (production techniques and product features) why such layered manufacturing techniques may indeed be used to yield prototypes.

BIBLIOGRAPHY

Beaman, Joseph J., et al. (1997). *Solid Freeform Fabrication: A New Direction in Manufacturing with Research and Applications in Thermal Laser Processing.* Boston: Kluwer.

Benedict, G. F. (1987). *Nontraditional Manufacturing Processes.* New York: Marcel Dekker.

Binnard, Michael (1999). *Design by Composition for Rapid Prototyping.* Boston: Kluwer.

Blanther, J. E (1892). Manufacture of contour relief maps. US Patent 473,901.

Boothroyd, Geoffrey, Knight, Winston A. (Oct. 1989). *Fundamentals of Machining and Machine Tools.* New York: Marcel Dekker.

Cheong, M. S., Cho, D.-W., Ehmann, K. F. (Oct. 1999). Identification and control for micro-drilling productivity enhancement. *International Journal of Machine Tools and Manufacture* 39(10):1539–1561.

Chryssolouris, George (1991). *Laser Machining: Theory and Practice.* New York: Springer-Verlag.

Chua, Chee Kai, Leong, Kah Fai (1997). *Rapid Prototyping: Principles and Applications in Manufacturing.* New York: John Wiley.

Crafer, R. C., Oakley, P. J. (1993). *Laser Processing in Manufacturing.* London: Chapman and Hall.

Dally, James W. (1990). *Packaging of Electronic Systems: A Mechanical Engineering Approach*. New York: McGraw-Hill.

DeBarr, A. E., Oliver, D. A. (1968). *Electrochemical Machining*. London: McDonald.

Doyle, Lawrence E., et al (1985). *Manufacturing Processes and Materials for Engineers*. Englewood Cliffs, NJ: Prentice-Hall.

Greco, A., Licciulli, A., Maffezzoli, A. (Jan. 2001). Stereolithography of ceramic suspensions. *Journal of Materials Science* 36(1):99–105.

Groover, Mikell P. (1996). *Fundamentals of Modern Manufacturing: Materials, Processes, and Systems*. Upper Saddle River, NJ: Prentice Hall.

Haberer, M (2001). Fibre-Resin Mixing and Layer Formation Subsystems for the Rapid Manufacturing of Short-Fibre-Reinforced Parts. M.A.Sc. thesis, Department of Mechanical and Industrial Engineering, University of Toronto, Toronto, Canada.

Hinczewski, C., Corbel, S., Chartier, T. (June 1998). Ceramic suspensions suitable for stereolithography. *Journal of the European Ceramic Society* 18(6):583–590.

Jacobs, Paul F. (1992). *Rapid Prototyping and Manufacturing: Fundamentals of Stereolithography*. Dearborn, MI: Society of Manufacturing Engineers.

Jacobs, Paul F. (1996). *Stereolithography and Other RP&M Technologies: From Rapid Prototyping to Rapid Tooling*. Dearborn, MI: Society of Manufacturing Engineers.

Jain, V. K., Pandey, P. C. (1993). Theory and Practice of Electrochemical Machining. New York: John Wiley.

Kalpakjian, Serope, Schmid, Steven R. (2000). *Manufacturing Engineering and Technology*. Upper Saddle River, NJ: Prentice Hall.

Ketting, H.-O., Olsen, F. O. (Nov. 1996). Comparison of CO_2 laser cutting with different laser sources. *J. of Welding in the World* 37(6):288–292.

Kochan, D. (1993). *Solid Freeform Manufacturing: Advanced Rapid Prototyping*. New York: Elsevier.

Kodama, H. (1981). Automatic method for fabricating a three dimensional plastic model with photohardening polymer. *Rev. Sci. Instru.* 1770–1773.

Kornienko, Oxana (1998). *A NURBS-Based Approach to Direct Slicing of Solid Models for Rapid Prototyping*. University of Toronto: Department of Mechanical and Industrial Engineering.

Kozak, J., Dabrowski, L., Lubkowski, K., Rozenek, M., Slzlawinski, R. (Nov. 2000). CAE-ECM system for electrochemical technology of parts and tools. *Journal of Materials Processing Technology* 107(1–3):293–299.

Kuhn, A., Blewett, I. J., Hand, D. P., French, P., Richmond, M., Jones, J. D. C. (Oct. 2000). Optical fibre beam delivery of high-energy laser pulses: beam quality preservation and fibre end-preparation. *J. of Optics and Lasers in Engineering* 34(4–6):273–288.

Kulkarni, P., Dutta, D. (Sep. 1996). An accurate slicing procedure for layered manufacturing. *J of Computer-Aided Design* 28(9):683–697.

McGeough, J. A. (1988). *Advanced Methods of Machining*. London: Chapman and Hall.

Metev, Simeon M., Veiko, Vadim P. (1998). *Laser-Assisted Microtechnology*. New York: Springer-Verlag.

Naguib, Ihab M. (1997). Determination of Optimal Building Orientation and Direct Slicing of Glass-Fibre-Reinforced Stereolithography Parts. University of Toronto, Toronto, Canada: M.Eng. project, Department of Mechanical and Industrial Engineering.

Pham, D. T., Gault, R. S. (1998). A comparison of rapid prototyping technologies. *International Journal of Machine Tools & Manufacture* 38:1257–1287.

Rozenberg, L. D., et al. (1964). *Ultrasonic Cutting*. New York: Consultants Bureau.

Schey, John A. (1987). *Introduction to Manufacturing Processes*. New York: McGraw-Hill.

Sheats, James R., Smith, Bruce W., eds. (1998). *Microlithography: Science and Technology*. New York: Marcel Dekker.

Sommer, C., Sommer, S. (1992). *Wire EDM Handbook*. Houston, TX: Technical Advanced Publishing.

Tabata, N., Yagi, S., Hishii, M. (Mar. 1996). Present and future of lasers for fine cutting of metal plate. *Journal of Materials Processing Technology* 62(4): 309–314.

Thoe, T. B., Aspinwall, D. K., Wise, M. L. H. (Mar. 1998). Review on ultrasonic machining. *International Journal of Machine Tools and Manufacture* 38(4):239–255.

Thompson, Larry F., Willson, C. Grant, Bowden, Murrae J., eds. (1994). *Introduction to Microlithography*. Washington, DC: American Chemical Society.

Thompson, D. C., Crawford, R. H. (1997). Computational quality measures for evaluation of part orientation in freeform fabrication. *Journal of Manufacturing Systems* 16(4):273–289.

Tlusty, Jiri (2000). *Manufacturing Processes and Equipment*. Upper Saddle River, NJ: Prentice Hall.

Weller, E. J. (1984). *Nontraditional Machining Processes*. Dearborn, MI: Society of Manufacturing Engineers.

Wiedemann, B., Jantzen, H.-A. (June. 1999). Strategies and applications for rapid product and process development in Daimler-Benz AG. *Computers in Industry* Vol. 39(No. 1):11–25.

Wilson, John Fay (1971). *Practice and Theory of Electrochemical Machining*. New York: Wiley-Interscience.

Yevko, V. (1997). Cladding Formation in Laser-Beam Fusion of Metal Powder. M.A.Sc. thesis, Department of Mechanical and Industrial Engineering, University of Toronto, Toronto, Canada.

Yevko, V., Park, C. B., Zak, G., Coyle, T. W., Benhabib, B. (Dec. 1998). Cladding formation in laser-beam fusion of metal powder. *Journal of Rapid Prototyping* 4(4):168–184.

Yilbas, B. S. (Oct. 1997). Parametric study to improve laser hole drilling process. *Journal of Materials Processing Technology* 70(1–3):264–273.

Yu, L. M. (Jan. 1997). Three-dimensional finite element modelling of laser cutting. *Journal of Materials Processing Technology* 63(1–3):637–639.

Zak, G. (1999). Rapid Layered Manufacturing of Short-Fibre-Reinforced Parts. Ph.D. diss., Department of Mechanical and Industrial Engineering, University of Toronto, Toronto, Canada.

Zak, G., Haberer, M., Park, C. B., Benhabib, B. (2000). Prediction of mechanical properties in short-fibre-reinforced, layer-manufactured composites. *Journal of Rapid Prototyping* 6(2):107–118.

Zak, G., Sela, M. N., Yevko, V., Park, C. B., Benhabib, B. (Aug. 1999). Layered-manufacturing of fiber-reinforced composites. *ASME Transactions, J. of Manufacturing Science and Engineering* 121(3):448–456.

Zaw, H. M., Fuh, J. Y. H., Nee, A. Y. C., Lu, L. (May 1999). Formation of a new EDM electrode material using sintering techniques. *Journal of Materials Processing Technology* 89–90:182–186.

10

Assembly

The assembly of parts and subassemblies to form a product of desired functionality may involve a number of joining operations, such as mechanical fastening, adhesive bonding, and welding. Although assembly processes are not value-adding operations and are commonly seen as necessary but "wasteful" tasks, most of today's products are not manufacturable as single entities. Complexities of products range from a few parts in a piece of furniture to several million parts in commercial aircraft. Thus while some products are passed on to customers as a collection of individual parts for their assembly by the user, with an incentive of reduced price, most products have to be assembled prior to their sale because of either their complex and long assembly process or the specialized tools needed for their joining that are not usually owned by the perspective customers.

Since an assembly process may add significant cost to the fabrication of a product, different manufacturing strategies have been adopted over the past century for increased assembly efficiency (Chap. 1). These cost-cutting measures have included the use of mass production techniques for reduced setup and fixturing costs, as well as specialization of human operators on one or two specific joining tasks; the use of automation for highly repetitive operations; and more recently the use of modular product design for simplification of assembly.

Assembly relies on the interchangibility of parts concept introduced in the mid-1800s. Individual parts' dimensions must be carefully controlled, within their tolerance levels, so that they can be assembled without further rework during their joining. This is a paramount issue in the batch production of goods (i.e., more than one of a kind) and even more important when in the future individual components that wear out must be replaced with off-the-shelf parts. Systems operating at remote locations requiring replacement parts cannot be expected to be returned to a service location for custom fitting of broken or worn parts.

Design for assembly was discussed in Chap. 3 in the context of minimizing cost, satisfying disassembly requirements for maintenance and repair, and even in the context of being environmentally friendly. It was argued that (1) minimization of parts would reduce assembly cost, (2) reduction of permanent joints would ease maintenance, and (3) lesser variety in materials would facilitate recycling.

The objective of this chapter is to address a variety of representative methods for different types of joining operations available to a manufacturer in the fabrication of multicomponent products. These include mechanical fastening, adhesive bonding, welding, brazing, and soldering. Automation issues pertinent to these processes will be briefly discussed in their respective sections. The chapter will be concluded with a detailed review of two specific assembly applications: automatic assembly of small mechanical parts and automatic assembly of electronic parts.

10.1 MECHANICAL FASTENING

Joining of mechanical components through fasteners (screws, bolts, rivets, etc.) is most desirable when future disassembly of the product is expected for maintenance, or when other joining techniques, such as welding or adhesive joining, are not feasible. Several factors affect the number, type and locations of fasteners used in assembling two or more parts: strength of the joints (tensile or compression), ease of disassembly, and appearance.

The location and number of fasteners to be used in a mechanical joint is primarily a function of the strength level we wish to achieve, subject to geometrical constraints (e.g., minimum wall thickness, distance from edge, and potential creation of stress concentrations). The strength of such joints can normally be calculated analytically (for example, through the area of contact of the number of threads on a fastener). Though, in some cases, empirical methods may have to be employed for more reliable estimations.

Product appearance also influences the locations of the fasteners and their types. However, we must be very conscious of ease of assembly and disassembly when making such placement decisions. Designers and

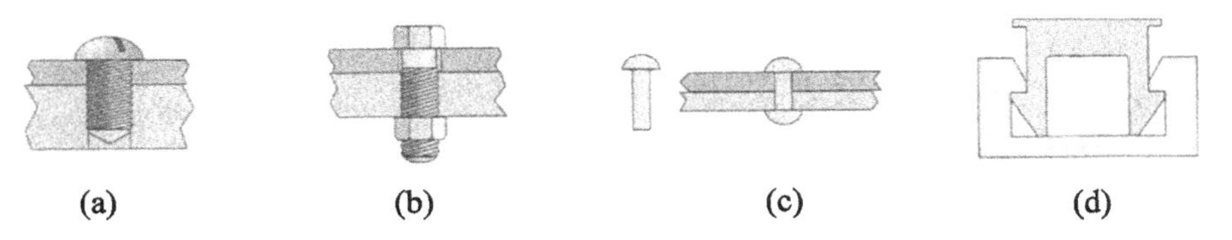

FIGURE 1 Mechanical assembly using (a) a screw, (b) a bolt, (c) a rivet, and (d) a snap-fit joint.

manufacturing engineers must not place fasteners at hard-to-reach places simply for aesthetic purposes, especially if the product is to be assembled by the customer, who may not have a large variety of tools at his or her disposal for fastening.

Mechanical fasteners can be mainly categorized as threaded fasteners, (nonthreaded) rivet-type fasteners, snap-fit fasteners, and interference-type fasteners (Fig. 1). Threaded fasteners can be further divided into two types: self-tapping screws, which do not require the parent component to have already been drilled and tapped, and bolts (and some screws) that either require threads in one (or both) of the parts to be assembled or utilize nuts.

10.1.1 Threaded Fasteners

Although screws and bolts may be fabricated in a variety of sizes and shapes, due to interchangeability requirements, designers should utilize standard fasteners, as opposed to requiring special-purpose screws or bolts at (relatively) high costs.

Tension fasteners are available in a number of different head styles, each suitable for a specific task. The two most common ones are briefly reviewed below (Fig. 2).

Pan and truss heads: Both of these head shapes are very popular and come in a variety of drive types (Phillips, Robertson, etc.). Although the truss head normally has a larger bearing area, both types of screws will fail first in the threaded area, as opposed to the head.

Hex head: Such screws have hexagonal external head shapes, though some may have hexagonal internal, socket-drive shapes. They are characterized by their large load-carrying ability, as well as the readily available (industrial) tightening tools. Socket screws with hex drives are normally used for high-strength and high-tolerance applications.

Compression fasteners normally are setscrews that are headless. They are utilized to locate and immobilize one part with respect to another.

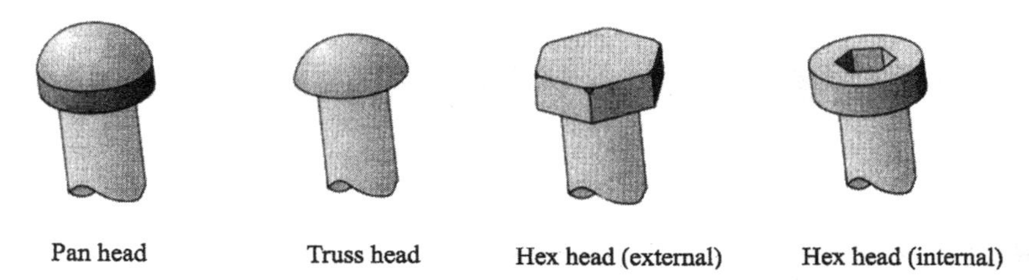

FIGURE 2 Screw and bolt heads.

Common setscrew point shapes include the cone point, oval point, flat point, and dog point (Fig. 3).

Although most threaded fasteners are fabricated using a cold-forming operation (Chap. 7), some are machined on dedicated, specialized automatic thread-cutting lathes (Chap. 8). Steel is the most commonly used material for fasteners due to its high strength, good resistance to environmental conditions, and low cost. Some nonferrous fastener materials include brass, bronze, aluminum, and titanium. Due to its high strength-to-weight ratio, titanium fasteners are often utilized in aerospace and sports assemblies, where higher costs are not prohibitive.

Fastener corrosion is the most problematic issue in mechanical assembly and presents users with high maintenance costs (e.g., bolts used in automobile assemblies, especially those used in corrosive environments, such as Canada and Northern European countries). Two common corrosion protection mechanisms are sealing (using plastic- and rubber-based sealant) and electroplating. Zinc, cadmium, nickel, copper, tin, and brass are the most frequently utilized plating materials. Electroplating of threaded fasteners (typically, about 0.01 mm) is a complex process and must be carefully planned for in dimensioning of holes and threads. For example, during electroplating, coated material tends to build up more on the thread "crests" (peak of the thread) and less in the "roots."

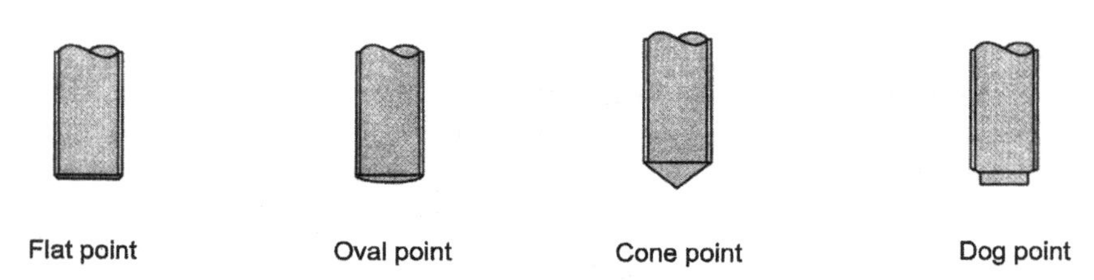

FIGURE 3 Setscrew geometries.

10.1.2 Rivets

Riveting is a highly effective joining process for the fastening of two segments with a permanent joint. Permanency implies that the only way of removing the joint for disassembly is by destroying the joint. Riveting is very commonly used in the joining of thin-walled structures, such as the fuselages and wings of aircraft.

The riveting process comprises two primary operations: Placement of the (unthreaded) rivet in the hole and deforming the headless end of the rivet (normally through an "upsetting" operation) to form a second head and thus a tight connection. The four primary rivet geometries are shown in Fig. 4. As with threaded fasteners, the designer must choose the minimum number of rivets (not to overfasten) and place them optimally to avoid stress concentrations, which is especially critical in thin-walled parts.

Riveting materials include:

Low- and medium-carbon steels: The majority of rivets (above 90%) are made of such steels for their low cost, high strength, and easily formable characteristics. Typical applications include automotive assembly, photographic equipment, home appliances, and office hardware.

Copper alloys: Rivets of this material are used for appearance, good electrical conductivity, and corrosion resistance. Typical applications include electrical assemblies, luggage, and jewellery.

Aluminum: Rivets of this material have the lowest cost, a bright appearance, and corrosion resistance. Typical applications include transportation equipment, lighting fixtures, storm windows and doors, and toys.

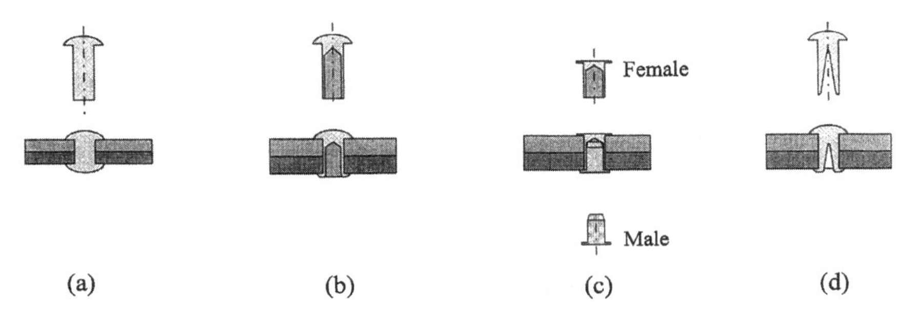

FIGURE 4 Rivet geometries: (a) solid, (b) tubular, (c) compression, and (d) split (or bifurcated).

10.2 ADHESIVE BONDING

Adhesives can be utilized for the joining of most engineering materials: metals, plastics, ceramics, wood, and paper. The joining process involves the placement of an adhesive filler material between the surfaces of two segments of a product (adherents) and the subsequent curing of the adhesives using an initiating mechanism: applications of heat and mixing of two or more reactive components. Some adhesives employ a solvent that evaporates or is absorbed by the adherents that are joined and leaves behind a dry hardened adhesive layer. The resultant joint is permanent and frequently cannot be broken without damage to one or both parts.

Adhesives can be traced back to the gluing of furniture and the use of sealings in a variety of forms in ancient civilizations. Their use in modern times, however, only became widespread with the availability of (organic) monomers at the end of the 19th century and the beginning of the 20th, accelerating after the 1940s. The first use of adhesives in manufacturing was in the bonding of load-bearing aircraft components during the 1940s. Since then, they have been used in the machine-tool industry, the automotive industry, the electronics industry, the medical industry, and the household products industry.

Some advantages of adhesive bonding are

Joining of dissimilar materials: Different materials or similar materials with different thermal characteristics (e.g., thermal-expansion coefficient) can be adhesive bonded.

Good damping characteristics: Adhesive bonded assemblies yield good resistance to mechanical vibration, where the adhesive acts as a vibration damper, as well as resistance to fatigue.

Uniform stress distribution: Broader joining areas yield better stress distribution, allowing the use of thinner assembly components and resulting in significant weight reductions.

Thermal and electrical insulation: Adhesives provide electric and thermal insulation, as well as resistance to corrosion.

Niche application: Adhesive bonding can be used in the joining of parts with complex shapes and different thicknesses that do not allow the use of other joining processes. It can also be used to yield visually attractive products with no visible joints or fasteners.

Naturally, as do other processes, adhesive bonding suffers from numerous drawbacks: parts' surfaces must be carefully prepared to avoid contamination; the joints can be damaged in the face of impact forces and weakened significantly at high temperatures ($>200°$–$250°C$); and actual bonding strength may not be accurately verifiable (i.e., a quality control problem). In the following subsections, some of these issues will be briefly addressed.

10.2.1 Joint Design and Surface Preparation

The first step in joint design for adhesive bonding is to understand how adhesives behave under mechanical loading. Tensile loads that may induce peeling or cleavage must be avoided owing to low cohesive strength of adhesives. In contrast, adhesive joints can resist high shear and compression loads, when the joint overlap area is sufficiently large to allow distribution of the applied load. However, high pure shear forces applied for long periods of time may eventually damage the joint.

Empirical data have shown that there is an upper limit to the degree of overlapping of the joints for increased load carrying. Beyond a certain limit, one cannot increase the strength of a joint simply by increasing the length of the overlap. In reference to joint thickness, although engineering intuition

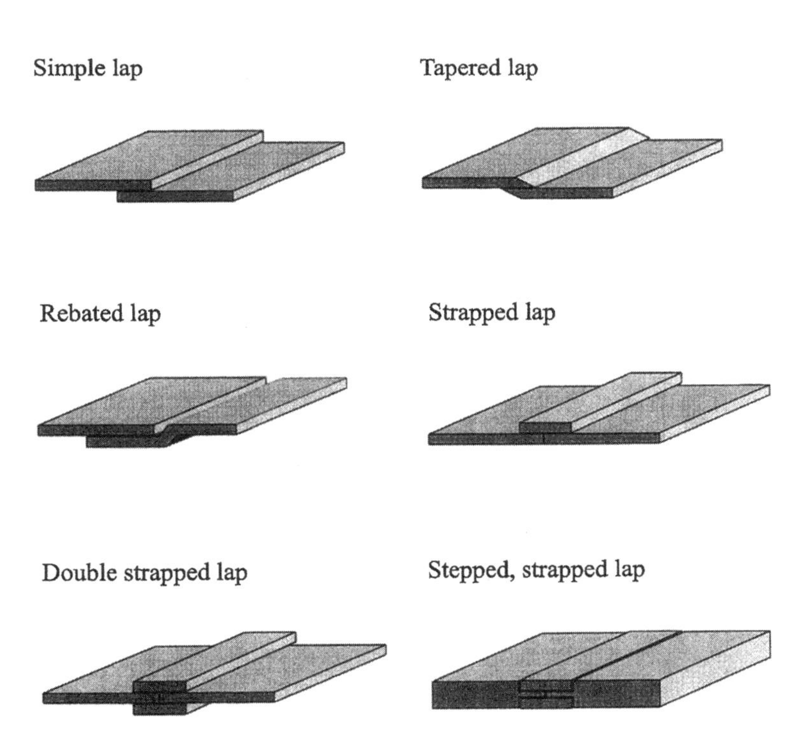

FIGURE 5 Adhesive bonding lap joints.

would advocate the minimization of bond thickness, several studies have shown that some level of increase may actually strengthen the joints.

Most adhesive bonding joints are of the lap-joint type and its variants, some of which are shown in Fig. 5. The simple lap joint requires a toughened adhesive that will not experience a brittle failure due to distortions occurring under shear loading. Otherwise, the geometry of the joint or its configuration, through the addition of third-party segments, must be varied for optimal distribution of loads.

Butt joints (end-to-end contact) are normally viewed as poor forms of adhesive bonding, unless large contact areas are created. Some design guidelines for good adhesive bonding are illustrated in Fig. 6.

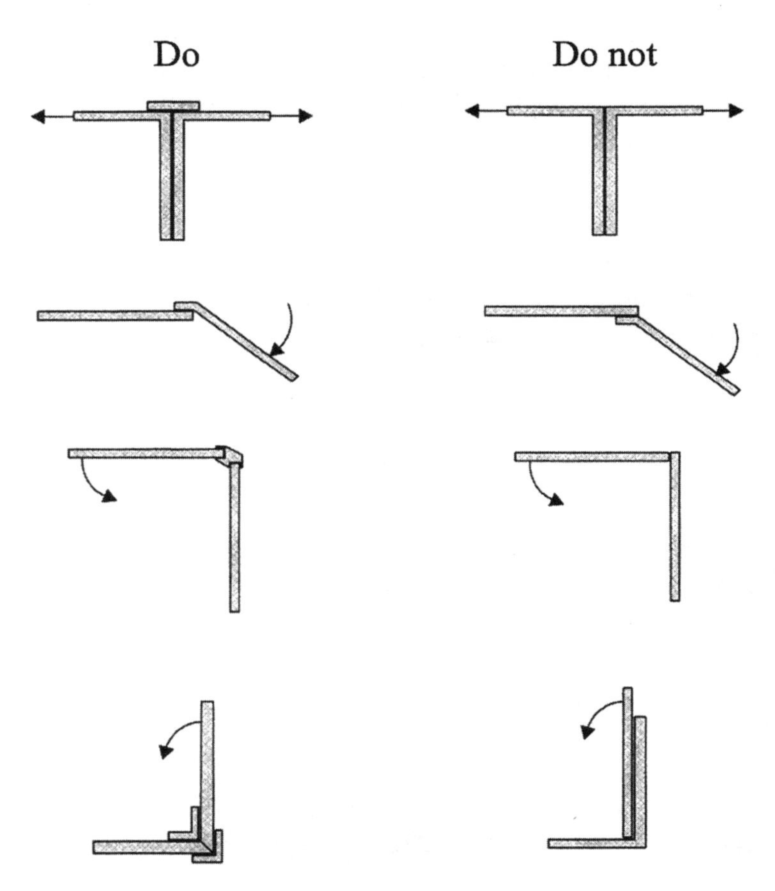

FIGURE 6 Guidelines for adhesive bonding.

Surface preparation is the most important step in adhesive bonding. Contaminants present on the adherents poorly affect wetting and cause premature failure of the joint. There are three common techniques for surface cleaning: solvent degreasing through wiping or vapor degreasing, chemical etching or anodizing, and the use of surface primers. Naturally, the optimal technique(s) selected for surface preparation is a function of the materials of the adherents. However, all surface preparation activities, regardless of adherents' materials, must be quickly followed by the deposition of the adhesive and subsequent bonding operation.

10.2.2 Adhesives and Bonding Techniques

Adhesives can be broadly classified into two groups: organic and inorganic. The former can be further classified into natural (e.g., dextrin, rubber) and synthetic (e.g., acrylic, epoxy, phenolic) types. In this section, we will briefly review several organic synthetic adhesives. Inorganic adhesives (cement, silicate, solder, etc.) will not be addressed herein.

Epoxy adhesives: These (thermoset) adhesives normally have two parts: the epoxy resin and its hardener. They are commonly used on large aluminum objects. Single-part, temperature-hardened epoxy adhesives can also be found in use in the manufacturing industry

Acrylic adhesives: This class contains a variety of (thermoset) subspecies: anaerobic, cyanoacrylate, and toughened acrylics. Anaerobic adhesives are one-part, solventless pastes, which cure at room temperature in the absence of oxygen. The bond is brittle. Cyanoacrylate adhesives are also one-part adhesives that cure at room temperature ("crazy glue"). Toughened acrylic adhesives are two-part, quick-setting adhesives that cure at room temperature after the mixing of the resin and the initiator. Overall, acrylic adhesives have a wide range of applications: metals in cars and aircraft, fiberglass panels in boats, electronic components on printed circuit boards, etc.

Hot-melt adhesives: These (thermoplastic) single-part materials include polymers such as polyethylene, polyester, and polyamides. They are applied as (molten) liquid adhesives and allowed to cure under (accelerated) cooling conditions. Owing to their modest strengths, they are not widely used for load-carrying applications.

The methods of applying adhesives onto prepared surfaces vary from industry to industry and are based on the type of adhesive materials: manual brushing or rolling (similar to painting), silk screening (placement of a metal screen on designated surfaces and deposition through cutouts on the screen), direct deposition or spraying using robot-operated pressure guns, and slot coating (deposition through a slot onto a moving substrate—"curtain coating").

10.2.3 Industrial Applications

Industrial examples of adhesive bonding include the following.

Automotive industry: Sealants and adhesives are widely used in the manufacturing of automobiles in temporary or permanent roles. On the welding line, examples include front hoods and trunk lids, hemming parts of door bottoms, front and rear fenders, and roof rails. On the trim line, examples include door trims, windshields, windows, wheel housings, and weather strips. Other automotive examples include (neoprene and nitrile rubber) phenolic adhesives used in the bonding of brake linings to withstand intermittent high shear loads at high temperatures, drain holes on body panels, and of course carpet fixing.

Machine-tool industry: Retaining adhesives have been utilized by machine-tool builders for strengthening a variety of bushings and bearings that are press-fitted against loosening due to intense vibrations (e.g., in chucks with hydraulic clamping mechanisms and in the spindle). Adhesives are also commonly used on a variety of body panels.

Other industries: Epoxy phenolics have been used in (aluminum-to-aluminum) bonding of honeycomb aircraft and missile parts onto their respective skins, in solar cells in satellites, and even in resin–glass laminates in appliances. One- or two-part epoxy adhesives have also been used in joining cabinet, telephone booth, and light fixture parts. Other examples include small electric armatures, solar heating panels, skis, tennis rackets, golf clubs, beverage containers, medical skin pads, loudspeakers, shoes, and glassware.

10.3 WELDING

Welding is a joining operation in which two or more segments of a product are permanently bonded along a continuous trajectory or at a collection of points through the application of heat. In certain cases, pressure can be utilized to augment the thermal energy expended. Most solid materials (metals, plastics, and ceramics) can be welded, though, with different difficulty levels.

The existing tens of different welding techniques can be grouped into two major classes: fusion welding and solid-state welding. The former class uses heat to create a molten pool at the intended joint location and may utilize a filler material to augment the existing molten pool for larger gaps, stronger bonds, difficult joint geometries, etc. The latter class utilizes heat and/or pressure for the welding process, though when heat is utilized the temperature is kept below the melting point of the segments to be joined.

In this section, our focus will be on fusion welding. Solid-state welding operations, such as friction welding and ultrasonic welding, will not be addressed herein. Also, joining operations involving brazing and soldering (often discussed with welding) will be addressed separately in Sec. 10.4.

Although welding became a widely used joining process after the introduction of electrical power in the mid-1880s, its origins can be traced to Egypt (1300s B.C.) and Rhodes (300s B.C.). These were rudimentary forge-welding processes using pressure and heat. The first (fusion-based) arc and resistance welding processes were commercialized during the period of 1880 to 1890 in England and the U.S.A. The use of oxyacetylene torches in arc welding, which utilize oxygen and acetylene for achieving a high-temperature flame, can be traced back to the period of 1900 to 1905. Today's most popular welding processes, however, were only developed during the period 1940–1960: gas tungsten arc welding (GTAW) in the late 1940s, gas metal arc welding (GMAW) in the mid-1950s, plasma welding in the late 1950s, and laser welding in the early 1970s.

As will be discussed below, welding is a complex manufacturing operation in which many process variables (such as feed rate and electrode angle) must be carefully controlled during the joining of two parts. The welding environment is quite hostile to humans owing to process temperatures, light emissions, and gases utilized. Thus in the past three decades two important trends have emerged: extensive use of robotic welders and widespread utilization of laser welding. Today, industrial robots have almost completely replaced human welders on manufacturing shop floors owing to their resistance to hostile environments and their capability of yielding repetitive high-quality welds.

10.3.1 Arc Welding

The most common fusion welding process, arc welding, involves the use of electric arc for the generation of extreme temperatures at a localized point (up to 10,000 to 20,000°C) in the joining of conductive metal workpieces. The electric arc is generated as (AC or DC) current passes between the welding electrode and the metal workpiece separated by a controlled short distance (Fig. 7). The welding electrode can be a consumable one, which itself acts as a filler material, or a nonconsumable one, which acts only as an arc generator and melter of an externally supplied filler material (normally in the form of a continuous wire). The extreme heat generated serves two complementary purposes: melting the regions of both workpieces at the closest vicinity of the joint and melting the filler material, thus forming a mixed continuous molten pool, which when solid forms a very strong bond. The primary contributors to the transfer of molten filler material (mostly as

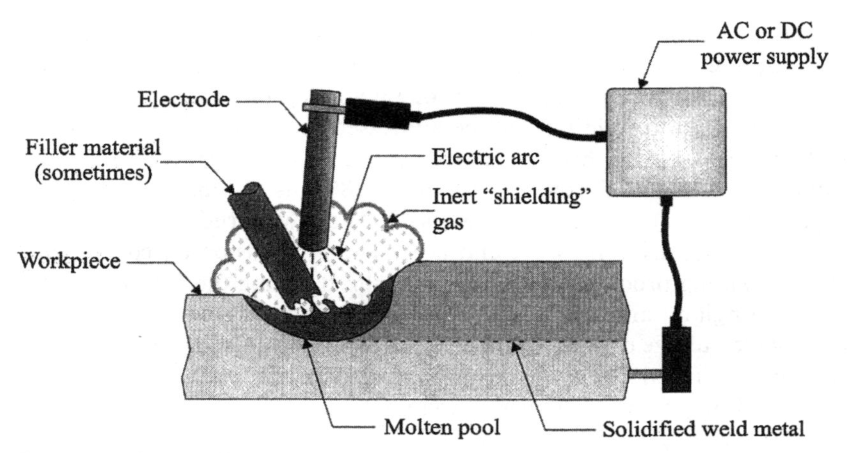

FIGURE 7 Arc welding.

drops or globules) into the welding zone include gravity, electromagnetism, and surface tension.

Since at high temperatures metals are very reactive to oxygen, hydrogen, and nitrogen present in the surrounding air, the welding region must be protected against such reactions through the use of inert gases such as argon and helium.

Weld Joints

The most common joints in arc welding are shown in Fig. 8. (a) Butt joints are connections between two sheets or plates; the weld penetrates through the parts' thickness. It is advised to configure the edges of both parts appropriately for a good flow of molten metal. (b) Corner joints are connections between two edges (30° to 150° relative inclination). (c) T-joints are orthogonal connections between two workpieces. (d) Lap joints are the fillet welding of two overlapping surfaces along one of the edges.

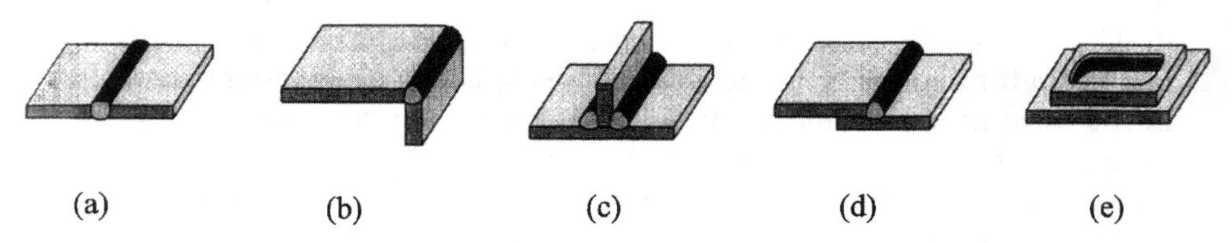

FIGURE 8 (a) Butt; (b) corner; (c) T; (d) lap; (e) slot weld joints.

These joints are commonly used for thin-sheet products. (e) Slot joints are connections between two parts through the fillet welding of the inner edge of the slot (periphery welding).

Shielded Metal Arc Welding

Shielded metal arc welding (SMAW) is primarily a manual process; a consumable electrode is advanced in the direction of the desired joint trajectory while being fed into the welding region according to the rate of consumption. The electrode is the metal filler melted into the joint (Fig. 9). This electrode is coated with a material that melts concurrently with the "stick," providing the weld region with necessary protective gas against reaction with the surrounding air. The power source can be either AC or DC. The electrical current is in the range of 150 amps (for AC) to 400 amps (for DC) at voltage levels of 15 to 45 volts. The diameter of the rod/stick (1.5 to 10 mm) dictates the current level—thinner rods require less current. Naturally, these rods have to be repeatedly replenished, making SMAW a process that is difficult to automate.

Gas Metal Arc Welding

In gas (shielded) metal arc welding (GMAW), also known as metal inert gas (MIG) welding, a protective region is established around the weld pool through a continuous supply of inert gas. The consumable (bare) wire is fed

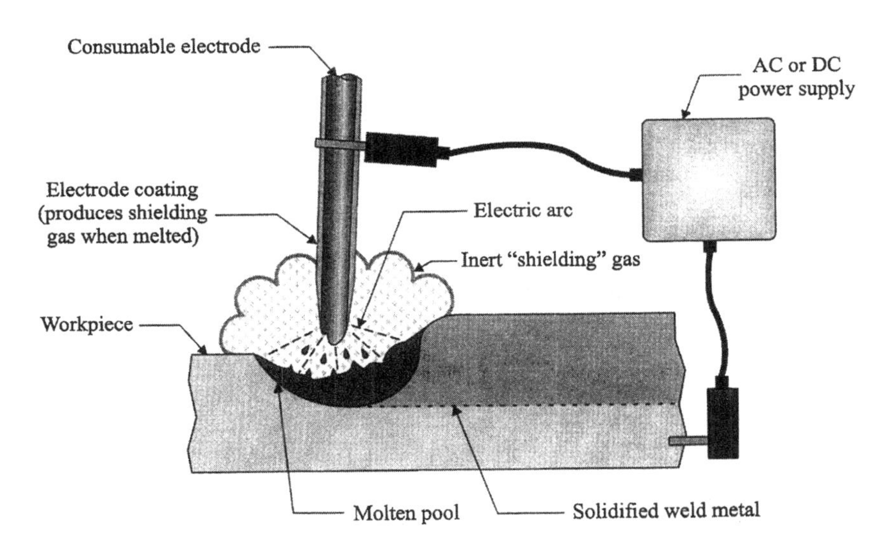

FIGURE 9 Shielded metal arc welding.

into the weld pool through a wire guide within a nozzle that supplies the shielding gas as well (Fig. 10). GMAW can be utilized for the welding of different metals, at high speeds (up to 10 m/min) and with minimum distortion. Owing to its continuous wire-feed feature (supplied in reels), GMAW is suitable for robotic applications.

The selection of filler wires, in diameters of 0.6 to 6.4 mm, depends on the material of the two parent workpiece segments. Frequently, steel wires are coated with copper for better conductivity, reduction of feeding friction, and minimization of corrosion while kept in stock. The selection of inert shielding gas also depends on the welding material: argon (Ar) is very suitable for nonferrous metals and alloys; additions of 12% oxygen into argon yields higher arc temperatures and increases wetting—this mixture is suitable for stainless steel welding; mixtures of argon and helium are suitable for aluminum, magnesium, nickel and their alloys; and high-purity argon

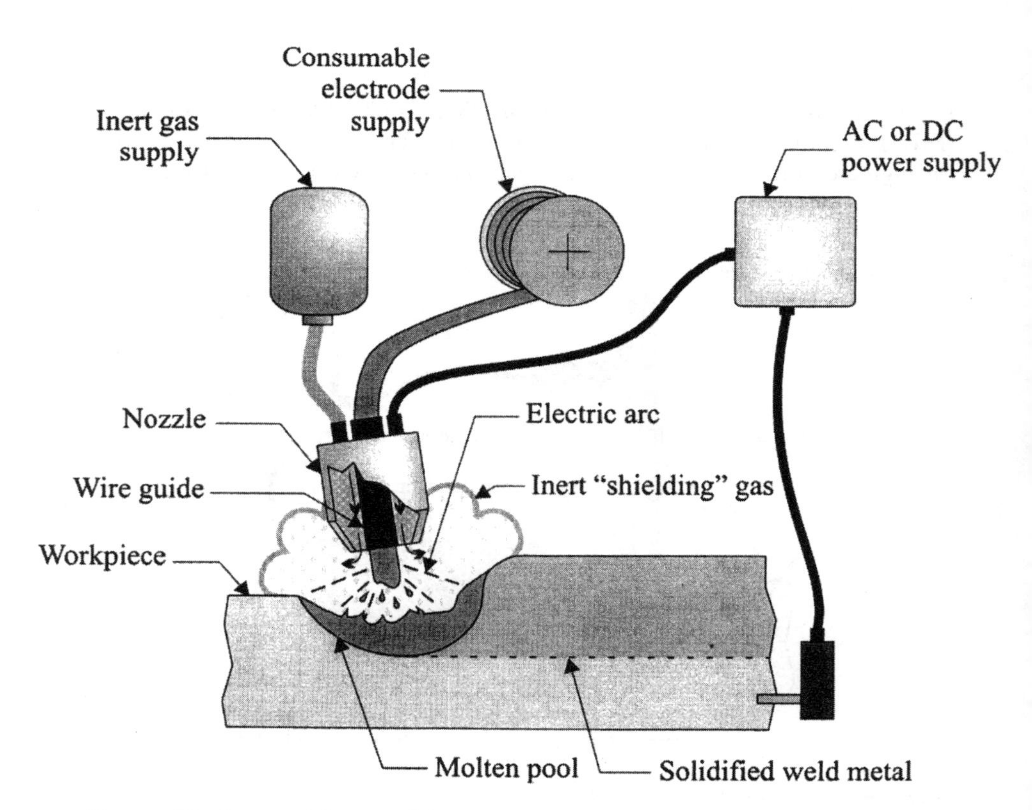

FIGURE 10 Gas metal arc welding.

gas is necessary for highly reactive metals such as titanium, zirconium, and their alloys.

Numerous operating parameters must be controlled when welding with a shielding inert gas: welding current influences the deposition speed and the shape of the weld—increased current increases the size of the weld pool and penetration, while the width of the bead remains practically unchanged; arc voltage also influences deposition—increased arc voltage yields enlarged bead width, while having diminished penetration; welding speed directly influences the quality of the weld joint—excessive speed yields nonuniform welds, while slow speed prevents deeper penetration. One must carry out extensive experimentation for combinations of different materials and shielding gases in order to obtain the optimal values for the three important process parameters: welding speed, arc voltage, and welding current.

Gas Tungsten Arc Welding

Gas tungsten arc welding (GTAW), also known as tungsten inert gas (TIG) welding, utilizes a nonconsumable electrode for the generation of very high welding temperatures. Since the filler wire no longer constitutes an electrode that has to be kept at a distance from the workpiece, it can be fed directly into the weld pool with no spatter to yield high quality weld joints (Fig. 11). Although targeted mostly for nonferrous metals, GTAW can be utilized for the welding of all metals, even for the joining of

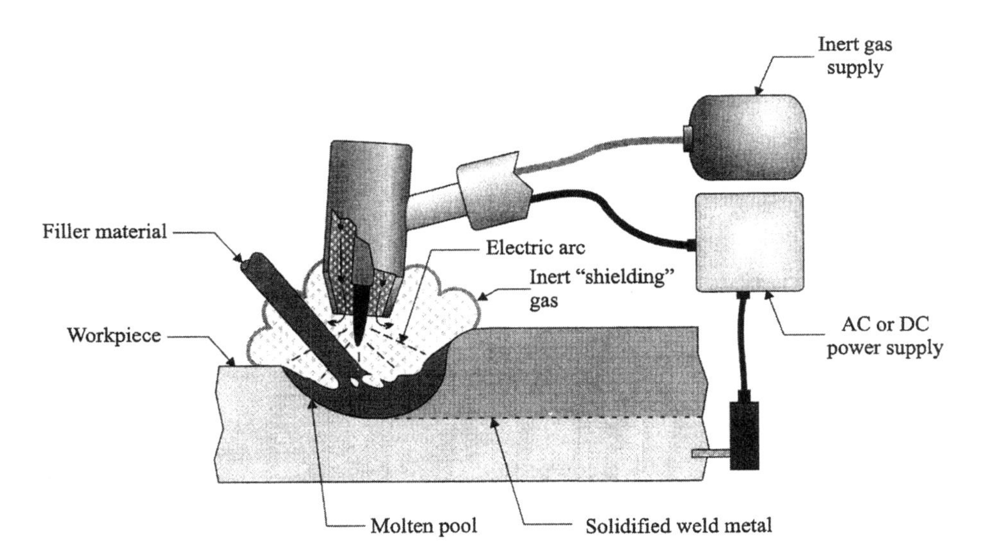

FIGURE 11 Gas tungsten arc welding.

dissimilar metal segments. Argon is utilized for the welding of most metals, with the exception of some alloys of aluminum and copper, for which helium is the recommended shielding gas. Mixtures of argon and helium are also commonly used in high-speed GTAW—the addition of helium improves penetration.

Tungsten electrodes range in diameter (1 to 8 mm) and allow welding currents of up to 800 amps. However, owing to very high electrode temperatures, the welding torch is equipped with a ceramic nozzle to hold the tungsten electrode and a water-based cooling system. Occasionally the tungsten electrodes may have to be ground for the welding of thin materials.

10.3.2 Spot Welding

In resistance spot welding, joining is achieved through a localized heating effect occurring due to an electrical current encountering resistance during its flow. When a large current (up to 100,000 amps) at a low voltage (up to 10 volts) is passed through two very highly conductive electrodes in contact with a pair of lower-conducting plates, heat is generated between the two plates along the line of the current, causing local melting. The localized welding of the two materials is normally augmented by applying pressure also directly along this line of current (Fig. 12).

Spot welding of thin-walled plates (less than 3 to 5 mm) using welding guns is the predominant joining operation in car bodies (up to 1,500 to 3,000 joints per vehicle), typically performed by industrial robots. Although such welds are extremely strong, one must always carefully control the welding cycle: the actual weld time, when current is

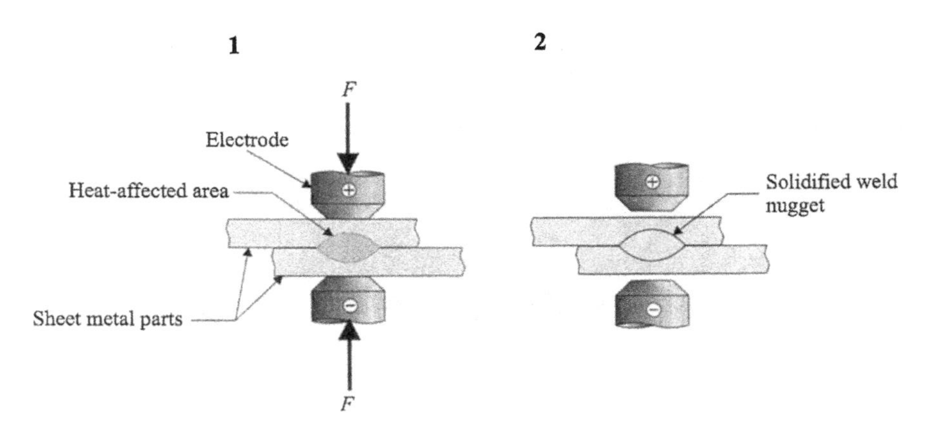

FIGURE 12 Spot welding.

applied under constant pressure, and the cooling period, when the current is off but the pressure is maintained for a little while more. Also, the electrodes (with water-cooling systems) should have the following properties: high electrical conductivity, high thermal conductivity, high resistance to mechanical deformation, and low adhesiveness to metals being welded.

10.3.3 Laser-Beam Welding

Laser-beam welding is a thermal process that utilizes a high-energy coherent light beam to melt particles on the surfaces of two adjacent workpieces for their permanent joining. As first discussed in regard to laser-beam machining in Sec. 9.2, the term laser is an acronym—light amplification by stimulated emission of radiation. A laser source converts electrical energy into a high-energy light beam, which is subsequently converted into heat as it is absorbed by the recipient material. In laser beam welding, this heat generation leads to the formation of a liquid melt pool. As the laser spot travels along a trajectory (i.e., continuous welding), the liquid pool advances forward, while the past location of the spot solidifies and forms a permanent joint between the two parts.

Laser welding offers several unique advantages over other traditional welding processes: it can produce a high-intensity spot at remote, difficult-to-reach locations (or even along continuous trajectories); it yields minimal distortions and high-quality uniform joints; it facilitates the welding of dissimilar materials without the use of filler materials; and it allows high welding speeds. As in laser beam machining, however, laser beam welding may be more challenging for highly reflective material, such as aluminum and copper.

Lasers

The three primary classes of lasers are gas, liquid, and solid. All lasers operate in one of the two temporal modes: continuous wave and pulse. The two most commonly used lasers in welding are as follows.

Nd:YAG: The neodymium-doped yttrium aluminum garnet ($Y_3Al_5O_{12}$) laser is a solid-state laser. Although very low in efficiency, its compact configuration, ease of maintenance, and ability to deliver light through a fiber-optic cable makes it an excellent choice for welding. In pulsed mode, a Nd: YAG laser can deliver an output power of up to 50 kW at a frequency of up to 500 Hz.

CO_2: The carbon-dioxide laser is a gas laser that can deliver an output power of up to 25 kW in continuous wave mode. It is the most efficient commercial laser (up to 10%)—though still a very inefficient power

source, when efficiency is measured as the ratio of output power to input power. CO_2 lasers, however, cannot be coupled to fiber-optic systems.

Continuous Welding

Most of the joint configurations shown in Fig. 8 for continuous arc welding can be achieved using laser beam welding, commonly with CO_2 lasers. As in laser beam machining, the laser spot can be focused onto the surface region of the weld joint using moving optics systems for faster welding speeds or by utilizing a moving workpiece system. In all cases, however, a shielding gas must be supplied to the welding region (Fig. 13).

Spot Welding

Spot welding was the first attempt in joining metals via a laser light source. Today, (lap joint–type) spot welds can be achieved by using both Nd:YAG and CO_2 lasers by creating a localized melting point between two parts in contact. Nd:YAG lasers are suitable for spot welding due to the very high intensity heat impulses they can generate at high frequencies. Such pulsed mode lasers have also been used in the welding of small-diameter wires or very thin walled sheets of metal—"microwelding."

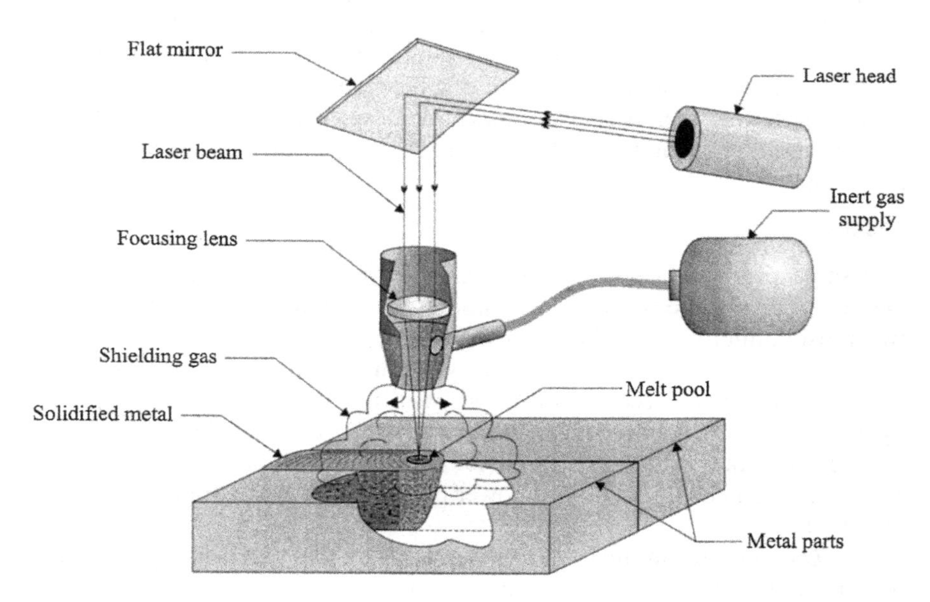

FIGURE 13 Laser welding.

Applications

Probably the best known application of laser welding is the spot welding of the two blades on the Gillette Sensor Razor (a total of 13 welds). The fabrication line for this product utilized thirty Nd:YAG lasers with fiber-optic light-delivery systems yielding a production rate of three million welds per hour. Other less known examples of laser welding include gears, steering units, engine parts and body parts for the automotive industry, TV picture tubes, food mixer parts and pen cartridges for the consumer goods industry, and heat-exchanger tubes for the nuclear industry.

10.3.4 Weldability and Design for Welding

Most engineering materials (metals, plastics, and ceramics) can be welded—some with more difficulty than others, using one of the available welding techniques. Due to the melting and solidification cycle and the resultant microstructure changes, one must carefully monitor all the welding process parameters, including shielding gases, fluxes, welding current and voltage, welding speed and orientation, and preheating and cooling rates.

In terms of different materials: carbon and low-carbon alloy steels are weldable with no significant difficulties—thicknesses of up to 15 mm are more easily weldable than thicker workpieces that require preheating (to slow down the cooling rate); aluminum and copper alloys are difficult to weld because of their high thermal conductivity and high thermal expansion; titanium and tantalum alloys are weldable with careful shielding of the weld region; thermoplastics (such as polyvinychloride, polyethylene, and polypropylene) are weldable at low temperatures (300 to 400°C), though glass-reinforced plastics are not generally weldable; ceramics (SiO_2-AlO_2) have also been welded in the past using CO_2 and Nd:YAG lasers with some preheating.

Welding defects can be classified as external (visible) and internal defects. Some of these are listed below (Fig. 14):

Misalignment: It is an external defect caused by poor preparation.

Distortion: It is an external defect caused by residual stresses due to unsuitable process parameters.

Incomplete penetration: It is an internal defect caused by excessive weld speed, low weld current, too small a gap, or poor preparation.

Undercut: It is an internal defect—a groove that appears at the edge of the joint, caused by high current or voltage, irregular wire speed, or too high welding speed.

Porosity: It is an internal defect—in the form of isolated or grouped bubbles, caused by an insufficient flow of gas, moist or rusty base metals, or entrapment of gases.

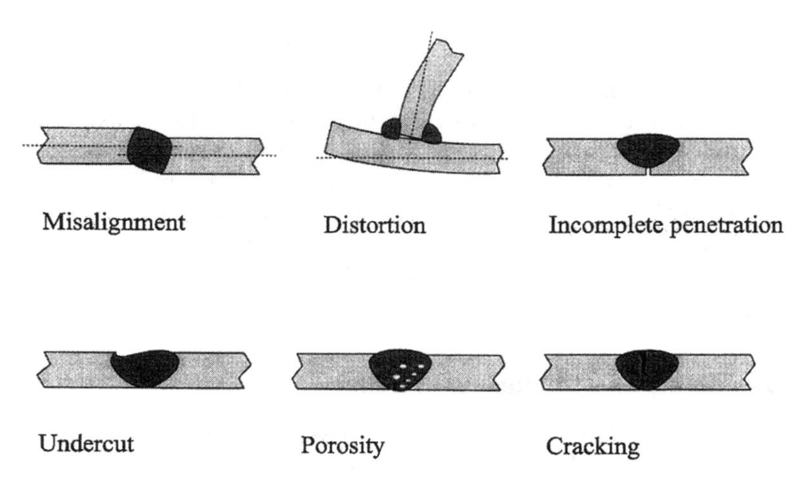

FIGURE 14 Welding defects.

Cracking: It is an internal defect—localized fine breaks that may occur while the joint is hot or cold, caused by hydrogen embrittlement, internal stress, lack of penetration, excessive sulphur and phosphorus content in base metal, or rapid cooling.

In addition to the process parameters that must be controlled to avoid welding defects, a designer may consider the following additional guidelines: weld locations should be chosen to maximize strength and avoid stress concentrations, though some awareness of intended use and appearance is important; careful edge preparation must be employed if unavoidable; and welding should be minimized owing to potential dimensional distortions.

10.4 BRAZING AND SOLDERING

Brazing and soldering are similar joining processes: a filler metal is melted and deposited into a gap between segments of a product. Unlike in welding, the base materials (similar or dissimilar) are not melted as part of the joining process. In brazing, the filler material normally has a melting point above 450°C (840°F), but certainly lower than that of the base materials, whereas, in soldering, the filler materials have melting points well below 450°C. Capillary forces play an important role in both processes in the wetting of the joint surfaces by the molten (fluid-state) filler material and thus the flow of the liquid metal into the gaps between the two base segments.

The use of low-melting-point metals in joining operations has been around for the past 3,000 years. The primary advantages of such processes have been joining of dissimilar materials joining of thinned walled and/or complex geometry parts that may be affected by high temperatures (such as those in welding), achievement of strong bonds (stronger than adhesive bonding but weaker than welding), and adaptability to mass production and automation.

In the following two subsections, the fundamental issues in brazing and soldering will be presented.

10.4.1 Brazing

Brazing is a simple joining process, in which a liquid metal flows into narrow gaps between two parts and solidifies to form a strong, permanent bond. Ferrous and nonferrous filler metals normally have melting temperatures above 450°C, but below those of the two base materials, which do not melt during the joining process. Almost all metals, and some ceramic alloys, can be brazed using filler metals, such as aluminum and silicon, copper and its alloys, gold and silver and their alloys, and magnesium and nickel alloys.

The ability of the liquid filler metal to wet the materials it is attempting to join, and completely to flow into the desired gaps prior to its solidification, determines the success or failure of a brazing application. The most commonly utilized wetting metric is the contact angle (Fig. 15a). In 1805, T. Young concluded that for each combination of a solid and a fluid there exists a corresponding contact angle that is governed by the following (Young's) surface tension angle,

$$\gamma_S = \gamma_L \cos\theta + \gamma_{SL} \tag{10.1}$$

where γ is the surface tension (J/m), the subscripts S and L refer to solid and liquid surfaces, respectively, and SL refers to the solid–liquid interface. Acceptable wetting occurs when the contact angle is less than 90° (Figs. 15b and 15c).

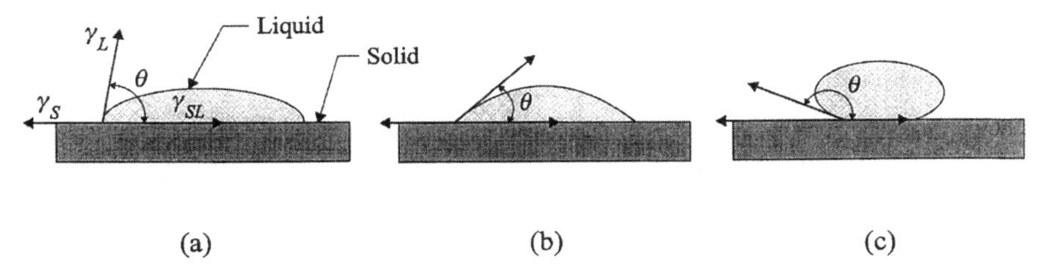

Figure 15 (a) Contact angle θ; (b) $\theta < 90°$, wetting; (c) $\theta > 90°$, no-wetting.

Besides surface tension, the following parameters affect the brazing process: temperature and duration of melting, surface preparation, joint design and gap dimensions, source and rate of heating, and, naturally, base material and filler metal characteristics.

In the past three decades, the brazing process has been successfully automated and applied in numerous industries. Several such examples are listed:

Cemented carbides onto cutting tools' metal shanks.
Ceramic (automotive) bladed–turbocharger hubs onto metal shafts.
Ceramic-on-metal joining in microelectronic products.
Metal-on-metal (automotive) pipes.

Joint Design in Brazing

The brazing metal filler is normally applied to a preheated joint through the melting and deposition of a rod or a wire. However, commonly, the brazing metal can be placed in the immediate vicinity of the joint prior to heating, in the form of preformed rings, disks, slugs, etc. Subsequent capillary forces that develop during heating draw the molten filler metal into the intended clearances (Fig. 16). Bond integrity and strength depend on joint geometry, clearances, and surface cleanliness.

Butt and lap joints are the two primary brazing joint configurations (Fig. 17). Butt joints are simple in design and preparation. However, in such joints, all of the load is transmitted in the undesirable tensile stress form, where the thinnest section of the joint dictates the strength of the joint. Lap joints' strengths do not depend on the cross sections of the components, and the load is normally transmitted in the desirable shear stress form. In both configurations, however, the clearance must be carefully controlled—beyond an optimal gap, the capillary forces may not be enough to uniformly distribute the filler metal fluid through the joint. Although joint clearances are functions of the filler and base materials, the empirical, ideal gap has been traditionally defined as 0.05 to 0.15 mm (up to 0.25 mm for precious metals).

Materials and Environment in Brazing

The brazing of all metals and ceramics depends on the wetting of the filler material at relatively high temperatures. Formation of unwanted oxides at such high temperatures, however, may impede the ability of the filler material to wet the joint surfaces. Use of a suitable flux complemented with an inert gas atmosphere dissolves and/or prevents the formation of oxides and promotes wetting by lowering the surface tension of the filler

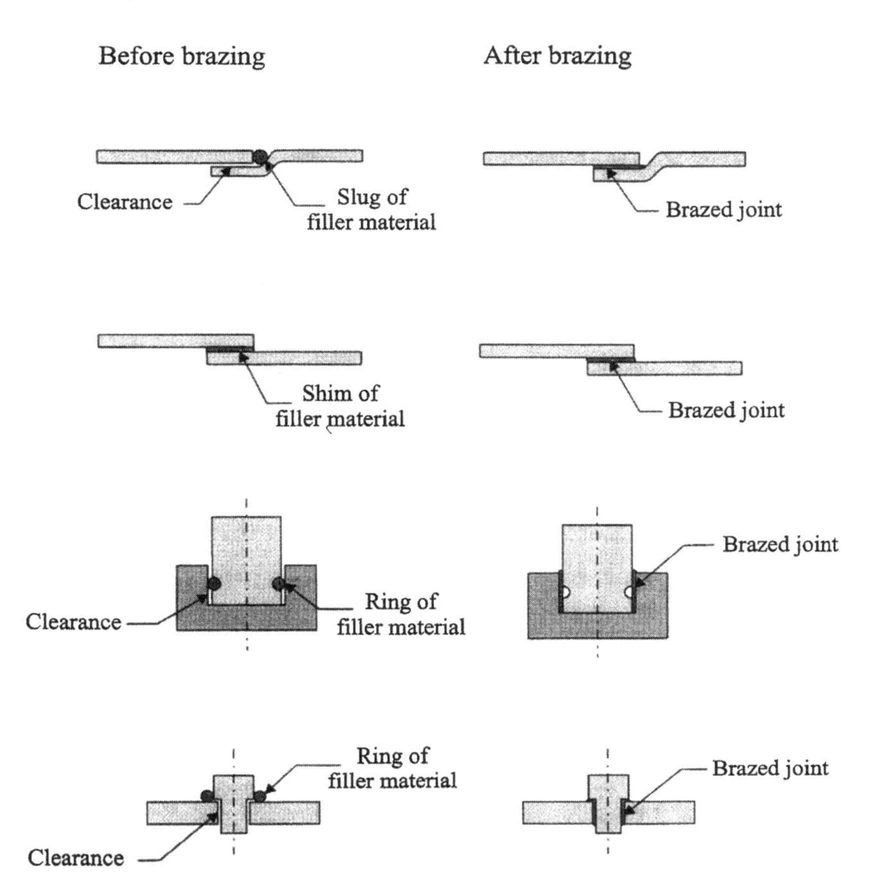

FIGURE 16 Preapplication of filler metal.

metal. Common fluxes used for brazing include chlorides, fluorides, borates, and alkalis.

Filler materials commonly used in brazing are

Aluminum–silicon: This group of alloys can be used as fillers for aluminum (and its alloys) base materials at melting temperatures of 500°–600°C—heat exchangers, aircraft parts, car radiators, etc.

Copper and copper–zinc: This group of alloys can be used as fillers for ferrous-base materials at melting temperatures of 750 to 1700°C.

Nickel and nickel alloys: This group of alloys can be used as fillers for stainless steel, nickel, and cobalt-based alloys at melting temperatures of 950 to 1200°C.

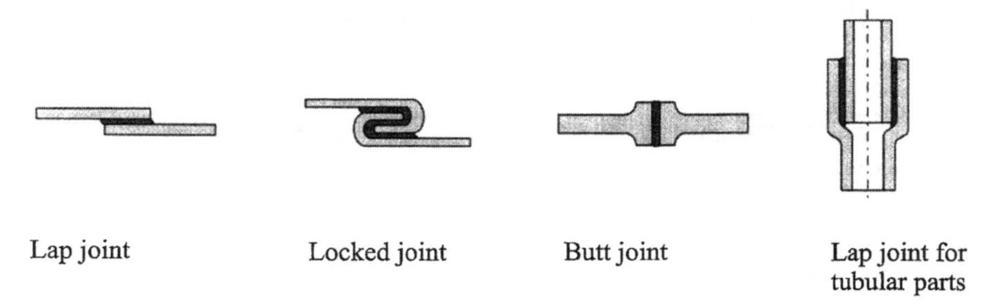

Figure 17 Brazing-joint configurations.

Silver-copper: This group of alloys can be used as fillers for titanium, ceramics, steel, and nickel at melting temperatures of 620 to 850°C—honeycomb structures, tubing, etc.

Brazing Methods

Manual torch brazing is the simplest and most commonly used technique, primarily reserved for one-of-a-kind (repair or prototyping) jobs. Other brazing techniques that allow automation for batch or continuous processing include

Furnace brazing: Parts with preplaced filler segments are brazed in (electric, gas, or oil-heated) furnaces that typically employ a conveyor for (time-controlled) continuous through motion of the parts to be joined.

Induction and resistance brazing: Electrical resistance is utilized to melt preplaced filler materials quickly.

Dip brazing: Complete immersion of small parts into a (constant-temperature) molten filler material vat provides wetting and filling of the joints.

10.4.2 Soldering

Soldering is a joining process in which two segments of a product are bonded using a liquid filler material (solder) that rapidly solidifies after deposition. As in brazing, the process occurs at the melting temperature of the solder (typically, below 315°C), which is significantly below the melting temperature of the base material. Wetting of the joint surfaces by the liquid solder and its flow into the desired gaps of optimal clearances due to capillary forces is a paramount issue in soldering. Due to wetting and bonding of the joints at low temperatures, soldering is a desirable joining process for applications with no significant load carrying situations, such as soldering of electronic components on printed circuit boards (PCBs).

Joint design guidelines for soldering are very similar to those established for brazing. Of the two most common joint configurations, butt and lap joints, the lap joint is the preferred one because of its strength (Fig. 18). Since joint strength is directly related to overlap area, designers must carefully configure lap joints for achieving uniform solder flow into the gaps. A common solution to such flow problems, however, is the preplacement of the solder prior to heating. Preplacement can be achieved using solid preforms of solders, such as washers, wire rings, discs, or powder form solder suspended in a paste.

The soldering process starts with a careful preparation of the surface—removal of contaminants and degreasing. An appropriate soldering flux is then applied to prevent oxidization and facilitate wetting. Inorganic fluxes include hydrochloric and hydrofluoric acids and zinc chloride and ammonium chloride salts. Organic fluxes include lactic and oleic acids, aniline hydrochloride halogens, and a variety of resins. Fluxing is followed by the joining process, where molten solder is directly applied to the joint area or heating is applied to melt the preplaced solder. Once the joint is cooled, all flux and solder residues must be removed.

Soldering Materials

For successful soldering, the base metal, the solder, and the flux materials must be chosen concurrently. Although most metals can be soldered, some are easier to solder than others: copper and copper alloys are the easiest base metals to solder, so are nickel and nickel alloys, while aluminum and its alloys, titanium, beryllium, and chromium are not normally soldered. The primary reason for our inability to solder materials in the last group is the difficulty encountered in removing the oxide film (i.e., fluxing) at the low temperatures of soldering.

Most commonly used solders are tin–lead alloys, with occasional addition of antimony (less than 2 to 3%). The equilibrium diagram of the

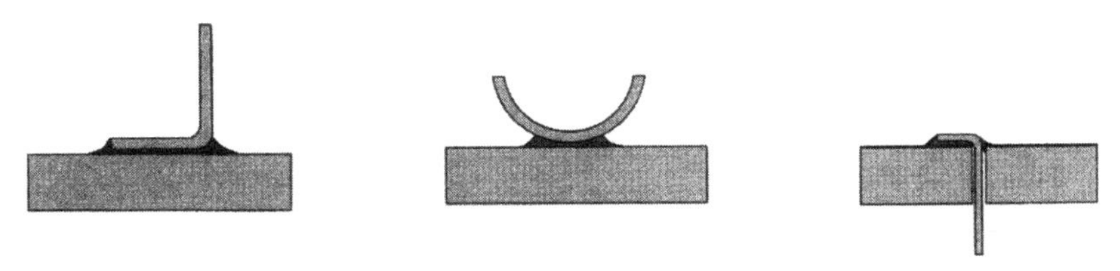

FIGURE 18 Soldering lap joint configurations.

tin–lead alloy is shown in Fig. 19. As can be noted, the eutectic composition of this alloy comprises approximately 62% tin and 38% lead, with a discrete melting point of 183°C. (The word eutectic derives from the Greek word *eutectos* meaning easily meltable.) On both sides of this composition, the alloys go through a transition phase (in "pasty" form) when being converted from solid to liquid.

Since tin is an expensive material, manufacturers may choose to use lower percentages of tin in the tin–lead alloy, at the expense of higher melting points. The electronic industry prefers to use the eutectic composition due to its rapid solidification at the lowest melting point level. Other solder materials include tin, silver, tin–zinc, tin–bismuth, and cadmium–silver. Tin–silver, for example, would be used for applications (intended product usages) with high service temperatures, since they have higher melting points (221°C) than do most tin–lead solders (183°C).

Soldering Methods

Soldering is a highly automated method of joining metal components. Most soldering methods can be classified according to the application of heat: conduction (e.g., wave soldering), convection (e.g., reflow soldering) and radiation (e.g., infrared and laser beam soldering).

Wave soldering: In this automatic method, a pump located within a vat of solder (heated through conduction) creates an upward spout (a laminar

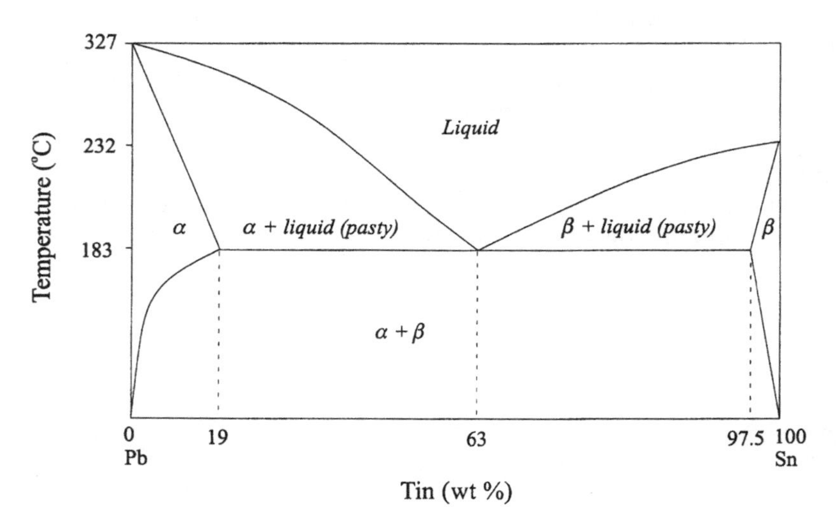

FIGURE 19 Equilibrium diagram for tin–lead alloy.

flow wave). Exposed metal joints are passed over this wave in a continuous motion, where liquid solder attaches itself to the joint due to capillary forces (Fig. 20). Wave soldering is one of the most common techniques used in the electronics industry, especially, for the joining of through hole components.

Reflow soldering: Remelting of preplaced/predeposited solder between two surfaces, for forming the intended joint, using a convection-type heat source is normally called reflowing. In the electronics industry, reflowing is commonly utilized for the soldering (via wetting of predeposited solder paste) of surface-mount devices on PCBs. (Occasionally, the term reflow soldering is erroneously utilized for soldering in infrared ovens).

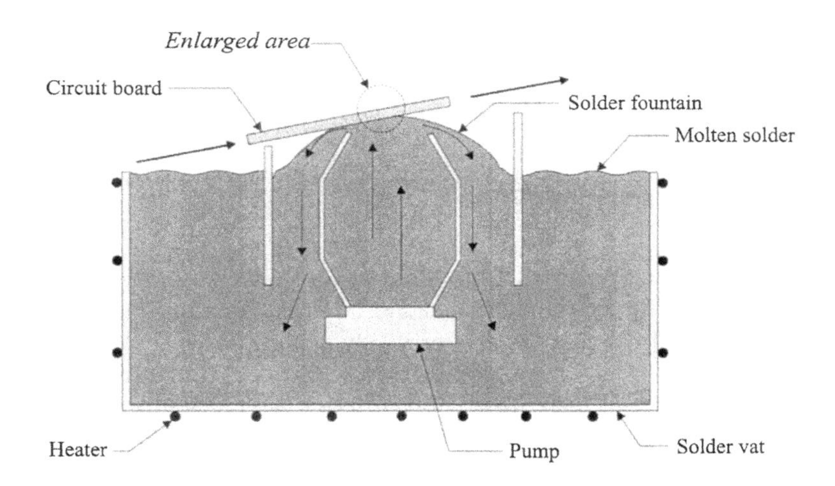

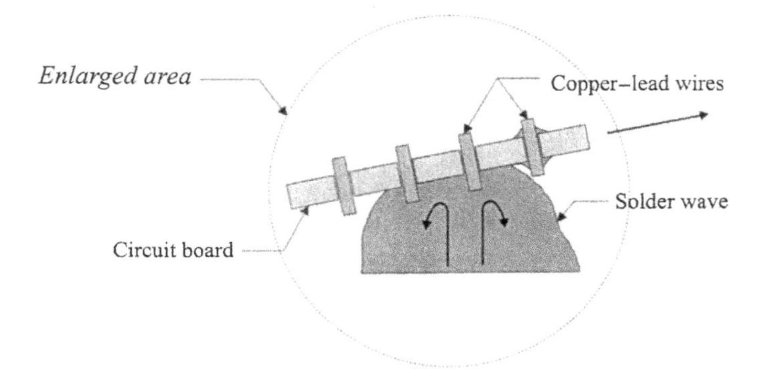

FIGURE 20 Wave soldering in electronics industry.

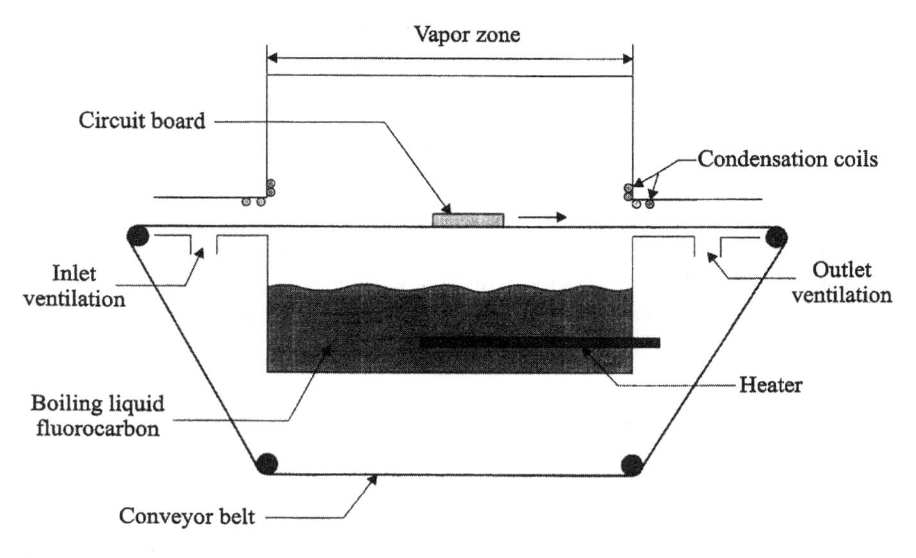

FIGURE 21 Vapor-phase soldering.

Infrared reflow soldering: In infrared ovens, remelting of preplaced solder is achieved via electromagnetic radiation in the far range of the 0.75 to 1000 μm infrared wavelength. As one would expect, however, only 20 to 30% of heat in such ovens is provided by radiation. Convection provides the remaining percentage of heat, thus eliminating potential shadowing problems. Some infrared ovens also utilize inert gas atmospheres.

Vapor phase soldering: This convection based heat transfer method (developed in 1973 by Western Electric Company, U.S.A.) provides soldering ovens with excellent temperature control and uniformity of heating. A liquid vat (normally with fluorinated hydrocarbon fluid) is utilized to generate vapor with good oxidation resistance and fill a chamber, through which a product (usually PCBs) with preplaced solder moves.

The vapor, when in contact with the product, raises the temperature of the solder to the boiling temperature of the liquid in the vat and allows it to reflow and form the necessary joints (Fig. 21). As the joints are formed, the product is retrieved from the oven for fast solidification.

10.5 ELECTRONICS ASSEMBLY

The use of electronic products during the period 1950–1980 was primarily restricted to industrial purposes, such as control of manufacturing machines, aircrafts, and large-scale computers. The computer revolution that started in the early 1970s brought electronic products into many households, such as hand-held calculators, telephones and answering machines, microwave ovens, and eventually personal computers. Today, most modern household appliances and automobiles have extensive numbers of electronic components in them. (This trend eventually led to the development of a new engineering field, mechatronics, broadly defined as the use of electronic components in the operation and control of mechanical systems.)

The manufacturing of an electronic product starts with the fabrication of its individual components, such as resistors, capacitors, and integrated circuits (ICs). The production of an IC, in turn, starts with the production of a single-crystal silicon ingot (up to 150 mm diameter and sometimes more than 1 m long solid cylinder), which is subsequently sliced into thin disc-shaped wafers (approximately 0.5 mm in thickness).

Through traditional processes such as lithography and etching (Chap. 9), thousands of ICs can be built on a single wafer and then separated for individual (or group) packaging. Packaging refers to the preparation of these devices for future connection onto PCBs, for example, attachment of heads and encapsulation in ceramic carriers. Two common technologies for component attachment onto a PCB (i.e., electronics assembly) are through-hole mounting, also known as pin-in-hole (PIH), and surface-mount technology (SMT).

10.5.1 Component-to-Board Connections

PCBs, also known as printed-wiring boards (PWBs), have the following primary functions: to provide a mounting surface for most of the components and allow component-to-component connections through a (printed) wiring system. Most PCBs are laminates of one or more layers of copper claddings (copper foils) separated by dielectric laminates (most commonly, polymers reinforced with glass cloth, cotton fabric, paper, etc.) (Fig. 22).

Lithography and etching are the primary processes used in the fabrication of a desired connection configuration on a copper foil (attached

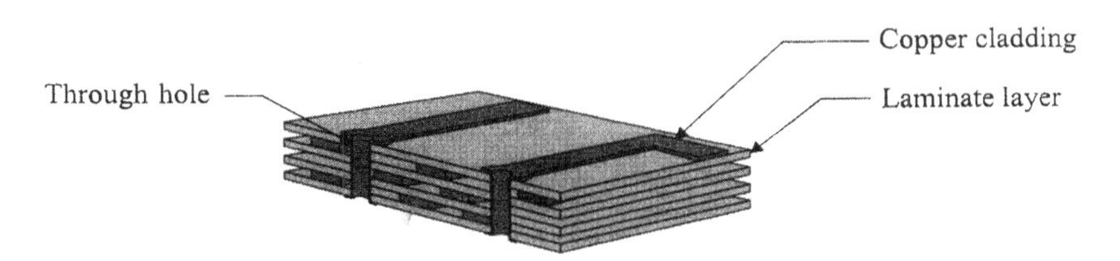

FIGURE 22 Printed-circuit-board configuration.

to an underlying dielectric laminate). Holes on PCBs (for interlayer connections and PIH component attachments) are normally punched or drilled. The design of the connection configurations and the spacing of components must be carefully considered in the fabrication of optimal size PCBs, which should also allow efficient assembly of components (also known as the "population" of a board) and testing of functionality.

As mentioned above, two common technologies used in component-to-PCB connections are through-hole and surface-mount technologies. In PIH connections, a connection is established by soldering the leads (pins) of the components onto the PCB by depositing molten solder at the holes and allowing it to wet the complete region due to capillary forces and solidify (Fig. 23a). In SMT, the components are placed on holeless PCB regions configured with landing areas for electrical connections to the components. These "lands" have already been equipped with predeposited solder paste prior to the placement of the components. Some SMT-connection (joint) configurations are shown in Fig. 23b.

The two most common SMT (IC-package) connections have long been the gull-wing and the J leads (Fig. 23). However, owing to the intense pressure for miniaturization of components since the early 1990s, manufacturers developed a very effective connection technology, named ball-grid-array (BGA) packaging. The BGA technology was to provide the electronics industry with a capability for more than 1,000 I/O connections per device, though recent BGA devices provide less than half of this targeted number.

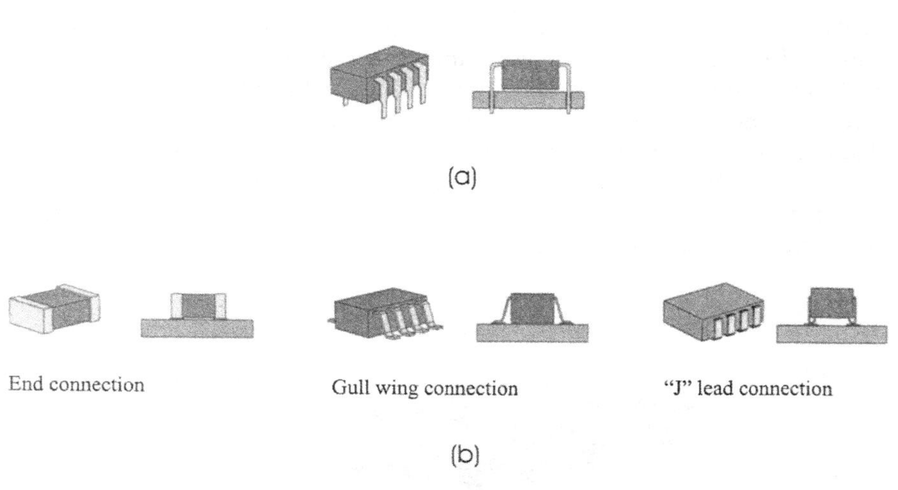

FIGURE 23 (a) Through-hole; (b) surface-mount connection.

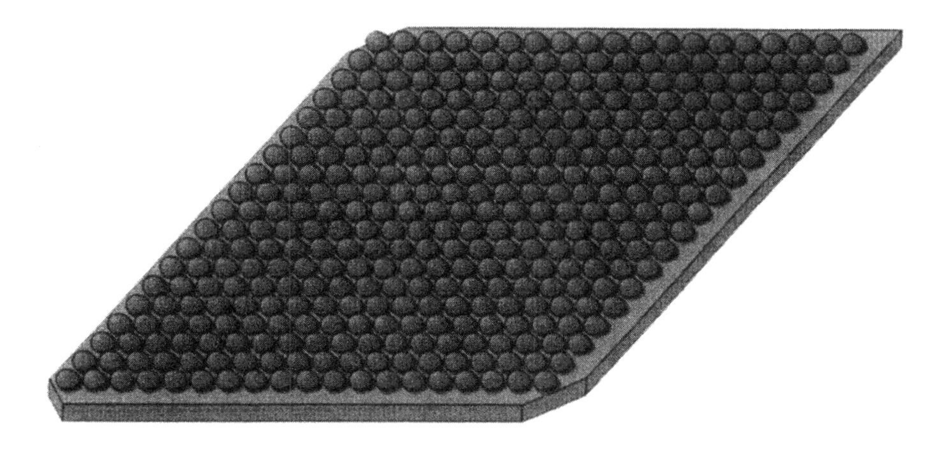

FIGURE 24 Ball-grid array I/O connections.

The contacts are normally less than 0.8 mm solder spheres attached to the bottom of the device in an array form (Fig. 24).

Beside their primary advantage of high I/O intensity, BGA devices also provide ruggedness and excellent thermal energy management. Their primary disadvantage is the severe difficulty of inspection, which may make them unattractive to small companies. BGA connections can only be inspected via computed tomography–type technologies that are capable of collecting x-ray- or ultrasound-based cross-sectional information about the flow of solder. This quality-control issue will be further addressed in Chap. 16 of this book.

A variation of BGA technology has been the utilization of cylindrical solder columns instead of spheres.

10.5.2 Assembly Process

In this section, our emphasis will be on the assembly process of SMT components, which differs from PIH component assembly, primarily on the placement subprocess of the individual components. SMT components rely on adhesives for remaining securely attached to the board prior to soldering, while the leads of PIH components must be slightly bent after insertion into their respective holes for secured attachment.

SMT technology was originally applied in military and aerospace industries in the mid-1960s for the major reason of yielding significantly higher densities than those offered by PIH technologies. These were ribbon-leaded flat packs comprising a ceramic package with flat ribbon leads (similar to today's gull-wing leads). The next generation SMT components, hermetic leadless chip carriers (HCC or LCC) became available in the early 1970s. However, these components faced compliancy (stress-relief) problems when placing ceramic components directly on organic (polymer-based) boards. The problem was overcome in the late 1970s with the introduction of plastic SMT packages. Today, SMT-based products can be found in all industries: automotive, aerospace, medical, and so on.

The typical assembly process of a two-sided PCB with SMT components is shown in Fig. 25. The primary steps in this process are solder paste application, adhesive application (if necessary), component placement, reflow soldering, solvent cleaning, and quality control.

Solder-paste application: Solder paste is, normally, a homogeneous, viscous material containing solder powder and flux material. Besides its primary function of joining the SMT components onto the PCB, it can also act as an adhesive that holds the component in place until soldering. The application of this paste onto the PCB can be achieved via screen (or stencil) printing, syringe dispensing, or pin transfer. Screen printing is the most widely utilized technique, where (in off-contact printing) the solder paste is flooded onto and forced through a screen onto the PCB (Fig. 26). The process is highly automatable.

Adhesive application: Syringe-dispensing or pin-transfer techniques can effectively apply adhesives to necessary points (dots) on the PCB.

Component placement: The pick-and-place operation of components refer to (normally vacuum-based) picking of a component from a feeder,

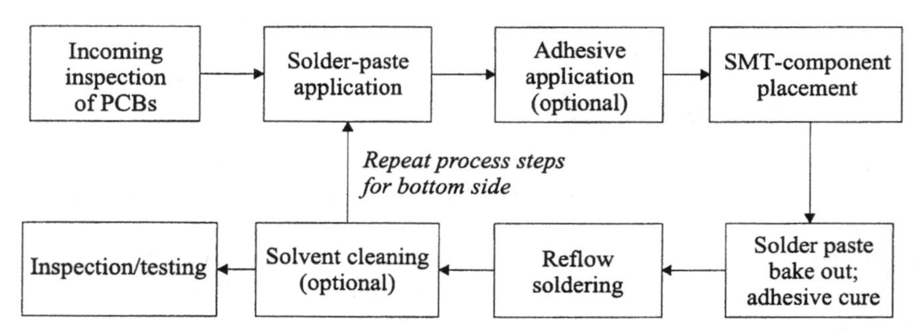

FIGURE 25 SMT assembly process.

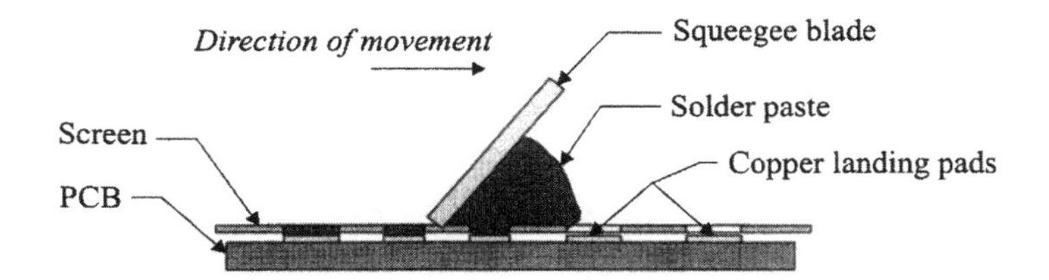

FIGURE 26 Screen printing of solder paste.

aligning it with its designated placement position and orientation, and releasing it. In sequential placement machines, this cycle is repeated for every component. Two variations of this machine are the fixed-PCB and moving-placement-head configuration versus moving-PCB and fixed- (but maybe rotating about the Z axis) placement-head configuration. Such gantry-type, X–Y planar, robotic placement machines can place up to 30,000 parts per hour, when placing components such as resistors and capacitors presented to the placement head on tapes and reels (Fig. 27). (Much lower rates would be achieved by special-purpose industrial robots utilized for the placement of irregular geometry components.)

Recent versions of placement machines include two placement heads working in concert for faster placement rates, especially for difficult-to-place components, which are fed using single-line magazines or array-type trays. Prior to z-axis motion placement of components with tight tolerances, most machines rely on position and orientation feedback of placement location achieved through a machine-vision system for minor adjustments in the x–y placement of the component.

Soldering: Prior to soldering, in most cases the populated board is subjected to sufficient heating for baking of the flux (solder) binder and the curing of the adhesives for better attachment, with an added side benefit of reducing potential gas escape during soldering. Most soldering equipment utilized in the electronics industry is of the conveyer type owing to mass production requirements. Reflow soldering using vapor-phase and infrared ovens is very common (Sec. 10.4). Wave soldering is applied to PIH components.

The above assembly process is concluded by a cleaning stage, prior to the inversion of the PCB for the assembly of components on the other side (for two-sided boards). The primary cleaning task is the removal of flux residues after the soldering process (especially critical for SMT components). If left on the board, such contaminants could lead to electrical and mechanical failures. Water-based washing is the most common cleaning technique.

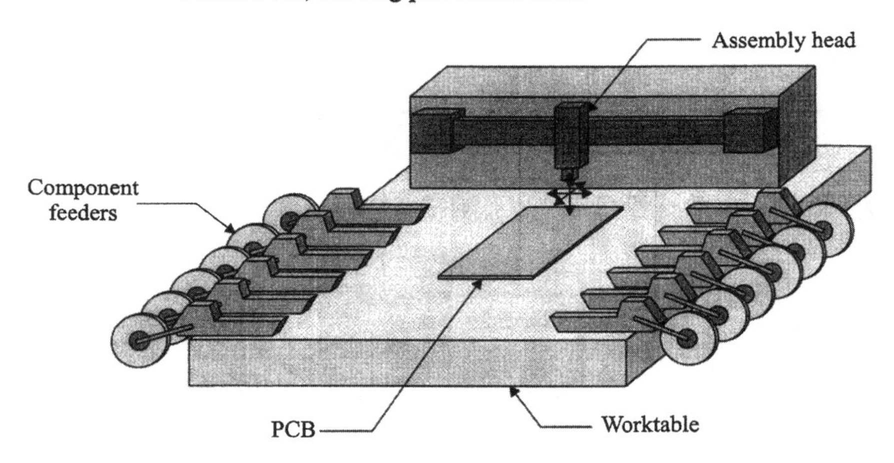

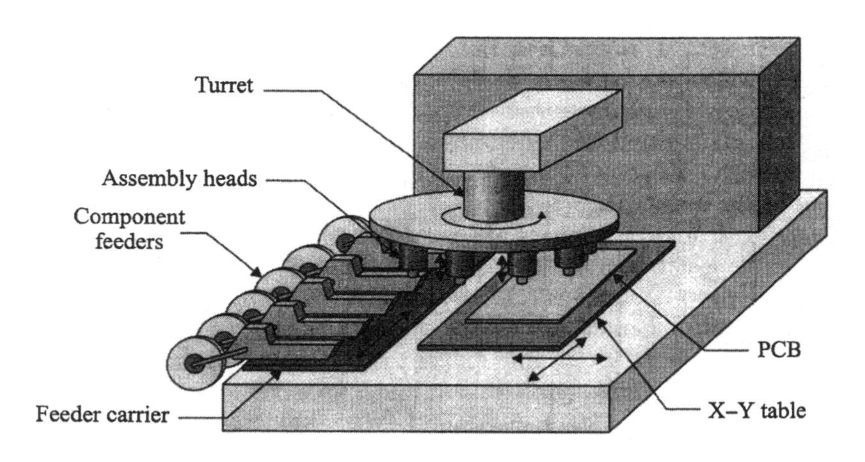

FIGURE 27 Component placement machines.

10.6 AUTOMATIC ASSEMBLY OF SMALL MECHANICAL COMPONENTS

The automatic assembly of small mechanical (versus electronic) components can be traced to the earlier part of the 20th century, though the widespread utilization of such nonprogrammable automation only started in the second half of the century, with the design and configuration of dedicated, single-purpose assembly machines. Then came the pioneering works of G. Boothroyd and P. Dewhurst, who challenged engineering designers to "think automation" when designing for assembly and to configure product geometries suitable for automatic assembly that does not rely on any sensory feedback. The majority of the literature on automatic assembly is in one way or another based on their work, and this section is no exception.

The primary joining operation relevant to automatic assembly discussed in this section is mechanical fastening, though parts positioned and oriented using the techniques described in the following subsections could be beneficial to other joining operations, such as adhesive bonding and soldering.

The first challenge in automatic assembly is the sorting and orientation of small mechanical components that arrive at the assembly area in unsorted bins. This operation is loosely defined as feeding. Its output is normally a single line of single-orientation components ready to be transported to the assembly area (from a remote but not too distant location) via feedtracks. The next challenge is the presentation of these components to pick-and-place manipulators for their assembly on pallets placed on an indexing machine or a conveyor.

10.6.1 Nonvibratory Feeding

Nonvibratory feeding is the use of devices that do not employ vibration as part of their sorting process; they rely on simple reciprocating or rotary-motion-based mechanisms. The basic requirement is that parts to be sorted must have a very limited number of stable orientations (preferably only two). A stable orientation refers to a static undisturbed orientation of a part when placed on a planar surface—for example, a rectangular prism would have three and a cylinder would have two stable orientations.

Although such feeders differ in their configurations, almost all have a limited-size hopper that acts as a large buffer from which the moving mechanism draws parts. A large number of examples can be found in the publications of Boothroyd et al. Only two cases are presented in this section to provide readers with sufficient understanding of the concept nonvibratory feeding.

Feeding of cylindrical parts: The feeder shown in Fig. 28 employs a one degree-of-freedom (dof) reciprocating mechanism with a semicylindrical concave-cavity (or a V-shaped-cavity) cross section. At every cycle, the reciprocating cavity dips into a "crowd" of cylindrical components and scoops up several of them during its upswing motion. Once at the limit of this upswing motion, the mechanism briefly stops to give a chance to the parts to slide down the tubular feedtrack, whose upper opening is aligned with the cavity of the reciprocating mechanism. As can be noted, parts can be engaged and slide down the tube only in one of their two possible stable orientations. Screws and rivets, and other similar geometry parts, can benefit from such a reciprocating nonvibratory feeder.

Feeding U-shaped parts: The feeder shown in Fig. 29 employs a one-dof rotary mechanism with multiple blades that have rectangular cross sections matching the inner dimension of the U-shaped part. At each cycle, every one of the blades of the rotary mechanism sweeps through the hopper (from the bottom, upward) and engages the U-shaped parts along its curved section. As the mechanism continues its motion, the parts slide down through the blade and move toward its end (straight-line) section. Eventually, the blade aligns itself with an inclined discharge rail and

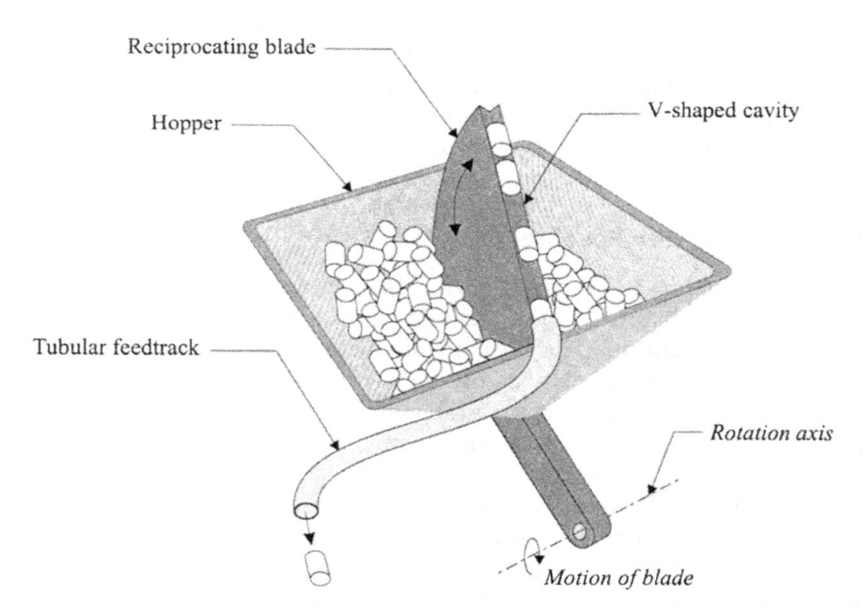

FIGURE 28 Feeding cylindrical parts.

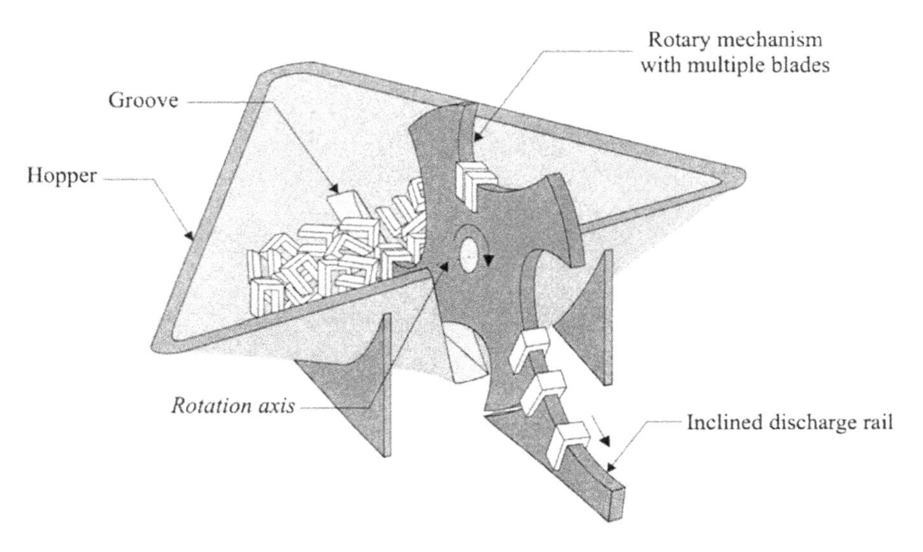

FIGURE 29 Feeding U-shaped parts.

transfers its load. The wheel must be slowed down or stopped completely at the discharge point.

10.6.2 Vibratory Feeding

Vibration has long been used by many industries for the separation and transportation of parts. In automatic assembly, these two objectives are achieved for small mechanical parts using vibratory bowls (Fig. 30a). This feeder device has two primary parts: a vibration-generator base and a replaceable bowl attached to this base through a number of leaf springs. Vibration is generated in two parts, whose combination moves the components upward along a spiral path along the outer wall of the bowl: a torsional vibration about the vertical axis (antigravity) and a translational vibration along this vertical axis (Fig. 30b).

Although parts hop along and move forward owing to the favorable vibration of the bowl, they may reach the end of the spiral path in any one of their stable orientations (preferably but not necessarily on a single line). Thus vibratory feeders must be equipped with orientation devices that would only allow parts with the desired orientation to exit the vibratory bowl. Most such devices employ a variety of attachments or cutouts for passive rejection of the incorrectly oriented parts (i.e., their return into the

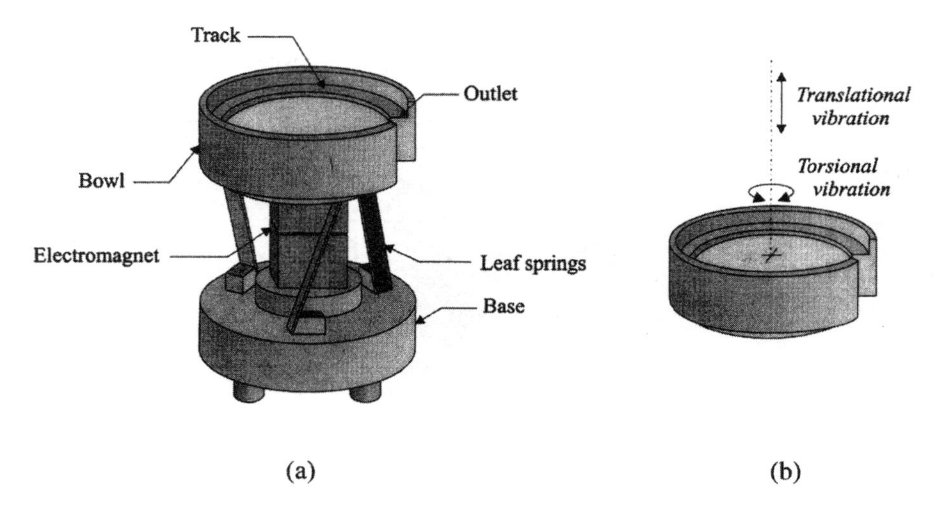

FIGURE 30 (a) Vibratory feeder; (b) vibration axes.

bowl). Naturally, since every stable part orientation would have an associated probability of occurrence, the efficiency of such orientation systems (i.e., the number of parts exiting it versus the number of parts returned back into the bowl) depends on the desired orientation chosen to exit the vibratory bowl. Orientation systems are built externally and normally welded onto the vibratory bowl, replacing an equivalent length section cut out of the last part of the spiral track.

Figure 31a shows an orienting device for screws that have six stable orientations (Fig. 31b). The device has several sections (attachments), each targeted for the rejection of one or more undesirable orientations back into the bowl. The first attachment rejects all parts that are standing up, orientations *a* and *b*; the second attachment narrows the path and thus rejects all parts with orientations *c* and *d*; the geometry of the last attachment not only rejects all parts that are not entering the cutout slot either with their head or tail first, namely all sideways parts, orientations *c* and *d*, that somehow made it to this point, but also it allows all parts that arrived with their head first to reorient themselves into the desired orientation of tail first, orientation *f*; the slot is large enough to let the threaded part of the screws fall through, but narrow enough to hold the head sections and allow the screws to move forward until they engage the discharge hole; once in the hole, the head of the screw must be correctly oriented to slide down, i.e., the cutout must engage the rail that is part of the tube—the torsional vibration

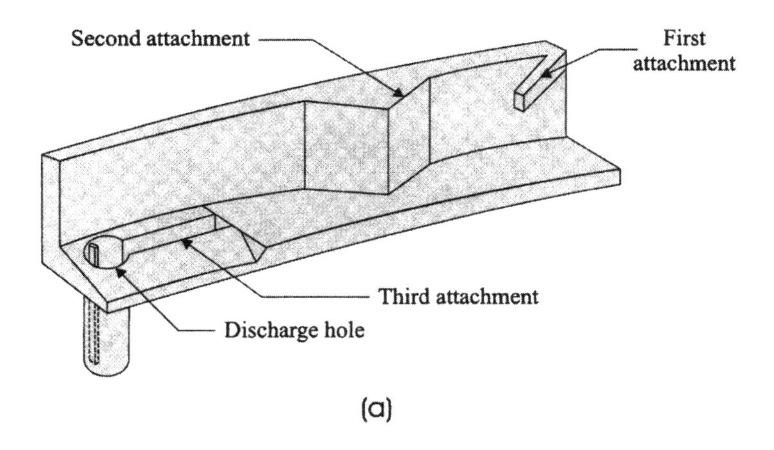

(a)

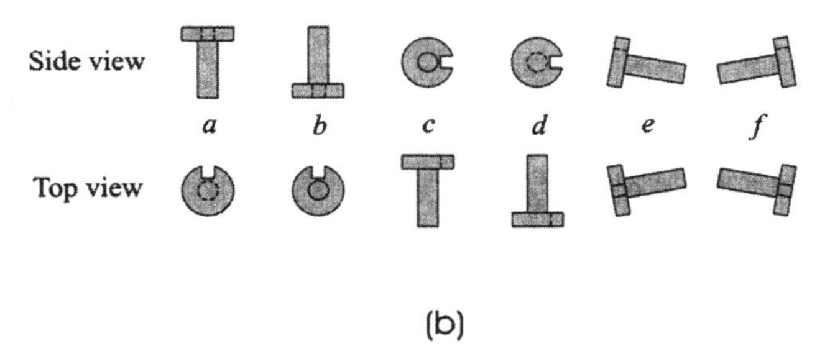

(b)

FIGURE 31 (a) Orientation device; (b) stable orientations.

of the bowl would cause the screw to rotate within the tube until this engagement occurs.

As mentioned earlier, the efficiency of the feeding/orientation system shown in Fig. 31 depends on the probability of occurrence of the individual six orientations. Let us say that they are $P_a = 1\%$, $P_b = 5\%$, $P_c = P_d = 12\%$, and $P_e = P_f = 35\%$, for a total of 100%. When 100 screws are dropped onto a hard flat surface, one would have an "*a*" orientation, five would have a "*b*" orientation, etc. Since the orientation device shown in Fig. 31 rejects all orientations *a* to *d*, accepts *f* as is, and reorients *e* into *f*, we would have a (35% + 35% =) 70% efficiency. Thus of 100 parts that randomly enter the orientation device, on average, 70 would exit it with the desired orientation.

Four other orientation devices are shown in Figs. 32a–32d.

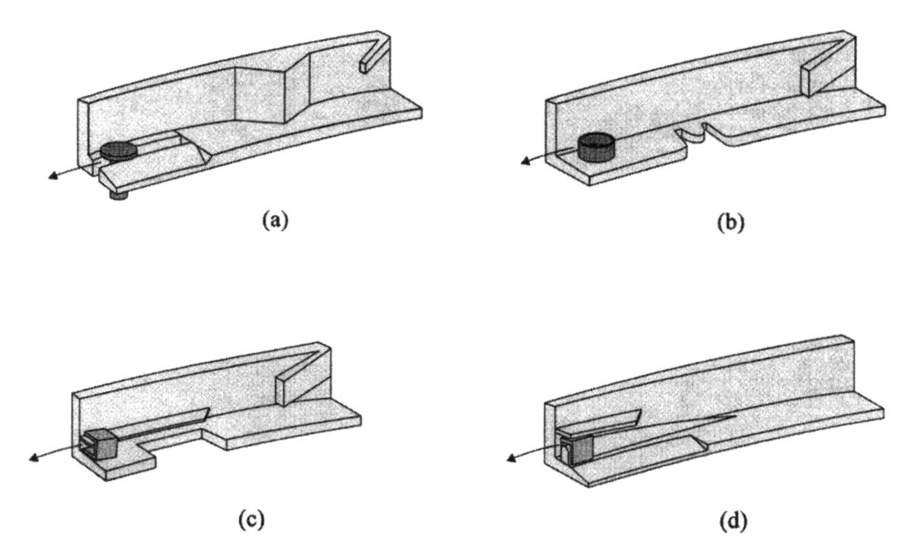

FIGURE 32 Four orientation devices.

10.6.3 Feedtracks and Escapements

Assembly workcell configurations may require sorting machines to be either in the periphery of the assembly area for easy access or noise requirements or as close to a robotic manipulator as needed. In both cases, parts exiting a sorting machine (vibratory or nonvibratory) must be maintained in a single line and desired orientation during their transfer to a designated discharge point. Feedtracks (or discharge rails) used for this purpose must be designed so that they are wide enough to allow fast and jam-free transportation, though narrow enough to prevent parts from changing their orientation during their motion.

Transportation along a feedtrack can be achieved by utilizing gravity or a powered arrangement. Most automatic assembly configurations elevate the sorting machines and allow parts exiting these feeders to slide down the rails via gravity (Fig. 33a).

Powered feedtracks can be of either the vibratory or the pneumatic type. The former vibrates, normally along a horizontal feed track for the forward hopping motion of parts, while the latter utilizes compressed air (input at one or many locations) to accelerate parts on a low-angle gravity track (Figs. 33b and 33c).

An alternative method of supplying an assembly station with sorted parts would be the use of magazines (i.e., finite length, finite capacity, and

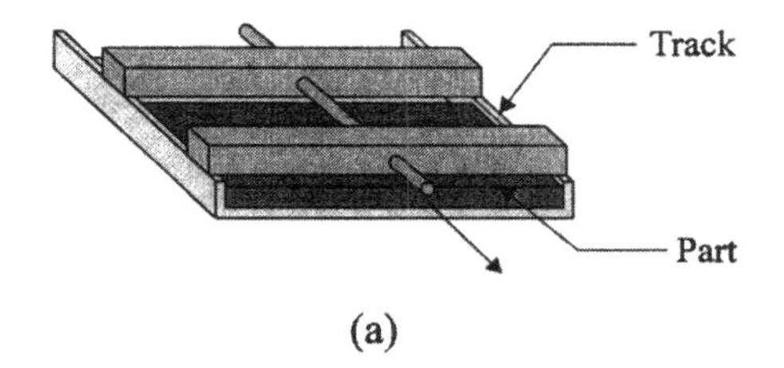

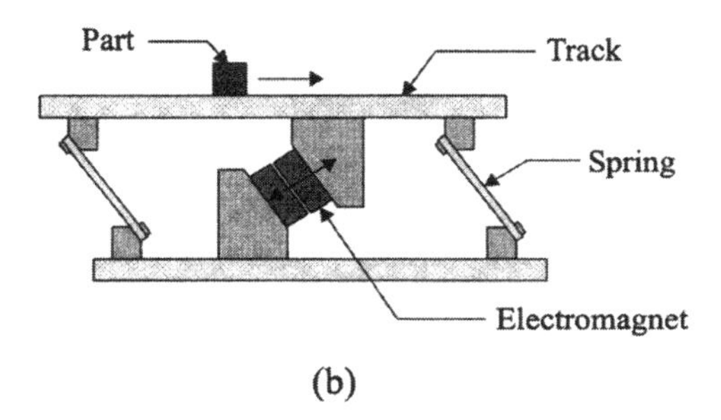

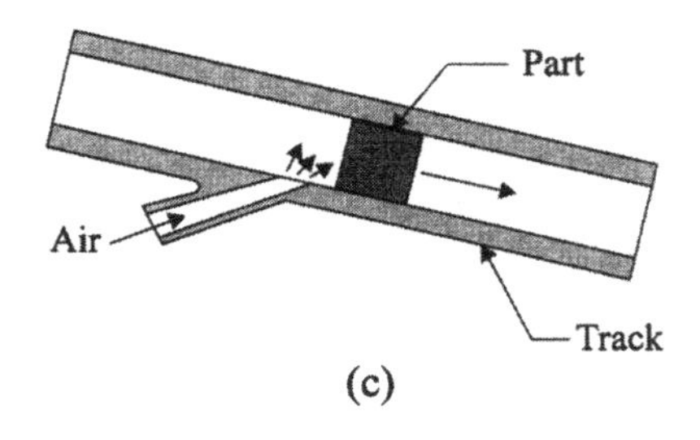

FIGURE 33 Feedtracks: (a) gravity based; (b) vibrational; (c) pneumatic.

transportable feedtracks). Magazines can be filled at remote locations by being attached to a sorting machine or directly to the production machine that is fabricating the parts, thus not necessitating the use of a subsequent sorter. These magazines would then be mounted onto proper fixtures at assembly areas and connected to escapement devices. The use of magazines is very common in the electronics industry, though these should be reusable to be cost effective.

Parts that reach the end of the feedtrack or a magazine must be stopped and presented to the assembler (a robotic manipulator or human operator) at required exact quantities and at exact times or instants. There

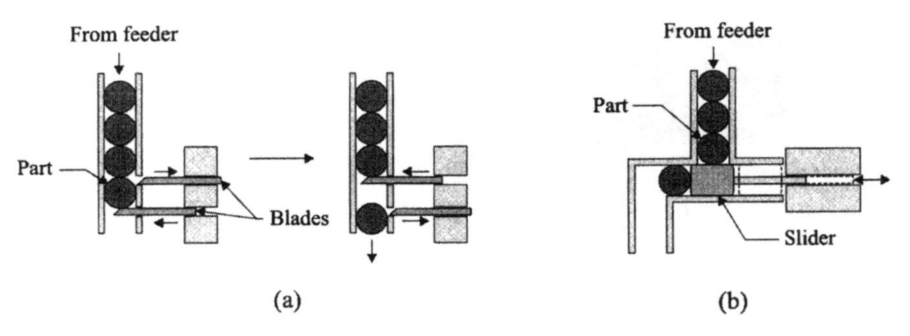

Figure 34 (a) Ratchet; (b) slider escapements.

are a variety of industrial escapements (metering devices) that allow the fulfilment of both of these objectives. The most commonly used escapements for vertically stacked parts are the ratchet and slider escapements (Fig. 34).

The ratchet escapement employs two knives (blades) that move in opposite directions: the lower one allows the passage of one (or many) parts, while the upper blade stops the slide of the rest of the waiting parts. The slider escapement utilizes a sliding separator that redirects one or more parts from a vertically stacked column of parts waiting to be dispatched in a feedtrack. For both ratchet and slider escapements, the actions of the dispatchers (blades, blocks, etc.) would be event controlled—i.e., they would execute their motions based on orders received from the controller of the robot that would notify the escapement about a need for the next set of parts (in contrast to a clock-based time control).

10.6.4 Transfer Equipment

Product assembly via most mechanical joining operations are carried out on pallets equipped with fixtures (Chap. 11) that are stationary or moving at a speed that can be matched by an assembler for on-the-fly assembly. Transfer equipment utilized in transporting these pallets from one assembly station to another (located within a closed vicinity of each other) can be achieved using indexing tables or pallet conveyers (Fig. 35).

A *rotary indexing table* is a circular plate with built-in multiple pallets that is rotated using a drive unit (e.g., ratchet drive, Maltese-cross drive, cam drive, rack-and-pinion drive). The table advances one step (station) forward when all assembly operations at individual stations have been completed—an event-based control.

A *pallet conveyor* is recommended as an intermittent transfer system for applications that require high assembly accuracies. Pallets placed on

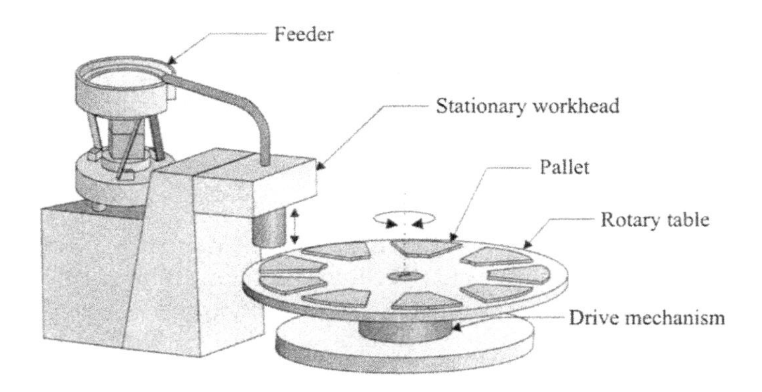

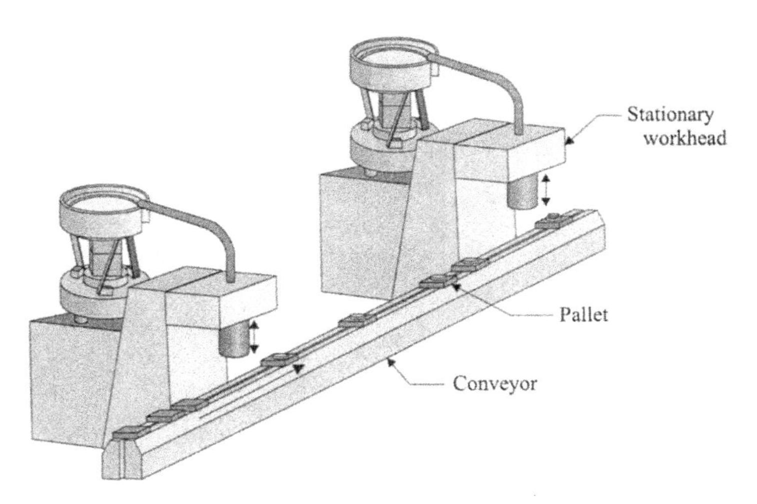

FIGURE 35 Transfer equipment.

such conveyors advance from one station to another by the continuous motion of the conveying medium (e.g., the belt). Once at the desired location, the pallet is stopped and disengaged from the moving conveyor and fixtured accurately by a pallet raiser. Upon termination of the assembly task, the pallet is returned onto the conveyor and allowed to proceed to the next station.

10.6.5 Positioners

In this section, the term "positioner" is utilized to describe nonreprogrammable robotic pick-and-place mechanisms. Such mechanisms are normally low-cost custom-built manipulators with one to three controlled axes of motion (dof). Frequently they are assembled into a desired configuration using a set of modular components (links, actuators, grippers, etc.). Their

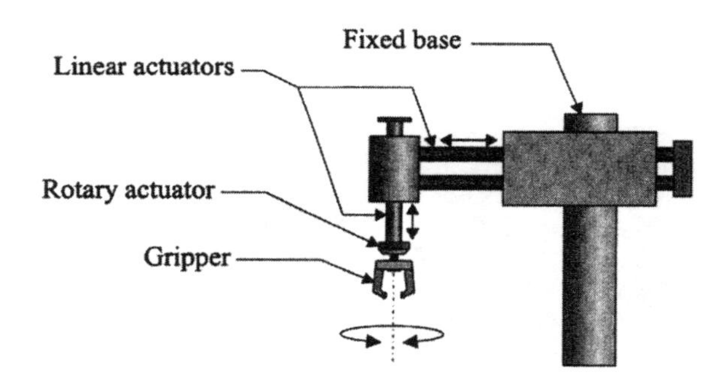

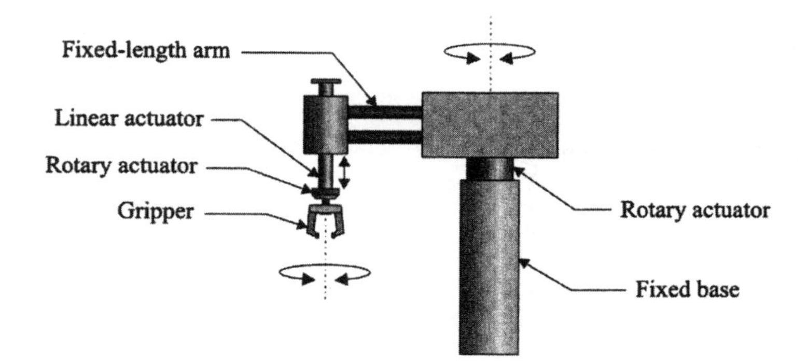

FIGURE 36 Pick-and-place manipulators.

objective is to move mechanical components from one location (e.g., feedtrack) to another (e.g., pallet) in a point-to-point motion at the highest possible speed. For cost effectiveness, the motions of the joints (for example, powered via pressurized air) are not controlled between the two end-of-motion points that are specified by hard stops (with possible built-in mechanical switches).

Two 2-dof pick-and-place mechanism configurations are shown in Fig. 36. Other reprogrammable robot configurations will be discussed in detail in Chap. 12 of this book in the larger context of material handling for a variety of manufacturing applications, including the assembly of small and large parts.

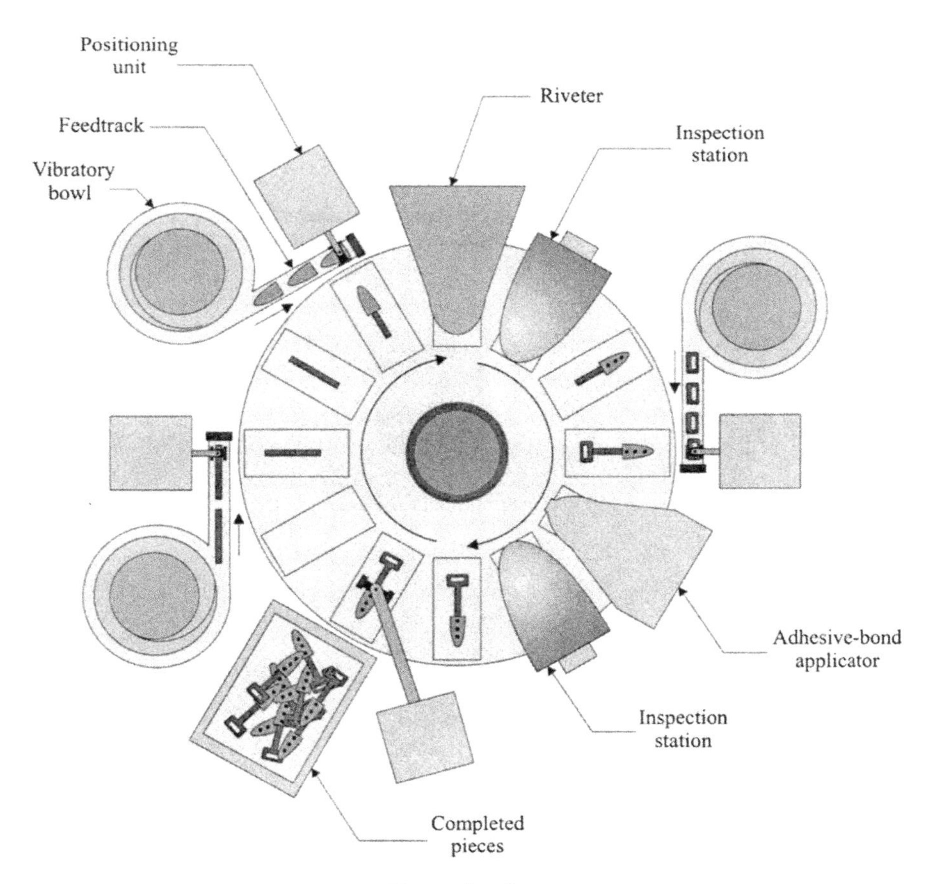

FIGURE 37 An automatic assembly workcell.

In conclusion to the description of the automatic assembly process (Secs. 10.6.1 to 10.6.5), an example assembly workcell that utilizes multiple vibratory bowl feeders, positioners, and numerous joining equipment to assemble a product on an indexing table is shown in Fig. 37.

10.6.6 Design for Automatic Assembly

Assembly has been defined in this chapter as a manufacturing operation that adds cost to a product more than it adds value. The first guideline of design for assembly, thus, has always been to minimize the number of parts that must be joined when assembling a product. Boothroyd et al. set the

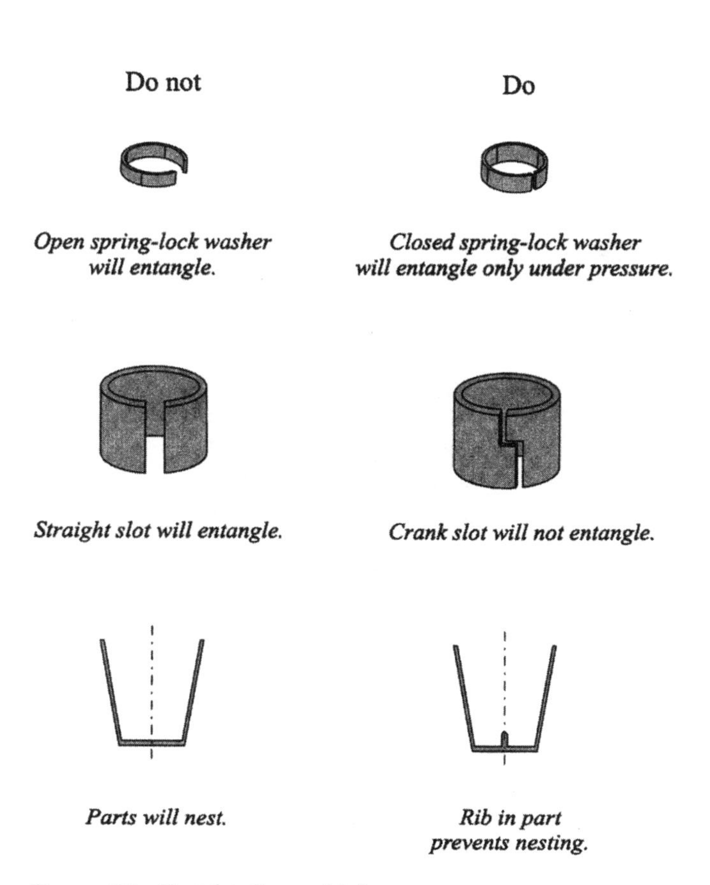

Figure 38 Part feeding guidelines.

following three criteria for whether a part needs to exist as a separate entity from others in the product:

Does the part need to have mobility after assembly?
Does the part need to be of different material or separated from others, owing to reasons such as electrical insulation?
Does the part obstruct or prevent the joining of other parts?

In this section, we will review an exemplary set of secondary guidelines that a product designer must be aware of when configuring part geometries for automatic assembly of small mechanical parts. The order of presentation will be in conformity with the typical order of tasks carried out sequentially: feeding, orienting, transporting, and positioning for joining.

Do not

Do

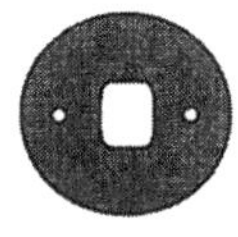

Circular shape makes it difficult to properly orient the holes.

Nonfunctional flat edges help in orienting the holes.

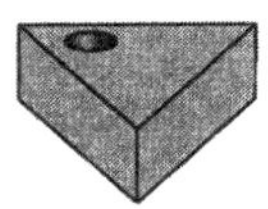

Triangular shape makes it difficult to properly orient the hole.

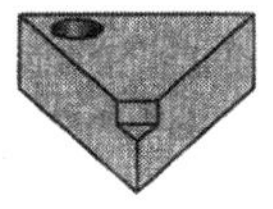

Nonfunctional shoulder helps orient the hole.

Tubular shape makes it difficult to properly orient head.

Double head allows two possible correct orientations.

FIGURE 39 Part orienting guidelines.

Feeding: Parts that arrive in bulk and deposited into (vibratory or non vibratory) feeders must be separated using vibration or mechanical mixing and be maintained free of each other prior to their orientation. Part geometries must not have features (projections, slots, etc.) that will cause tangling, nesting, jamming, etc. (Fig. 38). Nonfunctional features can be added to parts to facilitate feeding.

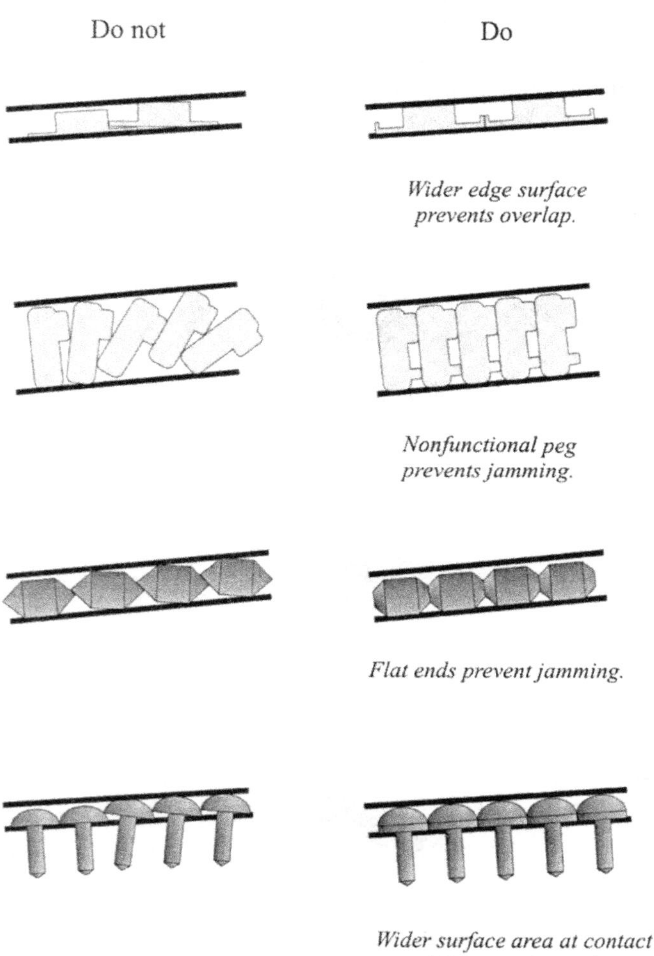

FIGURE 40 Part transportation guidelines.

Do not | Do

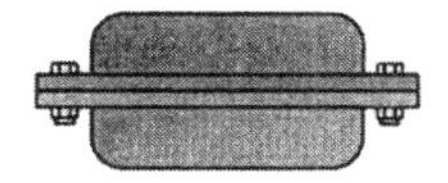

To fasten nut, assembly needs to be rotated.

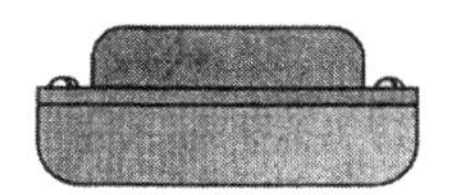

No rotation of assembly is required.

Part can jam when inserted into hole.

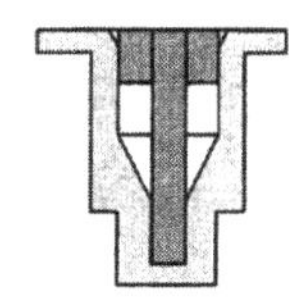

Tapering the hole facilitates part insertion.

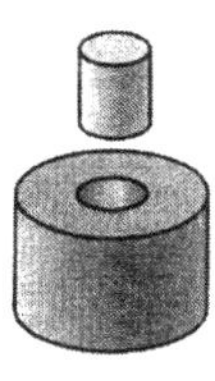

Parts are difficult to assemble.

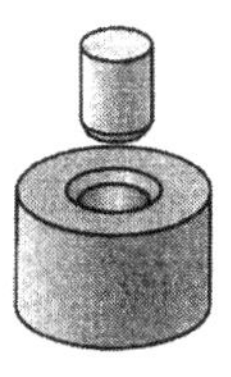

Chamfers assist assembly.

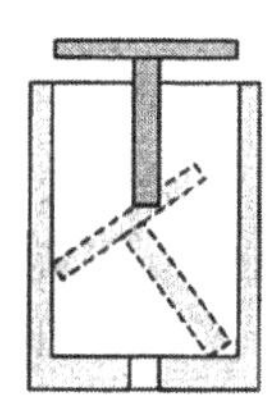

Parts are difficult to align properly; jamming can occur.

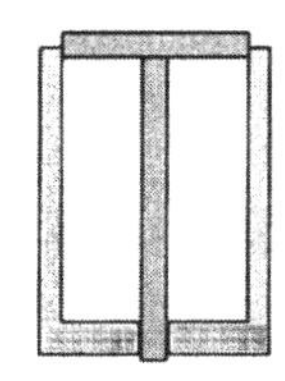

Changing part geometry allows it to be aligned properly before it is released.

FIGURE 41 Part positioning guidelines.

Orienting: Although mechanical part orienting systems can be manufactured to high tolerances, it is strongly advised to design parts with complete symmetry or exaggerated asymmetry. As in feeding, designers should consider adding nonfunctional features to a part to facilitate automatic orientation, especially in vibratory environments (Fig. 39). An added benefit for such an approach would be ease of recognition of correct assembly orientation by a human assembler.

Transporting: Feedtracks need to be designed and manufactured in their narrowest possible configurations in order to maintain part orientation during transportation. Thus parts must be designed to cooperate with this objective for jam-free transportation (Fig. 40).

Positioning and assembly: The primary requirement of automatic assembly is layered joining. That is, parts should be joined in a minimum number of different orientations. This will reduce costs of fixturing and positioning—requiring robotic manipulators with fewer dof. The second requirement is ease of insertion/mating—parts should have chamfers, self-locating features, tapers, etc. (Fig. 41).

For manual assembly, part designs should also consider the limited dexterity of the human hand (parts should not be too small or slippery) and the safety of the operator (not too sharp).

REVIEW QUESTIONS

1. Why are most products assembled by manufacturers, as opposed to being sold unassembled, even though assembly is seen as a cost-adding process?
2. Define interchangeability of parts for profitable manufacturing.
3. Define the three primary principles of design for assembly.
4. When would mechanical fastening be preferable to other joining methods?
5. Discuss the two conflicting interests in choosing the locations of mechanical fasteners: marketability (aesthetics) versus ease of assembly.
6. Many fasteners used in product assembly could have an unnecessarily large number of threads, thus prolonging the joining process. Argue both sides of this issue. You may further refer to Chap. 7 for similar concerns in die setups (quick die exchange).
7. What are the two common corrosion protection mechanisms for fasteners?
8. What does permanency imply in riveting, in contrast to in adhesive bonding, in soldering, etc.?
9. What are the primary advantages/disadvantages of adhesive bonding? Discuss its potential use in aviation industries versus riveting.

10. Discuss the following two issues in adhesive bonding: overlapping of the joints and joint thickness.
11. Why is surface preparation the most important step in adhesive bonding?
12. Define fusion welding versus solid-state welding.
13. Why should welding regions be shielded using inert gases?
14. Compare shielded metal arc welding (SMAW) to gas metal arc welding (GMAW).
15. Why would one use nonconsumable electrodes in arc welding (e.g., GTAW)?
16. Discuss the advantages of laser-beam welding.
17. Why should brazing and soldering not be considered welding operations? What is the primary difference between brazing and soldering?
18. Describe the wetting process and its importance in brazing and soldering.
19. What is the purpose of using a flux in the soldering process?
20. What is the eutectic composition of the tin–lead alloy and why do electronics manufacturers prefer to use it for soldering?
21. What is the principal similarity/difference between reflow soldering and infrared (reflow) soldering?
22. Describe the ball-grid array (BGA) packaging technology used in the electronics industry.
23. What are the primary steps of assembly of small mechanical components when using nonprogrammable (vibratory or nonvibratory) feeding devices?
24. How do parts move forward in the spiral track of a vibratory feeder?
25. What are stable part orientations? How would one use probability data about stable part orientations in designing orienting systems for vibratory feeders?
26. Discuss the design of effective feedtracks for optimal part transportation.
27. Discuss the use of magazines in parts assembly.
28. Discuss the three criteria proposed by Boothroyd et al. in evaluating whether a part needs to exist as a unique component or whether it could be joined to another in order to reduce the number of parts in a product.

DISCUSSION QUESTIONS

1. Assembly is often viewed as a cost-adding operation as opposed to fabrication, which is viewed as a value-adding operation. Discuss this issue in the context of several exemplary products: computers and automobiles that are preassembled versus furniture that is sold for "some assembly" by the customers.
2. The majority of mechanical joining techniques yield nonpermanent joints (i.e., those that can be removed without destroying the joining

elements and/or damaging the joined components). Discuss the advantages/disadvantages of such joints (when compared to permanent joints, such as rivets, adhesive joints, soldered joints, welded joints, etc.) in various operational conditions (static versus dynamic).

3. Welding, soldering, and painting are a few manufacturing operations that rely on the maintenance of consistent and repeatable process parameters. Discuss the use of automation (including the use of industrial robots) for these and other processes that have similar requirements as replacements for manual labor.
4. Nonfunctional features on a product facilitate their manual (or even automatic) assembly. Discuss some generic feature characteristics for such purposes. Furthermore, examine some consumer products whose daily use could be facilitated by such features (e.g., near-square cookie boxes with not-connected lids, door handles, etc.).
5. Process planning in assembly (in its limited definition) is the optimal selection of an assembly sequence of the components. For example, solving the traveling salesperson problem in the population of electronics boards. Although there exist a number of search techniques for the solution of such problems, they could all benefit from the existence of a good initial (guess) solution. Discuss the role of group technology (GT) on the identification of such initial (guesses) sequences of assembly.
6. Analysis of an assembly process via computer-aided modeling and simulation can lead to an optimal process plan with significant savings in assembly time and cost. Discuss the issue of time and resources spent on obtaining an optimal plan and the actual (absolute) savings obtained due to this optimization: for example, spending several hours in planning to reduce assembly time from 2 minutes to 1 minute. Present your analysis as a comparison of one-of-a-kind production versus mass production.
7. Several fabrication/assembly machines can be physically or virtually brought together to yield a manufacturing workcell for the production of a family of parts. Discuss the advantages of adopting a cellular manufacturing strategy in contrast to having a departmentalized strategy, i.e., having a turning department, a milling department, a grinding department, etc. Among others, an important issue to consider is the transportation of parts (individually or in batches).
8. Human factors (HF) studies encompass a range of issues spanning from ergonomics to human–machine (including human–software) interfaces. Discuss the role of HF in the autonomous factory of the future, where the impact of human operators is significantly diminished and emphasis is switched from operating machines to supervision, planning, and maintenance.

9. During the 20th century, there have been statements and graphical illustrations implying that product variety and batch size remain in conflict in the context of profitable manufacturing. Discuss recent counterarguments that advocate profitable manufacturing of a high variety of products in a mass-production environment. Furthermore, elaborate on an effective facility layout that can be used in such environments: job shop, versus cellular, versus flow line, versus a totally new approach.

BIBLIOGRAPHY

Barnes, T. A., Pashby, I. R. (2000). Joining techniques for aluminum space frames used in automobiles. *Journal of Materials Processing Technology* 99(1):72–79.

Boothroyd, Geoffrey (1992). *Assembly Automation and Product Design*. New York: Marcel Dekker.

Boothroyd, Geoffrey, Dewhurst, Peter, Knight, Winston (1994). *Product Design for Manufacture and Assembly*. New York: Marcel Dekker.

Brandon, David G., Kaplan, Wayne D. (1997). *Joining Processes: An Introduction*. New York: John Wiley.

Brindley, Keith (1990). *Electronics Assembly Handbook*. Oxford: H. Newnes.

Capillo, Carmen (1990). *Surface Mount Technology: Materials, Processes, and Equipment*. New York: McGraw-Hill.

Cornu, Jean (1988). *Advanced Welding Systems*. Vols. 1–3. Bedford, UK: IFS.

Dally, James W. (1990). *Packaging of Electronic Systems: A Mechanical Engineering Approach*. New York: McGraw-Hill.

Davies, Arthur C. (1992). *The Science and Practice of Welding*. Vols. 1–2. New York: Cambridge University Press.

DeLollis, Nicholas J. (1970). *Adhesives for Metals: Theory and Technology*. New York: Industrial Press.

Doko, T., Takeuchi, H., Ishikawa, K., Asami, S. (Aug. 1991). Effects of heating on the structure and sag properties of bare fins for the brazing of automobile heat exchangers. *Furukawa Rev.*, (9):47–54.

Doyle, Lawrence E., et al (1985). *Manufacturing Processes and Materials for Engineers*. Englewood Cliffs, NJ: Prentice-Hall.

Duley, W. W. (1999). *Laser Welding*. New York: John Wiley.

Dunkerton, S. B., Vlattas, C. (1998). Joining of aerospace materials—an overview. *International Journal of Materials and Product Technology* 13(1/2):105–121.

Geary, Don (2000). *Welding*. New York: McGraw-Hill.

Gibson, Stuart W. (1997). *Advanced Welding*. Basingstoke, UK: Macmillan.

Groover, Mikell P. (1996). *Fundamentals of Modern Manufacturing: Materials, Processes, and Systems*. Upper Saddle River, NJ: Prentice Hall.

Hahn, O., Peetz, A., Meschut, G. (Sep.–Oct. 1998). New technique for joining mobile frame constructions in car body manufacturing. *J. of Welding in the World* 41(5):407–411.

Haviland, Girard S. (1986). *Machinery Adhesives for Locking, Retaining, and Sealing*. New York: Marcel Dekker.

Higgins, A. (2000). Adhesive bonding of aircraft structures. *International Journal of Adhesion and Adhesives* 20(5):367–376.

Humpston, Giles, Jacobson, David M. (1993). *Principles of Soldering and Brazing*. Materials Park, OH: ASM International.

Kalpakjian, Serope, Schmid, Steven R. (2000). *Manufacturing Engineering and Technology*. Upper Saddle River, NJ: Prentice Hall.

Lane, J. D. (1987). *Robotic Welding*. Bedford, UK: IFS.

Lee, Lieng-Huang (1991). *Adhesive Bonding*. New York: Plenum Press.

Lees, W. A. (1989). *Adhesives and the Engineer*. London: Mechanical Engineering Publications.

Lindberg, Roy A. (1976). *Welding and Other Joining Processes*. Boston, MA: Allyn and Bacon.

Lotter, Bruno (1989). *Manufacturing Assembly Handbook*. London: Butterworths.

Manko, Howard H. (1992). *Solders and Soldering: Materials, Design, Production, and Analysis for Reliable Bonding*. New York: McGraw-Hill.

Manko, Howard H. (1995). *Soldering Handbook for Printed Circuits and Surface Mounting: Design, Materials, Processes, Equipment, Trouble-Shooting, Quality, Economy, and Line Management*. New York: Van Nostrand Reinhold.

Messler, Robert W. (1999). *Principles of Welding: Processes, Physics, Chemistry, and Metallurgy*. New York: John Wiley.

Nicholas, M. G. (1998). *Joining Processes: Introduction to Brazing and Diffusion Bonding*. Boston, MA: Kluwer.

Nof, Shimon Y., Wilhelm, Wilbert E., Warnecke, Hans-Jürgen (1997). *Industrial Assembly*. New York: Chapman and Hall.

Oates, William R., Connor, Leonard P., O'Brien, R. L., eds. (1996). *Welding Handbook*. Miami, FL: American Welding Society.

Oystein, Grong (1997). *Metallurgical Modelling of Welding*. London: Institute of Materials.

Packham, D. E., Harlow, Burnt Mill (1992). *Handbook of Adhesion*. New York: John Wiley.

Park, K. H., Barkley, J. M., & Woody, J. R. (1986). Vacuum brazed aluminum radiator for Ford Topaz. SAE Technical Paper Series, # 860644 (10 pages).

Parmley, Robert O. (1997). *Standard Handbook of Fastening and Joining*. New York: McGraw-Hill.

Pecht, Michael G. (1993). *Soldering Processes and Equipment*. New York: John Wiley.

Pender, James A. (1986). *Welding*. Toronto, ON: McGraw-Hill Ryerson.

Petrie, Edward M. (2000). *Handbook of Adhesives and Sealants*. New York: McGraw-Hill.

Rahn, Armin (1993). *The Basics of Soldering*. New York: John Wiley.

Ryan, E. J., King, W. H., Doyle, J. R. (March 1979). Brazing tomorrow's aircraft engines. *Weld Design and Fabrication* 52(3):108–114.

Sadek, M. M. (1987). *Industrial Applications of Adhesive Bonding*. New York: Elsevier.

Schey, John A. (1987). *Introduction to Manufacturing Processes*. New York: McGraw-Hill.

Schneberger, Gerald L. (1983). *Adhesives in Manufacturing*. New York: Marcel Dekker.

Schwartz, Mel M. (1995). *Brazing: For the Engineering Technologist*. London: Chapman and Hall.

Speck, James A. (1997). *Mechanical Fastening, Joining, and Assembly*. New York: Marcel Dekker.

Solberg, Vern (1996). *Design Guidelines for Surface Mount and Fine Pitch Technology*. New York: McGraw-Hill.

Strauss, Rudolf (1998). *SMT Soldering Handbook*. Oxford: H. Newnes.

Thwaites, Colin J., Barry, B. T. K. (1975). *Soldering*. Oxford: Oxford University Press.

Tlusty, Jiri (2000). *Manufacturing Processes and Equipment*. Upper Saddle River, NJ: Prentice Hall.

Traister, John E. (1990). *Design Guidelines for Surface Mount Technology*. San Diego, CA: Academic Press.

Wegman, Raymond F. (1989). *Surface Preparation Techniques for Adhesive Bonding*. Park Ridge, NJ: Noyes.

Woodgate, Ralph W. (1996). *The Handbook of Machine Soldering: SMT and TH*. New York: John Wiley.

11

Workholding—Fixtures and Jigs

Workholding in manufacturing is the immobilization of a part (workpiece) for the purpose of allowing a fabrication or an assembly process to be carried out on it. The term fixturing is also commonly used to describe workholding. Design of workholding devices normally falls within the domain of expertise of tool designers, who decide what fabrication or assembly tools to use as well as what fixtures or jigs to employ. The overall objective is to increase productivity through increased rates of manufacturing: utilize tools with appropriate lengths of life and fixtures/jigs with optimum accuracies.

A jig is a workholding device, primarily used in hole fabrication, for locating and holding a workpiece and guiding the production tool (e.g., a bushing for guiding a drill bit and thus preventing slippage and vibrations during the engagement of the tool with the workpiece). A fixture, on the other hand, is a workholding device used in machining and assembly for securely locating and holding the workpiece without providing a built-in guidance to the manufacturing tool. Both types of devices, jigs and fixtures, must provide maximum accuracy (including measures to prevent incorrect workholding) and be designed for ease of mounting and clamping of the workpiece by humans or robots.

Design of a workholding device requires a careful examination of the workpiece (geometry, material, mechanical properties, and tolerances), the fabrication processes (tool paths, machining/assembly forces, and environment, e.g., coolant liquids), and the specific machines to be utilized. An additional issue to be considered is the target setup cost that can be afforded. Owing to their complexity and high accuracy, workholding devices can be very expensive. Normally, these one-of-a-kind devices are expected to be used for a large number of workpieces in mass production, in order to minimize their per part cost. In case of small-batch or one-of-a-kind product manufacturing, modular fixtures that can be reconfigured according to the part geometry at hand should be utilized.

Although modular fixtures have been in existence since the 1940s, their primary users until the early 1980s were the machine-tool manufacturers, who fabricated small-batch or one-of-a-kind lathes, milling machines, and so on. With the widespread utilization of flexible manufacturing strategies in the past two decades, such reconfigurable devices have become very attractive in group-technology-based workcells for the fabrication of a family of similar parts. In parallel to industrial advancements on the design of modular fixtures, numerous academic research centers have also developed (1) reconfigurable and programmable ("flexible") fixtures for use in automated environments, and (2) computer-aided design tools for the efficient design of fixtures (one-of-a-kind or modular) in concurrent engineering environments.

In this chapter, following the description of general workholding principles and basic design guidelines for jigs and fixtures, we will review the use of such devices in manufacturing, in the form of dedicated or modular configurations. We will also present a brief discussion on the computer-aided design aspects of fixture/jig development.

11.1 PRINCIPLES OF WORKHOLDING

The design of a workholding device is governed by the geometry of the workpiece and the dynamics of the manufacturing process in which it is expected to participate. The fixture/jig must be able to hold the workpiece in place (i.e., preventing motion and deflections) while it is subjected to external forces. These forces are most prominent in metal cutting operations and might cause the workpiece to break away from the workholding device or to fracture if it were not supported suitably. Thus locating and clamping will be discussed below in greater detail.

11.1.1 Locating

A solid body has six degrees of freedom (dof) of mobility in unconstrained three-dimensional space: three degrees of translational movement freedom (D_x, D_y, D_z) and three of rotational movement freedom (R_x, R_y, R_z)

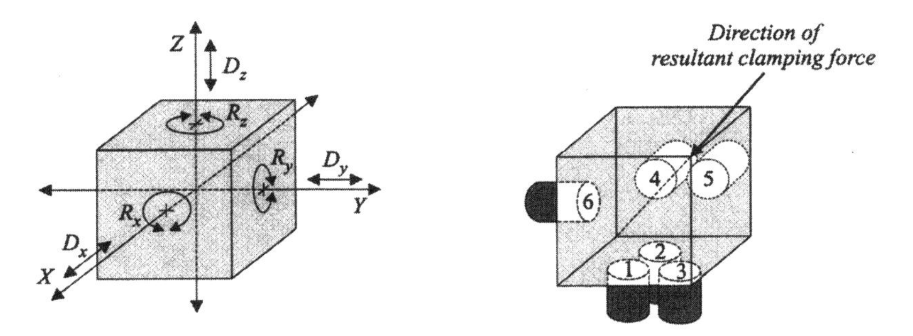

FIGURE 1 (a) Mobility of a solid body; (b) the 3-2-1 principle.

(Fig. 1). The objective of a workholding device is to eliminate all mobility and simultaneously provide adequate support to the workpiece to counteract external forces.

Three-dimensional mobility can be prevented by utilizing six points of constraint, by the 3-2-1 rule (Fig. 1b): three points (1, 2, and 3) provide a planar constraint, eliminating two rotational (R_x and R_y) and one translational ($-D_z$) dof, two additional (orthogonal) points (4 and 5) eliminate one more rotational (R_z) and one more translational ($-D_x$) dof and, finally, a sixth point (6) totally constrains the workpiece by eliminating the last translational dof ($-D_y$). Naturally, as seen from Fig. 1b, this immobility can be achieved only if the workpiece is pushed against these support points and held in place by a clamping device.

For the best possible accuracy, locators should contact the workpiece on its most accurate surfaces (versus unmachined, rough surfaces). Although point contact would yield best positioning accuracy, most locators have planar contact surfaces, in order to minimize damage to the workpiece due to potentially high-pressure contact points. Redundancy in locating should be avoided, unless necessary for safety reasons or to prevent deflections. Distribution and configuration of the locators is an engineering analysis issue: mechanical stress analysis should be carried out for the optimal placement of locators (Sec. 11.4).

Locators are manufactured separately from the main body of the workholding device (e.g., a mounting plate) using tool-quality-hardness steel for minimum wear. They are normally fabricated to exact specifications as fixed dimension components or as adjustable height locators. Some exemplary locators are shown in Fig. 2.

Locators may be placed on the periphery of the object or underneath it and, occasionally, fitted into existing holes on the workpiece. One must note that, for example in machining, locators should not be mounted directly on

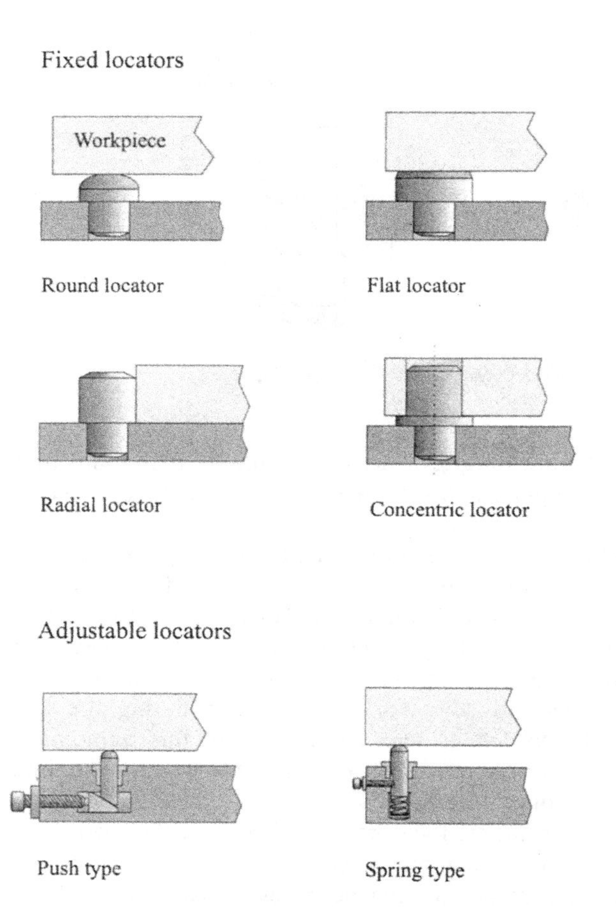

Figure 2 Fixed and adjustable locators.

the machine tool's table but on the workholding device's body, which is subsequently secured onto the machine tool's table.

11.1.2 Clamping

The role of a clamping device is to apply sufficient force on a workpiece to maintain its absolute immobility during the manufacturing process. Clamping forces should be sufficiently high not to allow any loosening due to potential vibrations and be directed toward support points (in the most solid sections of the workpiece) to prevent distortion or damage. Forces generated during manufacturing, however, should be counteracted by the fixed parts of the workholding device (locators and the base plate) and not by the clamps.

As with locators, clamps must allow for rapid loading/unloading of the fixture/jig and normally be located in the periphery for minimum interference with the manufacturing operations. The five basic classes of clamping are briefly described below (Fig. 3):

Strap clamps: The basic configuration comprises a bar, a heel pin, and a lever or a threaded rod. These clamps are the simplest to use and are found in most workholding devices.

Screw clamps: The moment developed by a screw is utilized to hold the workpiece in place. Although simple to use, these clamps are slower to operate than others.

Cam clamps: Cam-shaped levers are utilized in fast-operating clamping for direct or indirect application of pressure on the workpiece. Cam-action clamps would be susceptible to vibrations during the manufacturing operation.

Toggle clamps: Toggle-action clamps have the ability quickly and completely to move away from the workpiece once unlocked. The

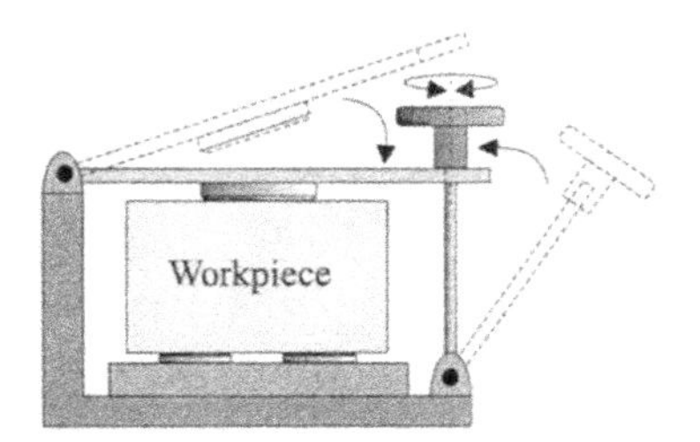

Strap clamp

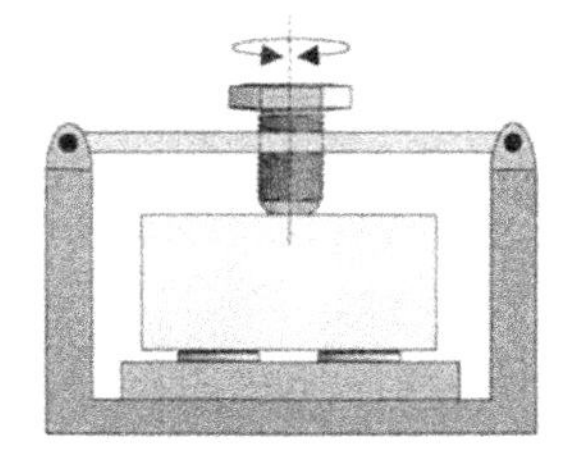

Screw clamp

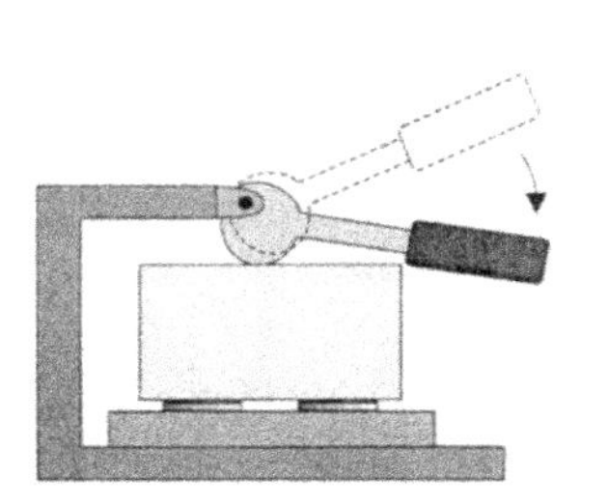

Eccentric-cam clamp

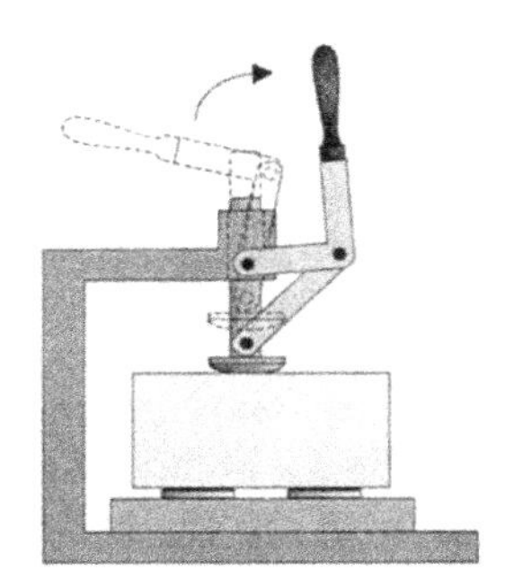

Toggle clamp

FIGURE 3 Clamps.

two common configurations used in manufacturing applications are the ones with the hold-down and straight-line-push actions.

Almost all clamping devices can be power activated using a hydraulic or electrical power source and occasionally a pneumatic power source. The obvious advantage of power activation is usefulness for automation.

Commercially available chucks (for lathes) and vises (for milling machines) are also considered as general-purpose clamping devices. Both devices can be configured for manual operation or automatic clamping. There also exist magnetic and vacuum chucks and vises for nonmechanical clamping of workpieces that would not be subjected to large forces during manufacturing.

11.1.3 Workholding Device Design

The mechanical design of a fixture/jig is a complex engineering task that includes all the typical steps of a traditional design process: synthesis, analysis, and prototyping. A tool designer can utilize the techniques addressed in Chap. 3 for effective fixture/jig design (e.g., axiomatic design theory, group technology, etc.). The outcome of this process is a specific fixture/jig configuration (layout), individual component designs, and a corresponding workpiece loading/unloading procedure.

Prior to the configuration of a suitable workholding device, however, the following issues must be addressed: the necessity of multiple fixtures/jigs owing to workpiece geometry complexity, the number of workpieces per fixture/jig, the determination of suitable surfaces on the workpiece for locating and clamping, and the sequence of workholding steps. The fixture/jig configuration process would yield the following information:

Types of locators and clamps
Positions of locators and clamps
Clamping sequence and magnitudes of clamping forces

The detailed designs (geometry, dimensions, and tolerances) of individual workholding elements are determined by workpiece geometry, contact information (point, line, or plane contact between the locators and workpiece surfaces), expected frequency of utilization (e.g., batch production versus mass manufacturing), availibility of off-the-shelf standard device geometries, mode of operation (manual versus automatic), and finally conditions of manufacturing (clean-room versus machining with coolants). Some jig and fixture design examples will be presented in the following sections.

11.2 JIGS

Jigs are workholding devices used for guiding hole-making tools into accurately located workpieces. Although used for a variety of hole-making processes, such as boring, reaming, tapping, etc., the majority of jigs are utilized for drilling. A typical jig used in drilling would include a baseplate, or a box, with a number of locators and clamps for holding the workpiece and (hardened-steel) bushings corresponding to the number of holes to be drilled.

11.2.1 Jig Configurations

Jig configurations vary from simple template: type jigs (a flat plate with a number of built-in bushings), which would be directly placed on a workpiece and held down manually during drilling, to box: type jigs that would allow drilling in different angles.

Plate Jigs

Plate jigs are variations of template-type jigs that also incorporate clamping devices for accurately and securely holding the workpiece. Leaf jigs constitute the most common configuration (Fig. 4). A workpiece is mounted onto the bottom half of the jig, located accurately, and subsequently clamped in place by the lowering of the upper half of the jig. Cam-action type latches allow for fast loading/unloading cycles.

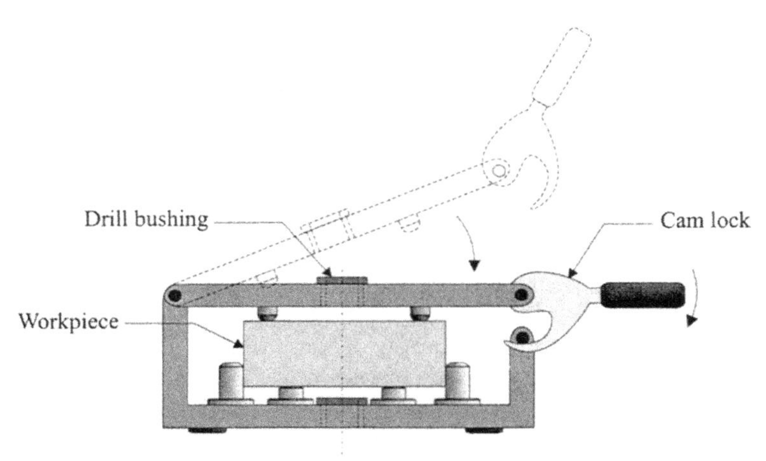

FIGURE 4 Leaf jig.

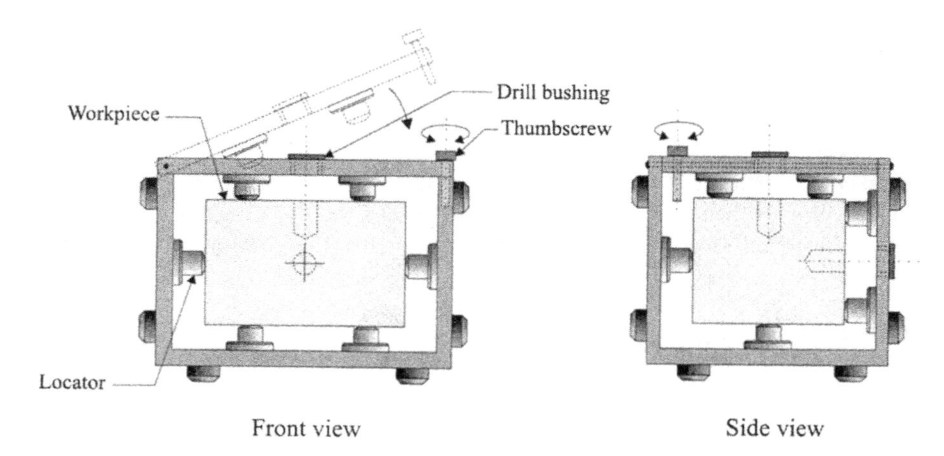

FIGURE 5 Box jig.

Box Jigs

Channel and box jigs are normally designed for complex part geometries and/or for manufacturing processes that would require drilling from a number of distinct angles, so one needs the part to be held accurately while repositioning the jig (Fig. 5). As in plate jigs, a number of locators placed on different walls of the box locate the workpiece securely while drilling is carried out. As in leaf jigs, the box is closed by a pivoting wall. Though common, placement of bushings on moving wall sections of the box jig should be avoided for better accuracy.

11.2.2 Bushings for Jigs

Drill bushings are normally manufactured from wear-resistant, hardened steel using precision finishing (grinding, or even lapping) to excellent concentricity. The most common types are press-fit, renewable, and liner bushings (Fig. 6):

Press-fit bushings are manufactured with or without "heads" and pressed directly into the jig plate for short production runs that would not require frequent changes of the bushings.

Renewable bushings slide into their respective locations in the jig plate with excellent fit and are held in place by a locking mechanism. These are typically used when multiple hole fabrication operations are performed on the same hole, which require different diameter bushings (e.g., accurate hole enlargement, tapping, etc.).

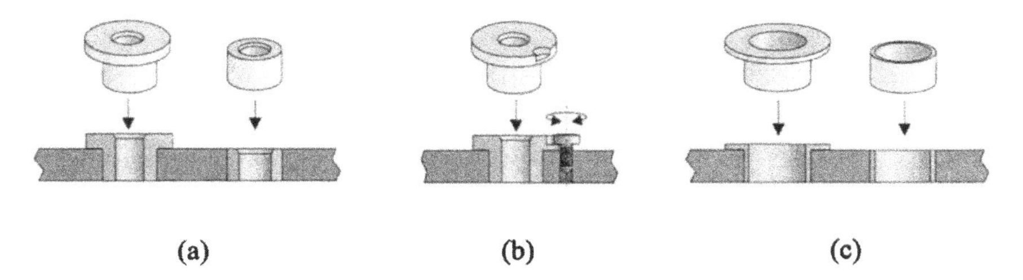

(a) (b) (c)

FIGURE 6 (a) Press-fit; (b) renewable; (c) liner bushings.

Liner bushings are employed for preserving the quality of the holes on the jig plate by being press-fitted into the holes and acting as "master" bushings into which the renewable bushings are fitted in turn. That is, they provide renewable bushings with high-accuracy, hardened holes to be fitted into.

11.3 FIXTURES

Fixtures are workholding devices utilized for locating, supporting, and clamping workpieces for fabrication and assembly tasks. Traditionally, they do not include special components, such as bushings, in order to guide tools. They do, however, employ components, such as tenons, for referencing purposes. Fixtures have been classified according to their configuration and/or according to the manufacturing task for which they are employed. In most cases, they are built to withstand external forces greater than those experienced by jigs, and to provide high positioning accuracy.

In this section, we will first briefly review dedicated fixture configurations that are typically used by most manufacturing applications, while discussing some applications' needs in more detail, and then discuss fixture modularity and reconfigurability, a topic of importance to flexible manufacturing.

11.3.1 Fixture Configurations

The majority of fixtures in use today are called dedicated workholding devices, since their configuration is fixed for one workpiece geometry, in contrast to modular fixtures, which can be assembled and disassembled according to the task at hand. Both dedicated and modular fixtures are normally built on a support plate using a variety of locators, supports, and clamping devices (Fig. 7). Occasionally, plates may be configured to provide an orthogonal wall of support (with respect to the machine table)

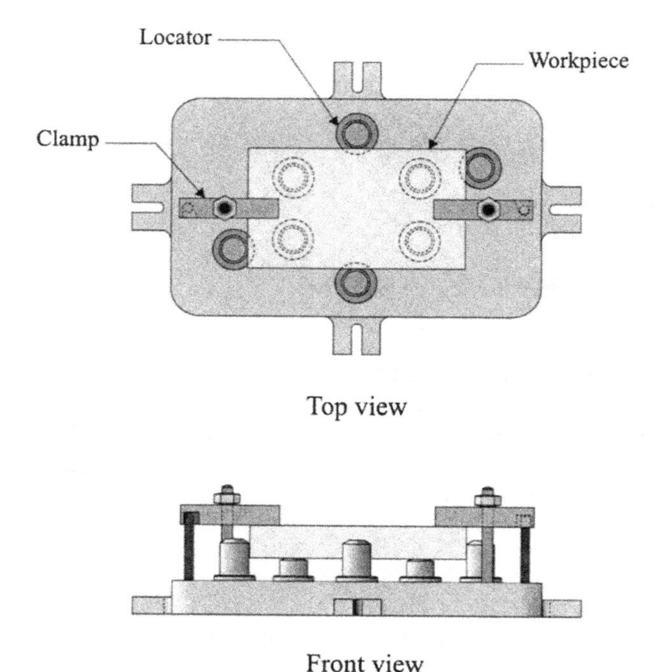

FIGURE 7 Plate fixture.

or even an arbitrary inclined wall of support (< 90°). In all cases, however, the fixture plate is constructed with special cut out slots for efficient mounting onto the worktables of manufacturing machines. Once mounted and secured via multiple bolts, they provide high rigidity. Tenons (square blocks) positioned underneath the plates fit into the narrow segments of the (reverse) T-slots of the worktables for improved accuracy in positioning.

Vise-held fixtures are small plate fixtures that are manually mounted onto the worktables of machines and fixed in place through the use of vises or chucks. They are normally targeted for light machining (low cutting forces).

Milling Fixtures

Milling is an intermittent cutting process, in which the (periodic) cutting forces can be very high (Chap. 8). The locators and supports of the fixture must be designed for these forces and configured to resist them while maintaining workpiece location accuracy and not allowing deflections. Tenons should be used to locate the fixture with respect to the worktable, and reference-setting blocks should be used to locate the fixture with respect to the cutting tool. Sufficient clearances must be incorporated for effective removal of chips and drainage of coolant liquid.

Turning Fixtures

The turning operation on a lathe subjects the workpiece, and thus the fixture holding it, to centrifugal forces in addition to the (continuous) cutting forces. Although the majority of workpieces can be directly mounted onto the (3- or 4-jawed) chuck of the lathe, those workpieces that cannot must be held by well-balanced fixtures, which may be in turn held in place by the chuck of the lathe or directly fastened onto the faceplace of the lathe (Fig. 8). An unbalanced fixture/workpiece assembly will cause vibrations, thus leading to cutting-tool chatter (Chap. 8). Balance can be achieved, when necessary, by the addition of nonfunctional weights to the fixture.

Assembly Fixtures

The primary objective of an assembly fixture is accurately to locate and clamp two parts prior to their joining operation (e.g., riveting, welding, etc., Chap. 10). Though rarely subjected to large fabrication forces, the clamping devices must provide sufficient reinforcement (especially in welding operations) while allowing fast loading/unloading cycles. Welding fixture designers should also consider the following workholding issues: protection of fixture components from sputters and heat; ensuring conduction of electricity and good grounding; proper heat dissipation control; and the use of

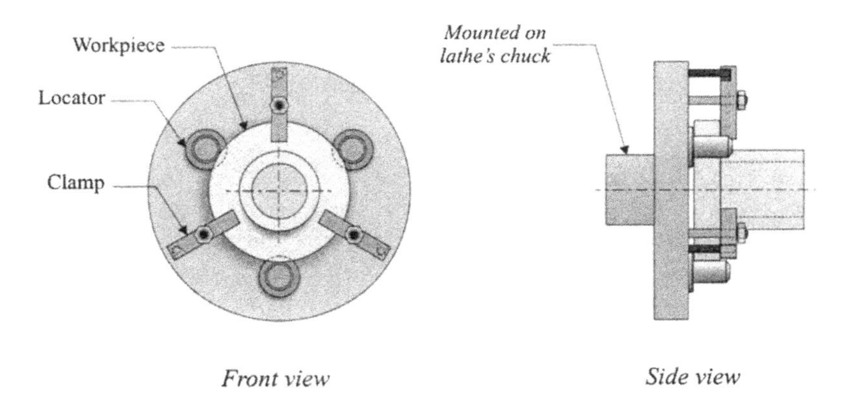

FIGURE 8 A turning fixture.

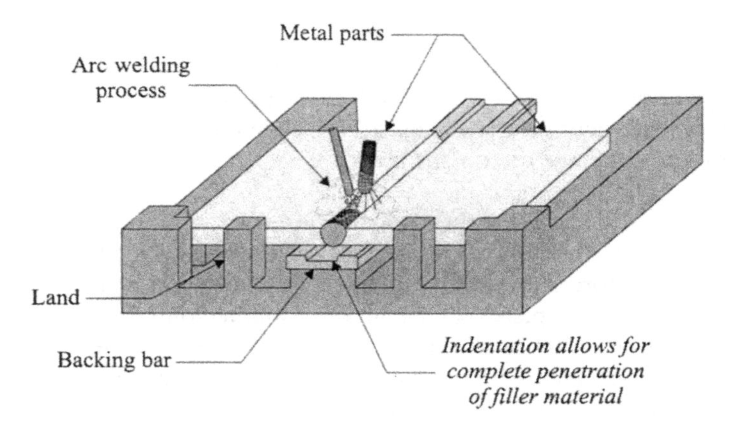

FIGURE 9 A welding fixture.

suitable backing bars (placed under the joints for arc welding) for complete penetration of filler material (Fig. 9).

11.3.2 Flexible Fixtures

Operational flexibility in manufacturing requires the use of flexible workholding devices that can be reconfigured for the latest workpiece at hand. Although the beginnings of such reconfigurable fixtures can be traced back to there early 1940s in Europe, in the form of modular devices, innovative fixture designs suitable for programmable automation have only been developed since the late 1970s and primarily by academics. However, despite a large number of such reconfigurable/programmable fixture design proposals, the manufacturing industry mostly still continues to use fully dedicated fixture configurations with only sparse efforts to use modular fixtures and very rarely any programmable devices.

Modular Fixtures

The rationale of using modular workholding devices is cost reduction by being able to accommodate multiple parts on a reconfigurable fixture, thus minimizing design and fabrication efforts for the fixture. Modular fixtures comprise a set of standard components (with variable dimensions), such as locators, V-blocks, clamps, and supports, which can be assembled on a base plate (with T-slots or holes) (Fig. 10). The assembly of the fixture can be carried out around an actual (reference) workpiece or using accurate measurement devices according to a plan, normally generated on a CAD workstation.

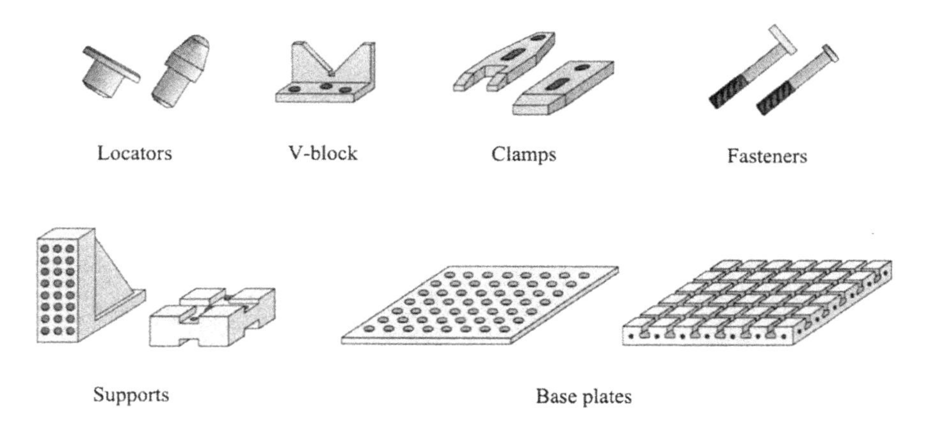

FIGURE 10 Modular fixture components.

As discussed above, modular fixtures can be utilized in flexible-manufacturing (or job-shop) environments for one-of-a-kind or small-batch productions. A typical application is the manufacturing (of the components) of machine tools themselves. Prototype production and pattern fabrication for casting are other common applications.

Modular fixtures are normally classified according to the geometry of their base plate: T-slot versus hole (or dowel-pin)-based systems (Fig. 11). The former systems were the first modular fixture configurations developed in order to duplicate the advantages of T-slot-based worktables on milling machines. Their primary advantage is the continuous variability/reconfigurability of individual components along the full range of the slots. However, all fixture components must be accurately placed on the plate and fastened down securely to counteract the cutting forces. Hole-based modular fixtures, on the other hand, can be easily reconfigured based on a CAD plan, and they provide higher stiffness. Furthermore, hole-based plates are easier to fabricate, though they provide a more limited reconfigurability owing to the discrete placement of the holes. Finally, one can note that there are hybrid plates that include holes and T-slots.

An important issue in modular fixtures is the size of the inventory of components. In order to accommodate workpieces of various sizes and shapes, the heights, widths, diameters, etc. of locators, supports, and clamps must also be variable. In most commercial modular fixture systems this variability is achieved by using add-on blocks for height variability and employment of a large number of locators and V-blocks of different sizes. The academic literature includes, however, designs of modular components whose dimensions can be continuously adjusted.

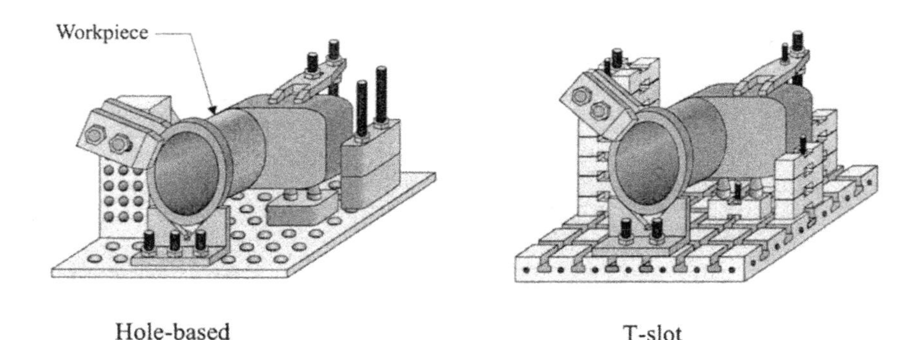

FIGURE 11 T-slot and hole-based modular fixtures.

Reconfigurable/Programmable Fixtures

The term reconfigurable fixtures has often been used interchangeably with modular fixtures that have limited ability to reconfigure. In this section, the former term is reserved for workholding devices whose locators, supports, and clamps can be adjusted in the continuous domain (versus in discrete increments) to adapt to the geometry of the workpiece.

The most commonly known reconfigurable fixture is the conformable clamping system developed by Cutkosky et al. for machining turbine-blade forgings (Fig. 12). The two primary characteristics of this system are (1) the use of variable-height (pneumatic) locators that fit the underneath profile of the turbine blade along a line, and (2) the use of a flexible belt that wraps around the upper profile of the turbine blade. The accurate positioning of the line of support and the exact height determination of each one of the

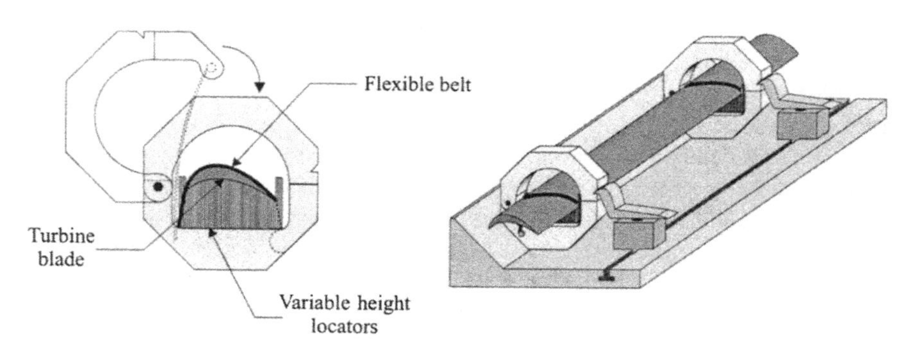

FIGURE 12 Conformable clamping system.

locators must be carried out with great care. The use of a master template has been proposed for this purpose.

An extension of the conformable clamping system from a 2-D line support to a 3-D surface support could be achieved via a "bed of nails," which would provide support to thick- and thin-walled surfaces. Such custom-made fixtures have been used in the aerospace industry for the drilling of thin-walled, large fuselage parts. Naturally, one may plan to use only a partial set of "nails" (locators) that would provide sufficient rigidity. The optimal number and locations of these locators can be determined using finite-element-based stress analysis tools.

An important issue to consider in workholding for flexible manufacturing is the intelligence of the fixtures. In this context, there have been only

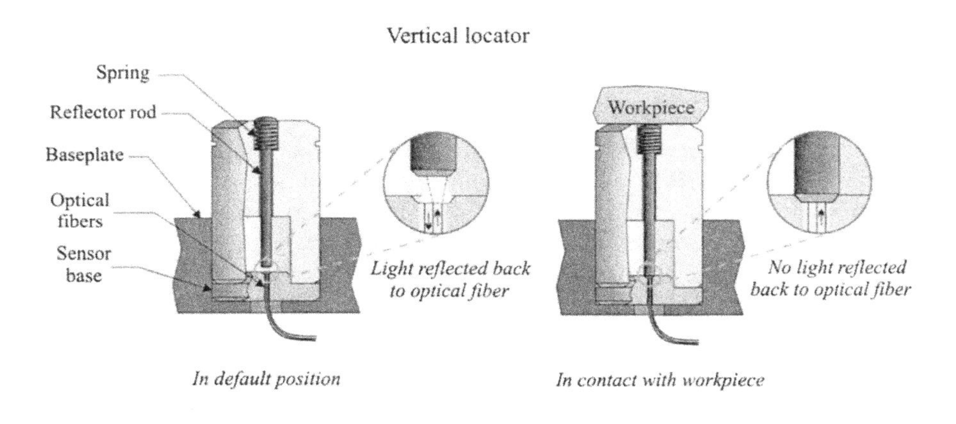

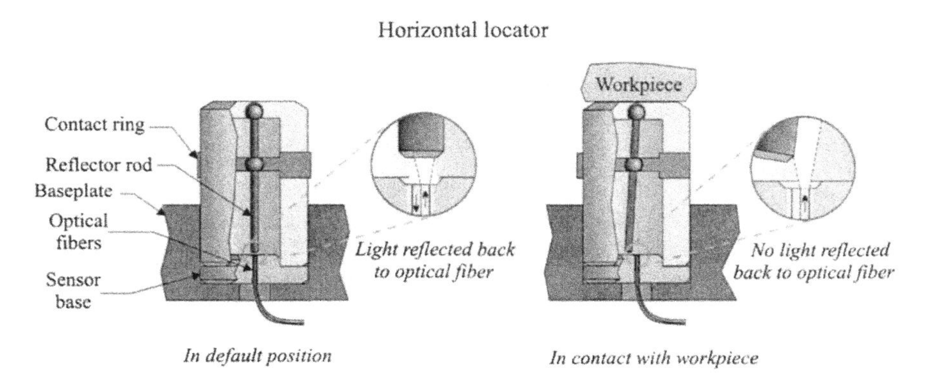

FIGURE 13 Intelligent locators for programmable fixtures.

a limited number of attempts to incorporate sensors into workholding devices in order to receive real-time feedback on the status of the fixturing process. Two challenges in programmable (intelligent) fixture development are (1) the detection of the accurate assembly of the reconfigurable fixture components on the baseplate and (2) the subsequent detection of the workpiece placement on the fixture and its clamping. Both of these challenges must be addressed without negatively affecting the accuracy of locating and clamping the workpiece. A variety of such fixture components were developed at the University of Toronto for a hole-based plate modular fixturing system (Fig. 13). This fixture is able to detect the presence of objects placed on it and activate clamps automatically for autonomous, computer-based workholding.

11.4 COMPUTER-AIDED FIXTURE DESIGN AND RECONFIGURATION

Fixture design for manufacturing may be a complex endeavor owing to accuracy needs in an environment of nontrivial tool paths and, where applicable, cutting forces. Commonly this task is carried out by an experienced and skilled tool designer. Given a workpiece geometry and manufacturing conditions, the designer is required to develop the most suitable fixture (dedicated or modular) and preferably a process plan for its fabrication. As in any product design, the tool designer should utilize existing design methodologies (Chaps. 3 and 4) and computer-aided-engineering (CAE) analysis tools (Chap. 5) in the design and reconfiguration of fixtures/jigs.

The role of computer-aided design (CAD) varies according to the fabrication strategy: for mass production, where dedicated fixtures would be utilized for long periods of time, the emphasis would be on design, whereas for small batch sizes or one-of-a-kind production, the emphasis would be on the reconfiguration of the modular workholding setup. In both cases, however, finite element analysis must be utilized for the prevention of potential workpiece deflections due to fabrication forces.

11.4.1 Design of Fixtures/Jigs

The most basic approach to fixture design is the utilization of a CAD package by a skilled tool designer for a design from scratch. The designer builds the fixture around the CAD model of the workpiece using a graphical user interface based on his or her past experience and knowledge of expected fabrication conditions. Some commercial software packages provide designers with a set of premodeled fixture components that they can retrieve from the database and modify them as necessary.

Group technology (GT) principles (Chap. 3) can be effectively utilized in the fixturing of workpieces with geometric similarity. The objective is to access fixture designs used in the past for workpieces that are similar to the workpiece at hand. The retrieval of the most appropriate/useful (past) fixture design can be achieved by the following sequential approach in a CAD environment, where all workpiece geometric models have been classified and coded:

1. Determine the GT code for the workpiece to be fixtured using the company's available classification and coding system.
2. Search the database of workpieces, for which there exist corresponding fixtures/jigs, to determine the most similar (past) workpiece geometry based on the GT code determined in Step 1.
3. Retrieve from the fixture database the (fixture) design corresponding to the (past) workpiece identified in Step 2.
4. Evaluate whether the most similar past fixture design could be effectively modified to yield a new design for the workpiece at hand. If the answer is no, then, we must return to Step 2 in order to determine and evaluate other similar designs, though the probability of finding a better past fixture design would be low, if the coding and classification system has functioned properly in the first iteration. After several evaluation iterations, Steps 2 to 4, if a suitable past fixture design can still not be identified, the designer must design a new fixture from scratch.If a retrieved fixture design is deemed to be suitable, the process continues.
5. Access all past information stored in the database regarding the past fixture design: reasoning behind the selection of specific locators/supports/clamps, etc., as well as the evaluation metrics for the specific fixture configuration chosen.
6. Modify the (retrieved) fixture design for the workpiece geometry and fabrication conditions at hand. This step is an iterative process itself, where different designs and configurations must be analyzed using CAE analysis tools (Sec. 4.2 below).
7. Store the new workpiece and fixture models and other pertinent data in their appropriate databases according to the GT code (of the new workpiece) for future use.

The above sequential process can be utilized for the design of dedicated fixtures as well as for the reconfiguration of modular fixtures.

There have been several attempts by academic investigators to develop CAD-based tools for the automatic synthesis of fixture designs (with almost no manual intervention). These systems utilize a variety of reasoning techniques (including heuristics and analytical models) to determine locating and clamping points on the workpiece, choose corresponding fixture component geometries, and assemble the fixture (in the CAD's virtual environment) for subsequent interference checks. Generative fixture design is another term used for such experimental design procedures.

11.4.2 Fixture Configuration and Analysis

Fixture configuration is commonly referred to as the process of determining the positions of locators and clamps for modular fixtures. However, as discussed earlier in this section, one must also select these positions with great care in the case of dedicated (nonreconfigurable) fixtures. The objective is engineering analysis for optimal fixture configuration.

Due to time-varying forces acting on the fixture, the problem at hand is a dynamic type, where fixture and workpiece behavior under loading must be analyzed. The analysis is almost always carried out using (numerical) finite element–based modeling owing to the complexity of workpiece geometry. The optimization process attempts to vary the fixture configuration in order to minimize deflections with preferred minimal clamping forces. Fixture configuration includes the following (optimization) variables: the number, types, and positions of locators, supports, and clamps and clamping forces. The problem is a mixed integer/continuous-variable type and must be solved by employing an appropriate search method (Chap. 5).

REVIEW QUESTIONS

1. Define workholding (fixturing) and state its primary objectives.
2. Define jigs versus fixtures.
3. Explain the 3-2-1 principle in workholding.
4. Why should locators be manufactured as entities separate from the body of the fixture/jig?
5. Define locating versus clamping. Why should manufacturing forces be directed toward support points and not be compensated by clamps (i.e., directed toward clamps)?
6. Why should clamps be power activated (versus being manual)?
7. Is the design process of a fixture different from that of the part it is manufactured to fixture? Explain.
8. Discuss the different classes of bushings available for jigs.
9. Discuss the use of tenons in the placement of fixtures onto machine worktables.
10. Compare the principal requirements for machining fixtures versus those for assembly fixtures.
11. Discuss the need for flexible fixtures in small-batch and/or one-of-a-kind manufacturing environments.

12. Compare modular fixtures versus reconfigurable/reprogrammable fixtures.
13. Compare the use of hole-based base plates versus T-slot-based ones in modular fixturing.
14. Why should fixtures/jigs be reprogrammable?
15. Discuss the use of computer-aided design (CAD) and engineering analysis (CAE) tools in fixture design. In your discussion, also refer to issues such as, group technology (GT), generative design, and so on.

DISCUSSION QUESTIONS

1. Discuss possible sensing technologies that can be incorporated into different workholding devices for the on-line monitoring and control of the manufacturing process, while the parts are being fabricated/assembled.
2. Fixtures can be designed for a specific range of metrics within the targeted family of products: (1) Those that allow reconfiguration via continuous and/or discrete incremental changes, or even through modularity of certain subcomponents, or (2) those that have been already manufactured in different dimensions, etc., for different product dimensions. Discuss these modes of fixture design in terms of manufacturing difficulties, durability, safety, cost, etc.
3. The use of design features has long been considered as improving the overall synthesis and analysis stages of products owing to the potential of encapsulating additional nongeometric data, such as process plans, in the definition of such features. Discuss feature-based design, in which the user, through some recognition/extraction process, can access and retrieve individual similar or identical features on earlier product/fixture designs and utilize them for the design of the fixture at hand.
4. Finite-element modeling and analysis (FEM/A) methods have been developed to cope with the engineering analysis of complex product geometries and/or physical phenomena. Discuss the use of FEA during the (iterative) fixture design process (i.e., synthesis analysis) for the determination of optimal design parameter values, for example in verifying part deflections under clamping and/or manufacturing forces.
5. Fixturing is a typical design process that involves the iterative synthesis and analysis stages, during which we would determine the optimal support and clamping positions for a workpiece at hand and accordingly configure, design and manufacture a mechanical fixture. Due to high accuracy requirements, the cost of a complex fixture can also be very high. This cost is normally amortized over a large number of identical parts in mass production cases. Discuss the utilization of

modular fixtures for small-batch and one-of-a-kind manufacturing cases. Address issues such as accuracy of components, ease of assembly, computer-aided planning of fixture configuration, and others.

6. Would several different GT based classification and coding systems be needed in a company for different objectives? That is, one system for product/fixture design, one system for manufacturing planning, and yet another for cost engineering.
7. In the near future, although the majority of engineering products will be modeled in the virtual (computer) space, representing the starting point of the design and manufacturing process, some products will still be crafted manually by artisans and/or industrial designers. Discuss the computer-aided design and manufacturing of fixtures for such products, whose features are not originally defined by exact mathematical relationships.
8. Machining centers increase the automation/flexibility levels of machine tools by allowing the automatic change of cutting tools via turrets or tool magazines. Some machining centers also allow the off-line fixturing of workpieces onto standard pallets, which would minimize the on-line setup time (i.e., reduce downtime of the machine): that is, while the machine is working on one part fixtured on Pallet 1, the next part can be fixtured on Pallet 2 and loaded onto the machine when it is has completed operating on the first part. Discuss the use of such universal machining centers versus the use of single-tool, single-pallet, unipurpose machine tools.
9. Job shops that produce one-of-a-kind products have been considered the most difficult environments to automate, where a product can be manufactured within a few minutes or may require several days of fabrication. Discuss the role of reconfigurable fixtures in facilitating the transformation of such manual, skilled-labor dependent environments to intensive automation-based environments.
10. Manufacturing flexibility can be achieved at three levels: operational flexibility, tactical flexibility, and strategic flexibility. Discuss operational flexibility. Is fixturing automation a necessary or a desirable tool in achieving this level of flexibility?

BIBLIOGRAPHY

Benhabib, B., Chan, K. C., Dai, M. Q. (1991) A modular programmable fixturing system. *ASME Journal of Engineering for Industry* 113(1):93–100.

Boyes, William E., ed. (1982). *Jigs and Fixtures*. Dearborn, MI: SME.

Boyes, William E. (1986). *Jigs and Fixtures for Limited Production*. Dearborn, MI: SME.

Chan, Ka-Ching (1989). Development of a Modular Programmable Fixturing System for Assembly. M.A.Sc. thesis, Department of Mechanical Engineering, University of Toronto, Toronto, Canada, 1990.

Chan, K. C., Benhabib, B., Dai, M. Q. (1990). A reconfigurable fixturing system for robotic assembly. *SME Journal of Manufacturing Systems* 9(3):206–221.

Cutkosky, M. R., Kurokawa, E., Wright, P. K. (1982). Programmable, conformable clamps. SME AUTOFACT Conference Proceedings, Philadelphia, PA, (11.51–11.58).

Drozda, Thomas, Wick, J. Charles, eds. (1998). Tool and Manufacturing Engineers Handbook: A Reference Book for Manufacturing Engineers, Managers, and Technicians. Society of Manufacturing Engineers, Dearborn, MI.

Gandhi, M. V., Thompson, B. S. (1986). Automated design of modular fixtures for flexible manufacturing systems. *SME Journal of Manufacturing Systems* 5(4):243–252.

Henriksen, Erik Karl (1973). *Jig and Fixture Design Manual.* New York: New York Industrial Press.

Hoffman, Edward G. (1980). *Jig and Fixture Design.* Albany, NY: Delmar.

Hoffman, Edward G. (1987). *Modular Fixturing.* Lake Geneva, WI: Manufacturing Technology Press.

Nee, Andrew Y. C., Whybrew, K., Kumar, A. Senthil (1995). *Advanced Fixture Design for FMS.* New York: Springer-Verlag.

Nee, John G., ed (1998). *Fundamentals of Tool Design.* Dearborn, MI: SME.

Pham, D. T., De Sam-Lazaro, A. (1990). AUTOFIX. An expert CAD system for jigs and fixtures. *Int. Journal of Machine Tools Manufacturing* 30(3):403–411.

Rong, Yiming, Zhu, Yaoxiang (1999). *Computer-Aided Fixture Design.* New York: Marcel Dekker.

Sela, M. N., Gaudry, O., Dombre, E., Benhabib, B. (1997). A reconfigurable modular fixturing system for thin-walled flexible objects. *Int. Journal of Advancd Manufacturing Technology* 13(9):611–617.

Smith, Graham T. (1993). *CNC Machining Technology: Volume 2, Cutting Fluids and Workholding Technologies.* New York: Springer-Verlag.

Thompson, B. S. (1984). Flexible fixturing—a current frontier in the evolution of flexible manufacturing cells. ASME Proceedings of the Winter Annual Meeting, 84-WA/PROD-16, New Orleans, LA.

12

Material Handling

Material handling is defined by the Materials Handling Institute as the movement of bulk packaged and individual goods, as well as their in-process and postprocess storage, by means of manual labor or machines within the boundaries of a facility. Although this field of study includes the handling of bulk (solid- or liquid-phase) material and individual goods, this chapter will only focus on the latter (i.e., "unit loads"), with a primary emphasis on material handling equipment, as opposed to facility planning and movement control.

Material handling does not add value to the product but only cost. Thus the objective of material handling is the efficient movement of goods for the on-time delivery of correct parts in exact quantities to desired locations in order to minimize associated handling costs. It is not uncommon to have parts/subassemblies moving around a plant several kilometers prior to their shipment. Manufacturing plants must therefore eliminate all unnecessary part movements, as well as in-process inventories, for just-in-time (JIT) production.

Material handling equipment can be classified according to the movement mode: above-floor transportation (e.g., belt conveyors, trucks, etc.), on-floor transportation (e.g., chain conveyors), and overhead transportation (e.g., cranes). In the following sections, we will review industrial trucks (including automated guided vehicles), conveyors, and industrial robots as the primary mechanized/automated material handling equipment. We will also briefly review the automated storage and retrieval of goods in high-density warehouses, as well as the important issue of automatic part identification (including bar codes). The chapter will be concluded with a discussion on automobile assembly.

12.1 INDUSTRIAL TRUCKS

Industrial powered trucks are the most versatile and flexible material handling devices in manufacturing. They can transport small or large loads over short distances in a plant with minimal restrictions on their movements. Powered trucks are generally classified into two broad categories: lift trucks and tow tractors.

12.1.1 Lift Trucks

Powered (lift) fork trucks are the most common industrial trucks used in the manufacturing industry for the transportation of parts placed on pallets (Fig. 1). The basic elements of a fork truck are (1) the mast assembly—a one-stage or multistage mechanism that lifts the forks, most commonly,

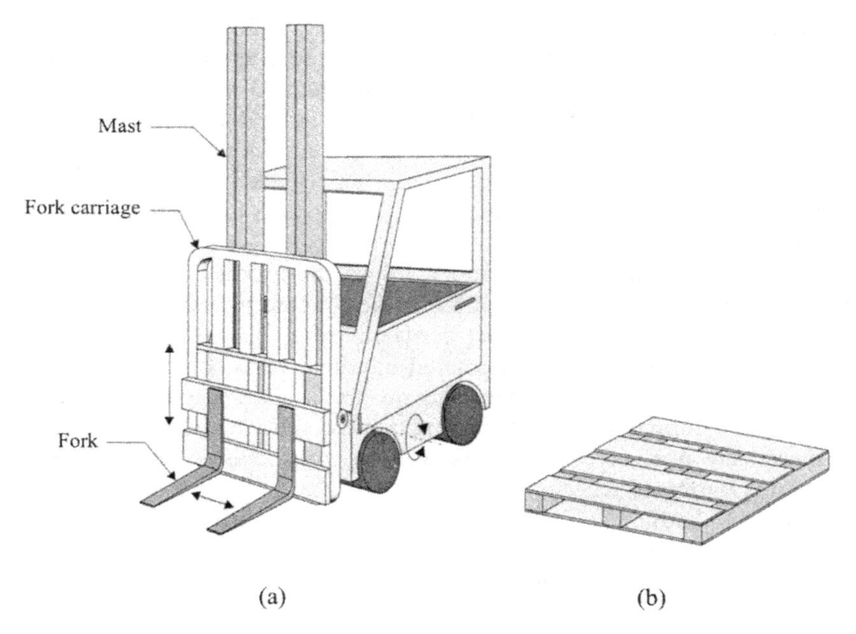

FIGURE 1 (a) A forklift truck; (b) a typical industrial pallet.

through hydraulic power; (2) the fork carriage—a carriage that is mounted on the mast, to which the forks are attached, with a primary objective of preventing loads from falling backward once they have been lifted off the ground; and, (3) the forks—the two forks can be of fixed configuration or with variable horizontal distance to accommodate varying load sizes. On most trucks, the mast assembly, including the forks, can be tilted backwards (4 to 12°) for increased security during motion.

The load capacity of forklift trucks is defined as the maximum weight carried at the maximum elevation of the forks. Typical forklifts can carry weights in the range of 1000 to 5000 kg at lift heights of up to 6 m and move at speeds of 5 to 10 km/hr. A large number of forklift trucks are counterbalanced at their rear for increased load capacity and more secure transportation, though at the expense of a larger footprint and potential difficulty working in confined spaces.

The primary advantage of forklift trucks is their path independence. A trained driver can transport parts by following the shortest available route, though their being driver-operated vehicles is also the greatest disadvantage of industrial trucks owing to increased costs.

12.1.2 Tow Tractors

A variety of wheeled vehicles are utilized in the manufacturing industry for towing (pulling) single- (or multi-)trailer cart attachments. Most tractors (like forklift trucks) are battery operated for maintaining clean-air environments within closed plants. The load carried can be palletized or (manually) placed directly on the trailer. The typical load-carrying capacity of electric-powered tow tractors ranges from 5,000 to 25,000 kg. As with forklift trucks, the primary disadvantage of tow tractors is their dependence on human drivers. In the following subsection, automated guided vehicles will be presented as a potential answer to this disadvantage.

12.1.3 Automated Guided Vehicles

The Materials Handling Institute defines an automated guided vehicle (AGV) as a driverless vehicle equipped with an on-board automatic-guidance device (electro-optical or electromagnetic) capable of following preprogrammed paths. The path information can be uploaded onto an on-board computer through (radio-frequency, RF) wireless communication or through a temporary (physical) connection to the plant's computer network. Reprogrammability of AGVs is their primary asset. Occasionally, AGVs are referred to as mobile robots owing to their reprogrammabilty. This is an erroneous classification, since AGVs mostly do not include a robotic manipulator arm capable of interacting with the environment.

The first AGV was developed by Barrett Electronics, U.S.A., in the early 1950s and installed at Mercury Motor Freight in 1954. These towing vehicles received poor acceptance by the manufacturing industry owing to their limited controllers and difficulty in their reprogammability. The subsequent period of 1960 to 1980, however, was marked by the introduction of a large number of AGVs in Europe, which was further accelerated during the following decade because of better (compact and reliable) on-board computers and electronics. As expected, the automotive industry (e.g., Volvo, Fiat) was the leader in the use of AGVs (more than 50% of over 10,000 installations).

General Motors (GM) has been the largest user of AGVs in North America since the late 1980s, with over 3,000 installations in their automotive plants. Most of GM's AGVs have been in their assembly lines (engine and body assemblies). Other main North American companies, on the other hand, have been utilizing AGVs since the early 1990s, largely in their warehousing/receiving/shipping activities, as opposed to on their factory floors.

AGVs have been intended to replace industrial trucks operated by human drivers. Thus they can be found in the following configurations: unit-load carriers, towing tractors, and forklift trucks (Fig. 2). The first two configurations are the more common types of AGVs in manufacturing environments. Despite their different intended usage, however, the operation of all AGVs is subject to a set of common constraints, some of which will be detailed below.

Basic vehicle design: AGVs are battery-powered wheeled vehicles that can automatically navigate through a network of paths (preinstalled) on the factory floor. The batteries can be recharged, commonly as the AGV is waiting between tasks (at distributed recharging locations) or when it is about to be depleted (at a central recharging location). The vehicles can be designed for forward-motion only or for forward and backward motions, as

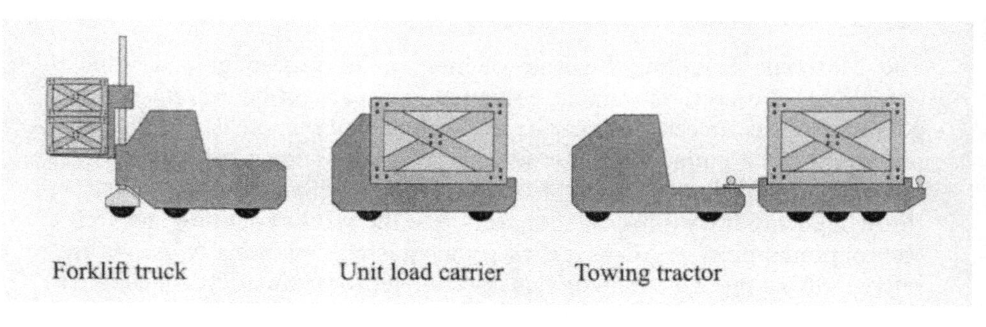

FIGURE 2 AGV types.

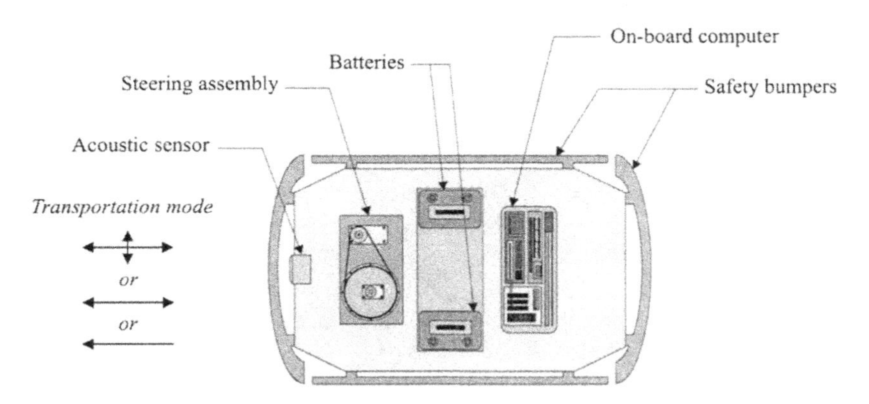

FIGURE 3 Basic AGV design.

well as for sideways motions. Steering is achieved by an on-board computer executing a prespecified motion trajectory, which is communicated to it from a central area controller, normally via RF. The vehicles are equipped with acoustic sensors to detect obstacles (for collision avoidance) as well as with safety bumpers that can detect contact and stop their motion in an emergency manner (Fig. 3).

Navigation guidance: Navigation guidance for AGVs can be in two forms: passive or active tracking of a guidepath. Both methods rely on noncontact tracking of a guidepath installed on or in the floor of the manufacturing plant. Optical passive tracking is the most economical and flexible method, where the guidepath is defined by a painted or taped-on strip. An optical detection device mounted underneath the vehicle follows the continuous guidepath (a collection/network of branched paths) and guides the vehicle to its destination (Fig. 4). Naturally, such a method can only be employed if the painted/taped-on guidepaths can be maintained reasonably well for prolonged periods of time.

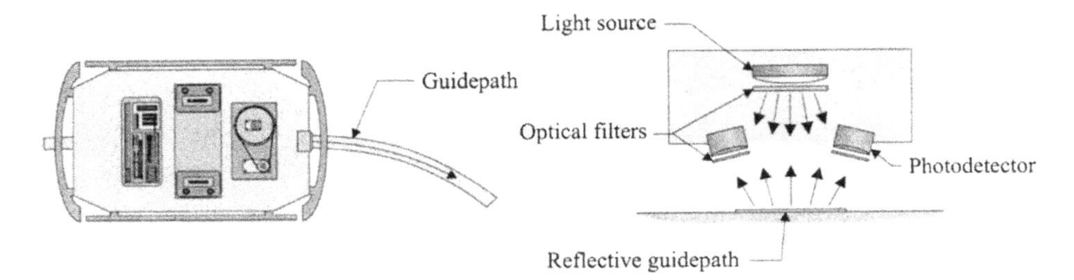

FIGURE 4 Optical guidance for AGVs.

Active tracking of a guidewire buried in a below-surface slot must be utilized for floors that cannot be easily kept clean. In this method, a network of paths is created using low-voltage, low-current-carrying wires that emit low-frequency (1 to 2 kHz) AC signals. In a networked environment, branching of a path can be specified by having the different paths emit different-frequency signals. Active tracking, however, is more expensive and less flexible than optical methods. Thus any reconfiguration of the factory floor would require reinstallation of a new network of buried wires (up to 40 mm deep).

Load carrying capacity: Most unit-load carrying AGVs have been designed to cope with weights in the range of 500 to 1,000 kg, though some custom-made vehicles can carry up to 50,000 kg. Tractor-type AGVs have been commonly designed to pull weights of up to 20,000 kg. AGVs can achieve typical speeds of up to 2 to 4 km/hr.

Applications

Several applications of AGVs for material handling in manufacturing environments are described below.

Car-engine assembly: A number of automotive manufacturers (including GM and Fiat) have utilized unit-load carrying AGVs for car-engine assembly (Fig. 5). On such assembly lines, AGVs transport in-process engines from one assembly station to another in an asynchro-

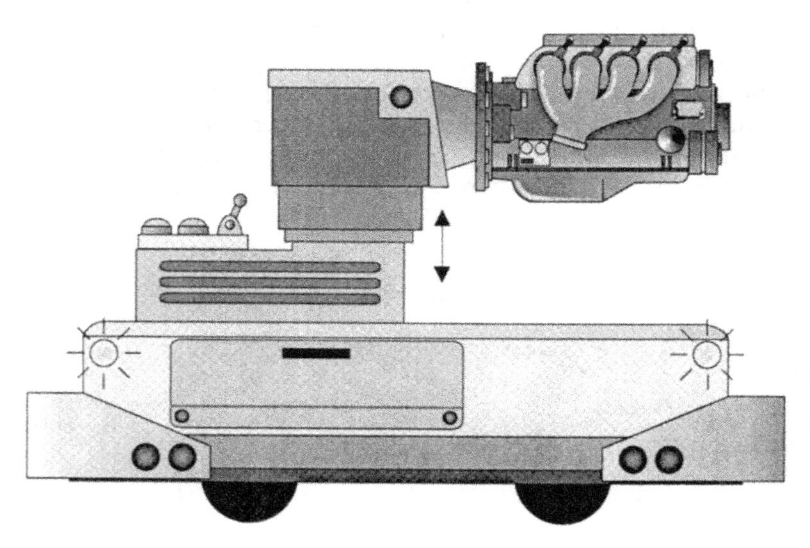

Figure 5 Car engine assembly.

nous manner—intermittent transfer. AGVs remain at a station until the specific assembly task is finished, which varies in time from engine to engine and from one operation to another. In a multi-car-model environment, bar codes placed on the engine or on the pallet indicate to the operator (or robot) the exact types of parts to be assembled.

Car-body assembly: Volvo has been the pioneer in replacing moving-conveyor-based assembly lines with AGV-based lines, where the entire car is assembled on a dedicated AGV that moves from one assembly station to another in an asynchronous mode (see Sec. 12.6). This original concept was also applied by GM in the late 1980s in their car assembly plants in Oshawa, Canada, and Lansing, U.S.A.

Electronics manufacturing: A number of electronics manufacturers (including Intel) have frequently used AGVs for the transportation of wafers, parts in magazines, printed circuit boards, etc., from one processing station to another and from storage to factory floor and vice versa. AGVs are very suitable to work in such clean-room environments in comparison to other material handling equipment.

Machining: AGVs, some with on-board robotic manipulators, have also been utilized by the metal-cutting industry in the transport of large workpieces, palletized batches of parts, and cutting tools. Most such AGVs are equipped with automatic transfer mechanisms that allow them to link to conveying devices (or other transfer mechanisms) for the automatic transfer of unit loads.

12.2 CONVEYORS

Conveyors are a broad class of material handling (conveying) equipment capable of transporting goods along fixed paths. Although conveyors are the least flexible material handling equipment (owing to their path inflexibility), they provide manufacturers with a cost-effective and reliable alternative. Conveying equipment is generally classified as above-floor conveyors versus on-floor or overhead tow-line conveyors. Both classes allow horizontal and inclined conveying, while tow-line type conveyors also allow vertical conveying (e.g., bucket elevators). In the following subsections, several examples of conveyors will be discussed with the emphasis being on conveying for manufacturing.

12.2.1 Above-Floor Conveyors

Above-floor conveyors have been also classified as package handling conveyors owing to their primary application of transporting cartons, pallets, and totes. On the factory floor, they are utilized to transport (palletized/

fixtured) workpieces (e.g., engine blocks, gearboxes, household items) from one assembly station to another. In a networked environment, where branching occurs, automatic identification devices must be utilized to route parts correctly to their destination along the shortest possible path.

Roller Conveyors

Powered roller conveyors are line-restricted conveying devices comprising a set of space rollers mounted between two side frame members and elevated from the floor by a necessary distance (Fig. 6). Rolling power can be achieved by having a moving flat belt underneath the rollers or a set of drive belts rotating the rollers individually, yielding speeds of up to 30 to 40 m/min.

Belt Conveyors

The early use of belt conveyors can be traced back to late 1800s in the mining industry. Today, the flat-belt version of such conveyors (versus the ones used in bulk-material transfer with side-inclined rollers—"troughing" idlers) are commonly used in the manufacturing industry for the transfer of individual (unpalletized) workpieces, as well as cartons/bins/etc. The highly durable, endless belt is placed in tension between two pulleys and normally operated in uni-directional motion (Fig. 7).

The belt is the most important and expensive component of a belt conveyor. A carcass, enclosed between top and bottom covers, provides the tensile strength necessary for conveying and absorbs the impact forces by workpieces being loaded onto the belt. The top cover protects the carcass against tear and wear and against high temperatures when needed (up to 200°C). Steel is commonly used in the construction of the carcass for high-tension applications.

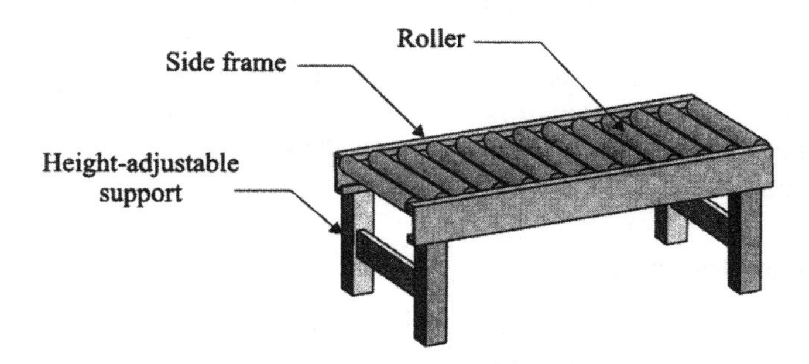

FIGURE 6 Roller conveyor.

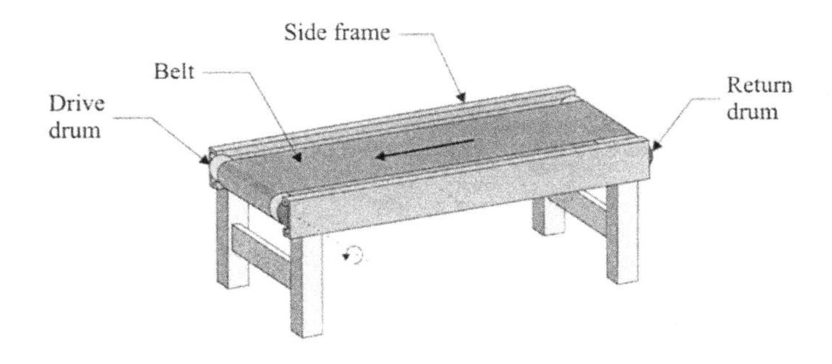

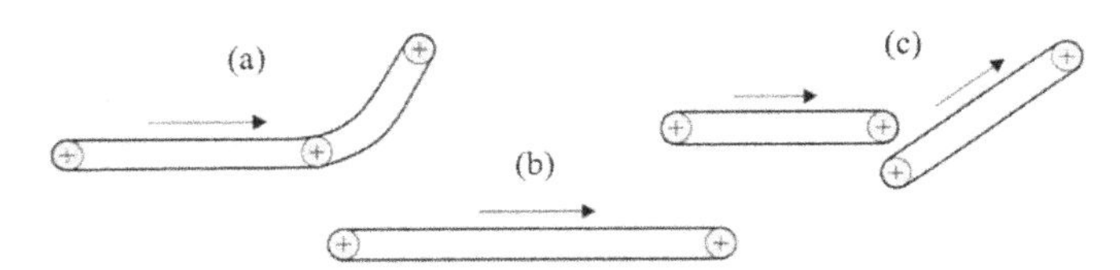

FIGURE 7 Belt conveyor.

Belt conveyors can be inclined up to 30 to 40° and operate at speeds of 10 to 40 m/min, over lengths of 20 to 30 m, while carrying loads of up to 800 kg/m.

12.2.2 On-Floor and Overhead Conveyors

On-floor towline conveyors provide manufacturers with versatile transportation systems for conveying goods unsuitable (large, irregular geometry, etc.) for above-floor conveyors. They normally comprise one, two, or multiple chains running in parallel tracks (in shallow trenches). Goods can be directly placed on the chains or on pallets. Towline carts of a variety of sizes and shapes have also been used in on-floor conveying using chain conveyors. Traditionally, chain conveyors have been configured to operate along straight lines, horizontally and at low speeds (typically, 1 to 5 m/min for large loads and less than 25 to 30 m/min for small loads).

Overhead conveyors maximize utilization of three-dimensional workspaces (Fig. 8). Although most are configured for the point-to-point transportation of unit loads directly mounted on the conveyor via hooks

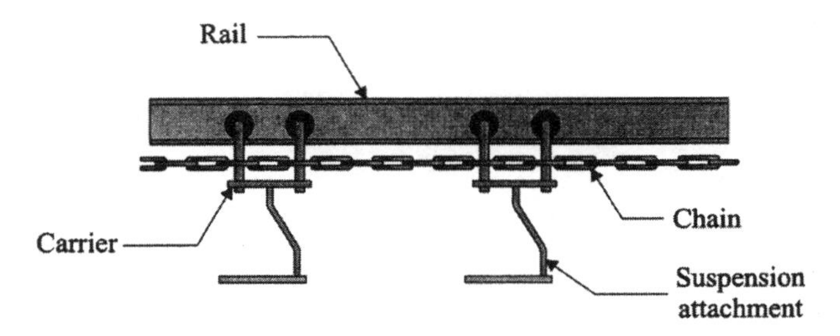

FIGURE 8 Overhead conveyor.

(e.g., automobile doors) or placed on suspension pallets, they can also provide a favorable environment for certain manufacturing applications, such as the on-the-fly spray-painting of workpieces. Overhead conveyors can operate horizontally or in inclined modes. The drive mechanisms employ chains or worm-screws. Occasionally, these conveyors also employ individually powered carriers capable of moving along monorails.

Overhead conveyors can reach speeds up to 80 to 100 m/min, though typically they operate in the range of 10 to 20 m/min.

12.3 INDUSTRIAL ROBOTS

Robotics is a multidisciplinary engineering field dedicated to the development of autonomous devices, including manipulators and mobile vehicles. In this section, our focus will be on robotic manipulators developed for industrial tasks. Although these devices are reprogrammable and multifunctional manipulators of goods and tools, the issues of planning, programming, and control will be addressed later in Chap. 14. Herein we will primarily address the mechanical design of manipulators and various applications of industrial robots in manufacturing environments.

The word robot has been often traced to the Czech word for forced labor mentioned by Karel Capek in his science fiction play Rossum's Universal Robots around 1921. The modern concept of industrial robotic manipulators was only introduced in late 1950s by G. C. Devol (U.S. Patent 2988237) and later championed by J. Engelberger—originators of the first industrial robot by Unimation Inc. in the 1959. The first installation of the Unimate robot for loading/unloading a die-casting machine at GM was in 1961. Today industrial robots can be found in almost all manufacturing applications, ranging from machine servicing to welding to painting.

12.3.1 Mechanical Design

An industrial robotic manipulator is typically an open-chain mechanism (fixed at one end to a base and free at the other end with an attached end-effector), whose mobility is defined by the number of independent joints in its configuration. As can human arms, these robotic mechanisms can manipulate objects/tools within a workspace defined by the geometry of the arm, via the end effector/gripper attached to the last link of the arm. The mobility of the end-effector is formally defined by the number of degrees of freedom (dof) of the robotic arm. In its most generic form, an end-effector can attain any position and orientation in three-dimensional Cartesian space (X, Y, Z, R_x, R_y, R_z), if the robotic arm has at least six independent dof. This mobility can also be described as the ability of the robot arm to relocate a Cartesian frame attached to its end-effector to any location (position and orientation) in its workspace with respect to a fixed "world" coordinate frame (Fig. 9).

Configurations

Industrial robots have been (mechanically) designed in several distinct configurations, each with a corresponding set of advantages. These robots can operate in point-to-point (PTP) or continuous-path (CP) motion mode: that is, they can move their end-effectors from one location (F_1) to another (F_2) in the fastest possible way without following any specified trajectory—i.e., PTP motion, or follow a specified continuous trajectory

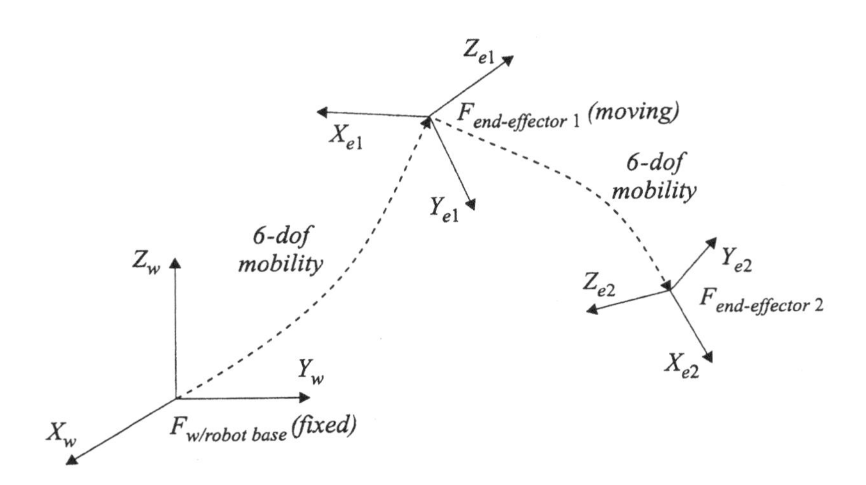

FIGURE 9 Six-dof mobility.

(e.g., a straight line) when moving their end-effector from F_1 to F_2—i.e., CP motion.

Classification of industrial robots is carried out according to their configuration based on the geometry of their workspace, primarily, defined by the first three joints:

Rectangular-geometry robots: There exist two primary robot configurations that belong to this group: Cartesian and gantry (Figs. 10a, 10b). Both configurations employ three linear joints assembled orthogonally for best achievable repeatability. The gantry type configuration is reserved for increased payload capacity due to its better structural stiffness and for better workspace utilization (e.g., electronic components assembly).

Rectangular-geometry robots can have up to three additional (closely located) rotary joints following the first three linear joints for a total of six-dof mobility.

Cylindrical-geometry robots: There exist two primary robot configurations that belong to this group: cylindrical and SCARA (Figs. 11a, 11b). The former has a sequence of rotary-linear-linear joints assembled orthogonally, while the latter has a sequence of rotary-rotary-linear joints. The cylindrical robot configuration was originally developed for fast peripheric access with good repeatability (due to the linear joints). The SCARA (selective compliance assembly robot arm) configuration, on the other hand, was developed in the late 1970s for vertical insertion of small parts (e.g., watch components) placed on a plane, thus requiring a maximum of four dof.

SCARA robots provide manufacturers with complete three-dof mobility in the part-placement plane and one-dof (vertical motion) linear mobility to reach the plane, for a total of four-dof mobility. In contrast, most cylindrical robots would have three additional (rotary) dof at their

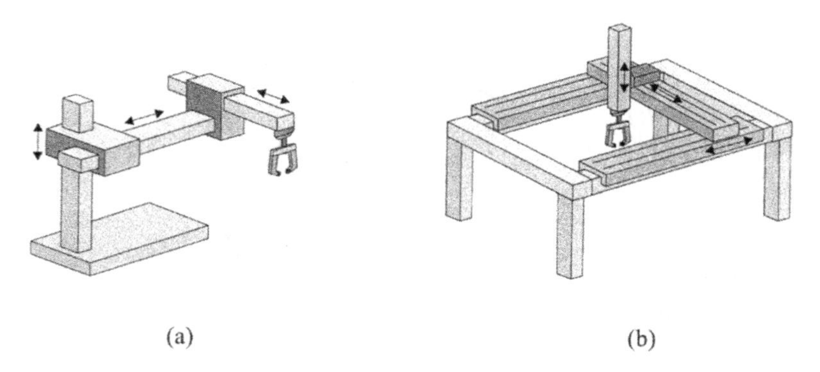

FIGURE 10 (a) Cartesian; (b) gantry robot configurations.

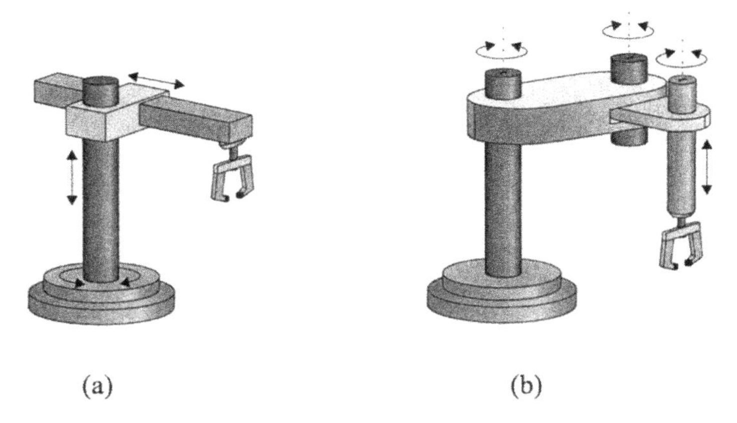

FIGURE 11 (a) Cylindrical; (b) SCARA robot configurations.

configuration end for a total six-dof mobility necessary for an arbitrary motion in three-dimensional space.

Spherical-geometry robots: There exist several robot configurations that yield spherical workspace geometries: spherical and articulated (anthropomorphic, human-like) types are the most common (Figs. 12a, 12b). The former (the first commercial robot configuration) has a sequence of rotary-rotary-linear joints assembled orthogonally to provide manufacturers with a fast reach-in/at capability along simple trajectories (e.g., machine loading/unloading). The latter articulated robot configuration provides maximum reachability among all available manipulator geometries (e.g., reaching into automobile bodies for spray painting) with a sequence of three rotary joints.

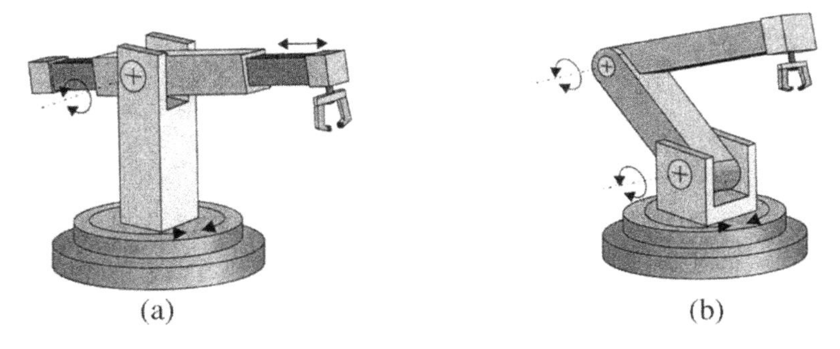

FIGURE 12 (a) Spherical; (b) articulated robot configurations.

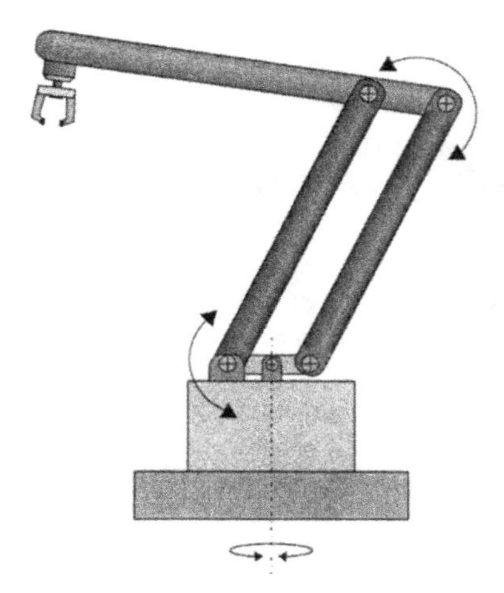

FIGURE 13 An articulated robot with parallelogram linkages.

Both robot configurations shown in Fig. 12 would have three additional (rotary type) joints for a total of six-dof mobility.

In comparison to other configurations, an articulated configuration robot would not be suitable for carrying large loads owing to its sequential employment of three rotary joints. In response to this drawback, a number of robot manufacturers redesigned their articulated configurations: better joint stiffness is achieved by employing two parallelogram linkages, for the "shoulder" and the "elbow" movements (Fig. 13).

Actuators

The three types of actuators used to power the primary (first three) joints of industrial robots are pneumatic, hydraulic, and electric. Pneumatic actuators are only suitable for light-load-carrying applications owing to the compressibility of air. As discussed in Chap. 10 (Sec. 10.6, Automatic Assembly), such actuators are frequently used in 1- to 2-dof pick-and-place mechanisms for the transportation of small parts. At the other extreme, hydraulic actuators are, normally reserved for heavy-load-carrying applications (up to 100 to 150 kg) owing to the incompressibility of hydraulic fluids. Electric motors are the most commonly used robotic actuators, despite their limited load-carrying capability, because of their low maintenance demand and low noise operation. Electric robots can carry loads up to 40 to 50 kg (but typically less than 10 kg). Most industrial robots can achieve end-effector speeds above 1 m/s, some achiev-

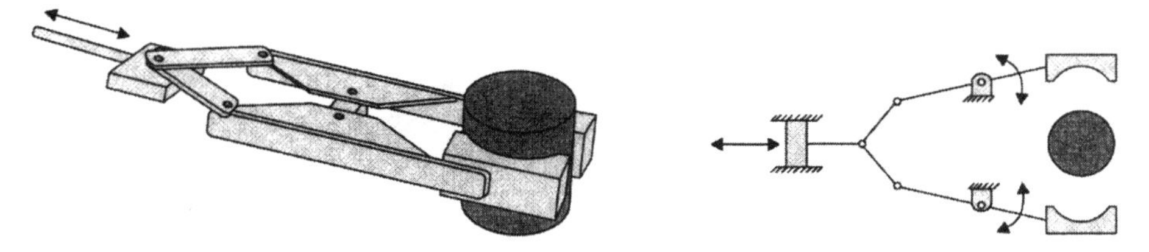

FIGURE 14 Slider-crank gripper.

ing up to 10 m/s in electronic component placement. End-effector repeatability of 0.005 to 0.1 mm can be achieved for linear electric-motor actuated robots.

Grippers

Industrial robotic applications require manipulators to be equipped with a variety of tools and grippers (end-effectors). Here our emphasis is on grippers designed for the handling of workpieces (i.e., their grasping, transportation, and placement). Stable grasping of objects can be achieved through mechanical grippers with one or multiple dof (depending on workpiece geometry complexity) or through vacuum/magnetic type gripping devices.

The slider-crank mechanism shown in Fig. 14 can be used for the grasping of cylindrical objects and, when adjusted accordingly, for the grasping of constant-width prismatic objects. This is a simple one-dof gripper, normally operated using pneumatic power (for lightweight objects) or hydraulic power. The gear-and-rack mechanism shown in Figure 15 can be very effectively used for the grasping of different width prismatic workpieces, without any adjustment, owing to the parallel motion of the jaws.

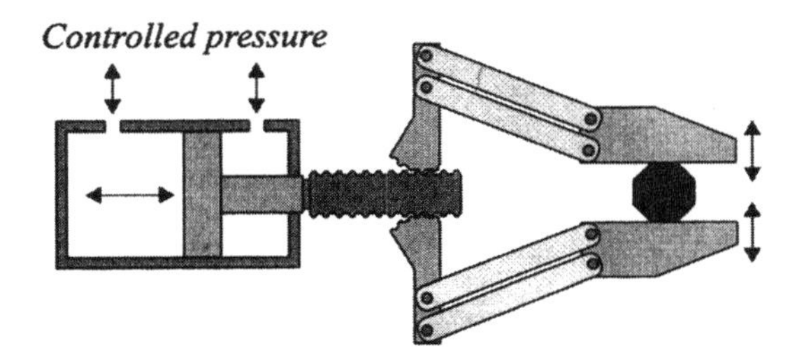

FIGURE 15 Gear-and-rack gripper.

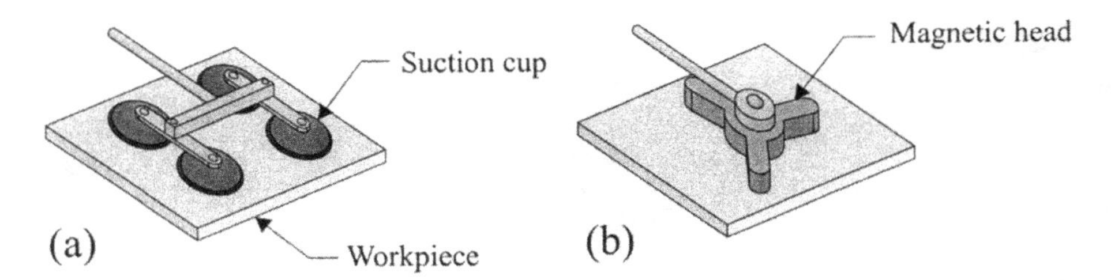

FIGURE 16 Grippers: (a) vacuum; (b) magnetic.

Vacuum type grippers can be used for the grasping of large nonferrous workpieces that cannot be handled using mechanical grippers (Fig. 16a). The workpieces must have appropriate flat surfaces, and a sufficient number of suction cups must be used for the successful grasping of heavy objects. For large ferrous objects, the use of magnetic grippers would be more suitable (Fig. 16b).

In a large number of applications, the industrial robot would be required to carry out an insertion operation—i.e., insert a peg held by the gripper into a hole on a fixtured object. In the absence of closed-loop force-feedback control, it would be recommended to use a passive alignment mechanism that would facilitate the insertion of the peg. One such facilitator is the remote–center compliance (RCC) device placed between the last link of the manipulator and the gripper (Fig. 17). The principle of an operation of an RCC device is to obtain compliance by employing flexible elements (spring, elastomer columns, etc.) between two (originally) parallel plates.

12.3.2 Applications

Industrial robots are automated mechanical manipulators that can be programmed to carry out a variety of material handling tasks, as well as production operations. As discussed earlier, the two primary end-effector-

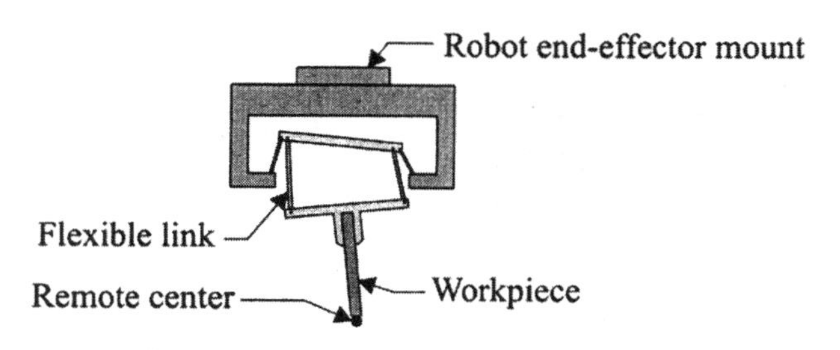

FIGURE 17 Remote-center compliance.

motion categories common to all industrial applications of robotic manipulators are PTP and CP. PTP motion implies the relocation of the end-effector from one position to another, such as in pick-and-place assembly operations, loading/unloading of machines, spot welding, riveting, and so on. CP motion, on the other hand, requires controlled movement of the end-effector (e.g., constant speed) along a prespecified trajectory defined in three-dimensional Cartesian space, such as in spray painting, arc welding, and so on.

Prior to a more detailed review of several industrial robot applications, a brief discussion of robot programming methods is provided below.

Robot Teaching

The majority of industrial robots can only be programmed after having been taken off the production line. The robot is taught to interact with its environment in show-and-teach mode and expected to operate subsequently in playback mode. In order to avoid this extremely time-consuming teaching process by carrying out an off-line teaching process, the following conditions must be satisfied:

The kinematic model of the robot describing the mobility of its end-effector (i.e., its Cartesian frame, $F_{\text{end-effector}}$, Fig. 9) with respect to its base (i.e., $F_{\text{robot-base}}$) must be known.

The locations of all devices (including the robot) must be accurately defined with respect to a fixed "world" coordinate frame and must not vary during the interactions of the robot end-effector with its environment.

The motion controller of the industrial robot must allow for off-line programming.

Only in a very limited number of occasions are all above three conditions satisfied. Thus in industry most robot teaching is still carried out by show-and-teach methods.

The most common robot teaching method relies on moving the end-effector to a series of points using a teach pendant (a sophisticated joystick) and asking the robot controller to memorize these locations in terms of robot joint positions. Then the operator compiles a program in a high-level (robot-dependent) language, specifying a set of time-dependent PTP motion segments or a set of CP trajectories (e.g., a straight line) between these points. A typical teach pendant would allow the programmer to move each joint individually or move the end-effector along individual Cartesian axes in the end-effector or in the world-frame coordinates in order to position the robot end-effector at a desired location.

Material Handling and Machine Loading

For a large number of industrial robot applications, the manipulator arm moves a workpiece from one location to another. The objective here is the safest and fastest transportation of the workpiece in PTP-motion mode with no particular emphasis on the trajectory followed by the end-effector. Typical tasks include picking and placing components from and onto conveyors, indexing tables, pallets, and bins, and loading and unloading lathes, milling machines, forging presses and casting machines (Fig. 18).

Today, a large number of manufacturing machines (milling machines, lathes, etc.) can be purchased with built-in robotic arms that allow the automatic loading and unloading of workpieces or even tools. On a number of occasions, material handling robots have also been mounted on AGVs for increased mobility and as an important step toward humanless autonomous factory floors.

Welding

Spot welding is one of the primary industrial robot applications especially in the automotive industry. Most commonly, the robot end-effector is a

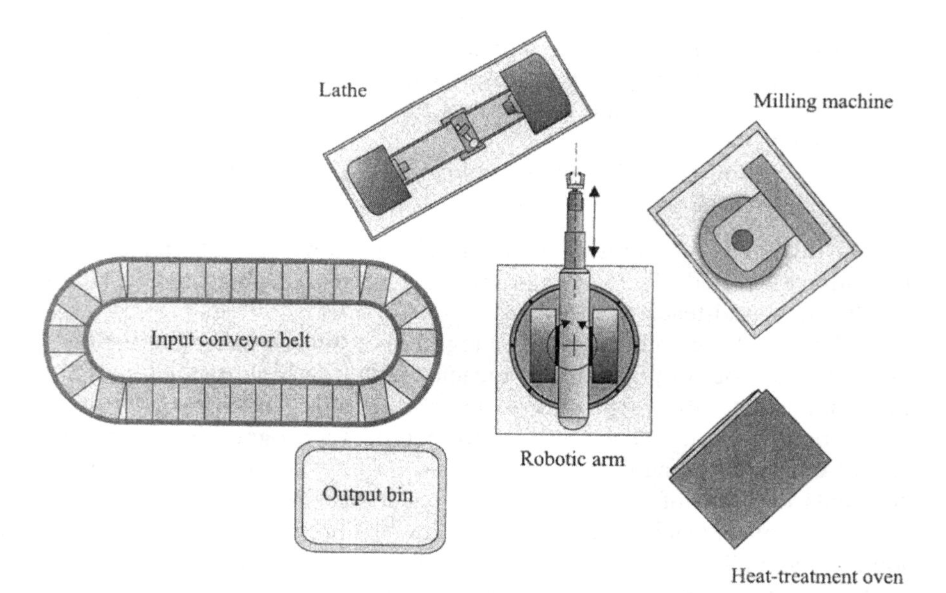

FIGURE 18 A robotic workcell.

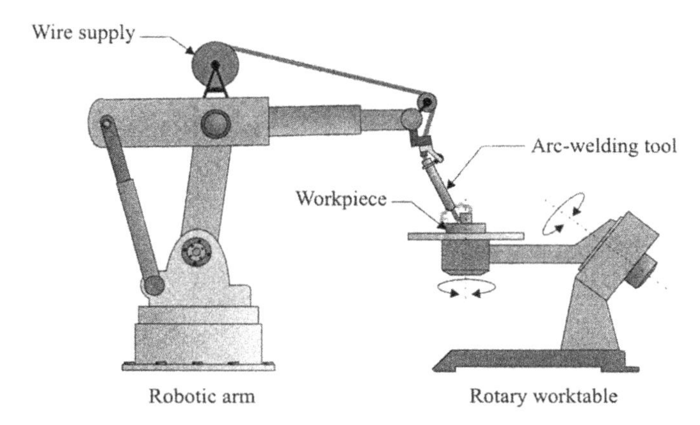

FIGURE 19 Robotic arc welding.

(resistance-weld) gun with two electrodes that joins two fixtured workpieces by squeezing their surfaces, while an electric current generates sufficient fusion heat (Chap. 10). The motion of such a robotic spot welder is of the PTP type—moving the welding gun from one spot to another. Occasionally, the industrial robot moves the two workpieces held in its gripper to a fixed spot welder, instead of moving the welding gun to fixed workpieces. Spot welding of automotive bodies (up to 1,500 to 3,000 spots per vehicle) is the most common example for robotic welding.

A large number of industrial robots have also been used in arc welding (Chap. 10). The robot end-effector is the arc welder fusing workpieces fixtured on stationary or rotary tables that themselves may have up to three rotational dof (Fig. 19). The robot end-effector is normally programmed to follow a CP trajectory at a constant speed. Most robot end-effectors for such arc welding tasks are equipped with proximity sensors to follow the seam (i.e., seam tracking). Typical product examples for robotic arc welding include water boilers, bicycle frames, gearbox casings, and car-seat frames.

Spray-Painting

Industrial robots dominate the spray-painting activities in manufacturing companies since they provide a clean-room environment and they are very repeatable. The CP motion trajectories followed by the robot end-effector (i.e., the spray gun), however, are difficult to teach to a conventional robot. Thus most robotic painters are taught through a lead-through technique.

An expert painter (literally) holds the end-effector of the robot and moves it through a desired complex trajectory, while the robot controller memorizes a large set of points on this trajectory. If the mechanical design of the robot does not allow such a lead-through motion because of the large gear ratios, these robots can be taught via a stripped-down version of the industrial robot (i.e., a "slave") with no gears or motors. Data collected using the slave robot is then transferred to the actual (master) robot, which is designated to carry out the spray-painting operation.

12.4 AUTOMATED STORAGE AND RETRIEVAL

The storage of goods until they are required for a manufacturing operation or shipment to a customer is commonly referred to as warehousing. The three common objectives of warehousing are ease of accessibility for random retrieval, effective protection of goods while they are stored and transported, and maximum utilization of space.

There are a variety of racks that can be used in the construction of high-density storage areas. Such structures could be as high as 20 to 30 meters and be totally automated in terms of storage of goods and their random retrieval based on an order issued by the warehousing computer—automated storage and retrieval systems (AS/RS).

Storage and retrieval equipment in high-density warehouses can be categorized into single-masted (single-column), double-masted, and human-aboard machines (stacker canes). Single-masted and double-masted machines are normally supported from the ceiling for accurate vertical alignment. All such machines are equipped with telescopic extraction devices for the loading/unloading of unit loads onto/from the racks based on an address defined by the warehousing computer (e.g., *X12, Y7, Z22* m) (Fig. 20).

12.5 IDENTIFICATION AND TRACKING OF GOODS

Effective material handling in flexible manufacturing systems requires automatic identification and tracking of goods that are stationary or in motion. This information must be transferred into a computer that oversees the transportation of goods in a timely manner. Some exemplary scenarios that necessitate automatic identification are listed below as a preamble to the descriptions of available technologies.

Unit loads (cartons, bins, pallets, etc.) moving on a conveyor network must be identified for correct branching.

A product arriving at an assembly substation must be identified for the correct assembly of parts by human or robotic operators.

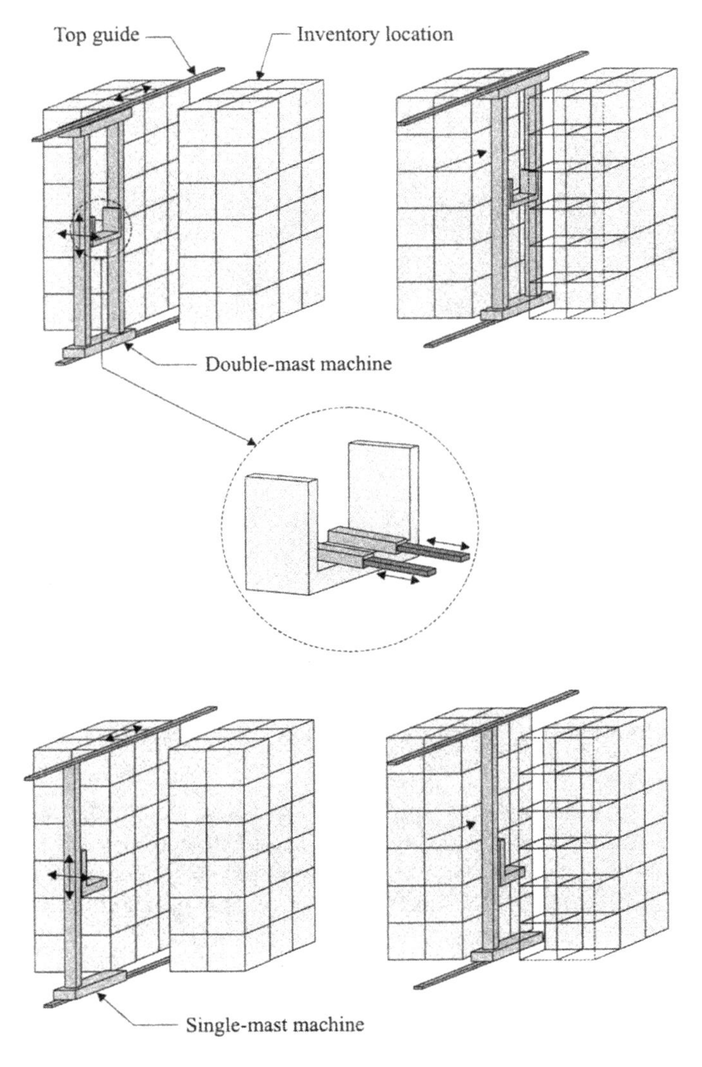

Figure 20 High-density warehousing.

A product arriving at a warehouse must be correctly identified for its automatic storage.

Automatic identification can be carried out by directly observing the geometry of the object or by indirectly reading an alphanumeric code attached to the object or onto the pallet/fixture carrying it. The former is, normally, carried out using a computer vision system or a collection of

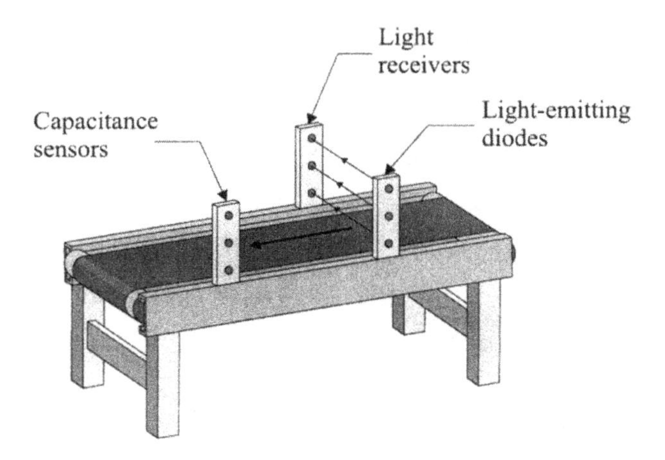

FIGURE 21 Direct object identification.

electro-optical or electromagnetic sensors that can detect a limited number of features on the object (Chap. 13).

Part identification through computer vision is a complex procedure and cannot be effectively utilized in high-volume, high-speed applications. Identification through individual noncontact sensors, on the other hand, can only differentiate among a limited number of object features, for example,

Electro-optical sensors mounted on both sides of a conveyor can differentiate among a number of different-height objects (Fig. 21).
Electromagnetic (e.g., capacitance) sensors mounted on one side of a conveyor can differentiate between conductive and nonconductive objects (metals versus plastics) (Fig. 21).

Bar Codes

Bar codes are the most commonly used identifiers of unit loads in manufacturing environments. Their primary advantage is the near-impossibility of incorrect identification—for most codes, a less than one in a million chance. Even if a bar-code scanner does not succeed in reading a code owing to improper printing or dirt, it will almost never read it as a different existing code.

A bar code is a collection of vertical printed bars (white and black) of two distinct thickness that form a constant-length string. Alphanumeric information is represented by assembling (combinatorically) different fixed-length subsets of vertical bars (i.e., characters) into a code. Almost all bar-code strings are provided with a check digit at their end for minimizing the occurrence of errors.

The 3-of-9 code is the most commonly used coding technique in manufacturing. A 3-of-9 bar-code character utilizes 3 wide and 6 narrow bars (black or white) to define a character. The symbology comprises the numbers 0 to 9, the letters A to Z, and six additional symbols, for a total of 43 characters. The Universal Product Code (UPC) symbology, on the other hand, is an all-numeric bar code, in which each digit is represented by 2 black and 2 white (narrow or wide) bars.

Bar codes printed on the highest possible quality printers are normally attached onto cartons/boxes that contain the goods or onto the pallets/fixtures that carry the goods and only very rarely onto the object itself. They must be placed on locations that are visible to the bar-code readers, preferably on the flattest parts of the goods/pallets/etc.

Bar-code readers (also known as "scanners") can be of the handheld type or the fixed-in-place type (e.g., stationed on the side of the conveyor). They must be placed at correct heights for effective reading. The reader scans through the bar code horizontally utilizing a light beam of circular cross section (reflected back for interpretation) with a diameter that is much smaller than the height of the bars. This relative dimensionality allows the bar-code reader to scan codes that are not well placed—misaligned and/or above or below their expected location (Fig. 22).

For increased effectiveness of bar-code reading (i.e., increased robustness to label misalignments), there are omnidirectional moving-beam scanners that use multiple light-beam scanners: one light beam is perpendicular to the motion direction of the good, while two other light beams

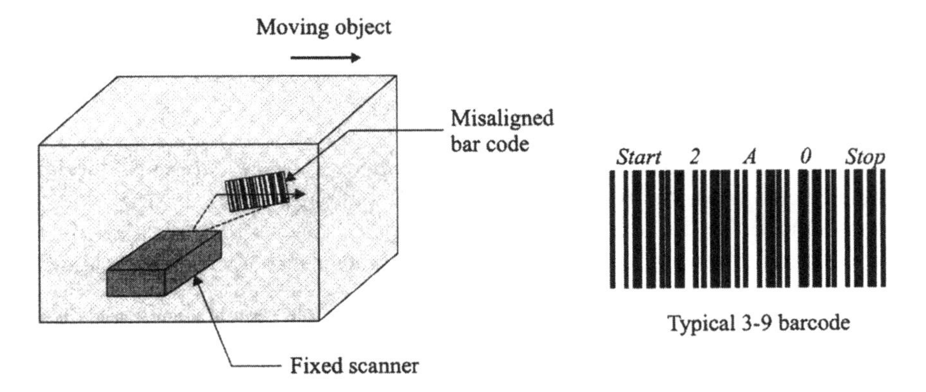

FIGURE 22 Bar-code scanning.

scan parallel to the motion direction of the good. There are also fixed-position moving-beam scanners that can project a scan line onto the good's surface in a rotating pattern to allow maximum flexibility of placing bar-code labels.

In certain circumstances, when the environmental and product constraints do not allow the use of optical devices, manufacturers have to use identifiers based on electromagnetic or RF-emission technologies (e.g., magnetic cards, tags). Such active (battery-powered) tags (transponders) are normally attached to the carriers of the goods (boxes, bins, pallets) that are reusable through reprogramming of the tag or the identifier (or both) for new goods.

12.6 AUTOMOBILE ASSEMBLY

Final assembly of automobiles has long been seen as the most complex manufacturing activity to automate because of high dexterity requirements, even long after the introduction of robotic manipulators on the factory floors. The typical car body is not assembly friendly; it requires individual components and subassemblies to follow complex paths to be fitted into their necessary locations. As discussed in Chap. 1, although the transfer of car bodies has been highly mechanized for their mass production since the 1910s, most direct assembly operations are still carried out manually.

12.6.1 Brief History

Today, the North American customer may erroneously assume that the name bearer of an automobile make manufactures and assembles the vehicle. In practice, however, a car maker is only primarily responsible for the design and assembly of the vehicles—the manufacturing of the individual components has long been the responsibility of independent or subsidiary suppliers (with some limited exceptions). Therefore major car makers are continually faced with low profit margins and thus must control assembly costs very carefully.

Historically, assembly plants have been set up as single-car-model plants, with most having large presses for the fabrication of the "white body" prior to its welding and painting on the same premises. The beginnings of these plants in the late 1880s and early 1900s were marked by the utilization of craft-production strategy; i.e., individual vehicles were assembled at specific fixed locations by highly skilled craftspeople who machined, matched, and assembled components with no great concern for interchangibility. Although this strategy was rapidly discontinued in favor of Ford's moving assembly line in the 1910s, one can still find recent examples of such neo-craft systems in the production of low-volume sports cars or luxury cars.

The early part of the 20th century was marked by the total dominance of conveyor lines in the transfer of vehicles from one assembly station to another. The vehicle was brought to the employee but almost always kept in motion, in order to maintain targeted productivity numbers. The 1960s witnessed the mechanization of two additional activities, complementary to mechanical assembly: painting and spot welding, which were eventually totally automated in the 1970s with the introduction of industrial robots and highly repeatable positioning mechanisms.

The period of 1975–1995 witnessed many innovations on the assembly floors, ranging from common-sense production strategies to modern material handling systems. The era was marked by intense competition between the Japanese auto makers and their U.S. counterparts, as well as drastic management changes in Europe. In hindsight, however, many concluded that these intense and varied changes had little impact on the productivity of assembly operations. The most important experiment was the concept of teamwork-based assembly introduced by Volvo at its plant in Uddevalla. The basic premise was to empower teams in decision making and subsequently increase their motivation in the faster assembly of high-quality cars (Sec. 12.6.3 below). Although concepts developed in this plant were adopted by others, including GM, the Uddevalla plant was closed in 1993.

12.6.2 Strategies for Automation

Although automation has been traditionally introduced in North America, as turnkey systems for improved productivity, in some European countries it has been introduced onto factory floors as human-motivating automation. The latter strategy aims at alleviating stress and boredom caused by repetitive jobs and then increasing employee satisfaction and job attractiveness. Companies such as Volvo that have adopted this strategy see automation as a tool that makes the assembly work station more human-friendly, where the human operator is expected to be always in control. The high-tech automation strategy adopted by companies such as GM, VW, and Fiat, on the other hand, emphasizes the eventual elimination of human operators from the factory floors.

Car assembly operations range from engine mounting, to suspension mounting, to windshield placement, to bumper mounting, to battery placement, to spare tire storage, and to a variety of wire harnessing operations. The characteristics of each are different in terms of weight, bulkiness, shape complexity, and alignment tolerances. Thus while some can be (and need to be) automated, most assembly tasks are not easy to automate with current technologies.

Statistics collected during the period of 1985 to 1995 indicated that less than 30% of assembly tasks have been automated. Among all automated tasks, welding is the leading operation (80 to 100% automated), while painting is the second leading automated task (100% of electrocoating and 70 to 100% of topcoat painting). In contrast to welding and painting, normally less than 2% of component insertion/mounting tasks have been automated: among those that are automated, windshield placement and sealing is the most automated task, while others include, rear-glass placement/sealing, engine mounting, suspension assembly, and spare tire placement. Some of these are further discussed below:

Body assembly: Levels of automation and technologies used in spot welding automation are very similar in the world auto industry. Although Japanese plants tend to have framing stations, where the roof, sides, and underbody of the vehicle are joined at once, U.S. and European plants usually have sequential lines. Furthermore, Japanese plants tend to automate a large portion of their arc welding operations, which may not be the case with U.S. plants.

Painting: The electrocoating process, the first coat of paint applied, primarily for rust protection, is always automated. The subsequent primer application and the last topcoat painting are also very widely automated. However, the interior painting of a vehicle is a challenging task and only infrequently automated (except in newer plants). Manual painting is usually required for finishing touch-ups at hard-to-reach interior places.

Final assembly: Despite massive investments in the past two decades, progress in the automation of final assembly has been very slow. However, modern plants do employ state-of-the-art conveying systems. For example, most plants utilize overhead conveyors with tilt-over mechanisms to allow efficient underbody work. For work on the upper body and interior of the car, most plants remove the doors at the start of the line and reattach them later. Most plants also utilize platform-type conveyors that carry the vehicle only or both the vehicle and the assembly person. Some newer plants allow for variable-speed conveying, for more efficient assembly, according to the task at hand. In carrying out the assembly tasks, the employees are provided with a variety of assistance tools for lifting, placing, bolting, and so on.

12.6.3 The Uddevalla System

Volvo's Uddevalla plant was initially planned to be a complete auto assembly plant with a body welding shop and paint shop. However, in

1989, the plant opened only as a final assembly plant, which eventually was shut down in 1993 because of a cyclic downturn in demand. Some have argued that a contributing factor to the closure was the lack of body and paint shops. However, all agree that the radical changes experimented with in this plant have been adopted by other Volvo plants and other auto makers, so Uddevalla's death has not been the end of the story!

As discussed above, the assembly strategy adopted by the designers of the Uddevalla plant was based on human-motivating automation, which attempted to make assembly work attractive by moving from a moving-line-type assembly to stationary assembly. Thus first came the development of such an implementable assembly strategy. The concept necessitated the empowerment of a small team (planned to be 5 to 10 people) to assume total assembly responsibility for the vehicle. However, due to efficiency limitations, the assembly of the vehicle had to be planned as a division into 8 sequential stations (as opposed to a single assembly station), where workers were to carry out assembly tasks for a duration of 20 minutes on each car (an upper time limit to human efficiency in learning and performing a sequence of tasks). Each station was to be supplied with complete component kits (prepared at a central location) for each 1/8th of the vehicle, respectively.

Next came the difficult task of factory layout. The first planned cost-effective layout comprised parallel assembly shops, each for complete car assembly. Each shop was to house four stations allowing eight separate teams to work in parallel on four stationary cars. In each shop, labor teams specializing on 1/8th of the assembly were to move from one station to another and complete the assembly of four cars at a time. Each team was to comprise 8 to 10 people. Stations were to be equipped with lifting and tilting mechanisms for maximum utilization of space.

The eventual shop layouts, however, employed only four teams (versus eight) working sequentially at four stations within the shop, for a total assembly time of almost two hours, repeated four times a day. The teams worked in an asynchronous manner deciding when to start to work on the next car and even occasionally switching tasks among themselves between cars.

Prior to its closure, Uddevalla had the following statistics: complaints per car reduced from 1.32 to 0.87, productivity of 32 hours per car assembly versus 42 in other European plants and above 50 in North American plants; 25% lower tooling costs per car and lead-time reductions of almost 50% on delivery of ordered cars.

Although the plant no longer exists in its intended form, many European and Japanese auto makers have adopted the principles of Uddevalla—a combination of low-cost technology and skilled labor for human-oriented automotive assembly. The use of AGVs, instead of conveyor lines, in automotive assembly has also been commonly attributed to the Uddevalla experience.

REVIEW QUESTIONS

1. What is the fundamental objective of material handling? Is this objective compatible with just-in-time manufacturing? Explain.
2. Why can industrial trucks be considered "flexible" equipment?
3. Define automated guided vehicles (AGVs). Can AGVs be considered industrial robots?
4. What has been the primary purpose of developing and using AGVs in industrial settings that typically employ industrial trucks, such as lift trucks and tow tractors?
5. Describe navigation guidance for AGVs.
6. Why should one consider overhead conveying?
7. Compare the use of AGVs versus conveyors in asynchronous manufacturing environments. Consider car engine assembly as an example application.
8. What is an industrial robot?
9. Define mobility for open-chain robotic (spatial) manipulators.
10. Define point-to-point (PTP) versus continuous-path (CP) motion for industrial robots.
11. Why can rectangular geometry robots (with linear joints) provide users with better end-effector accuracy than that achievable with nonrectangular geometry robots (with rotary joints)? Discuss the use of parallelogram linkages in this context.
12. Describe the use of a remote-center compliance device in robotic component insertion processes.
13. Discuss the limitations of robot teaching on the implementation of industrial manipulators. Consider the assembly of a multicomponent toy car as an example.
14. Why are industrial robots ideal for use in welding and painting applications?
15. What is automated storage and retrieval?
16. What are bar codes and how can they be used in manufacturing environments?
17. Discuss the use of nonoptical goods tracking devices in manufacturing environments.
18. Why has automobile assembly been so intensely studied? Discuss its historical development in the 20th century.

19. What are the most commonly automated tasks in automobile assembly?
20. Briefly review the Uddevalla automobile assembly system and discuss its innovations.

DISCUSSION QUESTIONS

1. Computers and other information management technologies have been commonly accepted as facilitators for the integration of various manufacturing activities. Define/discuss integrated manufacturing in the modern manufacturing enterprise and address the role of computers in this respect. Furthermore, discuss the use of intranets and extranets as they pertain to the linking of suppliers, manufacturers, and customers.
2. Manufacturing flexibility can be achieved at three levels: operational flexibility, tactical flexibility, and strategic flexibility. Discuss operational flexibility. Is material handling automation a necessary or a desirable tool in achieving this level of flexibility?
3. Explain the importance of investigating the following factors in the establishment of a manufacturing facility: availability of skilled labor, availability and closeness of raw materials and suppliers, closeness of customers/market, and availability of logistical means for the effective distribution of products.
4. Discuss strategies for retrofitting an existing manufacturing enterprise with automation tools for material as well as information processing. Among others, consider issues such as buying turn-key solutions versus developing in-house solutions and carrying out consultations in a bottom-up approach, starting on the factory floor, versus an top-to-bottom approach, starting on the executive board of the company and progressing downward to the factory floor.
5. Several fabrication/assembly machines can be physically or virtually brought together to yield a manufacturing workcell for the production of a family of parts. Discuss the advantages of adopting a cellular manufacturing strategy in contrast to having a departmentalized strategy, i.e., having a turning department, a milling department, a grinding department, and so on. Among others, an important issue to consider is the transportation of parts (individually or in batches).
6. In the factory of the future, it is envisioned that production and assembly workcells would be frequently reconfigured based on the latest manufacturing objectives without the actual physical relocation

of their machines/resources. Discuss the material handling options that should be available to the users of such workcells, whose boundaries may exist only in the (computer's) virtual space.

7. Discuss the advantages of utilizing reusable pallets and other temporary storage means in a manufacturing environment. Propose a number of features that would improve the usability, transportation, storage, removal, and so on of such pallets/boxes/magazines/etc.
8. AGVs have been accepted as being more flexible and less space restrictive than networks of conveyors. Discuss this comparison and extend it to include human-operated material handling vehicles (e.g., forklifts) in manufacturing environments as well as overhead conveying (i.e., three-dimensional material handling solutions.
9. Welding, soldering, and painting are manufacturing operations that rely on maintaining consistent and repeatable process parameters. Discuss the use of automation (including the use of industrial robots) for these and other processes that have similar requirements as replacements for manual labor.
10. The majority of industrial robots can be programmed using a high-level language that allows users to define end-effector trajectories between a priori taught points in Cartesian space. Lack of earlier standardization in NC machine controllers that have been frequently used as the basis for robot controllers, however, has also led to the absence of a single programming language common to all industrial robots. In trying to protect their markets, industrial robot makers have developed proprietary programming languages. Discuss the potential negative impact of the proliferation of programming languages on the use of industrial robots in manufacturing environments.
11. The necessary programming of robot task space locations by physically moving the robot's end-effector to these positions, while it is taken off the manufacturing line, has severely limited their use to mass-production environments. Thus although industrial robots provide a high level of automation, they cannot be time efficiently programmed and used for one-of-a-kind or small-batch productions. Discuss potential remedies that would allow robots to be programmed for their next task, while they are performing their current task.
12. Industrial robots have traditionally been designed and marketed as "all-capable" (generic) manipulators. That is, they have the necessary mobility to manipulate objects in three-dimensional space (position and orientation), the necessary workspace to replace a human operator, the necessary load carrying capability for a large number of applications, and so on. In most manufacturing environments, however, such robots would be overqualified. Discuss an alternative approach to industrial

robot design and marketing that is based on modularity and reconfigurability, a possibility of manufacturing industrial robots that are made to order based on the employment of standardized modular components: cut-to-length links and properly sized actuators assembled to the extent of the mobility (frequently less than six degrees of freedom) that is required and in an optimal geometrical architecture.

13. Industrial robots are normally sold with no end-effectors (grippers or other tools). Discuss the advantages/disadvantages of using generic humanlike multifingered hands versus simple single-purpose, limited-mobility (possibly, quick-exchange) end-effectors.

14. Industrial robots have been often labeled as being deaf and blind operators with no tactile feedback detection capability. Discuss in general terms what would be the benefits of having a variety of visual and nonvisual sensors monitoring the robot's working environment and feeding back accurate and timely information to the motion controller of such manipulators.

15. Human operators have been argued to be intelligent, autonomous, and flexible when compared to industrial robots. Discuss several manufacturing applications where one would tend to utilize human operators rather than industrial robots (even those supported by a variety of sensors) in the context of these three properties.

16. Industrial robots have been often designed to replace the human operator in manufacturing settings. The past several decades have shown us, however, that there still exist significant gaps between humans' and robots' abilities, primarily owing to the unavailability of artificial perception technologies. Compare humans to pertinent anthropomorphic robots in terms of the following and other issues: mechanical configuration and mobility, power source, workspace, payload capacity, accuracy, communications (wireless!), supervisory control ability, sensory perception, ability to process data, coping with uncertainties, and working in hazardous environments.

17. Discuss the need of having effective tracking means, distributed throughout a manufacturing facility, that would provide users (and even customers) with timely feedback about the status and location of products in motion. Include several examples of such sensing and feedback devices in your discussion. Furthermore, discuss the advantages/disadvantages of using wireless solutions.

18. It has been long argued that manufacturing inventories hide production problems, and that their elimination would expose the sources of chronic problems. Despite this understanding, what manufacturing scenarios would still require the use of large (automated or manually operated) warehouses?

19. Material handling has been always argued to be a cost adding (versus value adding) activity in manufacturing. Despite this acceptance, however, many manufacturers, including numerous vehicle assemblers, continue to utilize several-kilometer-long paths that their product must travel prior to their shipment. Discuss the reasons that fabricators and assemblers of many manufactured products continue to utilize long material handling paths. Consider several examples and propose remedies. In your discussion include a comparison of the assembly concepts of bringing the work to the operator versus bringing the operator to the work, as well as a comparison of the fabrication concepts of departmentalized facility layout (i.e., milling department, grinding department, heat-treatment department, etc.) versus flexible manufacturing workcell layout.
20. The 20th century has witnessed a historical trend in the strong reduction of manual labor in the agricultural industry with the introduction of a variety of (mechanized) vehicles, irrigation systems, crop treatment techniques, and so on. Discuss the current trend on continuing reduction in the (manual) labor force involved in materials handling activities. Can you identify similarities to what has happened in the agricultural industry (and even in the book publishing, textile, and other industries in earlier centuries) to what may happen in the manufacturing industry in the 21st century?

BIBLIOGRAPHY

Apple, James M. (1977). *Plant Layout and Materials Handling*. New York: Ronald Press.

Asfahl, Ray C. (1992). *Robots and Manufacturing Automation*. New York: John Wiley.

Stephen, Cameron, Penelope, Probert, eds. (1994). *Advanced Guided Vehicles*. River Edge, NJ: World Scientific.

Eastman, Robert M. (1987). *Materials Handling*. New York: Marcel Dekker.

Engstrom, T., Medbo, L., Jonsson, D. (July 1994). Extended work cycle assembly—a crucial learning experience. *International Journal of Human Factors in Manufacturing* 4(3):293–303.

Engstrom, T., Medbo, L., Jonsson, D. (1998). Volvo Uddevalla plant and interpretations of industrial design processes. *Journal of Integrated Manufacturing Systems* 9(5–6):279–295.

Groover, Mikell P. (2001). *Automation, Production Systems, and Computer-Integrated Manufacturing*. Upper Saddle River, NJ: Prentice Hall.

Fayed, M. E., Skocir, Thomas S. *Mechanical Conveyors: Selection and Operation*. Lancaster, PA: Technomic.

Hammond, Gary (1986). *AGVS at Work: Automated Guided Vehicle Systems*. Bedford: UK: IFS.

Hollier, R. H., ed. (1987). *Automated Guided Vehicle Systems*. Bedford, UK: IFS.
Holzbock, Werner G. (1986). *Robotic Technology, Principles and Practice*. New York, NY: Van Nostrand Reinhold.
Hunt, V. Daniel (1988). *Robotics Sourcebook*. New York: Elsevier.
Koren, Yoram (1985). *Robotics for Engineers*. New York: McGraw-Hill.
Kulwiec, Raymond A., ed. (1985). *Materials Handling Handbook*. New York: John Wiley.
Lundesjo, Gregor, ed. (1985). *Handbook of Materials Handling*. New York: Halsted Press.
Miller, Richard Kendall, Subrin, Rachel, ed. (1987). *Automated Guided Vehicles and Automated Manufacturing*. Dearborn, MI: Society of Manufacturing Engineers.
Mulcahy, David E. (1999). *Materials Handling Handbook*. New York: McGraw-Hill.
Müller, Willi (1985). *Integrated Materials Handling in Manufacturing*. Bedford, UK: IFS.
Nayak, Nitin, Asok, Ray (1993). *Intelligent Seam Tracking for Robotic Welding*. New York: Springer-Verlag.
Nof, Shimon Y., ed. (1999). *Handbook of Industrial Robotics*. New York: John Wiley.
Pham, D. T., Heginbotham, W. B. (1986). *Robot Grippers*. Bedford, UK: IFS.
Rembold, Ulrich (1990). *Robot Technology and Applications*. New York: Marcel Dekker.
Rivin, Eugene I. (1988). *Mechanical Design of Robots*. New York: McGraw-Hill.
Shimokawa, K., Jürgens, U. Fujimoto, T., eds. (1997). *Transforming Automobile Assembly: Experience in Automation and Work Organization*. New York: Spring-Verlag.
Todd, D. J. (1986). *Fundamentals of Robot Technology: An Introduction to Industrial Robots, Teleoperators and Robot Vehicles*. London: Kogan Page.

Part III

Automatic Control in Manufacturing

Automatic control in manufacturing refers to forcing a device or a system achieve a desired output in an autonomous manner through intelligent instrumentation. Control is carried out at multiple levels and at different modes. At the lowest level, the control of individual devices for the successful execution of their required individual tasks is achieved in the continuous-time domain. At one level above, the control of a system (e.g., a multidevice manufacturing workcell), for the correct routing of parts within it, is achieved in an event-based control mode. In both cases, however, automatic control relies on accurate and repeatable feedback received from individual device controllers and a variety of sensors.

In Chap. 13, the focus is on the description of various sensors that can be used for automatic control in manufacturing environments. A brief generic introduction to the control of devices in the continuous-time domain precedes the discussion of various pertinent analog- and digital-transducer based sensors (e.g., motion sensors, force sensors). Machine vision for two-dimensional image analysis is also addressed in this chapter. A variety of actuators are described in the conclusion of the chapter as the "executioners" of closed-loop control systems.

In reprogrammable flexible manufacturing, it is envisaged that individual machines carry out their assigned tasks with minimal operator intervention. Such automatic device control normally refers to forcing a servomechanism to achieve (or yield) a desired output parameter value in the continuous-time domain. In Chap. 14, our focus will thus be on the automatic control of two representative classes of production and assembly machines: material removal machine tools and industrial robotic manipulators. For the former class of machines, numerical control (NC) has been the norm for the control of the movement of the cutting tool and/or the workpiece since the early 1960s. In this context, issues such as motion trajectory interpolation, *g*-code programming, and adaptive control will be discussed in this chapter.

The planning and control of the motion of industrial robots will also be discussed in Chap. 14. Robotic manipulators can be considered the most complex assembly devices in existence. Thus solutions valid for their control would be applicable to other assembly machines. Regardless of their geometry classification (serial or parallel), industrial robotic manipulators carry out tasks that require their end effector (gripper or specialized tool) to move in point-to-point or continuous-path mode, just as do NC machine tools. Unlike NC motion interpolation for machining, however, trajectory planning for industrial robots is a complex matter owing to the dynamics of open-chain manipulators moving payloads in three-dimensional Cartesian space subject to gravitational, centrifugal, and inertial forces. In this context, the following issues are discussed in Chap. 14: robot kinematics/dynamics, trajectory planning and control, and motion programming.

In a typical large manufacturing enterprise, there may be a number of flexible manufacturing systems (FMSs) each comprising, in turn, a number of flexible manufacturing workcells (FMCs). An FMC is a collection of production/assembly machines, commonly configured for the manufacturing of families of parts with similar processing requirements, under the control of a host supervisor. The focus of Chap. 15 is thus the autonomous supervisory control of parts, flow within networked FMCs; in contrast to time-driven (continuous-variable) control of the individual devices in a FMC, the supervisory control of the FMC itself is event driven.

There are three interested parties to the FMC-control problem: users, industrial controller developers, and academic researchers. The users have been always interested in controllers that will improve productivity, in response to which industrial controller vendors have almost exclusively relied on the marketing of programmable logic controllers (PLCs). The academic community, on the other hand, has spent the past two decades developing effective control theories that are suitable for the supervisory control of manufacturing systems. In Chap. 15, we will thus first address two of the most successful discrete-event system control theories developed by the academic community: Ramadge-Wonham automata theory and Petri-nets theory. The description of PLCs, used for the autonomous DES-based supervisory control of parts flow in FMCs, will conclude this chapter.

Quality control refers to the establishment of closed-loop control processes capable of measuring conformance (as compared to desired metrics) and varying production parameters, when necessary, to maintain steady-state control. The final manufacturing issue thus addressed in this part of the book, in Chap. 16, is quality control with specific emphasis on on-line statistical control (versus postprocess sampling). Quality management strategies and measurement technologies targeted specifically to quality control are addressed in Chap. 16 as a preamble to a discussion on common statistical tools, such as statistical process control. A brief discussion of ISO 9000 is also presented in this chapter.

13

Instrumentation for Manufacturing Control

In flexible manufacturing systems (FMSs), control is carried out on multiple levels and in different modes. On the lowest level, our interest is in the control of individual devices (e.g., milling machine, industrial robot) for the successful execution of their required individual tasks. One level above, our concern would be with the control of a collection of devices working in concert with each other [e.g., a multidevice flexible manufacturing workcell (FMC)]. Here, the primary objective is the sequencing of tasks through the correct control of part flow. In both cases, however, automatic control relies on accurate and repeatable feedback, in regard to the output of these processes, achieved through intelligent instrumentation.

Automatic device control normally refers to forcing a servomechanism to achieve (or yield) a desired output parameter value in the continuous-time domain. Requiring a milling machine to cut through a desired workpiece contour is a typical manufacturing example. Motion sensors measuring the displacement and speed of the individual axes of the milling machine table provide the closed-loop control system with necessary feedback about the process output. Automatic supervisory control of FMCs, on the other hand, means forcing the system to behave within legal bounds of task sequencing based on observable events that occur within the system. This type of event-based control is primarily achieved based on feedback information received from individual device controllers and device-independent (workcell) sensors.

The principal element of any sensor is the transducer—a device that converts one form of energy into another (e.g., light into electrical current). The combination of a transducer and a number of signal-conditioning and processing elements forms a sensor. In this chapter, the focus is on the description of various sensors that can be used for automatic control in manufacturing environments. A brief generic introduction to the control of devices in the continuous-time domain will precede the discussion of various pertinent manufacturing sensors. The control of machine tools and robots will be discussed in greater detail in Chap. 14; an in-depth discussion of event-based manufacturing system control is presented in Chap. 15. Quality control issues will be addressed in Chap. 16.

13.1 PROCESS CONTROL AND CONTROLLERS

Closed-loop (feedback) control continuously adjusts the variable parameters of a process in order to yield an output of desired value. As shown Fig. 1, the actual output parameter value, c, is measured via a sensor and fed back to a comparator (summing junction) for the computation of the error, e, with respect to the desired output value, r. Based on this error value, $e = r - c$, a controller decides on an appropriate corrective action and instructs an actuator (or multiple ones) to carry out this response.

For a dynamic process, all process variables would be functions of time, where the primary objective of the control system is to reduce the output error to as close as possible to zero in the fastest manner. Although different controller designs will achieve this objective in varying transient-response ways, all must thrive to yield stable systems with minimum steady-state errors.

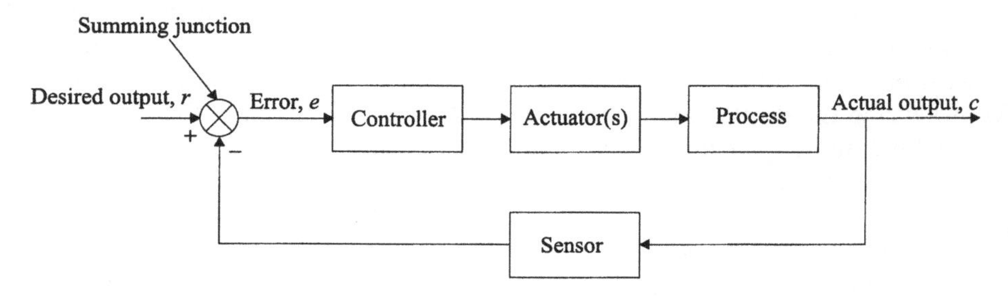

FIGURE 1 Closed-loop control block diagram.

Controllers have often been classified as analog versus digital. Analog systems are, naturally, more prone to electronic noise than their digital counterparts which utilize analog-to-digital-to-analog (AD/DA) converters for analog inputs/outputs.

In digital control, the digital processor (a computer) can be used in two different configurations:

Supervisory control: A microprocessor (computer) is utilized as a (digital) monitoring device and provides the control system with new desired output values (Fig. 2a). The control is still analog in nature. The microprocessor can be used to control several systems.

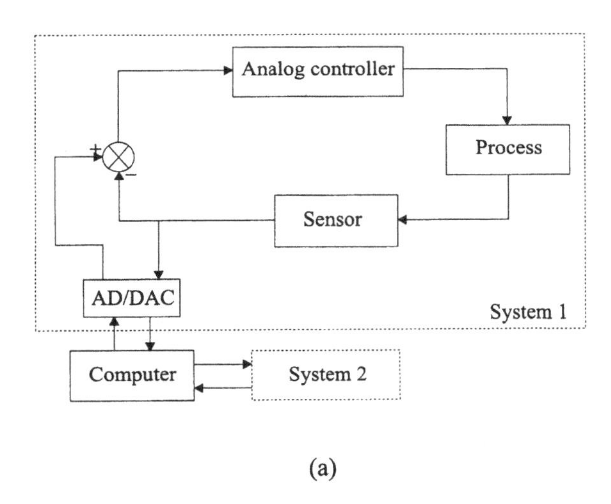

(a)

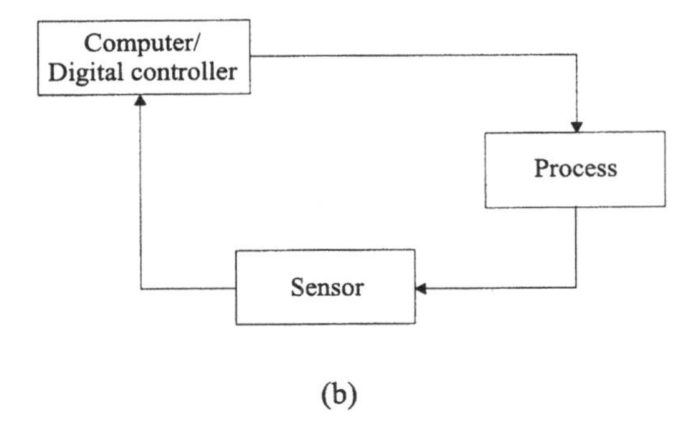

(b)

FIGURE 2 Digital control: (a) supervisory; (b) direct.

Direct control: A microprocessor replaces completely the analog controller and the comparator as the sole control device. All (computer) inputs and outputs are digital in nature (Fig. 2b).

13.1.1 Controller Modes

Continuous controllers manipulate the (input) error signal for the generation of an output signal in several different modes, most commonly relying on proportionality:

Proportional-integral (PI) control: This composite control mode uses the following typical expression for determining the output signal value, p:

$$p = K_p e + K_p K_i \int_0^t e \, dt + P_o \tag{13.1}$$

where K_p and K_i are the proportional and integral gains, respectively, and p_o is the controller output with no error. The integral mode of the composite signal eliminates the inherent offset (residual error) that would have been produced by the proportional mode of control. PI controllers may yield large overshoots owing to integration time.

Proportional-derivative (PD) control: This composite control mode utilizes a cascade form of the two individual proportional and derivative control modes:

$$p = K_p e + K_p K_d \frac{de}{dt} + p_o \tag{13.2}$$

where K_d is the derivative gain. The derivative mode of a composite controller responds to changes in the error (the rate of change)—it is a predictive action generator.

Proportional-integral-derivative (PID) control: This three-mode composite controller is the most commonly used controller for industrial processes:

$$p = K_p e + K_p K_i \int_0^t e \, dt + K_p K_d \frac{de}{dt} + p_o \tag{13.3}$$

13.1.2 Controllers

Electronic analog controllers that use analog (current) signals are commonly employed in the automatic control of manufacturing devices. Op-amp circuits form the backbone of these controllers. Error signals are computed by measuring voltage differences and used for determining the output current signal of the controller, where gains are defined by specific resistor and capacitor values.

Digital controllers are computers that are capable to interact with external devices via I/O interfaces and AD/DA converters. Their reprogrammability with appropriate software greatly enhances their usability for automatic control. The primary advantages of using digital controllers include ease of interface to peripheral equipment (e.g., data storage devices), fast retrieval and processing of information, capability of using complex control laws, and transmission of noiseless signals.

13.2 MOTION SENSORS

Motion control is of primary interest for the majority of manufacturing processes: automatic control of a milling operation requires precise knowledge of the motion of the table, on which the workpiece is mounted; industrial robots need to know the exact location of a workpiece prior to its grasping; and so on. Motion sensors can provide the motion controllers of such manufacturing equipment with displacement, velocity, and acceleration measurements. Mostly, they carry out their measurement tasks without being in contact with the object.

Motion sensors use a variety of transducers that yield analog output signals. Electromagnetic, electro-optical, and ultrasonic transducers are the most common ones and will be discussed individually below. Some digital transducers will also be presented in this section.

13.2.1 Electromagnetic Transducers

The majority of electromagnetic-transducer-based noncontact sensors are used in manufacturing environments as detectors of presence, as opposed to absolute or relative measurement of motion, owing to their low-precision yield. Such sensors, although frequently called proximity (i.e., distance and orientation) sensors, simply detect the presence of an object in their close vicinity. Some exemplary sensors are briefly described below:

Potentiometers: Resistive-transducer-based contact displacement sensors are often referred to as potentiometers, or as pots. The transducer of a potentiometer, a wire or a film, converts mechanical displacement into voltage owing to the changing resistance of the transducer (Fig. 3).

Potentiometers can be configured to measure linear or rotary displacements. In both cases, however, owing to their contact mode, they add inertia and load (friction) to the moving object whose displacement they are measuring.

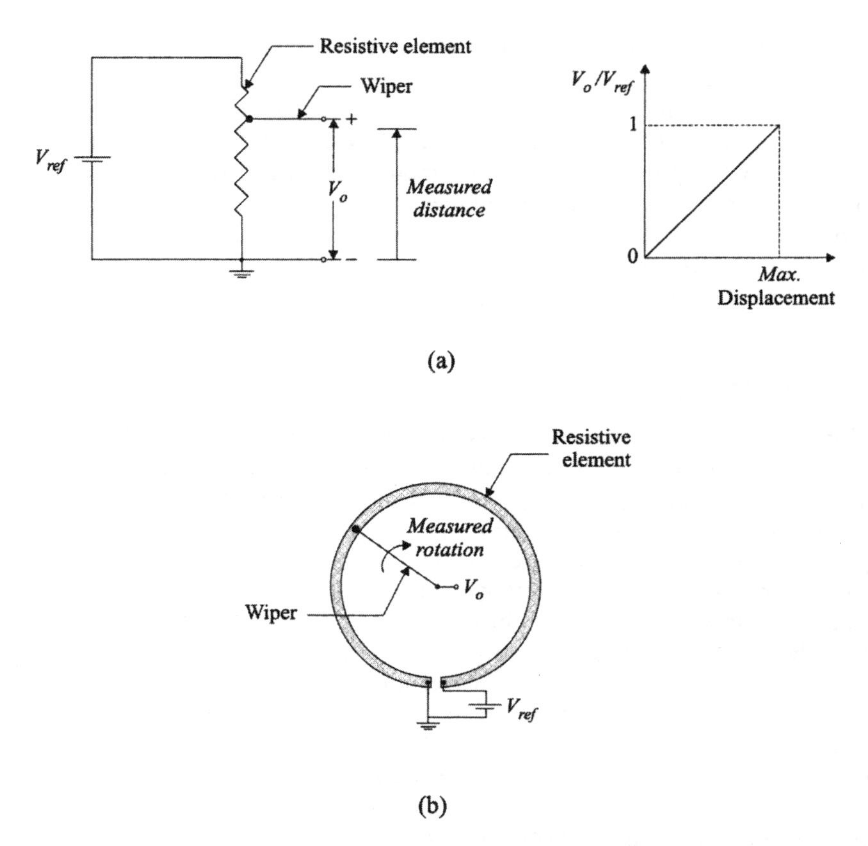

FIGURE 3 Resistive transducers for (a) linear; (b) rotary motions.

LVDT: The linear variable-differential transformer (LVDT) is a passive inductive sensor utilized for the contact measurement of linear displacement. This variable-reluctance transducer comprises a moving core that varies the magnetic flux coupling between two or more coils (Fig. 4). When the core is placed in the center, the output voltage is zero since the secondary voltages are equal and cancel each other. As the core is displaced in one direction or another, a larger voltage is induced in one or the other secondary coil, thus producing a voltage differential as a function of core displacement.

There also exist rotary variable-differential transformers (RVDTs) for rotational displacement measurements.

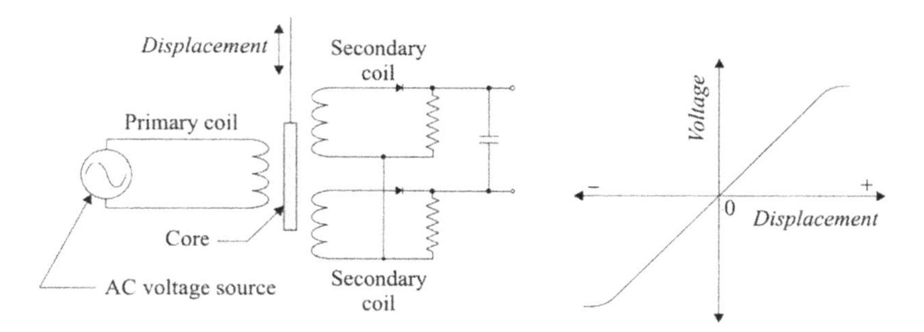

FIGURE 4 Linear variable-differential transformer.

Transverse inductive sensors: Inductive transducers can be configured to act as proximity or presence detection sensors, when only one coil is used. The flux generated by the coil is disturbed by a magnetic object in the close vicinity of the transducer (10 to 15 mm) (Fig. 5). Although the displacement of the object can be related to the amount of flux change, such sensors are rarely used for absolute (precision) measurements of displacement.

Capacitive sensors: Variations in capacitance can be achieved by varying the distance between the two plates of a flat capacitor. In capacitance displacement sensors for conducting material objects, the surface of the object forms one plate, while the transducer forms the other plate

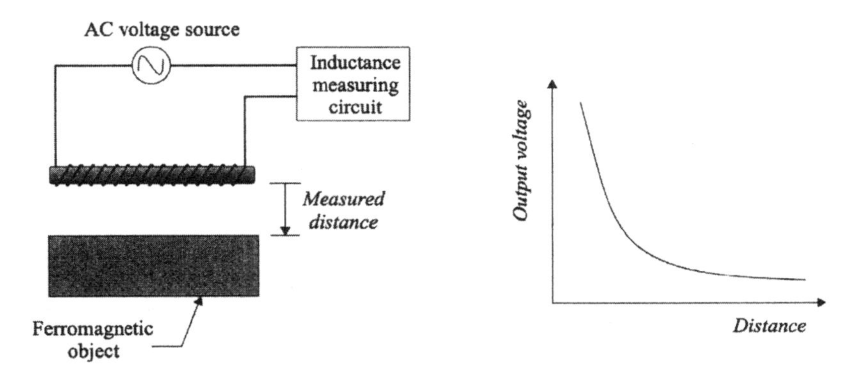

FIGURE 5 Inductive proximity sensor.

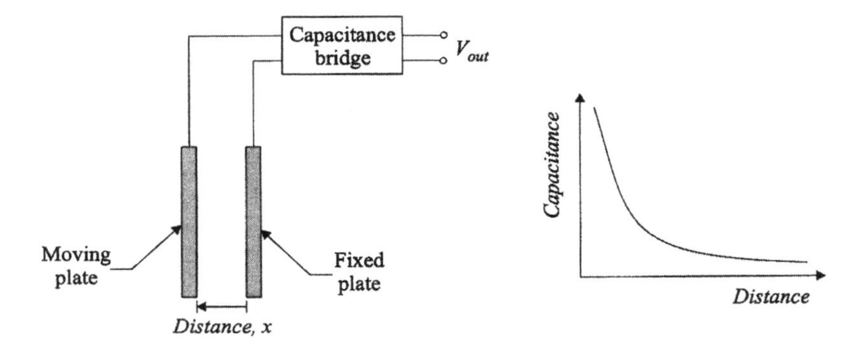

FIGURE 6 Capacitive sensors.

(Fig. 6). For dielectric objects, the capacitive sensor would have two live electrodes—the object does not need to be part of the sensing system (also called fringing capacitance).

As with transverse inductive sensors, the precise measurement of absolute motion is a difficult task for capacitive sensors. Thus they are commonly used only for the detection of the presence of conductive or dielectric objects close to the sensor (up to 30 to 40 mm).

13.2.2 Electro-Optical Transducers

Electro-optical-transducer-based sensors developed over the past three decades allow noncontact displacement measurement of the highest possible precision—for example, less than half a light wavelength for interferometers. Such sensors are also often used in the manufacturing industry for simply checking for the presence of an object. The common principle of all electro-optical sensors is the controlled emission of light, its reflection from the surface of an object, and the analysis and interpretation of the reflected light for absolute or relative position and, in some instances, orientation measurements.

Light Sources

The majority of electro-optical sensing devices in manufacturing utilize coherent or noncoherent light in the infrared range (0.76 to 100 μm wavelength). In some applications, the utilization of light in the visible range (0.4 to 0.76 μm wavelength) might be sufficient. Typical light sources include incandescent lamps, solid-state lasers, and light-emitting diodes (LEDs), the last developed in the early 1960s. LEDs are transducers that convert electrical current into light—namely, the opposite of light-detecting transducers.

Light Detectors

There are a variety of electro-optical transducers that can detect electromagnetic radiation based on the interaction of photons with semiconductor materials. These devices are often called photodetectors:

Photodiodes: These detectors operate in two distinct modes, photoconductive and photovoltaic. In the former mode, radiation causes change in the conductivity of the semiconductor material in terms of change in resistance. Photovoltaic detectors, on the other hand, generate a voltage proportional to the input light intensity. Photodiodes' primary advantage is their fast response time (as low as a few nanoseconds).

Phototransistors: These detectors produce electrical current proportional to input light intensity. Phototransistors provide higher sensitivity (i.e., higher current) than do photodiodes but operate at much lower response times (milliseconds versus nanoseconds).

Optical Fibers

Optical fibers allow remote placement of sensors that employ electro-optical transducers, as well as access to hard-to-reach places. They can be either plastic or glass and are normally protected by a cladding layer against potential damage and/or excessive bending. Fiber-optic cables can be easily coupled to light-emitting or light-receiving diodes—that is, they can be used to collect light reflected from a surface (normally, within a 20 to 30° cone) as well as emit coherent or noncoherent light onto desired surfaces.

Amplitude Modulation Proximity Sensors

In amplitude modulation electro-optical sensors, the magnitude of the light reflected from a surface can be utilized to determine the distance and orientation of the object. Such sensors usually comprise one light source and several photodetectors. Many utilize plastic optical fibers (typically, having a 0.3 to 2 mm core size) to reflect and collect light from objects' surfaces (Fig. 7a). The intensity of the light reflected from the surface is not a monotonic function of the distance. Thus the minimum operating distance of the transducer (x_{min}) is usually limited to a value that will guarantee a monotonic response (Fig. 7b).

For the measurement of surface orientation, a symmetrical three-fiber constellation can be used (Fig. 8a). In this configuration, the emitter is at the center and the two receivers are positioned symmetrically on either side. The light intensities detected by the receivers of this sensor, as a function of the surface orientation, are shown in Fig. 8b.

One must note that, although orientation measurements are not affected by variations in distance due to the normalization effect by the

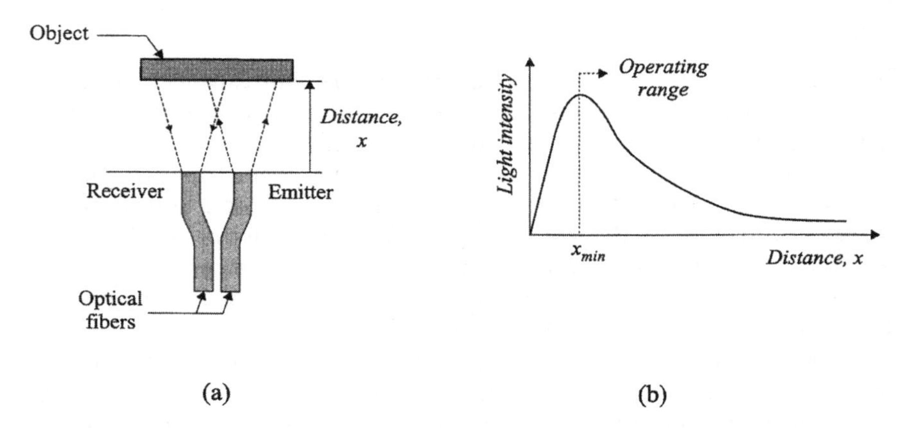

FIGURE 7 (a) Y-guide transducer; (b) Y-guide response for distance measurement.

symmetrical receivers, distance measurements are significantly affected by the orientation of the surface. Accordingly, in proximity sensing, an object's surface orientation is first estimated and subsequently the distance is determined. The accuracies of the measured distance and the orientation angle can be further improved by an iterative process.

A primary disadvantage common to all amplitude-modulation sensors is their dependence on the material of the object's surface. All distance/orientation versus light-intensity relationships must be calibrated with respect to the specific objects at hand.

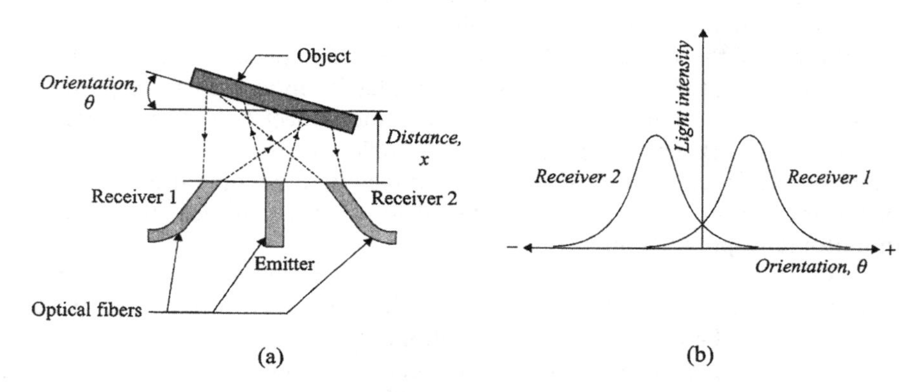

FIGURE 8 (a) Typical receiver pair constellation for orientation measurements; (b) the light intensity detected by each receiver as a function of orientation (θ).

Phase Modulation Proximity Sensors

A phase modulation proximity sensor usually comprises two light sources and one or more photodetectors. The light sources are driven by modulated sinusoidal signals having a 90° phase relationship (Fig. 9). The signal detected by the receiver is a superposition of the two reflected signals. The signal attenuation is a function of the geometrical and electrical parameters of the sensor, the reflectivity characteristics of the object's surface, and the surface's distance and orientation with respect to the sensor.

Triangulation Proximity Sensors

Triangulation proximity sensors can be used to determine the position of an object by examining the geometrical attributes of the reflected and incident light beams. In its basic configuration, a triangulation sensor comprises a laser light source and a linear array of photodetectors (Fig. 10). A narrow light beam reflected from the object's surface is detected by several of these detectors; the one detector that receives the maximum light intensity is considered as the vertex of the base of the triangle shown in Fig. 10. The geometry of the ray trajectory, then, provides the basic information for the estimation of the object's distance (x).

It is accepted that a triangulation sensor has the following properties:

The influence of irregularities, reflectivity, and orientation of the object is negligible.

The distance measurement is not affected by illumination from the environment and luminance of the object. Their influence is eliminated by comparison of two sensor signals obtained in successive on-and-off states of the light source.

The sensor's physical configuration can be sufficiently small for use in manufacturing applications.

Interferometers

All the above-mentioned electro-optical sensors can be configured to provide highly repeatable measurements, though the precision of measurements is inversely proportional to the operational range of these sensors. There are commercial sensors that can provide 2 to 5 μm precision, whose range, however, is less than 10 to 15 mm. Increased operational ranges would come at the expense of measurement precision (e.g., 10 to 20 μm precision for a 50 to 100 mm range). Laser interferometers capable of providing distance measurements with less than half a wavelength precision are thus the preferred choice of electro-optical sensors for high-precision applications (e.g., milling machines).

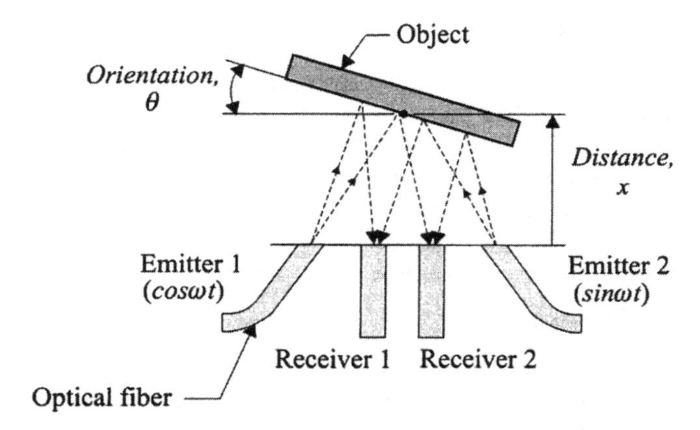

FIGURE 9 The basic phase modulation proximity sensor configuration.

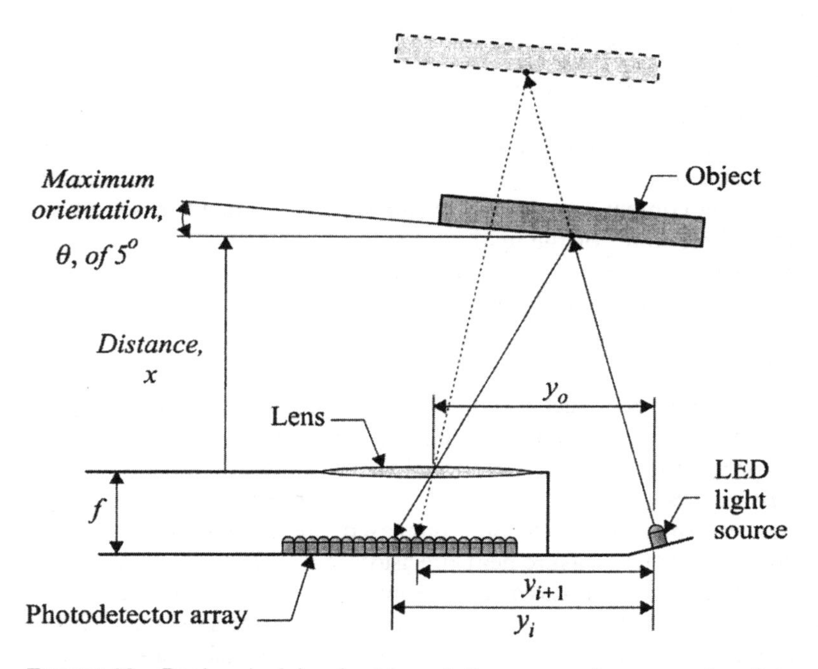

FIGURE 10 Basic principle of a triangulation sensor for measuring distance.

Interference of two light beams separated by a phase of 180° yields a total black fringe. Interferometers utilize this principle by superimposing two light beams, one reflected from a fixed mirror and one from the surface of a moving object, and count the fringes to determine the distance traveled by the object (Fig. 11). The distance traveled by the object is measured as a multiple of half wavelengths of the light source used. Modern interferometers can measure relative phase changes in a light wave to a precision of as a low as 1/52nd of the wavelength.

Nonproximity Sensors

LEDs and photodetectors can be arranged into a variety of configurations for the detection of presences of objects, finding their edges, etc. (Fig. 12). These sensors do not attempt to find the distance or orientation of the object's detected surface or its edges.

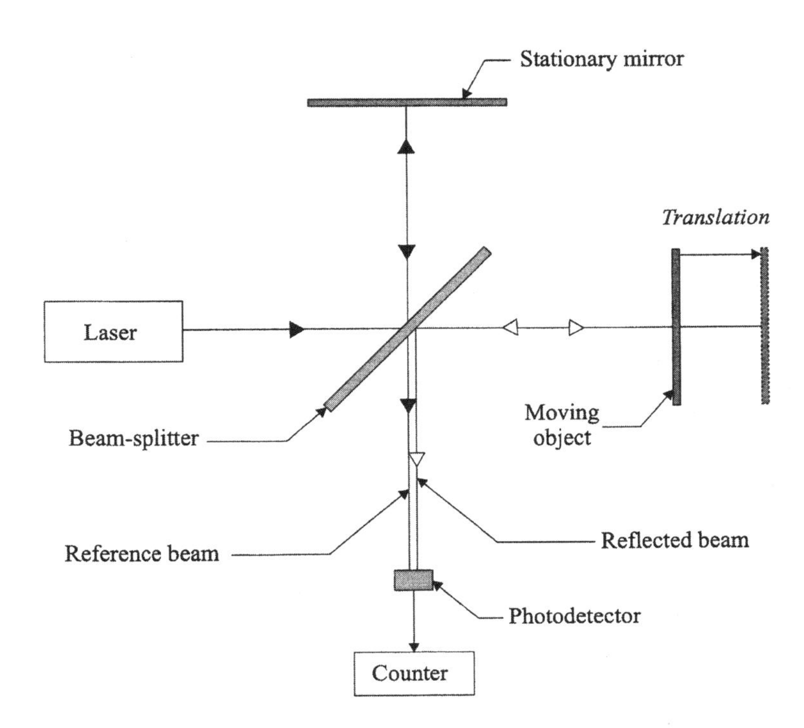

FIGURE 11 Interferometry.

13.2.3 Ultrasonic Transducers

Ultrasonic transducers can be configured using several emitters and receivers to measure the relative proximity of an object's surface. Acoustic waves above the 20 kHz range are labeled as ultrasonic—beyond the capability of humans to perceive. In industrial settings, ultrasonic transducers can emit acoustic waves with a frequency of as high as 200 kHz, and they can be used both as emitters and as receivers.

There are several time-of-flight methods for distance measurements:

Pulse-echo: The time elapsed from the emission of an ultrasonic acoustic pulse to the reception of the returned echo (reflection from the object's surface) can be utilized to calculate distance, when one logically assumes a constant signal velocity over relatively short distances (less than a few meters) (Fig. 13).

Phase-angle: The phase angle between emitted and received acoustic waves can be used to measure a distance normally less than the length of the ultrasonic wave.

Frequency modulation: Frequency modulated signals reflected from an object's surface, with no change in signal shape, except for the frequency shift, can be utilized accurately to calculate distance.

Piezoelectric transducers, which convert mechanical displacement into electrical current and vice versa, are the most commonly used devices in ultrasonic sensors. Ceramics and some polymers can be polarized to act as natural piezoelectric materials (e.g., natural crystals). Other ultrasonic transducers include electrostatic (i.e., plate capacitors with one free and one fixed plate), magneto-restrictive (based on dimensional changes of ferromagnetic rods), and electromagnetic (e.g., loudspeakers and microphones).

In some cases, ultrasonic transducers can also be utilized to detect the presence of objects that could not be achieved with electromagnetic or electro-optical sensors owing to large distances and reflectivity problems.

13.2.4 Digital Transducers

Transducers that output data in digital form, as discrete pulses or coded information, are classified as digital transducers. Such sensors' output can be directly interpreted by microprocessor-based controllers (with no need for analog-to-digital data conversion). Digital counters must be utilized when the output signal is in pulse form.

In this section, our focus will be on two popular digital transducers, encoders and tachometers, for displacement and velocity measurements, respectively.

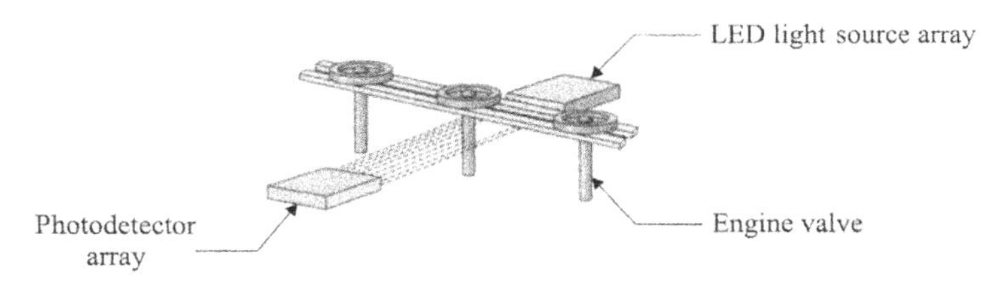

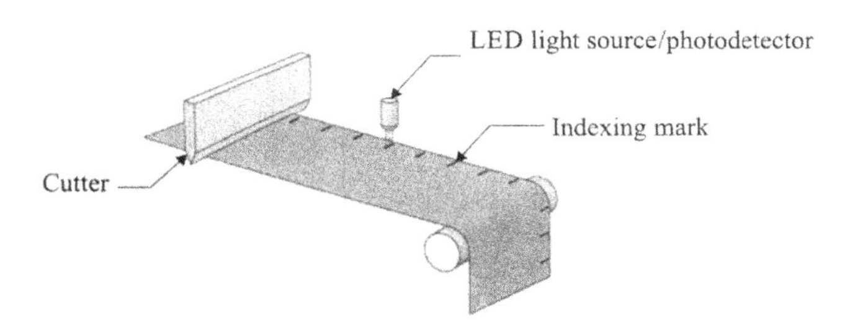

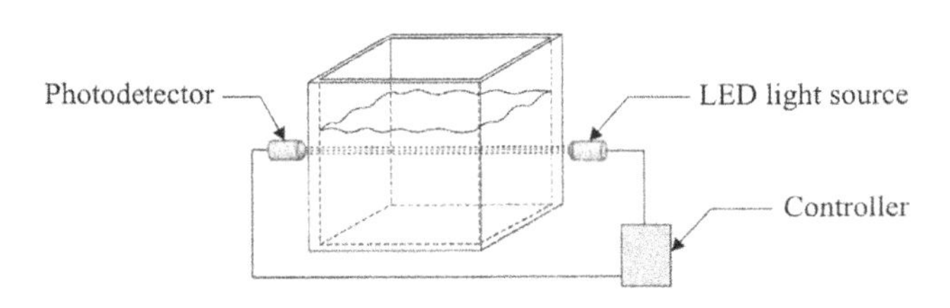

FIGURE 12 Some industrial applications of electro-optical proximity sensors.

Encoders

Digital encoders can be configured to measure linear or rotary displacement. They utilize physical contact, magnetic disturbance or optics for the detection of movement. Optical encoders are most commonly used owing to their high-accuracy manufacturability. They can be in incremental form (pulsed information) or absolute form (coded information). All, however,

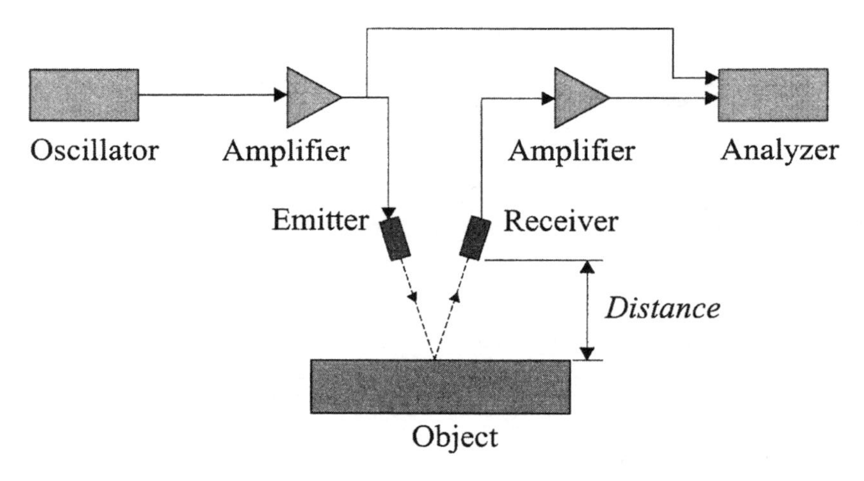

FIGURE 13 Ultrasonic sensor.

comprise two basic components: a marked grating (scale) component and a detection system.

Rotary encoders employ a disk-shape grating device with radial markings (also called "shaft encoders"), while in linear encoders the (linear) scale comprises one or more sets of parallel lines. The former will be discussed first because of their use in almost all motion-control systems (linear or rotary motion).

Optical rotary encoders use one of three methods to detect the motion of the (grating) disk:

Geometric masking is based on allowing light to pass through unmasked slits (grating) and be detected by photodetectors on the other side of the disk (Fig. 14).

Moiré fringes are generated by employing two disks with similar periodic patterns in (rotary) motion with respect to each other.

Diffraction effects, due to coherent light passing through a pattern of slits (a few wavelengths wide), can be utilized for very-high-precision encoding.

Figure 15 illustrates the basic principle of the incremental quadrature rotary encoder with four distinct outputs, $n = 4$. Two photodetectors are placed one half slit-width apart, thus detecting the rotary motion of the outer ring with a 90° phase, while a third detector registers a reference signal per each revolution of the disk by noting a reference slit on an inner

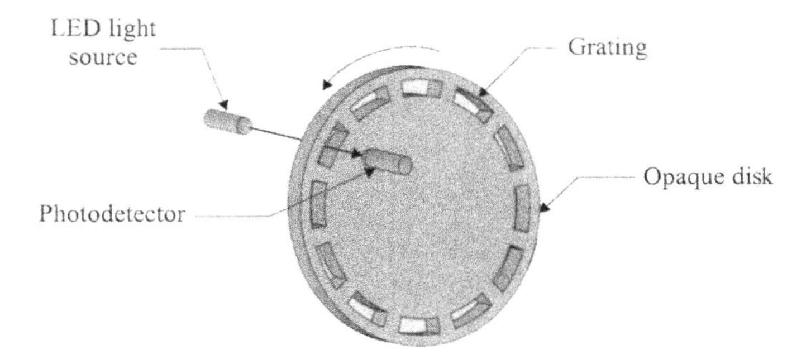

FIGURE 14 Geometric masking for rotary encoding.

track on the disk. A counter counts the number of slits detected for relative displacement calculations.

Figure 16 shows the basic principle of an absolute encoder with four rings, yielding sixteen distinct outputs, $n = 16$, employing the Gray coding scheme (versus the binary coding scheme) (Table 1). The use of the Gray code allows transition of only one bit of data between two consecutive sector numbers, thus minimizing ambiguity and requiring simpler electronics. Each of the four tracks (rings) is monitored by a separate photodetector as in the case of incremental encoders.

Optical linear encoders are normally of incremental form and utilize geometric masking as the encoding technique. Some commercial linear encoders also utilize the principles of diffraction for higher resolution measurements.

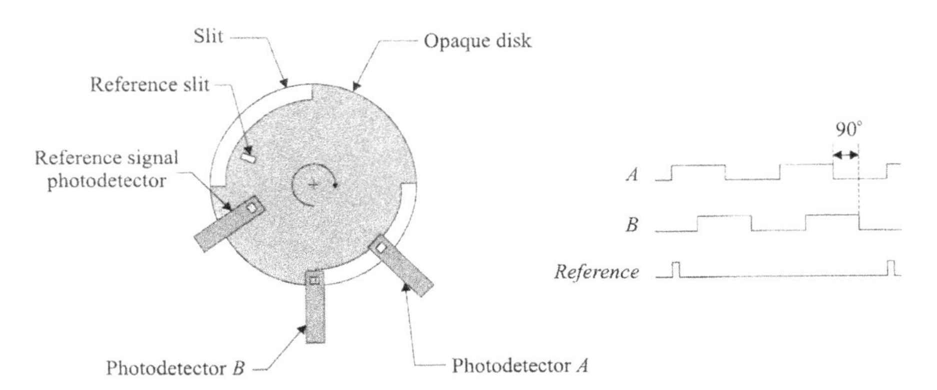

FIGURE 15 Incremental rotary encoding.

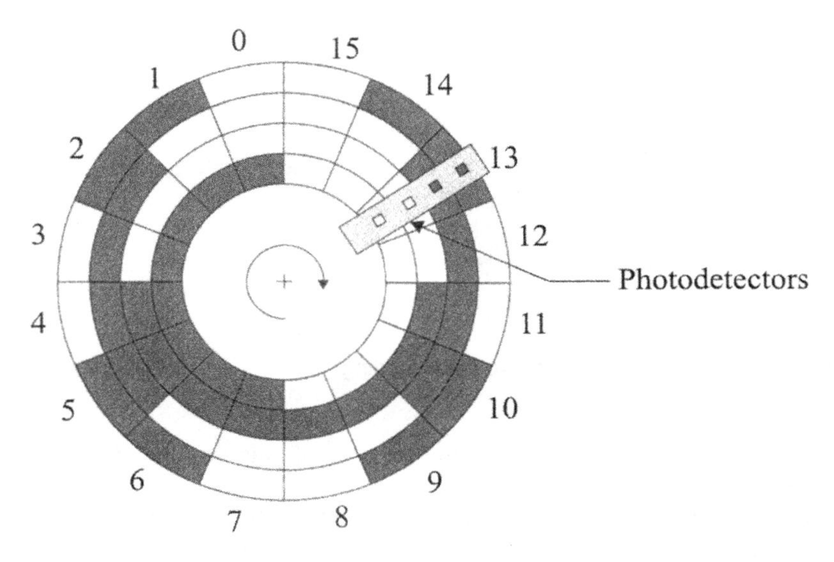

FIGURE 16 Absolute rotary encoding.

Rotary incremental optical encoders can have from fifty thousand up to several millions of steps per revolution, while their absolute counterparts can have from 512 up to 131,072 steps per revolution. Linear encoders can have a resolution from 0.005 μm to 5 μm.

Digital Tachometers

Angular speed can be measured using a tachometer, which is normally considered an analog transducer that converts mechanical energy into electromagnetic energy. A DC tachometer generates a voltage proportional to the speed of a rotating coil coupled to a shaft, whose rotational speed we want to determine. An AC tachometer generates a voltage with a frequency proportional to the rotational speed of a rotor.

Electromagnetic tachometers can also be configured to generate a pulsed-output signal; for example, the rotational speed of a ferromagnetic

TABLE 1 'Gray Coding for Four-Bit Words

0	1	2	3	4	...	13	14	15
0111	0110	0100	0101	0001	...	1100	1110	1111

gear would induce pulsed signals as the individual teeth pass in close proximity to a magnetic sensor (Fig. 17a). The frequency of the output voltage is directly proportional to the speed of the gear.

An electro-optical version of a pulsed tachometer is shown in Fig. 17b. This configuration forms the basis of digital tachometers: as in quadrature rotary encoders, a couple of pulsed signals, with a 90° phase shift, are counted for velocity measurement. Based on this principle, one may simply use an encoder for measuring both displacement and velocity.

13.3 FORCE SENSORS

Most manufacturing operations involve direct interactions between a tool and a workpiece. It is expected that the mechanical fabrication device exerts force on a workpiece in a tightly controlled fashion. Instruments for detecting and measuring such interactions are classified as force, torque, and tactile sensors. The most commonly used transducers for these sensors are strain gages, piezoelectric films, piezoresistive films or strips, and capacitive detectors. In this section, our focus will be only on the first three types of transducers.

13.3.1 Strain Gage Transducers

Strain gages, whose resistance changes under deformation, are utilized in the majority of force sensing applications owing to their simplicity. They are manufactured in the form of flat coils bonded onto a nonconducting elastic sheet of paper or plastic. The flat coil is normally a metallic element (e.g.,

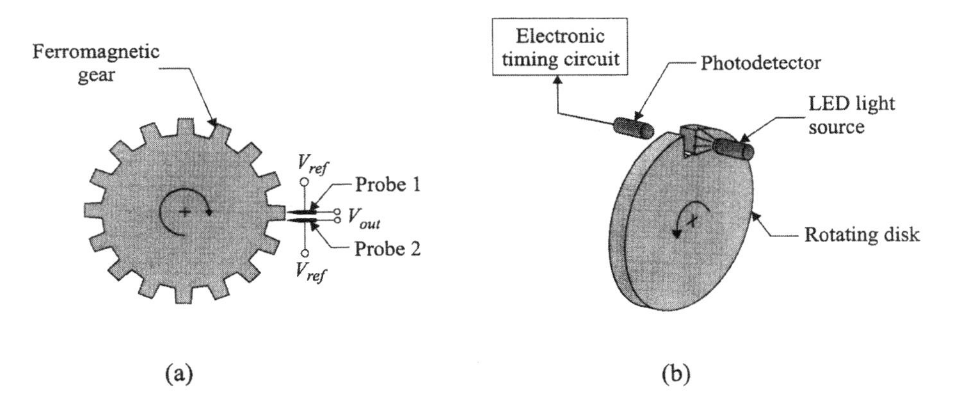

FIGURE 17 (a) Electromagnetic; (b) electro-optical pulsed tachometer.

copper–nickel alloy), though there are strain gages that are semiconductor elements. Strain gages are cemented onto the surface of an object whose strain we want to measure, as part of a bridge type circuitry. During this placement of the strain gages, we must ensure appropriate directionality. Most strain gages are designed to measure strain in one direction, although they can be configured for multidirectional measurement as well (Fig. 18).

Based on the principle of force-inducing strain, which in turn can be measured by strain gages, a large number of force and torque sensors (also known as load cells) have been built and commercialized in the past several decades. The most commonly used load cell is a rigid structure that elastically deforms under applied force. Cantilever-beam type cells are utilized for low-load cases, while ring-shaped cells are designed for large forces (Fig. 19a, 19b, respectively). Load cells placed on torsion members of mechanical devices can also effectively measure torque.

Strain gages have been occasionally used in the construction of tactile sensors for the detection and measurement of distributed forces along a two-dimensional contact surface between an object and the transducer. Such sensors are also capable of detecting and measuring slippage.

13.3.2 Piezoelectric and Piezoresistive Transducers

Piezoelectric materials generate an electric charge when subjected to an external force. The electric charge is collected by a capacitor and used to measure the magnitude of the force. Common piezoelectric materials are quartz crystals, lithium sulphate, lead zirconate titanate, and a number of synthetic materials (e.g., PVF_2).

There are also force-sensitive resistor (FSR) transducers used in the construction of force sensors. These materials, as do strain gages, alter their resistance to electrical current when subjected to an external force.

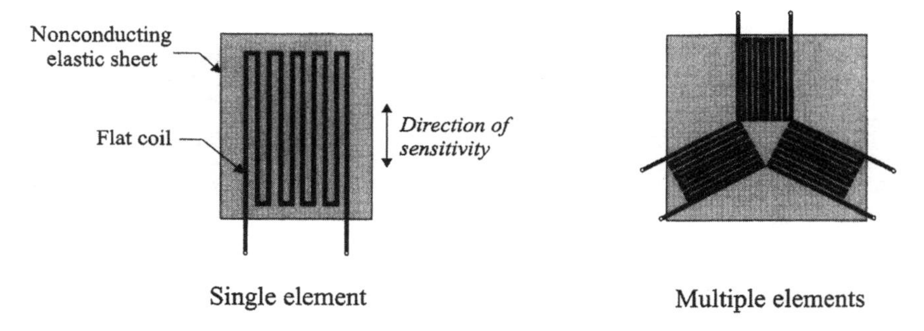

FIGURE 18 Strain gages: single- and multiple-element types.

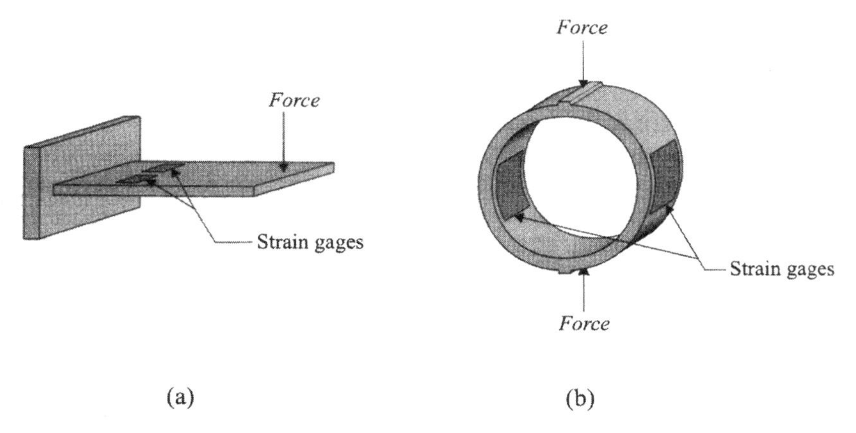

FIGURE 19 (a) Beam type; (b) ring type load cells.

They are fabricated using conductive elastomers such as silicon rubber and polyurethane impregnated with conductive particles/fibers (e.g., carbon powder). FSR-based transducers in the form of (overlapping) thin strings can be formed into a matrix configuration for tactile-sensing applications (Fig. 20).

13.4 MACHINE VISION

Machine vision is defined as the use of optical devices for noncontact acquisition of object/scene images and their automatic analysis for quality-

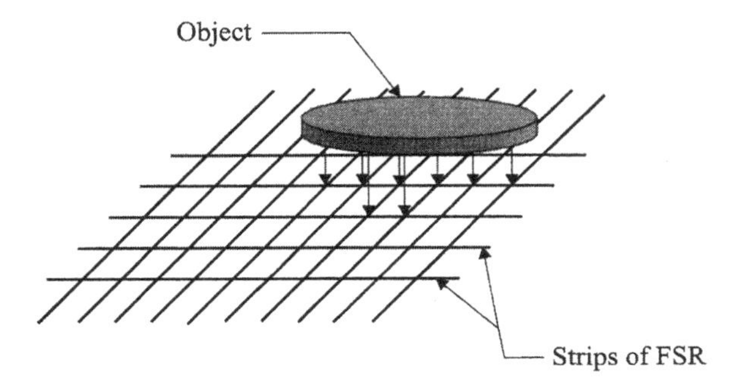

FIGURE 20 A force-sensitive resistive-based tactile sensors.

control or process-control purposes. In contrast to computer vision, targeted for three-dimensional imaging of unstructured environments, machine vision is utilized for two-dimensional imaging of objects in structured environments. Information about three-dimensional objects and/or scenes is interpreted from their two-dimensional images. Although range information can be deduced from multiple images, machine vision is primarily used for object recognition purposes and only in a limited number of cases for positional feedback to device controllers (e.g., position and orientation feedback provided to the controllers of electronic-component placement machines).

Machine vision applications can be traced back to the late 1930s and early 1940s, when analog systems were used in the U.S. for food sorting and inspection of refillable bottles for possible cracks. With the introduction of computers in the late 1950s and early 1960s into manufacturing environments, the utilization of machine vision rapidly expanded. By the early 1970s, several companies started to commercialize machine vision systems for inspection and control purposes. Diffracto, Canada, was one of these first such successful companies. Today, machine vision imaging principles are commonly used in the analysis of internal features of objects using x-ray based computed tomography, ultrasonics, and so on. Image acquisition is no longer restricted to the visible or even infrared wavelength region of the light spectrum.

Typical examples of machine vision use include high-speed bottle/can inspection, solder paste deposition inspection, dimensional verification of body in white in the automotive industry, label inspection, robot guidance in electronic component placement, welding seam tracking and mechanical assembly, and the dimensional measurement of machined parts for in-process control. The fundamental components of every machine vision system used in these and other applications are shown in Fig. 21. They include illumination devices, one or more imaging sensors (cameras) equipped with appropriate optical lenses/filters, image capture devices, and a computer. Image preprocessing/conditioning and image analysis/interpretation are tasks normally carried in software within the computer.

13.4.1 Image Acquisition

Image acquisition is to the capturing of light reflection from an illuminated scene by an electro-optical sensor and conversion of this data into a digital image. Machine vision systems must thus be provided with best possible illumination subsystems for the acquisition of sufficiently illuminated, shadow-free images with maximum contrast.

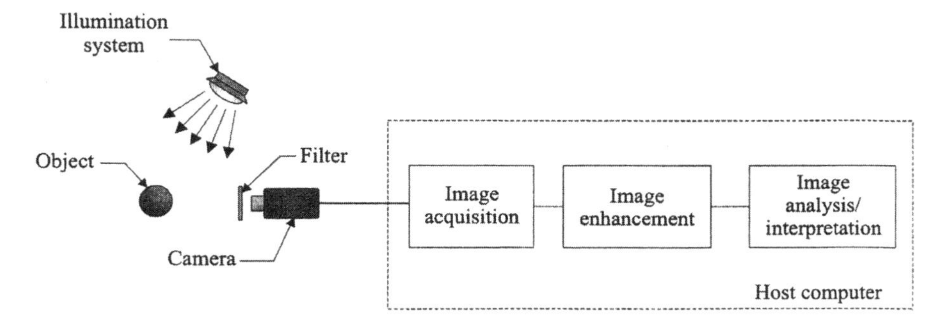

FIGURE 21 Basic machine vision system architecture.

Lighting

A variety of light sources (incandescent, quartz halogen, fluorescent, etc.) can be used to illuminate the scene in the form of diffused or collimated lighting (Fig. 22). The latter form is especially useful for backlighting; that is, providing the imaging sensor with a silhouette of the object. Fiber-optic cables are often used to illuminate hard-to-reach regions. Beam splitting can be also employed in providing an illumination direction in line with the optical axis of the sensing system (i.e., camera) (Fig. 22).

The image of an object is sensed as a mixture of the three primary colors red (R), green (G), and blue (B). All colors can be derived from these primary additive colors by varying their emission intensity (brightness): no light corresponds to black, while white is obtained by proportional mixing of red, green, and blue (Fig. 23a). Although most image sensors are based on the RGB color system, we may also note the existence of an alternative color space that is similar to human perception: hue, saturation, and brightness/intensity (HSB) (Fig. 23b). The hue component describes the color spectrum, the saturation component reflects the purity of the color, and the brightness component describes how close the color is to white (i.e., maximum brightness).

Cameras

Machine vision systems' reliability was significantly increased with the introduction of solid-state imaging sensors in the early 1970s: the charge-coupled device (CCD), charge-injection device (CID), metal-oxide semiconductor (MOS), and so on. CCD-based (matrix and linear) cameras have been the most widely utilized imaging sensors.

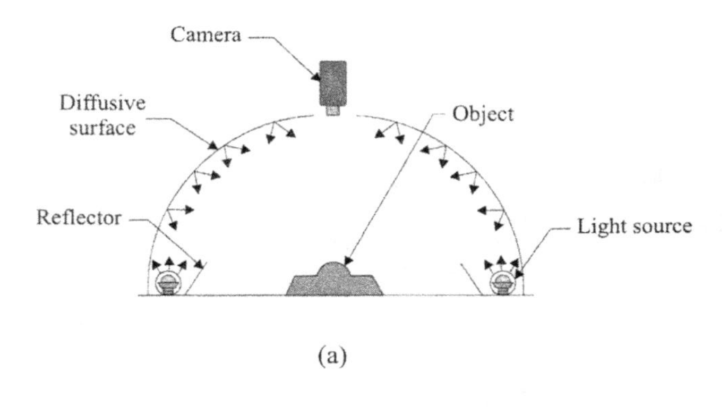

(a)

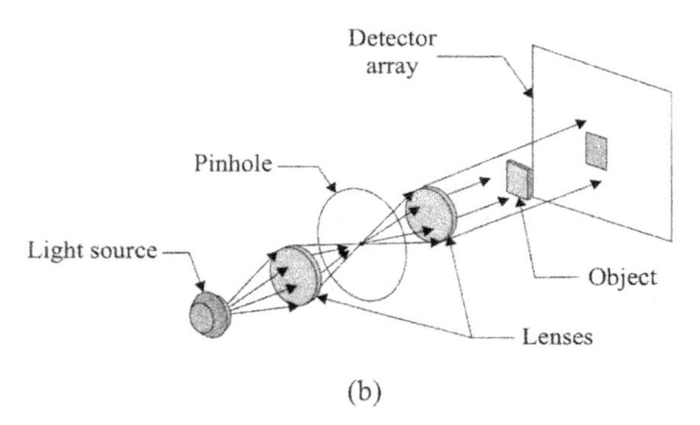

(b)

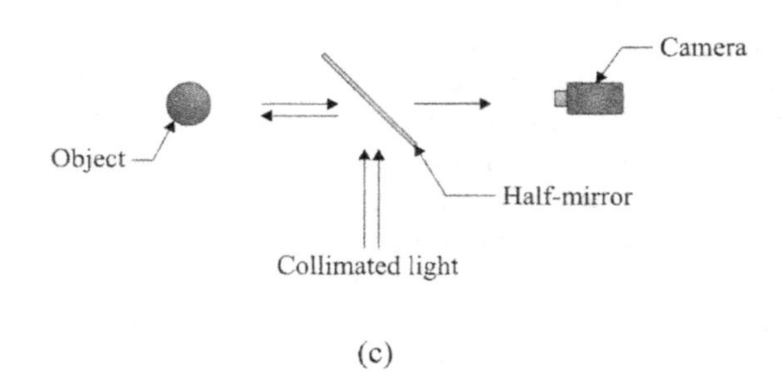

(c)

FIGURE 22 (a) Diffused; (b) collimated; (c) in-line lighting.

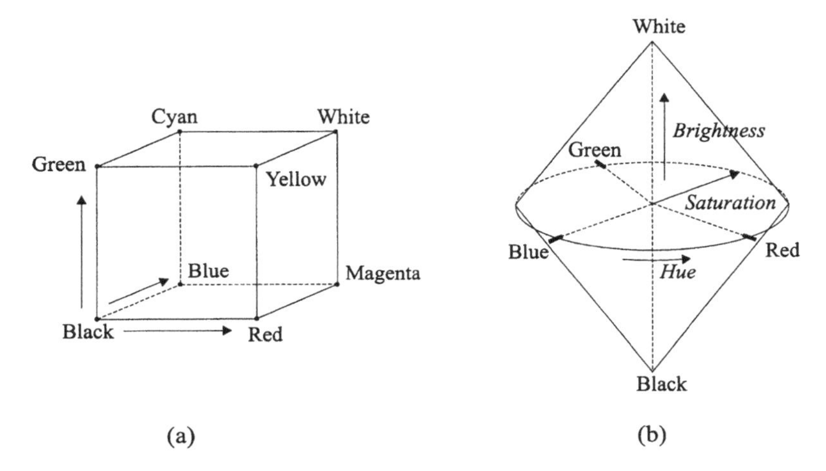

FIGURE 23 (a) RGB; (b) HSB color space.

CCD was invented in 1969 in the Bell Laboratories, U.S.A., by W. S. Boyle and G. E. Smith—a decade before the first CCD color camera was commercialized by Sony. The principle of operation of a CCD imaging sensor involves the following steps: conversion of light energy into electrical charge at discrete sites (pixels), transfer of the packets of charge through the silicon substrate, and conversion of individual charges into voltages by floating diffusion amplifiers.

Frame transfer–based CCDs utilize photogates (photoactive MOS capacitors) as photodetectors, while interline transfer CCDs use photodiodes (coupled to transfer gates) for photodetection and movement of charge. In the former type of CCDs, the whole frame of information is transferred simultaneously into a storage area, while in the latter type, the charge packets are transferred line by line. In both cases, however, the image is transferred out horizontally in a serial mode. The obvious advantage of frame transfer–based CCDs is the capture of the next image frame as soon as the previous frame is transferred to the storage area.

The latest use of CMOS (complementary MOS) technology in CCD cameras provides two further advantages: on-chip processing (leading to smaller sensors) and allowing random access to individual (active) pixels. The technology also prevents overflowing of charge into neighboring pixels ("blooming").

There are single- and three-array CCD cameras for the acquisition of color images. Single-array sensors use an arrangement of filters to detect the primary colors of collected light signals by individual (distinct) neighboring pixels. In most cameras, the number of pixels devoted to each color differ (most would have twice as many as green pixels than the red or blue pixels—e.g., for a horizontal line of 768 pixels, 384 would be devoted to the detection of green light, 142 to red, and 142 to blue, in an exemplary arrangement of GGRBGGRB...). In higher end cameras, three individual arrays are dedicated to detect red, green, and blue light.

The majority of CCD array cameras have an aspect ratio of 4:3, yielding pixel sizes in correspondence to the physical size of the overall array itself and the resolution of the array. For example for a 768×480 array, with a diagonal length of 4 mm, the pixel size would be 4.17×5 μm.

The NTSC (National Television Systems Committee) standard used in North America recommends video timing of 30 frames per second. The PAL (phase alteration line) and the SECAM (sequentiel color avec mémoire) standards, on the other hand, recommend a frequency of 25 Hz.

Cameras based on (noninterlaced) progressive-scan CCDs capture an image in its entirety instantaneously, in contrast to interlaced-scan CCDs that capture an image line by line. The former are thus very suitable for imaging moving objects, providing (an unsmeared) improved vertical resolution.

Image Grabbing

The last stage in image acquisition is the creation of the digital image by grabbing the analog image captured by the CCD camera (Fig. 21). For analog-output cameras, the analog image is grabbed by converting the voltage values of all the pixels into their corresponding digital levels, yielding a digital image. This digital image is then transferred into the memory of the computer (typically, the random-access memory RAM) for image analysis. Most commercial frame grabbers (also known as digitizers) can be mounted on the bus of the personal computer (PC).

A digital image would have a resolution of gray levels dictated by the resolution of the frame grabber (e.g., 256 gray levels per pixel for an 8-bit digitizer and 1024 gray levels for a 10-bit digitizer). For color cameras (versus monochrome cameras), we would need three separate digitizers, one per primary color, in order to carry out color imaging and analysis. A single digitizer linked to the green (or to the composite image) output of the color CCD camera could be used, if image analysis were to be carried out in the gray-scale mode.

13.4.2 Monochromatic Image Processing and Analysis

In this section, our focus is on the analysis of monochromatic images. Color image analysis is beyond the scope of this book. Analysis of digital images can be carried by directly using the acquired gray-scale representation or by first converting this representation into a binary (black and white) image and then using this limited representation for analysis. The objective of image analysis is either object identification and/or its dimensional verification or acquiring range information. The image analysis stage is normally preceded by the image preprocessing/enhancement stage.

Image Preprocessing

A digital image can be preprocessed for the removal of unwanted spatial frequencies—spatial filtering. Spatial convolution refers to spatial filtering through the use of neighborhood information obtained in the form of a window (or kernel) of size 3×3, 5×5, etc., whose center element coincides with the pixel we are examining.

The most common low-pass spatial filter is the mean filter used to yield a smoothing effect (attenuating the high-frequency elements), in which each pixel value is replaced by the average value of its neighbors within the kernel considered (Fig. 24):

$$h_{ij} = \frac{1}{N} \sum_{k=i-1}^{j+1} \sum_{l=j-1}^{j+1} g_{kl} \tag{13.4}$$

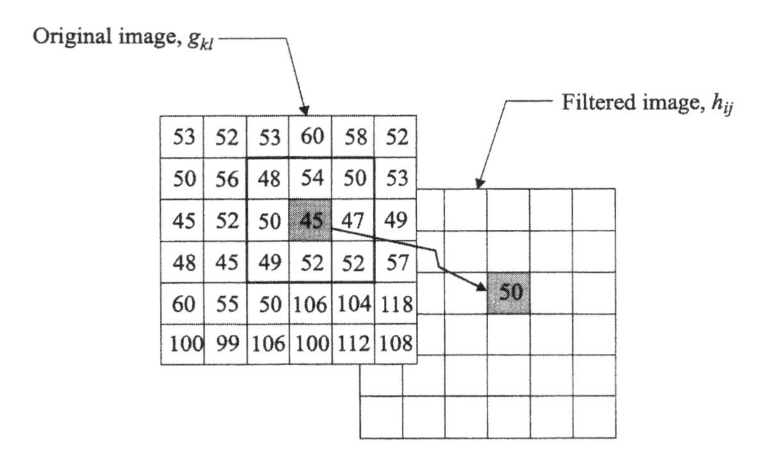

FIGURE 24 Mean filter.

where g_{kl} represents the original image, h_{ij} represents the new (filtered) image and N is the total number of pixels in the kernel considered (e.g., for a 3×3 kernel, $N = 9$).

A high-pass spatial filter yields the opposite of a low-pass filter by attenuating the low-frequency elements. The result is a sharper image with clearly defined edges. A common 3×3 high-pass filter kernel is given here:

−1	−1	−1
−1	9	−1
−1	−1	−1

Once an image has been preprocessed through spatial convolution, the next step is determining the edges of the object captured in the digital image. As mentioned above, since image analysis can be carried out using gray-scale or binary images, we must choose one or the other before proceeding to the edge detection step. A binary image, if desired, can be obtained through a process called thresholding.

Thresholding is normally applied for faster image analysis, when the object in the image is accentuated in contrast to the background. A gray-scale image is converted into a binary (0,1) image by assigning the value zero to all the pixels with a value lower than a chosen threshold and assigning the value one to all others. A suitable threshold value can be chosen by examining the histogram of the gray-scale image (Fig. 25a).

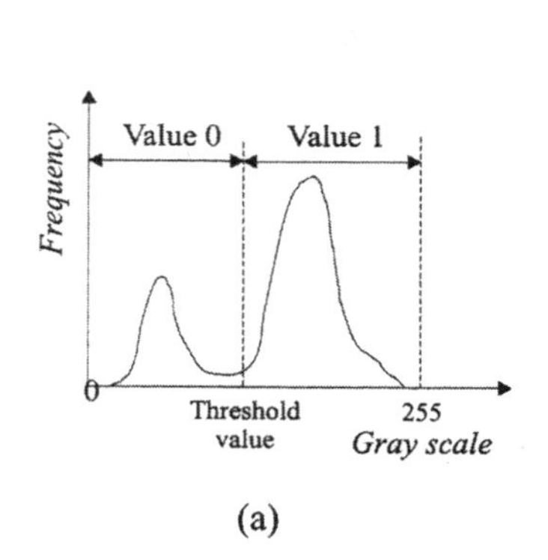

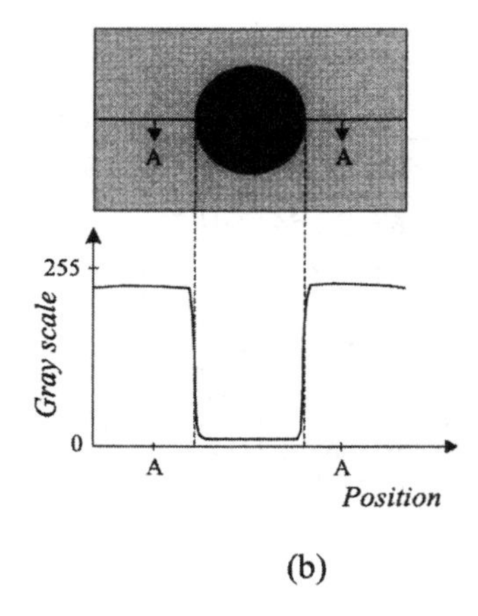

FIGURE 25 Thresholding.

Although for object identification, choosing the correct threshold value may not be that critical, for dimensional measurements, an incorrectly chosen threshold value would shift the edges of the object and thus yield an unreliable measurement (Fig. 25b).

For gray-scale image analysis, edge detection must be preceded by the application of an edge enhancer. One of the most commonly used edge enhancing filters is the Sobel operator. It uses two 3×3 kernels to calculate the new value of the pixel under consideration, h_{ij}:

$$h_{ij} = (S_x^2 + S_y^2)^{1/2} \tag{13.5}$$

where S_x and S_y are defined, respectively as

−1	0	1
−2	0	2
−1	0	1

1	2	1
0	0	0
−1	−2	−1

A thresholding process normally precedes the edge enhancement operation, such as the Sobel operator above, in order to enhance the detection of edges (Fig. 26).

The boundary of an object, whose edges have been determined using the tools described above, can be defined as a chain code. The objective is a concise representation to replace the boundary pixel coordinates with a sequence of numbers, each defining the relative location of the next boundary pixel. For example, the eight pixels neighboring a boundary point can be labeled by 1 to 8 (Fig. 27a). Using such a notation, an object boundary can be easily coded as a simple chain (Fig. 27b).

Image Analysis

Image analysis is the extraction of necessary information from one or more digital images of an object or scene, for the purposes of object identification, dimensional measurement, and/or range and orientation determination. In regard to object identification (recognition), we will restrict our attention to two object-identification methods that impose minimal constraints on the placement of objects in the scene (i.e., their relative position and orientation). Both methods assume that the image has been segmented and objects do not overlap.

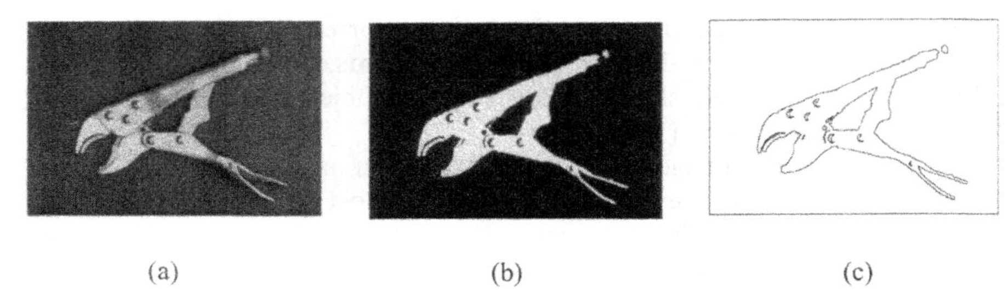

FIGURE 26 Edge detection process: (a) a gray-scale image; (b) a thresholded image; (c) object edges.

Model-based recognition: This method requires the modeling of objects in terms of several of their features that can be extracted from their acquired two-dimensional images. Features can include absolute measurements, such as area (total number of pixels included within the boundary of the object), perimeter, and maximum dimension; they can also include relative measurements or identification of certain geometrical features that would be scale-invariant (i.e., independent of the size of the image of the object), such as a circularity measure (area/perimenter^2), ratio of maximum dimension to minimum dimension, number of corners, number of holes, and so on.

The first task is to create individual models (feature vectors) for all the objects that we plan to recognize. The feature values in the vectors should correspond to the mean values of the potential measurements of dimensions,

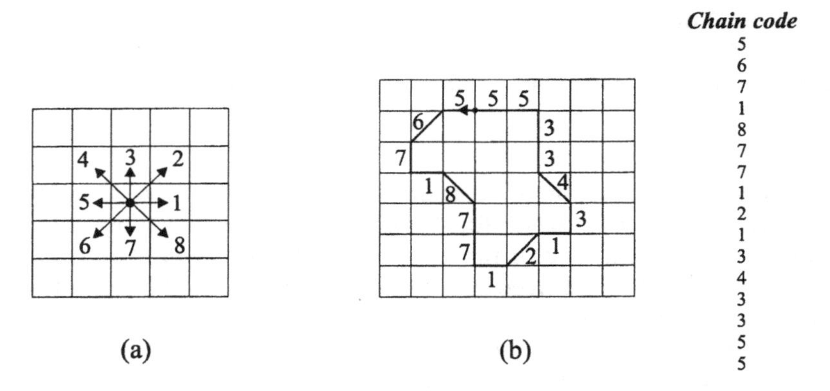

FIGURE 27 Chain coding; (a) directions; (b) example.

when these are distributed according to a Gaussian (normal) distribution owing to manufacturing variations (Chap. 16). Once the database of feature vectors is ready, on-line recognition can start.

For an object whose image we just acquired, the corresponding features are extracted and measured for the determination of the corresponding feature vector, U_i, $i = 1$ to m, where m is the number of features. This vector is then compared with all the feature vectors in our database, ${}^jO_i$, $j = 1$ to r, where r is the number of objects, for identifying a match. A match is established, by calculating a dissimilarity measure, D_j,

$$D_j = \sum_{i=1}^{m} (U_i - {}^jO_i)^2 \tag{13.6}$$

The smallest calculated D_j value, $(D_j)_{\min}$, indicates a potential match only if $(D_j)_{\min} \leq \varepsilon$, where ε indicates possible uncertainty due to imaging errors and/or manufacturing tolerances. Otherwise, the viewed object is labeled as unknown.

One should always attempt to form the smallest size feature vector (i.e., with minimum number of features) for the shortest possible processing time. However, the features chosen should provide maximum possible distinctiveness in order to prevent any potential ambiguity due to natural imprecisions (imaging errors dimensional tolerances, etc.). For example, in Fig. 28, four objects are first modeled using two features, Feature 1 and Feature 2. One notes that when using these two features the uncertainty regions of Object 1 and Object 2 overlap, thus yielding a potential ambiguity. In order to avoid this problem, the objects are remodeled using two new features, Feature 3 and Feature 4. These features do not yield ambiguity in the recognition of the four objects.

Shape recognition: An object's identity can be efficiently established by examining its outer boundary. Fourier transforms have been utilized to obtain approximate representations of objects' (periodic) contours and their subsequent identification through database comparison. Another popular method is the determination of a shape signature—"encoding." A polar-radii-signature-based encoding scheme is shown in Fig. 29. This scheme measures distances from the center of the object to all points (or a reduced representative set) on the boundary of the object yielding a unique signature. The centroid (center of mass) of any shape can be obtained by determining the balance point where the "mass" of the object's image is balanced with respect to both x and y axes.

As in model-based recognition, the shape signature of a viewed object must be compared with every one of the object boundaries stored in the database to determine a match. Since polar-radii signatures are not rotation

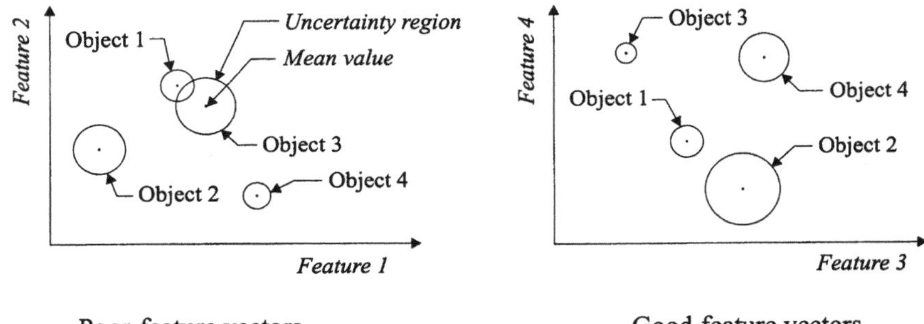

FIGURE 28 Two-dimensional feature vectors.

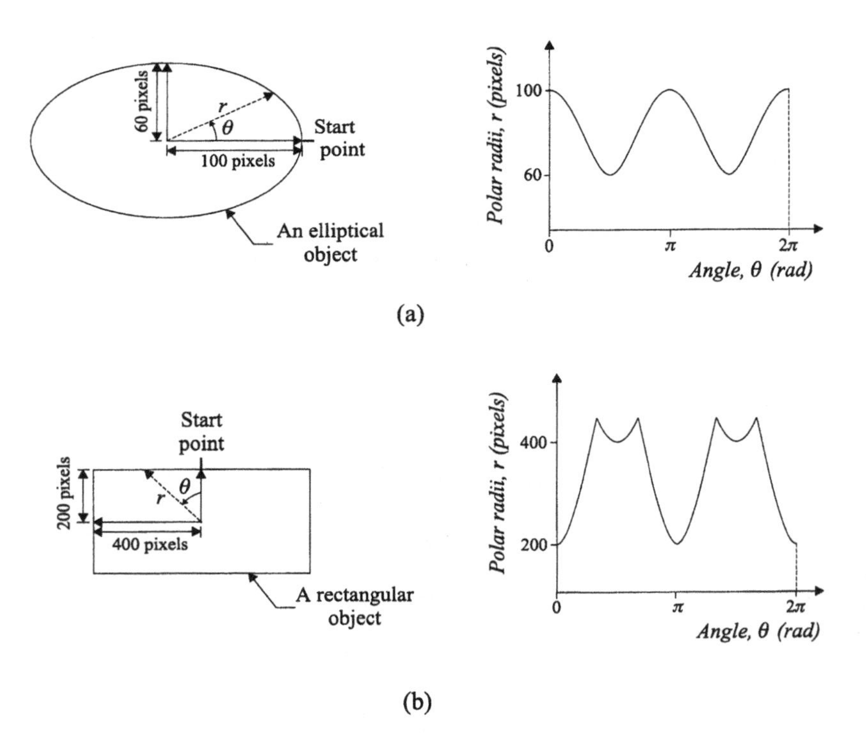

FIGURE 29 Polar radii signatures.

invariant, i.e., there is no established starting point for the generation of the signature, the alignment of signatures does present a challenge that must be addressed. Once an alignment is achieved, a dissimilarity measure can be calculated by summing up the absolute differences between the reference and the acquired signatures at every boundary point.

There are shape signatures that are rotation invariant—they utilize a characteristic ellipse unique to every shape in order to determine the starting point of the signature. The characteristic ellipse rotates together with the shape, where the intersection of its major axis with the shape boundary is always the same point.

13.4.3 Object Location Estimation

Machine vision can be used for estimating the position and orientation of objects, whose digital images have been acquired by one or more cameras. The task at hand is to relate the location of the object in the "image" plane to its actual location in a three-dimensional "world" coordinate frame. Such a process first requires the calibration of the camera used for image acquisition: determining the appropriate spatial transformation matrices by calculating the extrinsic (exterior orientation) and intrinsic (internal geometry) parameters of the camera. Once the camera has been calibrated, two images of the object must be acquired from different distances in order to calculate its range (one image could suffice, however, for determining its orientation).

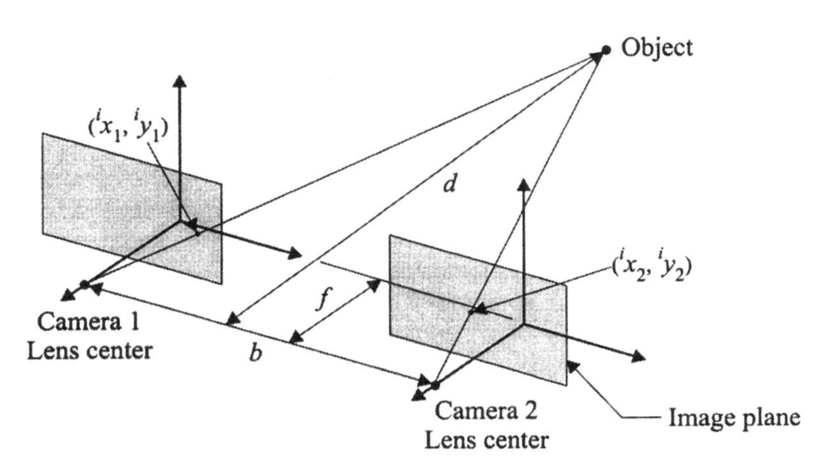

FIGURE 30 Stereo imaging.

A typical technique for range determination is stereo imaging (Fig. 30). The disparities of feature locations in two (separated) image planes, one of each camera, are utilized to extract depth information, d:

$$d = \frac{bf}{({}^{i}x_1 - {}^{i}x_2)} \tag{13.7}$$

where b is the separation distance of the two cameras (baseline), ${}^{i}x_1$ and ${}^{i}x_2$ are the x-axis image coordinates of the point feature as detected by Camera 1 and Camera 2, respectively, and f is the distance from the lens center to the image plane in both cameras.

13.5 ACTUATORS

Devices that execute motion commands generated by a controller are called actuators. Actuators convert energy input from an external source, typically electrical, hydraulic, or pneumatic, into mechanical (kinetic) energy and thus cause motion. Hydraulic actuators are normally reserved for large-load carrying applications, while pneumatic actuators are targeted for light-load carrying applications. Electrical actuators are the most versatile motion generators used for large-load carrying and minute-load carrying applications. They are quiet, easy to control, and easy to maintain, three favorable features in manufacturing applications. In a closed loop control environment actuator actions are monitored by sensors, like those mentioned in this chapter.

13.5.1 Electric Motors

Electrical actuators can be configured to provide rotary or linear motion. They are typically classified as AC (alternating current), DC (direct current) and stepper motors. Although a number of machines do use AC servomotors with voltage and frequency control via microelectronic drivers, the majority of manufacturing devices use DC motors and stepper motors:

Rotary-motion (brush) DC motors utilize a (stationary) stator and a rotating core element (armature) placed inside the stator. Power transfer to the armature (commutation) is achieved through brushes placed on the rotary armature. The primary disadvantages of these motors are rapid wear out of the brushes and noise. Brushless DC motors, first introduced in the early 1980s, overcome these problems by utilizing electronic switching of the electrical current. Such motors can provide large torques over a range of speeds reached almost instantaneously

Linear motion can be achieved using rotary DC motors by coupling them to leadscrew drivers. For short-distance movements and light-load applications, however, one may choose to use linear DC motors, where a stage moves in a (stationary) channel. Power brushes are mounted on the stage for supplying the necessary current.

The displacement and velocity of DC servomotors must be controlled in a cascade (two-loop) architecture, tachometers measuring velocity in the inner-loop and encoders measuring displacement in the outer loop. Although the motion of a DC motor can be controlled by varying the input voltage, precise control cannot be achieved owing to the varying torque–speed relationship. Thus a desirable control mode would be adjusting the input current and thus controlling the torque applied—the voltage is manipulated to maintain a desired current.

Stepper motors, developed in the 1920s, provide accurate positioning through a stepping motion. That is, they provide incremental displacement (comprising equal length steps) in response to a sequence of precisely controlled digital pulses. As in DC motors, rotary motion is achieved by utilizing a stator and a rotor: a step motion is accomplished by the alignment of rotor poles with starter teeth through sequential alternating energizing of the teeth (Fig. 31).

The stepper motor has been generally targeted for use in an open loop control environment. A required displacement is achieved by a corresponding command of the necessary number of steps. Velocity is specified by varying

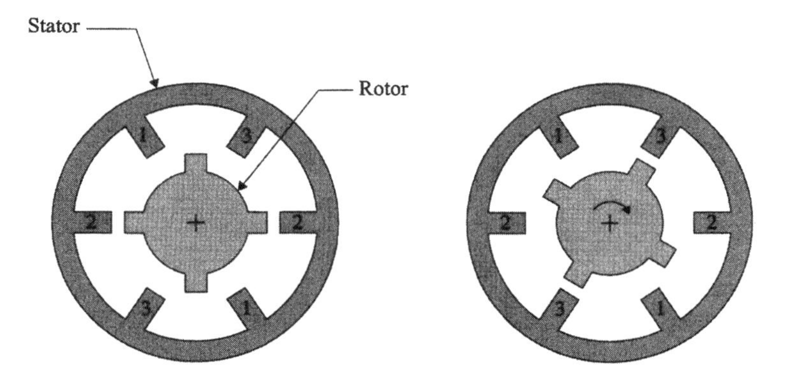

FIGURE 31 Step motion in variable reluctance stepper motor.

the rate of pulse input to the motor. Stepper motors, although desirable for high-positioning-accuracy tasks, would not be suitable for tasks with high torque/force resistance. In such cases, it would be necessary to utilize displacement sensors for feedback control to overcome step jumping (missing): rotary encoders for rotary and linear encoders for linear stepper motors.

13.5.2 Pneumatic and Hydraulic Actuators

Fluid power can be very effective in remote actuation. The power source is located a distance away from the output point, thus minimizing the weight carried by the manufacturing device. For example, an industrial robot with electric actuators incorporates the motors into the manipulator arm—the fifth joint carries the sixth motor, the fourth joint carries both the fifth and sixth motors, and so on. In fluid-power-based machines, on the other hand, the electrical pumps supplying the necessary pressurized fluid to all actuators are located in the exterior of the load-carrying mechanism.

Hydraulic actuators utilize incompressible (hydraulic) fluids to transfer energy to a desired output point. The incompressibility of the fluid allows them to output large forces/torques. Directional control valves and pressure control valves can be configured for the closed-loop control of hydraulic actuators. Linear cylinders can be utilized to output linear motion, while rotary pumps, such as the vane pump, can be utilized to output rotary motion (Fig. 32). In a vane motor, an eccentric rotor has chambers of unequal volume separated by spring-loaded vanes. Pressure differences between the chambers drive the rotor one way or another according to the direction of fluid input.

Pneumatic actuators are much easier to maintain than hydraulic actuators and are inexpensive. However, due to the compressibility of air, they cannot be used for large-load carrying applications and are only very

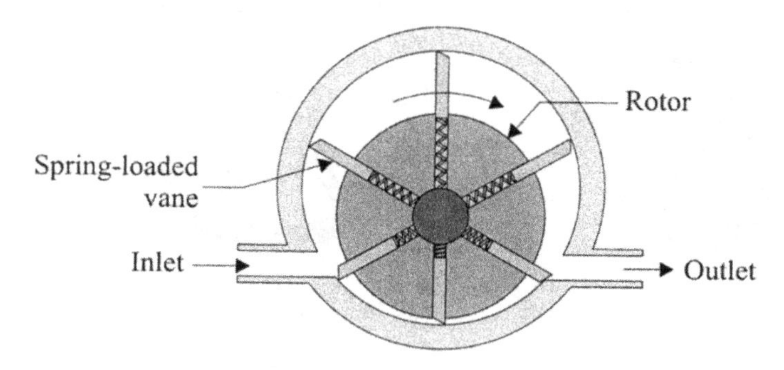

FIGURE 32 Vane motors.

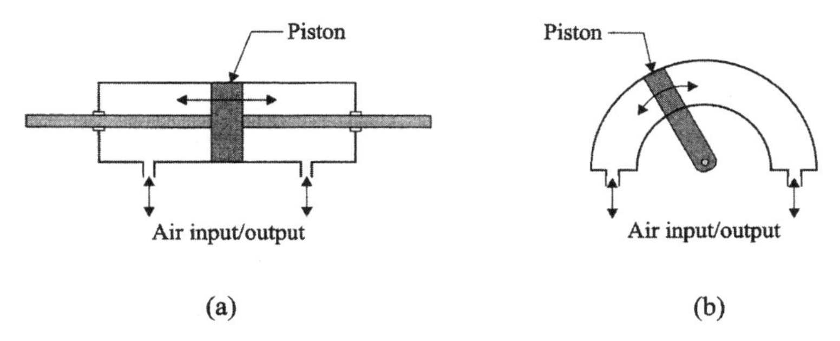

FIGURE 33 (a) Linear; (b) rotary pneumatic actuators.

infrequently used as servo systems (i.e., for motion control). They are primarily used in limit-to-limit motions (Fig. 33).

REVIEW QUESTIONS

1. Define the two primary levels of manufacturing control.
2. What is closed-loop process control?
3. Compare the different controller modes (i.e., PI, PD, and PID).
4. Compare analog controllers to digital controllers.
5. Describe the basic elements of a sensor.
6. Why is motion sensing of primary interest in many manufacturing processes?
7. Compare inductive transducers to capacitive transducers.
8. Noncontact sensing is considered a preferred mode of instrumentation for motion control. Discuss cases where a contact sensor, such as a linear variable-differential transformer (LVDT), would be the preferred choice.
9. Discuss amplitude-modulation-based (electro-optical) proximity sensing.
10. Compare amplitude-modulation-based proximity sensing to triangulation proximity sensing. Discuss advantages/disadvantages.
11. Explain the principle of interferometry and discuss some manufacturing application areas.
12. Compare ultrasonic proximity sensing to electro-optical proximity sensing.
13. Explain the principle of digital (absolute and relative) encoding and discuss some manufacturing application areas.
14. What are the most commonly employed transducers in force sensing?
15. Discuss machine vision versus computer vision.

16. Describe the primary steps of (digital) image acquisition and analysis.
17. Discuss the following issues in digital imaging: camera resolution, color versus monochromatic imaging, and the number of gray levels. In your discussion, you may refer to different manufacturing applications that would impact on the selection of these features.
18. What is image preprocessing? What are the advantages/disadvantages of image preprocessing?
19. For dimensional measurement purposes, would it be more accurate to utilize gray level images or (thresholded) binary images? Explain.
20. Discuss the use of model-based recognition versus (boundary) shape-based recognition in image analysis.
21. Discuss the use of DC motors versus stepper motors in manufacturing applications. Include some applications in your discussion.
22. Discuss the use of pneumatic actuators versus hydraulic actuators in manufacturing applications. Include some applications in your discussion.

DISCUSSION QUESTIONS

1. The factory of the future would be a totally networked enterprise. Information management in this enterprise would be a complex issue. In regards to planning, monitoring, and control, discuss the level of detail of information that the controllers (humans or computers) would have to deal with in such environments. For example, some argue that in a hierarchical information-management environment, activities are more of the planning type at the higher levels of the hierarchy and of the control type at the lower levels. It has also been argued that the level of details significantly decreases as you ascend the enterprise ladder.
2. The quality of a digital image is influenced by lighting conditions, electronic signal-transmission noise, and other image processing related problems. There exist numerous mathematical operators that can be utilized for the pre processing of such images. Discuss the advantages/disadvantages of using such operators in terms of specific manufacturing objectives: dimensional measurements versus pattern recognition.
3. In pattern recognition problems, an object (or a feature) can be identified by comparing its acquired image with a set of reference images in our database. Regardless of the analysis/identification technique utilized, two fundamental issues must be considered in the compilation of the database of reference object (image) models: Use of a dimensionally exact CAD-based object/feature model and the definition of statistical dimensional variations of the object's features that are

manufacturing-process dependent. Discuss both issues in the generation of a reliable set of reference object models for use in controllable/structured manufacturing environments. Furthermore, discuss how would one cope with (image) modeling problems in the absence of CAD-based object models.

4. Machine vision is a branch of computer vision targeted for inspection or guidance purposes in manufacturing environments. Such systems primarily operate in two-dimensional, static, uni color (gray scale) spaces for the autonomous measurement of geometric dimensions or recognition of patterns on viewed objects. The precision of the measurements is a direct function of the resolution of the camera's photodetector and the optics used. Discuss image processing time and measurement precision for several manufacturing applications. In this discussion, include a comparison on the suitability of machine vision for dimensional measurement versus pattern recognition in the context of manufacturing time and cost constraints.
5. The absence of accurate robot kinematic models, compounded with the absence of accurate geometric world models of their working environments, has often forced users to define Cartesian locations through manual teaching techniques (teleoperation). The robot end-effectors are moved to their desired destinations through teach pendants, and the controllers are required to memorize joint-encoder readings at these locations. The use of such playback-based robot motion techniques thus force objects always to be at their expected locations with very stringent tolerances. Discuss the potential of using a variety of visual and/or non visual task space sensors in controlling the robot motion that can lead to the relaxing of positioning requirements for objects that are static or those that are in motion.
6. Bin picking is a term used for the robotic grasping of a single component from an open bin that contains many randomly oriented, identical (and sometimes not identical) components. Discuss the difficulties associated with this operation from the machine vision point of view, assuming the robot can be directed to carry out such a picking operation, once it has been instructed about where the Cartesian location of the component is. Propose alternative solutions to robotic bin picking.
7. The supervisory control of a manufacturing process relies on timely and accurate sensory feedback. However, owing to the difficult production conditions (high temperatures, high pressures, physical obstructions, etc.), many production output parameters cannot be directly measured. Sensors instead observe and quantify certain physical phenomena (e.g., acoustic emissions) and relate these measurements to production

output parameters (e.g., tool wear) that we desire to monitor. Discuss the following and other issues in the context of effective autonomous process control: the availability of effective signal processing and pattern recognition techniques, the use of multiple sensors to monitor one phenomenon (i.e., sensor fusion), the decoupling of information obtained by a sensor whose outputs may have been influenced by several output parameters, and the availability of models that could optimally change a machine input parameter in response to changes monitored in one or more output parameters.

8. In machining, tool change (due to wear) may constitute a significant part of setup times. This is especially true in multipoint cutting, such as milling, where all the inserts have to be replaced together. Almost a century of work in the area of tool wear has yet to yield reliable models of the wear mechanisms, which would allow users to maximize the utilization of the tools and thus minimize the number of tool changes. Discuss the use of a variety of sensors and pattern recognition techniques for on-line intelligent machining in the absence of such models, or in support of approximate models.
9. Discuss the need of having effective tracking means, distributed throughout a manufacturing facility, which would provide users (and even customers) with timely feedback about the status and location of products in motion. Include several examples of such sensing and feedback devices in your discussion. Furthermore, discuss the advantages/disadvantages of using wireless solutions.
10. In the factory of the future, it is expected that manufacturing workcells will operate autonomously under the control of an overall computer-based supervisor. In such workcells, all devices, including fixtures and jigs, must be equipped with their own (device) controllers that can effectively communicate with the workcell supervisor. Discuss the integration of sensors and other communication components into intelligent fixtures that can be remotely controlled (via two-way communications) in the flexible manufacturing workcells of the future.

BIBLIOGRAPHY

Auslander, David M., Kempf, Carl J. (1996). *Mechatronics: Mechanical System Interfacing*. Upper Saddle River, NJ: Prentice Hall.

Barney, George C. (1988). *Intelligent Instrumentation: Microprocessor Applications in Measurement and Control*. New York: Prentice Hall.

Baxes, Gregory A. (1994). *Digital Image Processing, Principles and Applications*. New York: John Wiley.

Beckwith, Thomas G., Marangoni, Roy D., Lienhard, John H. (1993). *Mechanical Measurements*. Reading, MA: Addison-Wesley.

Boldea, I., Nasar, Sayed A. (1997). *Linear Electric Actuators and Generators*. New York: Cambridge University Press.

Bolton, W. (1999). *Mechatronics: Electronic Control Systems in Mechanical Engineering*. Harlow, UK: Longman.

Bonen, Adi (1995). Development of a Robust Electro-Optical Proximity Sensing System. Ph.D. diss., Department of Electrical and Computer Engineering, University of Toronto, Toronto, Canada.

Bonen, A., Saad, R. E., Bonen, A., Smith, K. C., Benhabib, B. (June 1997). A novel electro-optical proximity sensor for robotics: calibration and active sensing. *IEEE Transactions on Robotics and Automation* 13(3):377–386.

Boyle, W. S., Smith, G. E. (1970). Charge coupled semiconductor devices. *Bell Systems Technical Journal* 49:587–593.

Busch-Vishniac, Ilene J. (1999). *Electromechanical Sensors and Actuators*. New York: Springer-Verlag.

Davies, E. Roy. (1997). *Machine Vision: Theory, Algorithms, Practicalities*. San Diego, CA: Academic Press.

De Silva, Clarence W. (1989). *Control Sensors and Actuators*. Englewood Cliffs, NJ: Prentice Hall.

Figliola, Richard S., Beasley, Donald E. (2000). *Theory and Design for Mechanical Measurements*. New York: John Wiley.

Fraden, Jacob (1997). *Handbook of Modern Sensors: Physics, Designs, and Applications*. Woodbury, NY: American Institute of Physics.

Frank, Randy (2000). *Understanding Smart Sensors*. Boston: Artech House.

Fraser, Charles, Milne, John (1994). *Integrated Electrical and Electronic Engineering for Mechanical Engineers*. New York: McGraw-Hill.

He, D., Benhabib, B. (July 1998). Solving the orientation-duality problem for a circular feature in motion. *IEEE Transactions on Systems, Man and Cybernetics*, Part A 28(4):506–515.

Hobson, G. S. (1978). *Charge-Transfer Devices*. London: E. Arnold.

Holst, Gerald C. (1998). *CCD Arrays, Cameras, and Displays*. Bellingham, WA: JCD.

Horn, Berthold (1986). *Robot Vision*. Cambridge, MA: MIT Press.

Jain, Ramesh, Kasturi, Rangachar, Schunck, Brian G. (1995). *Machine Vision*. New York: McGraw-Hill.

Janesick, James R. (2001). *Scientific Charge-Coupled Devices*. Bellingham, WA: SPIE Press.

Johnson, Curtis D. (2000). *Process Control Instrumentation Technology*. Upper Saddle River, NJ: Prentice Hall.

Ejiri, Masakazu (1989). *Machine Vision: A Practical Technology for Advanced Image Processing*. New York: Gordon and Breach.

Lambeck, Raymond P. (1983). *Hydraulic Pumps and Motors: Selection and application for Hydraulic Power Control Systems*. New York: Marcel Dekker.

McMillan, Gregory K., ed. (1999). *Process/Industrial Instruments and Controls Handbook*. New York: McGraw-Hill.

Nachtigal, Chester L. (1990). *Instrumentation and Control: Fundamentals and Applications*. New York: John Wiley.

Partaatmadja, Ojong (1990). Development of an Electrooptical Distance Sensor and an Electrooptical Orientation Sensor. M.A.Sc. thesis, University of Toronto, Department of Mechanical Engineering, Toronto, Canada.

Partaatmadja, O., Benhabib, B., Goldenberg, A. A. (June 1993). Analysis and design of a robotic distance sensor. *J. of Robotic Systems* 10(4):427–445.

Partaatmadja, O., Benhabib, B., Kaizerman, E., Dai, M. Q. (April 1992). A two-dimensional orientation sensor. *J. of Robotic Systems* 9(3):365–383.

Phillips, Charles L. (2000). *Feedback Control Systems*. Upper Saddle River, NJ: Prentice Hall.

Rosenfeld, Azriel, Kak, Avinash C. (1982). *Digital Picture Processing*. New York: Academic Press.

Ristic, Rasko (1988). *An Electro-Optical Tactile Sensor Design*. University of Toronto, Toronto, Canada: Department of Mechanical Engineering.

Ristic, R., Benhabib, B., Goldenberg, A. A. (June 1989). Analysis and design of an electro-optical tactile sensor. *IEEE J. of Robotics and Automation* 5(3):362–368.

Saad, Ricardo E (1996). Development of a Photo-Elastic Tactile Transducer. Ph.D. diss., Department of Electrical and Computer Engineering, University of Toronto, Toronto, Canada.

Saad, R. E., Bonen, A., Smith, K. C., Benhabib, B. (1999). Proximity sensing for robotics. In: Webster, J. G. (Ed.), *The Measurement, Instrumentation, and Sensors Handbook*. CRC Press.

Saad, R. E., Bonen, A., Smith, K. C., Benhabib, B. (1999). Tactile sensing for robotics. In: Webster, J. G. ed. *The Measurement, Instrumentation, and Sensors Handbook*. CRC Press.

Saad, R. E., Bonen, A., Smith, K. C., Benhabib, B. (May 1998). Phase-lead reconstruction of a photoelastic tactile sensor. *J. of Robotic Systems* 15(5):259–280.

Safaee-Rad, Reza. An Active-Vision System for 3D-Object Recognition in Robotic Assembly Workcells. Ph.D. diss. University of Toronto, Toronto, Canada, 1991.

Safaee-Rad, R., Smith, K. C., Benhabib, B., Tchoukanov, I. (July 1992). Application of moments and fourier descriptors to the accurate estimation of elliptical shape parameters. *J. of Pattern Recognition Letters* 13:497–508.

Shigeyuki, Ochi, et al. (1996). *Charge-Coupled Device Technology*. Amsterdam, The Netherlands: Gordon and Breach.

Soloman, Sabrie (1999). *Sensors Handbook*. New York: McGraw-Hill.

Sun, Andy (1991). Design and Analysis of an Electrooptical Force/Torque Sensor: M.A.Sc. thesis, Department of Mechanical Engineering, University of Toronto, Toronto, Canada.

Tchoukanov, I., Safaee-Rad, R., Smith, K. C., Benhabib, B. (Dec. 1992). The angle-of-sight signature for 2D shape analysis of manufactured objects. *J. of Pattern Recognition* 25(11):1289–1305.

Tsai, R. Y. (Aug. 1987). A versatile camera calibration technique for high-accuracy 3D machine vision metrology using off-the-shelf TV cameras and lenses. *IEEE, Journal of Robotics and Automation* 3(4):323–344.

Tomkinson, Donald, Horne, James (1996). *Mechatronics Engineering*. New York: McGraw Hill.

Tzou, H. S., Fukuda, T. (1992). *Precision Sensors, Actuators, and Systems*. Boston: Kluwer Academic.

Van de Vegte, John (1994). *Feedback Control Systems*. Englewood Cliffs, NJ: Prentice Hall.

Vernon, David (1991). *Machine Vision: Automated Visual Inspection and Robot Vision*. New York: Prentice Hall.

Webster, John G., ed. (1999). *The Measurement, Instrumentation and Sensors Handbook*. Boca Raton, FL: CRC Press and IEEE Press.

Zuech, Nello (2000). *Understanding and Applying Machine Vision*. New York: Marcel Dekker.

14

Control of Production and Assembly Machines

In reprogrammable flexible manufacturing, it is envisaged that individual machines will carry out their assigned tasks with minimal operator intervention upon receipt of an appropriate high-level execution command. Such automatic device control normally means forcing a servomechanism employed by a production or assembly machine to achieve (or yield) a desired output parameter value in the continuous-time domain. In this chapter, our focus will be on the automatic control of two representative classes of production and assembly machines: material removal machine tools and industrial robotic manipulators. In Chap. 15, our attention will shift to the (higher-level) manufacturing system control that is based on discrete event system (DES) control theory, that is, the control of the flow of parts between machines.

14.1 NUMERICAL CONTROL OF MACHINE TOOLS

Material removal is achieved by the relative motion of a cutting tool with respect to a workpiece (Chaps. 8 and 9). In turning operations, the cutting tool can move in two orthogonal directions (feed and depth) and engage a rotating workpiece. The real-time control objective is to move the cutting tool along a prescribed path while controlling its position and velocity—the spindle rate is normally set to a fixed value. In three-axis milling operations, the workpiece can move in three orthogonal directions and engage a rotating cutting tool. The real-time control objective is to move the workpiece (via the motion of the worktable) along a prescribed path while controlling its position and velocity—the (tool holder's) spindle rate is normally set to a fixed value. In drilling operations, the workpiece can move in two orthogonal directions, in a plane perpendicular to the one-axis motion of the cutting tool. The real-time control objective is to move the workpiece from one point to another and translate the tool vertically according to the specific hole depth requirement while the workpiece is kept stationary—the (tool holder's) spindle rate is normally set to a fixed value.

14.1.1 Development of Machine Tool Control

The term numerical control (NC), synonymous with machine tool control, can be traced back to the development of the pertinent control technology in 1952 at the Massachusetts Institute of Technology (MIT), U.S.A. The Servomechanism Laboratory at MIT was contracted at the time by the Parsons Corporation to develop a universal control technology for machine tools through a US Air Force contract. The preliminary outcome of this research was a retrofitted vertical (tracer) milling machine, whose three motion axes could be simultaneously controlled by a hybrid (digital/analog) controller. A punched tape, coded with the sequence of machining instructions, was utilized to program the controller of this first NC machine tool.

The first commercial NC machine controllers were developed by four separate companies based on US Air Force contracts—Bendix, EMI, General Dynamics, and General Electric. Some claim that this diversification attempt and promotion of competition is the lead cause of still having different formats for NC programs and thus a lack of portability of a NC program from one controller to another.

In 1960s, NC controllers relied on dedicated digital hardware for the execution of simple motion commands (straight line and circular arcs). These machine control units (MCUs) allowed programmers to download a sequence of operations to be executed by the dedicated hardware–based (versus software-based) motion generators (interpolators) and controllers. Many of these controllers are still in use today, in the form of original equipment (older NC machine tools) or as customized controllers retrofitted on originally manual machines.

The mid and late 1960s were marked by the development and widespread use of mainframe computers (especially those by IBM). At the time, several large manufacturers attempted to network their individual NC machines under the umbrella of one (or more) such mainframe computers. The purpose was centralized control, where one computer assigned tasks and directly downloaded corresponding programs to the individual NC controllers. The term direct numerical control (DNC) was appropriately adopted for such configurations. The practice of DNC, however, was short lived owing to frequent down times of the main computer (not tolerable in manufacturing) and continued use of mass production strategies that did not require frequent changes in the programming of NC machines.

The term DNC has also referred in the past to attempts to control several machines using one centralized computer, where this controller downloaded step-by-step individual instructions to individual machines, as opposed to complete programs. Naturally, this practice had an even shorter life in manufacturing environments owing to frequent computer down times.

The term computerized numerical control (CNC) was introduced in the early 1970s with the development of minicomputer-based controllers for machine tool control. The early use of minicomputers was later replaced with the use of dedicated microprocessor-based NC controllers, as miniaturization rapidly allowed the packaging of CPU and memory devices with servo controllers into small controller units. Such controllers carry out motion planning and control functions in software, as opposed to via very restricted hardware circuits. The primary advantage of CNC machines, however, has been noted as their capability of allowing the adaptive control of machining operations. That is, CNC controllers can be appropriately programmed to vary the (input) process parameters, such as cutting speed and/or feed rate, in direct response to varying cutting conditions, such as tool wear and variable depth of cut that would cause undesirable increases in machining forces.

The factory of the future will be a networked environment, where production plans and control programs will be downloaded to appropriate CNC machines when needed (i.e., just-in-time control) (Fig. 1). Based on this

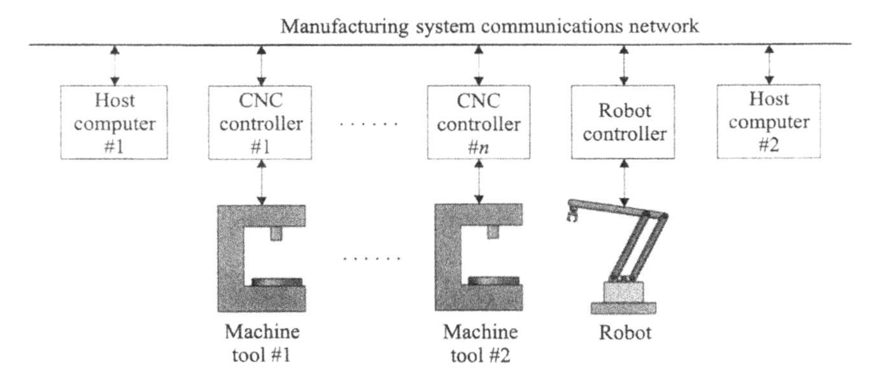

FIGURE 1 Distributed numerical control.

premise, the term distributed numerical control (DNC) has rapidly gained acceptance since the 1990s and was replaced the earlier acronym for direct numerical control. Although the current DNC architectures normally assume direct physical connection of CNC controllers to a centralized computer, in the near future there will be no such apparent connections. As shown in Fig. 1, all CNC controllers will have networking capabilities and receive commands and/or be downloaded programs over the communications network backbone of the factory.

14.1.2 Motion Control

Motion control in NC machines is achieved by issuing coordinated motion commands to the individual drives of the machine tool (Fig. 2). Almost all commercial NC machines employ DC or AC electrical motors that linearly drive stages/tables mounted on ball-bearing leadscrews. These leadscrews provide low-friction (no stick-slip), no-backlash motions with accuracies of 0.001 to 0.005 mm or even better. High-precision machines employ interferometry-based displacement sensors to provide sensory data to the (closed loop) controllers of the individual axes of the machine tool (Chap. 13). Rotational movements (spindle and other feed motions) are normally achieved using high-precision circular bearings (plain, ball, or roller).

Motion Types

Machine tools can be utilized to fabricate workpieces with prismatic and/or rotational geometries. Desired contours are normally achieved through a controlled relative motion of the cutting tool with respect to the workpiece. Holes of desired diameters, on the other hand, are normally achieved by

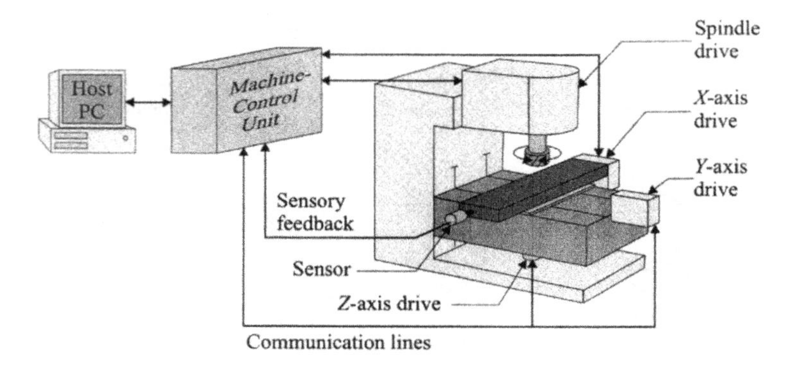

FIGURE 2 Overall NC machine tool control architecture.

holding the workpiece fixed and moving a rotating drill bit into the workpiece vertically. Correspondingly, NC motions have been classified as point-to-point (PTP) motion (e.g., drilling) and contouring, or continuous path (CP), motion (e.g., milling and turning).

In PTP systems, the workpiece is moved from one point to another in the fastest manner without regard to the path followed. The motion is of asynchronous type, where each axis accomplishes its desired movement independent of the others. For example, the $X-Y$ table of the drilling press would follow the path shown in Fig. 3a, where the Y axis continues its motion from Point A to the desired Point B, while the X axis remains stationary after it has already accomplished its necessary incremental motion. Once the table reaches Point B, the drill head is instructed to move in the Z axis, the necessary distance, and cut into the workpiece.

In CP systems, the workpiece (in milling) or the tool (in turning) follows a well-defined path, while the material removal (cutting) process is in progress. All motion axes are controlled individually and move synchronously to achieve the desired workpiece/tool motion (position and speed). For example, the $X-Y$ table of a milling machine would follow the path shown in Figure 3b, when continuously cutting into the workpiece along a two-dimensional path from Point A to Point B.

For both PTP and CP motions, the coordinates of points or paths can be defined with respect to a global (world) coordinate frame or with respect to the last location of the workpiece/tool: absolute versus incremental positioning, respectively. Regardless of the positioning system chosen, the primary problem in contouring is the resolution of the desired path into multiple individual motions of the machine axes, whose combination would yield a cutter motion that is closest possible to the desired path. This motion-planning phase is often called interpolation. In earlier NC machine controllers, interpolation was carried out exclusively in dedicated hardware

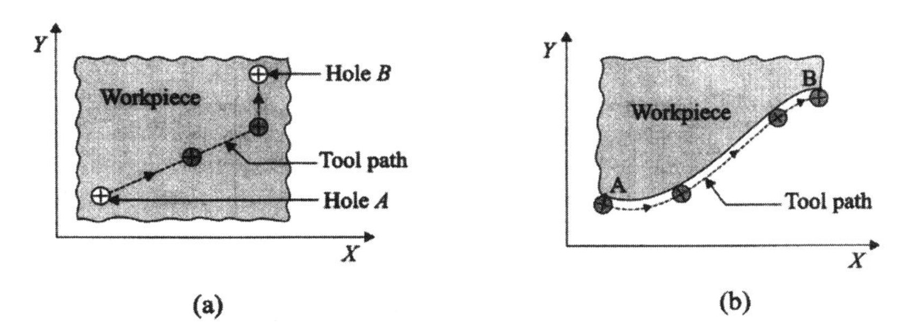

FIGURE 3 (a) Point-to-point; (b) continuous path motion.

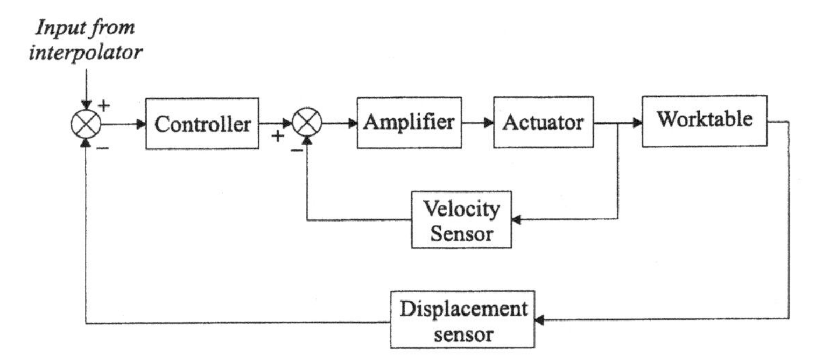

Figure 4 Closed-loop NC machine tool CP motion control.

boards, thus limiting the contouring capability of the machine tool to mostly straight-line and circular-path motions. In modern CNC machines, interpolation is carried out in software, thus allowing any desired curvature to be approximated by polynomial or spline-fit equations.

Closed-Loop Control

In PTP motion, individual axes are provided with incremental motion commands executed with no regard to the path followed. Although control can be carried out in an open-loop manner, encoders mounted on the leadscrews allow for closed-loop control of the motion (Chap. 13).

In CP motion, the interpolator provides individual axes with necessary motion commands in order to achieve the desired tool path (Fig. 4). Encoders and tachometers provide the necessary feedback information; interferometry type sensors can be used for high-precision displacement and velocity applications (Chap. 13).

Adaptive Control

Adaptive control of machine tools refers to the automatic adaptation of cutting parameters in response to changes in machining conditions (Fig. 5). A collection of sensors (acoustic, thermal, dynamic, etc.) are utilized to monitor cutting forces/torques, cutting temperatures, mechanical vibrations, acoustic emissions, in order to predict tool wear, the potential for tool breakage, chatter, and so on. A software-based adaptive controller utilizes the collected information in order to change feed rate and cutting velocity in real time and provide this information to the interpolator of the CNC controller for the generation of new motion commands (Fig. 4). Prediction techniques, such as neural networks, fuzzy logic, and heuristic rules, can be used in the calculation of new cutting parameters.

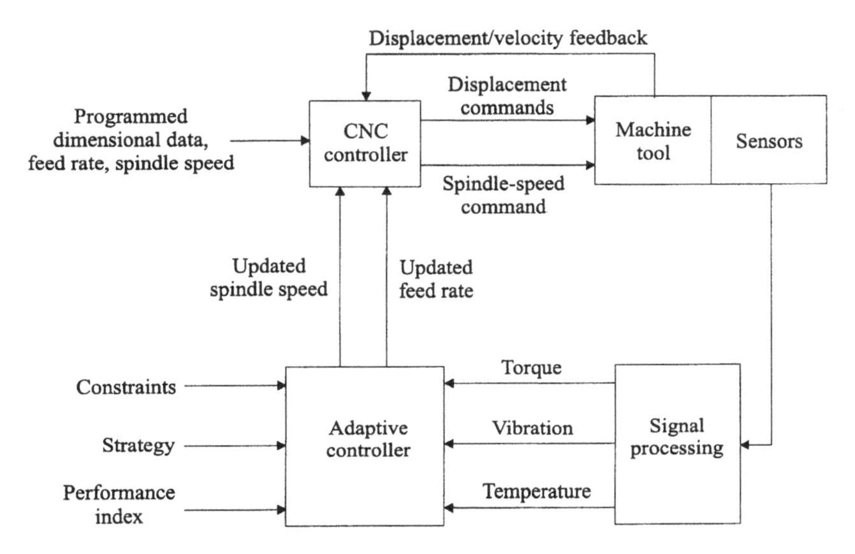

FIGURE 5 Adaptive control for machining.

From a commercial point of view, the primary objective of an adaptive control system should be to optimize a performance index, such as machining time or cost, subject to the capability limits of the machine tool and the dimensional constraints imposed on the workpiece. It would, for example, be desirable to adjust automatically the cutting parameters in real time for maximizing material removal rates.

Adaptive controllers capable of real-time optimization are still in their research phase owing to the high complexity of the machining process. Constraint-based adaptive controllers, however, are considered to be mature enough for commercialization. Such controllers adjust cutting parameters in real time in order to maintain cutting forces/torques, vibrations, temperature, and so on at or below their user-specified limits. For example, a machine tool's feed rate would be reduced in response to cutting-force increases due to tool wear, unexpected variations in workpiece hardness, and raw-material (stock) geometry, and so on.

Adaptive control is discussed below for two metal-cutting applications:

Adaptive control in turning: The cutting tool in NC lathes is mounted onto a stage whose motion is controlled in two orthogonal axes, the feed and depth-of-cut directions. Tool wear in turning is normally a continuous process leading to tool degradation in the form of flank wear and crater wear (Chap. 8). Although flank wear yields a continuous increase in cutting forces, initial crater wear can create favorable cutting conditions and lead to reduction in cutting forces. Beyond a crater-wear threshold, both wear mechanisms lead to gradual increases in cutting forces.

There exist a variety of force sensors, commercialized since the early 1970s (e.g., Kistler, Prometec, Montronix, and Sandvik), that can be easily mounted on the stage of the lathe, underneath the tool holder. Such instruments utilize piezoelectric or strain gages as the force detection transducers (Chap. 13). Cutting forces can also be evaluated by monitoring torque requirements on the drivers of the cutting tool stage and/or on the spindle motor. Such measurements, however, are only used as complementary information and not as sole indicators of force owing to difficulties in mathematical predictions of force directions and magnitudes.

Acoustic emissions from the cutting zone (low amplitude and high frequency) have also been sensed via piezoelectric detectors (microphones) for estimating tool wear. Continuous signals are generated in the shear zone and at the workpiece–tool and chip–tool interfaces, while discontinuous signals are generated by the breakage of the chips. The frequencies of these signals are much higher than other potential emissions in the surroundings, such as machine tool vibrations. A number of classical statistical pattern-recognition schema to have been developed by academic researchers during the 1980s and 1990s for identifying tool wear via acoustic emissions. However, in practice, acoustic sensors have only been used as early warning systems to indicate imminent failure of the cutting tool and not for continuous feedback to the adaptive controllers.

Adaptive control in milling: The cutting operation in milling is an intermittent process, where a cutting edge engages the workpiece periodically and remains engaged for a portion of the full rotation of the multitooth tool. Thus, besides the gradual tool wear, one must monitor for force and torque overloads, chatter-causing vibrations, and catastrophic tool failure. Force overloads at the engagement of the tool with the workpiece (especially in the case of small-diameter tools) can severely damage the tool and subsequently the workpiece.

As in turning, force sensors placed underneath the workpiece fixtures and torque sensors mounted on the spindle of milling machines can be effectively utilized to detect spindle stalls, cutting-force overloads, and tool wear/breakage. Acoustic sensors have also been used in milling to detect chatter—a self-excited vibration mechanism due to the regeneration of periodical waviness on the machined workpiece—by listening to emissions of increasing amplitude (Chap. 8).

As discussed above, many different sensors can be used to monitor the working condition of a machine tool for its adaptive control. For example, tool wear can be monitored using force sensors mounted under the tool holder (in turning) or under the workpiece (in milling), torque sensors mounted on the spindle motor, and acoustic sensors placed in close vicinity to the cutting interface. Naturally, each sensor outputs its conclusion based on its received and analyzed signals with an associated uncertainty. This uncertainty consists of components such as (Gaussian) random noise, (systematic) fixed errors due to inaccurate calibration, and limitations of the pattern analysis technique used in manipulating the collected data. The use of multiple sensors (multisensor integration) and the merging of their outputs (data fusion) can benefit the monitoring process by reducing the uncertainty level.

Multisensor integration is the choice of the number and the types of sensors for the task at hand and their optimal placement in the workspace for maximum accuracy. Two possible strategies for multisensor integration are (1) to select and configure a minimum number of sensors and utilize them continuously (for the entire duration of the process monitored), or (2) to select a large number of sensors (more than the minimum) and configure them in real time (i.e., select subsets of sensors) according to a criterion to best suit the needs of the monitoring objective as machining progresses. For the latter strategy, for example, we can use only force transducers at the beginning of cutting but activate and merge additional data received from acoustic sensors toward the end of the expected/predicted tool life.

Multiple sensors can provide a data fusion module with two types of information: (1) data about one feature observed by multiple sensors—redundant information, or (2) data about the subfeatures of one feature, in cases where no one single sensor can perceive the totality of the feature level—complementary information. The data collected can in turn be fused at multiple levels: signal level or feature level. Signal level data fusion is common for sensing configurations, multiple identical (redundant) sensors observing the same feature. A common problem at this level of fusion is the temporal and spatial alignment of data collected from multiple sensors (i.e., ensuring that all sensors observe the same feature at the same time—synchronization). At feature-level fusion, the primary problem is the spatial transformation of information for spatial alignment.

Common methods for signal-level data fusion include weighted averaging of measurements, recursive estimation of current and future measurements using the Kalman filter, hierarchical estimation using a Bayesian estimator for combination of multisensor data according to probability theory, Dempster–Shafer reasoning approach for combining only evident data (i.e., not assigning probabilities to unavailable data), fuzzy-logic reasoning via the assignment of discrete values (between 0 to 1) to different propositions—a multivalued-logic approach, and so on.

14.1.3 Programming of NC Machine Tools

The programming of a NC machine tool is preceded by the determination of a suitable (preferably optimal) process plan. A process plan specifies how a part is to be machined: the sequence of individual operations, the specific machine tools on which these operations are to be carried out, the machining parameters (e.g., feed rate, cutting velocity) for each operation, and so forth.

All NC machine tools are equipped with controllers that can interpret a machine language–based program and convert these instructions into motion commands of the numerically controlled axes. These machine language programs have been commonly referred to as *g*-code. Unfortunately, for historical reasons, different commercial NC controllers use similar but different *g*-codes.

During the period 1955 to 1958, the first high-level programming language for NC machine tools was developed under the coordination of researchers from MIT. This programming language (APT, automatically programmed tool) reached maturity in the early 1960s and served as a guideline for the development of many subsequent NC programming languages, such as EXAPT (extended subset of APT) developed by the Institute of Technology in Aachen, Germany, ADAPT (adaptation of APT) and AUTOSPOT (automatic system for positioning tools), both by IBM, U.S.A., among many others. A program written in one such high-level languages needs to be translated into the specific *g*-code of the NC machine tool to be utilized for the machining of the workpiece at hand.

Since the late 1980s, most commercial CAD software packages allow users to generate cutting tool paths automatically in an interactive manner, bypassing the generation of a high-level language program. The user can simulate the machining operation and, having been satisfied with the outcome, can request the CAD system to generate the corresponding *g*-code program (specific to the NC controller to be utilized) and directly download it to the NC machine tool over the communications network.

g-Code

A *g*-code program consists of a collection of statements/blocks to be executed in a sequential manner. Each statement comprises a number of "words"—a letter followed by an integer number. The first word in a statement is the block number designated by the letter N followed by the number of the block (e.g., N0027, for the 27th line in the *g*-code program). The next word is typically the preparatory function designated by the letter G (hence, the letter "*g*" in *g*-code) followed by a two-digit number. Several examples of G words are given in Table 1.

TABLE 1 Some *G* Words[a]

Code	Function	Code	Function
G00	Point-to-point motion	G20	Imperial units
G01	Linear-interpolation motion	G21	Metric units
G02	Clockwise circular-interpolation motion	G32	Thread cutting
G03	Counterclockwise circular-interpolation motion	G98	Per-minute feed rate
		G99	Per-revolution feed rate

[a] May be different for different NC controllers.

The preparatory function is followed by dimensional words designated by axes' letters X, Y, and Z with corresponding dimensions, normally expressed as multiples of smallest possible incremental displacements (e.g., X3712 Y-47000 Z12000; multiples of 0.01 mm) or in absolute coordinates (e.g., X175.25 Y325.00 Z136.50). The feed rate and spindle speed words are designated by the letters F and S, respectively, followed by the corresponding numerical values in the chosen units. Next come the tool number word designated by the letter T and the miscellaneous function word designated by the letter M (Table 2).

A typical *g*-code program block is

N0027 G90 G01 X175.25 Y325.00 Z136.50 F125 S800 T1712 M03 M08;

the statement Number 27 (N0027) specifies the use of absolute coordinates (G90), a linear interpolation motion (G01) from current location to a position defined by the X, Y, Z coordinates (X175.25 Y325.00 Z136.50), a feed rate of 125 mm/min (F125) along the path, a spindle speed of 800 rev/min (S800), tool number 1712 (T1712), a clockwise turn of the spindle (M03), and coolant on (M08).

TABLE 2 Some M Words[a]

Code	Function	Code	Function
M00	Program stop (during run)	M08	Coolant on
M02	End of program	M11	Tool change
M03	Spindle start clockwise	M98	Call a subprogram
M05	Spindle stop	M99	Return to main program

[a] May be different for different NC controllers.

APT Language

The APT language serves two purposes: it (1) provides the NC controller with the pertinent geometric description of the workpiece, and (2) instructs it to carry out a series of operations for the machining of this workpiece. It achieves these objectives by utilizing about 600 geometric and motion command words.

The typical APT statement comprises two segments separated by a slash. The APT word to the left of the slash is modified by the information provided on the right side of the slash.

Geometric statements: The geometry of the workpiece pertinent to its machining can be described by a collection of points, lines, and surfaces. A few exemplary ways of describing such entities are given here:

Point definition by its coordinates:

Point_Name = POINT/X, Y, Z coordinates

P7 = POINT/200, 315, 793

Point definition by the intersection of two lines:

Point_Name = POINT/INTOF, Line_Name_1, Line_Name_2

P11 = POINT/INTOF, L3, L7

Line definition by two points:

Line_Name = LINE/Point_Name_1, Point_Name_2

L3 = LINE/P9, P21

Line definition by a point and an angle with respect to an axis:

Line_Name = LINE/Point_Name, ATANGL, Angle_Value, Axis_Name

L7 = LINE/P8, ATANGL, -75, YAXIS

Defining a circle by its center and radius:

Circle_Name = CIRCLE/CENTER, Point_Name, RADIUS, Radius_Dimension

C3 = CIRCLE/CENTER, P14, RADIUS, 35

Defining a plane by its equation $ax + by + cz = d$:

Plane_Name = PLANE/a, b, c, d

PL1 = PLANE/7.5, -3.1, 0.3, 6.7

Defining a (circular) cylindrical surface by a tangent plane, along a given line, with a given radius:

Surface_Name = CYLNDR/Side_of_Plane, TANTO, Plane_Name, THRU, Line_Name, RADIUS, Radius_Dimension

CYL3 = CYLNDR/ZLARGE,TANTO, PL1, THRU, L7, RADIUS, 25

Motion statements: The relative movement of the tool with respect to the workpiece can be of PTP or CP (contouring) type. A few exemplary ways of describing motion are given here:

PTP motion commands:
  GOTO/Point_Name; Go to Point Point_Name.
  GODLTA/ΔX, ΔY, ΔZ; Move incrementally by (ΔX, ΔY, ΔZ).
CP motion commands: In APT programming, motion commands are based on the relative movement of the cutting tool with respect to a stationary workpiece. The tool's motion is restricted by three surfaces: The depth (part) surface, on which the tool-end moves, the tangent (drive) surface, along which the tool slides, and the constraint (check) surface, which defines the end of the motion (Fig. 6). Thus the contouring motion commands on a given part surface are defined by the drive-surface and check-surface planes:

$$\begin{bmatrix} \text{GOFWD} \\ \text{GOBACK} \\ \text{GOLFT} \\ \text{GORGT} \\ \text{GOUP} \\ \text{GODOWN} \end{bmatrix} / \ \text{Drive\_Surface}, \begin{bmatrix} \text{TO} \\ \text{ON} \\ \text{PAST} \\ \text{TANTO} \end{bmatrix}, \ \text{Check\_Surface}$$

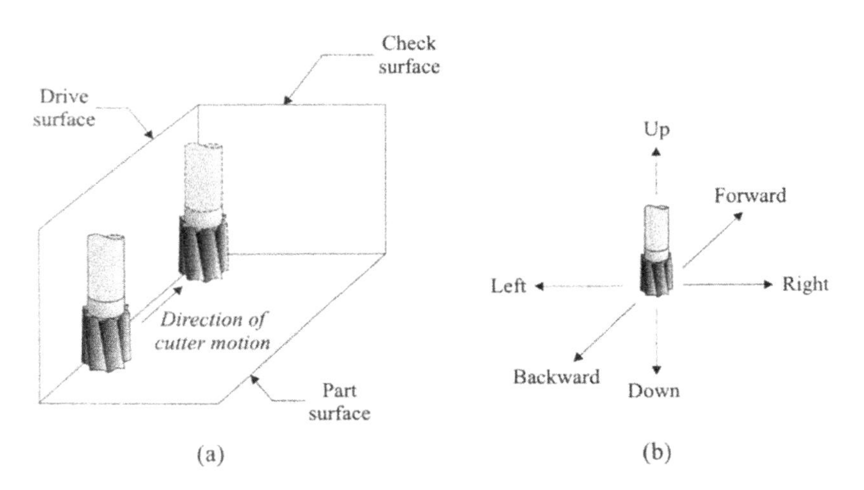

FIGURE 6 (a) Control surfaces; (b) control directions for contouring.

In three-dimensional machining, the contouring motion starts by relocating the tool from its current location to a point defined by the three surfaces constraining the motion of the tool:

FROM/Point_Name

$$\text{GO/}\begin{bmatrix}\text{TO}\\ \text{ON}\\ \text{PAST}\end{bmatrix}\text{, Drive\_Surface, }\begin{bmatrix}\text{TO}\\ \text{ON}\\ \text{PAST}\end{bmatrix}\text{, Part\_Surface, }\begin{bmatrix}\text{TO}\\ \text{ON}\\ \text{PAST}\end{bmatrix}\text{, Check\_Surface}$$

The following example program defines the two-dimensional contouring of the part profile shown in Fig. 7:

```
FROM/P1
GO/TO, L1, ON, PSURF, ON, L2
GORGT/L1, TANTO, C1
GOFRWD/C1, TANTO, L2
GOLFT/L2, PAST, L1
GOTO/P1
```

Other APT Statements:

MACHIN/Postprocessor_Name	; Machine-specific postprocessor
UNITS/MM or UNITS/INCHES	; Units
FEDRAT/Value_per_minute or Value_per_revolution	; Feedrate
SPINDL/Speed, CLW	; Spindle turns at "Speed" clockwise
COOLNT/ON	; Coolant on

Once an APT program is obtained, a processor is needed to generate the cutter location data (CLDATA) file. The CLDATA file is then utilized

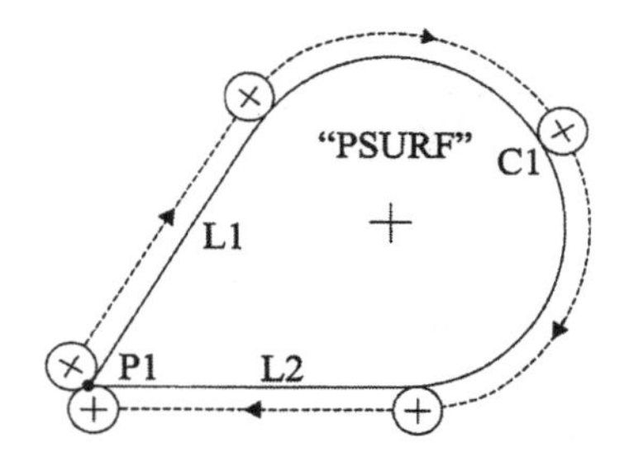

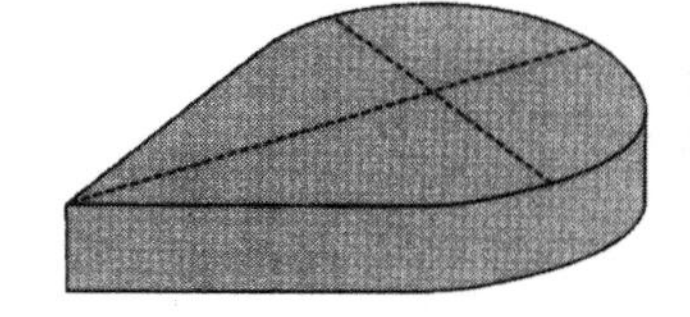

FIGURE 7 Contouring example.

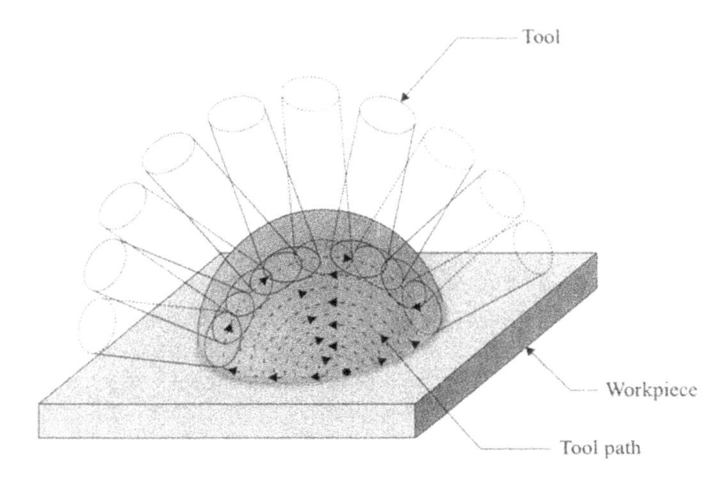

FIGURE 8 Tool path simulation.

as an input to a postprocessor specific for the NC machine tool at hand for the generation of the *g*-code. The first line in the APT program specifies this specific postprocessor (MACHIN/ Post_Processor_Name).

Computer-Aided NC Machine Programming

In the integrated and networked factory of the future, engineers will likely bypass the manual generation of NC part programs and exclusively adopt the rapidly developing computer-aided tools for the automatic generation of *g*-codes. Today an engineer can create the solid model of a stock, define cutting tool paths on this stock by specifying control surfaces corresponding to desired optimal cutting parameters, and prompt the software to generate corresponding a CLDATA file and subsequently to postprocess this file according to the NC controller at hand. This automatic process is then finalized by the downloading of the *g*-code to the controller of the CNC machine over the computer network.

Figure 8 shows a simulation example for a computer-aided generated tool path.

14.2 CONTROL OF ROBOTIC MANIPULATORS

As discussed in Chap. 12, robotic manipulators have been utilized in the manufacturing industry in a variety of applications, ranging from spot welding to spray painting, to electronic component assembly, and so on. The vast majority of these manipulators are open-chain mechanisms, comprising a set of links attached via revolute (rotary) or prismatic (linear) actuators in series. A number of closed-chain mechanism manipulators, comprising a set of links/actuators configured in parallel versus in series, have also been utilized in the manufacturing industry for high-precision tasks, but our focus herein will be on serial manipulators (Fig. 9).

Serial robotic manipulators have been configured in three distinct geometrical forms (Sec. 12.3.1, Figs. 12.10 to 12.13): rectangular, cylindrical, and spherical. This classification is based on the geometry of the workspace of the manipulator. For example, an articulated robot comprising a sequence of rotary joints can be classified as a spherical-geometry robot since its workspace is spherical in nature. Regardless of their geometric classification, industrial robotic manipulators carry out tasks that require their end-effector (gripper or specialized tool) to move in point-to-point (PTP) or continuous path (CP) mode. Thus, as NC controllers for machine tools, robot motion controllers must ensure specific trajectory following in real time, as defined by the trajectory planning module of the controller.

Unlike in NC motion interpolation for machining, with the exception of five-axis machining, trajectory planning for industrial robots is a complex matter owing to the dynamics of open-chain manipulators moving payloads in three-dimensional Cartesian space subject to gravitational, centrifugal, and inertial forces. Thus in this section, robot motion planning and control will be addressed in the following order: kinematics/dynamics, trajectory

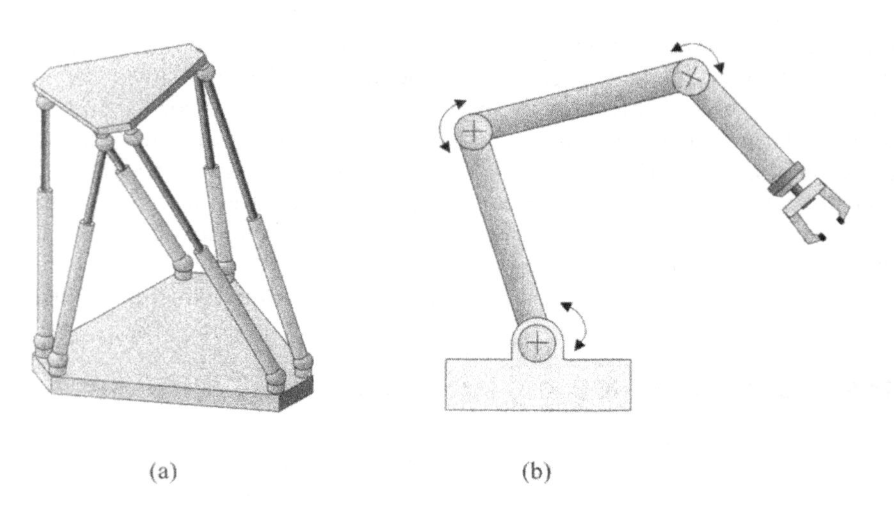

FIGURE 9 (a) A parallel; (b) a serial manipulator.

planning, and control, (Section 14.2.1). Robot programming techniques will be discussed in the subsequent subsection.

14.2.1 Motion Planning and Control

The first challenge in robot motion control is the transformation of a desired task space motion command into corresponding joint space (actuator) motion commands for the individual joints of the manipulator. For example, given the manipulator's latest stand (configuration) and a desired incremental end-effector translational motion of (ΔX, ΔY, ΔZ), while maintaining a constant orientation, the task at hand is to determine corresponding joint motions, $\Delta\theta_i$, $i=1, n$, where n is the degrees of freedom (dof) of the robot.

A transformation of positional/velocity/acceleration information between task space (normally, Cartesian) and joint space coordinates can only be achieved via the kinematic model of the manipulator. A dynamic model of the manipulator, however, is needed for calculating available joint torques/forces in response to load carrying task space requirements, especially when one attempts to minimize the required effort or motion time along a given end-effector path—trajectory planning.

The majority of industrial robots employ closed loop controllers designed to drive the individual actuators of the manipulator, when executing a desired task space trajectory converted into individual joint commands by a trajectory planning module (Fig. 10).

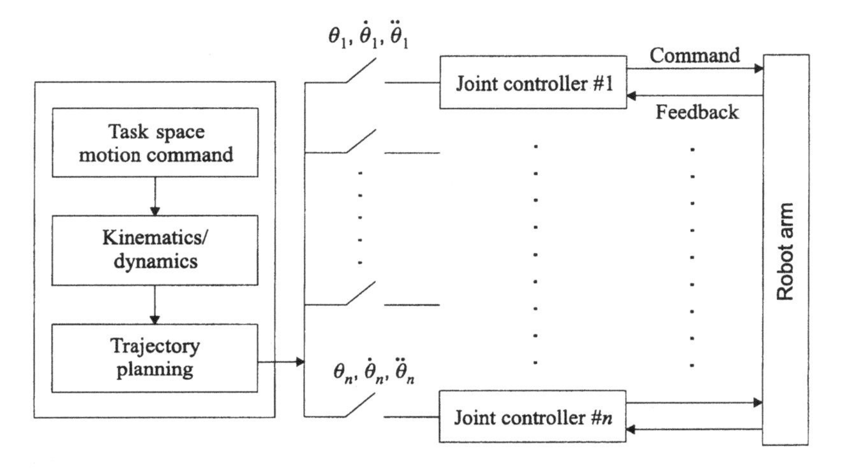

FIGURE 10 Robot motion control.

Robot Kinematics

The objective of a kinematic model is to relate the motion of the robot end-effector in task space coordinates to joint space coordinates: the individual motions of the joints. This is a typical rigid body motion description in three-dimensional space.

Homogeneous transformations (4 × 4 matrices) have been often used to describe the motion of a rigid body, defined by a Cartesian frame, with respect to a fixed coordinate system. The following four matrices describe a translation of (d_x, d_y, d_z) and a rotation of θ with respect to the X, Y, and Z axes, respectively:

$$\text{Trans}\ (d_x, d_y, d_z) = \begin{bmatrix} 1 & 0 & 0 & d_x \\ 0 & 1 & 0 & d_y \\ 0 & 0 & 1 & d_z \\ 0 & 0 & 0 & 1 \end{bmatrix} \tag{14.1a}$$

$$\text{Rot}\ (x, \theta) = \begin{bmatrix} 1 & 0 & 0 & 0 \\ 0 & \cos\theta & -\sin\theta & 0 \\ 0 & \sin\theta & \cos\theta & 0 \\ 0 & 0 & 0 & 1 \end{bmatrix} \tag{14.1b}$$

$$\text{Rot}\ (y, \theta) = \begin{bmatrix} \cos\theta & 0 & \sin\theta & 0 \\ 0 & 1 & 0 & 0 \\ -\sin\theta & 0 & \cos\theta & 0 \\ 0 & 0 & 0 & 1 \end{bmatrix} \tag{14.1c}$$

$$\text{Rot}\ (z, \theta) = \begin{bmatrix} \cos\theta & -\sin\theta & 0 & 0 \\ \sin\theta & \cos\theta & 0 & 0 \\ 0 & 0 & 1 & 0 \\ 0 & 0 & 0 & 1 \end{bmatrix} \tag{14.1d}$$

Any rigid body motion can be described by the above matrices multiplied in the sequence of their application. For example, a translation of the object frame from F_1 to F_2 (d_x, d_y, d_z), followed by an arbitrary rotation, θ, about the new X_2 axis, would yield a new frame location, F_3, which could be followed by an arbitrary rotation, ψ, about the new Z_3 axis to yield F_4. The last object location defined by F_4 with respect to its initial location F_1 would then be defined by

$${}^1T_4 = \text{Trans}(d_x, d_y, d_z)\ \text{Rot}(X_2, \theta)\ \text{Rot}(Z_3, \psi)$$

In the case of open chain (serial) manipulators, a (rigid body) frame attached to the end of a link is moved in space by a joint located at the start of the link. We must sequentially combine the individual motions of every moving link, from the end-effector all the way to the base, in order to obtain the overall kinematic model of the robot. A commonly used notation for this purpose was developed by J. Denavit and R. S. Hartenberg in the early 1950s—now called the D–H transformation.

According to D–H notation, a rotary joint causes the following transformation to a frame attached to the end of the link it is driving (Fig. 11):

$$A_i = \begin{bmatrix} \cos\theta & -\sin\theta\cos\alpha & \sin\theta\sin\alpha & a\cos\theta \\ \sin\theta & \cos\theta\cos\alpha & -\cos\theta\sin\alpha & a\sin\theta \\ 0 & \sin\alpha & \cos\alpha & d \\ 0 & 0 & 0 & 1 \end{bmatrix} \tag{14.2}$$

where θ is the variable rotation of the ith joint about its Z-axis and (α, a, d) are constant offsets. Similarly, the following transformation matrix describes

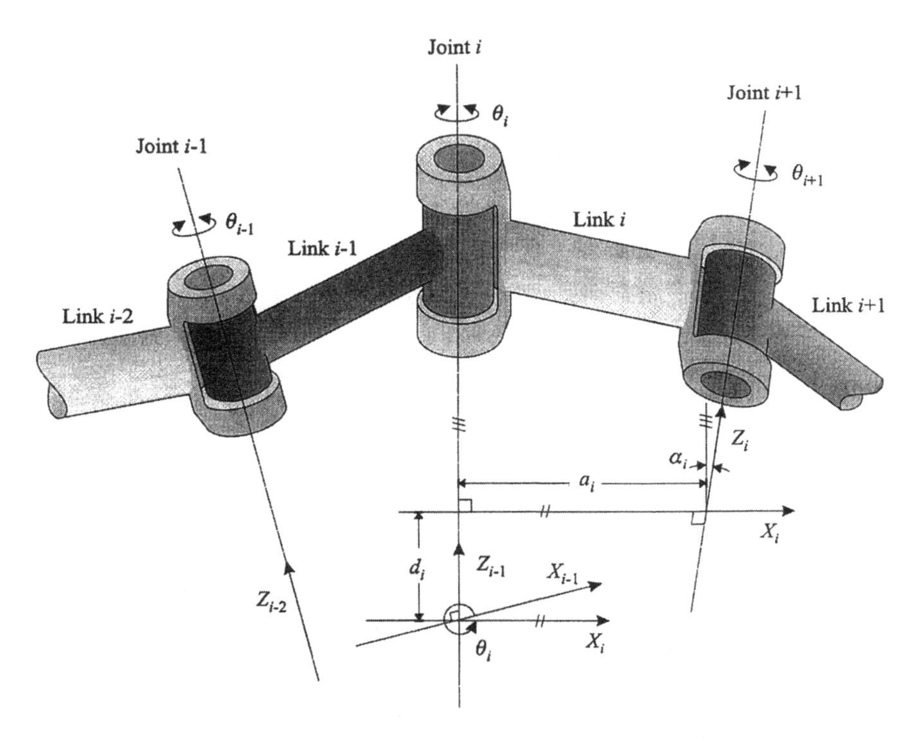

FIGURE 11 D–H notation for a rotary joint.

the displacement of a frame attached to the end of the link driven by a prismatic (linear) joint (Fig. 12):

$$A_i = \begin{bmatrix} \cos\theta & -\sin\theta\cos\alpha & \sin\theta\sin\alpha & 0 \\ \sin\theta & \cos\theta\cos\alpha & -\cos\theta\sin\alpha & 0 \\ 0 & \sin\alpha & \cos\alpha & d \\ 0 & 0 & 0 & 1 \end{bmatrix} \tag{14.3}$$

where d is the variable displacement of the ith joint along its Z axis, and θ, α are constant offsets.

For an n dof manipulator, the transformation matrix relating the motion of the end-effector, defined by the frame, F_e, with respect to a fixed

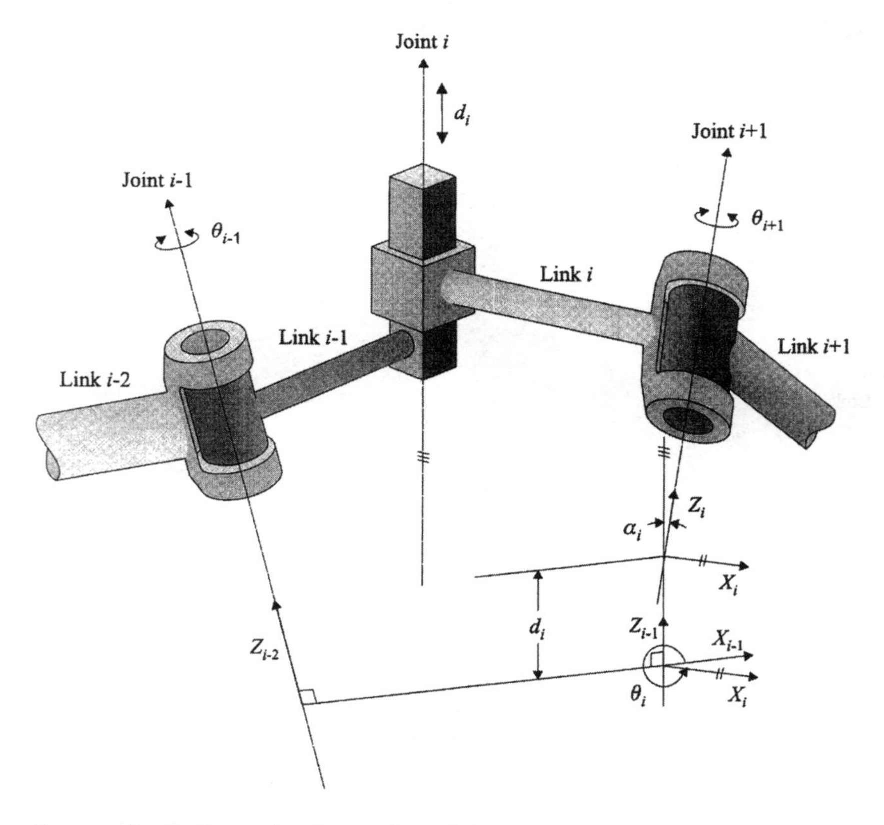

FIGURE 12 D–H notation for a prismatic joint.

frame at the base of the robot, F_0, is obtained by multiplying sequentially all the pertinent A matrices,

$$^0T_e = A_1 A_2 \cdots A_{n-1} A_n = f(\theta_i, \alpha_i, a_i, d_i) \tag{14.4}$$

where ^0T_e is a function of the joint variables and constant offsets, $i = 1, n$ (Fig. 13).

The term direct kinematics is used for serial manipulators to describe the process of obtaining the location of the end-effector, F_e, with respect to the robot's fixed base frame, F_0, by solving the Eq. (14.4) for a set of joint variable values. This answers the question, where is the end-effector, having moved the joints of the robot by arbitrary amounts? Correspondingly, the term inverse kinematics is used (for serial manipulators) to describe the process of obtaining the individual joint displacement values for a specific location of the end-effector: this process would require the inversion of Eq. (14.4). Inverse kinematics is a nontrivial process owing to the presence of inverse trigonometric terms in the kinematic models of the majority of industrial robots that employ rotary joints.

In the context of kinematics, PTP motion planning requires us to find the robot configuration, defined by the variable joint displacement values, corresponding to the location to which we want to move the end-effector, and planning appropriate joint trajectories from the current robot configuration to that point (Fig. 14a). CP motion planning, on the other hand, requires us to define a given continuous path in terms of a sufficient number of representative points, carry out inverse kinematics, just as in PTP motion planning, to determine the corresponding robot configurations, and plan

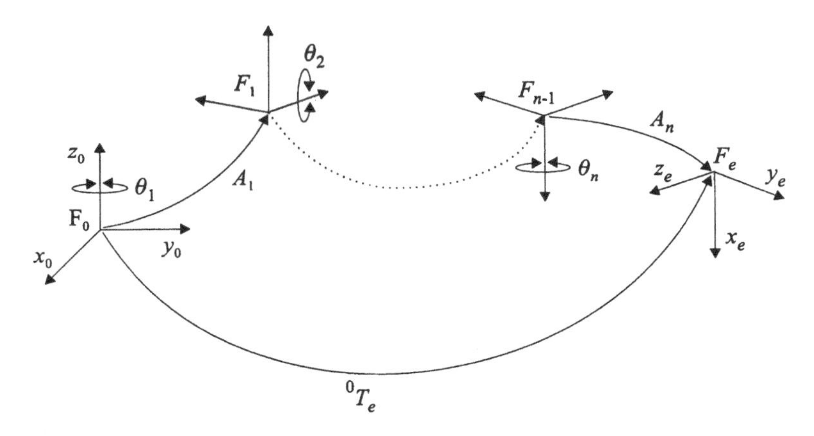

FIGURE 13 An n-dof robot transformation.

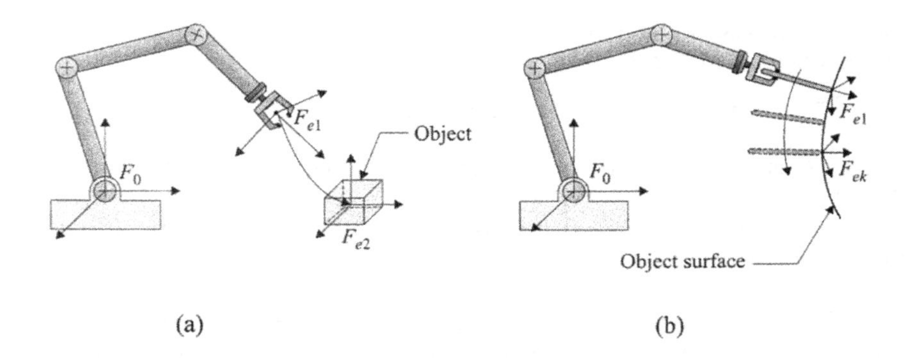

FIGURE 14 (a) PTP motion to grasp object; (b) CP motion to arc weld.

individual synchronized joint trajectories to obtain the desired (smooth) CP motion (Fig. 14b).

The kinematic model of a serial manipulator can be differentiated to yield a relationship between the joint and the Cartesian end-effector velocities. This relationship can be expressed in a matrix form as

$$\mathbf{V} = \mathbf{J}(\boldsymbol{\theta})\dot{\boldsymbol{\theta}} \tag{14.5}$$

where **V** is the Cartesian end-effector velocity vector comprising both linear and angular velocity components, $\boldsymbol{\theta}$ is the (generic) joint velocity vector, and $\mathbf{J}(\boldsymbol{\theta})$ is the Jacobian matrix expressed as a function of the instantaneous robot configuration (i.e., the joint displacement values, $\boldsymbol{\theta}$).

The (serial) robot's Jacobian matrix can also be utilized to express the relationship between joint torques/forces and the static Cartesian end-effector forces/moments:

$$\boldsymbol{\tau} = \mathbf{J}^{\mathrm{T}}\mathbf{F} \tag{14.6}$$

where $\boldsymbol{\tau}$ is the generic joint torque vector, **F** is the generic end-effector static force vector, and $\mathbf{J}^{\mathrm{T}}$ is the transpose of the robot's Jacobian matrix.

As an example, the Jacobian matrix of a SCARA robot shown in Fig. 15 is

$$\mathrm{J} = \begin{bmatrix} 0 & a_1\cos\theta_1 & a_1\cos\theta_1 + a_2\cos(\theta_1+\theta_2) & 0 \\ 0 & a_1\sin\theta_1 & a_1\sin\theta_1 + a_2\sin(\theta_1+\theta_2) & 0 \\ 0 & 0 & 0 & 1 \\ 0 & 0 & 0 & 0 \\ 0 & 0 & 0 & 0 \\ 1 & 1 & 1 & 0 \end{bmatrix} \tag{14.7}$$

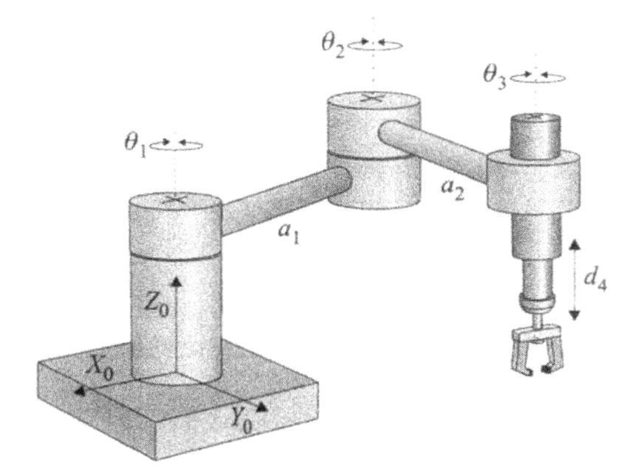

FIGURE 15 A SCARA geometry robot.

The Jacobian in Eq. (14.7) can be expressed as a square matrix by eliminating the fourth and fifth rows, which indicate that the end-effector does not have more than the only angular velocity, ω, with respect to the Z axis.

Robot Dynamics

The dynamic model of a robotic manipulator, that is, its equations of motion, can be obtained from the common laws of physics and Newtonian, Eulerian, and Lagrangian mechanics. The objective at hand is to relate the available joint torques/forces to the motion of the robot end effector. The problem is complicated due to the multibody structure of the (serial) open chain manipulator that is, the motion of the end effector is governed by the individual motions of a series of interconnected (rigid body) links, each driven by an actuator. The pertinent literature includes detailed algorithms for the real-time calculations (solutions) of the dynamic model. Herein, our focus will be only on the overall discussion of the typical robot dynamic model and not on its solution for the calculation of joint torques, for example, for a desired output (i.e., end-effector velocity, acceleration, and force output).

The two common approaches to modeling the dynamic behavior of a serial manipulator are the Lagrange–Euler and the Newton–Euler formulations. The latter method can be adapted to a recursive solution of the robot dynamics using the d'Alembert principle of equilibrium for each link individually.

The overall Newton–Euler dynamic model of the robot in joint-space coordinates can be expressed as

$$\mathbf{M}(\boldsymbol{\theta})\ddot{\boldsymbol{\theta}} + \mathbf{C}(\boldsymbol{\theta}, \dot{\boldsymbol{\theta}})\dot{\boldsymbol{\theta}} + \mathbf{G}(\boldsymbol{\theta}) = \boldsymbol{\tau} \tag{14.8}$$

where $\boldsymbol{\theta}$, $\dot{\boldsymbol{\theta}}$, $\ddot{\boldsymbol{\theta}}$, and $\boldsymbol{\tau}$ are the (generic) joint displacement, velocity, acceleration, and torque vectors, respectively, $\mathbf{M}$ is the inertia matrix, $\mathbf{C}$ is the Coriolis matrix, and $\mathbf{G}$ is the gravity vector. As is clearly apparent from Eq. (14.8), the torque requirements are a function of the instantaneous robot configuration. For example, when accelerating around its base joint, a robot would require less torque if the arm were retracted (i.e., minimum distance from the shoulder), as opposed to being fully stretched.

Trajectory Planning

As discussed above, a robotic manipulator is required to move either in PTP or in CP motion modes in task space. However, robot motion control necessitates that commands be given in joint space to yield a required task space end-effector path. We must plan individual joint trajectories—joint displacement, velocity, and acceleration as a function of time, to meet this objective. PTP and CP motions are treated separately below:

Point-to-Point Motion: In PTP motion, the robot end-effector is required to move from its current point to another: a point is defined by a Cartesian frame location (position and orientation—"pose") with respect to a global (world) coordinate system (Fig. 14a). Although the path followed is of no significance, except for avoiding collisions, all manipulator joints must start and end their motions synchronously. Such a strategy would minimize acceleration periods and thus minimize joint torque requirements—a phenomenon not considered in NC machining due to the absence of significant inertial forces.

Trajectory planning for PTP motion starts by determining the joint displacement values at both ends of the motion corresponding to the two end-effector Cartesian frames F_{e1} and F_{e2}—inverse kinematics, ${}^{1}\theta_i$ and ${}^{2}\theta_i$, $i = 1, n$, for an n-dof manipulator. The next step involves determining individual joint trajectories: a vast majority of commercial robots only utilize kinematics to determine these trajectories; only a very few utilize the dynamic models of the manipulators. We will address both approaches below.

The dynamic model of the robot, Eq. (14.8), clearly indicates that the availability of joint torques to maximize joint velocities and accelerations is a function of the instantaneous robot configuration and the geometry and mass of the object carried. In the absence of dynamic model utilization during trajectory planning, one must therefore assume some logical limits for the joint velocities and accelerations. Normally, worst-case values are utilized for these limits, i.e., assuming that the robot configuration is in its most unfavorable configuration and carrying a payload of maximum mass. Based on these limits and user-defined joint trajectory velocity profiles, trapezoidal, or parabolic, we must first determine the slowest joint, #k, i.e., the joint that will take the longest time to accomplish its motion, $\Delta\theta_k = {}^2\theta_k - {}^1\theta_k$. The time required to achieve $\Delta\theta_k$ is defined as the overall robot motion time for the end-effector to move from frame F_{e1} to F_{e2}. All other joints are slowed down to yield a synchronous motion for the robot that ends at time t_f.

Over the past three decades, many joint trajectory velocity profiles have been proposed. The two most common ones are shown in Fig. 16a and 16b for the slowest joint, θ_k.

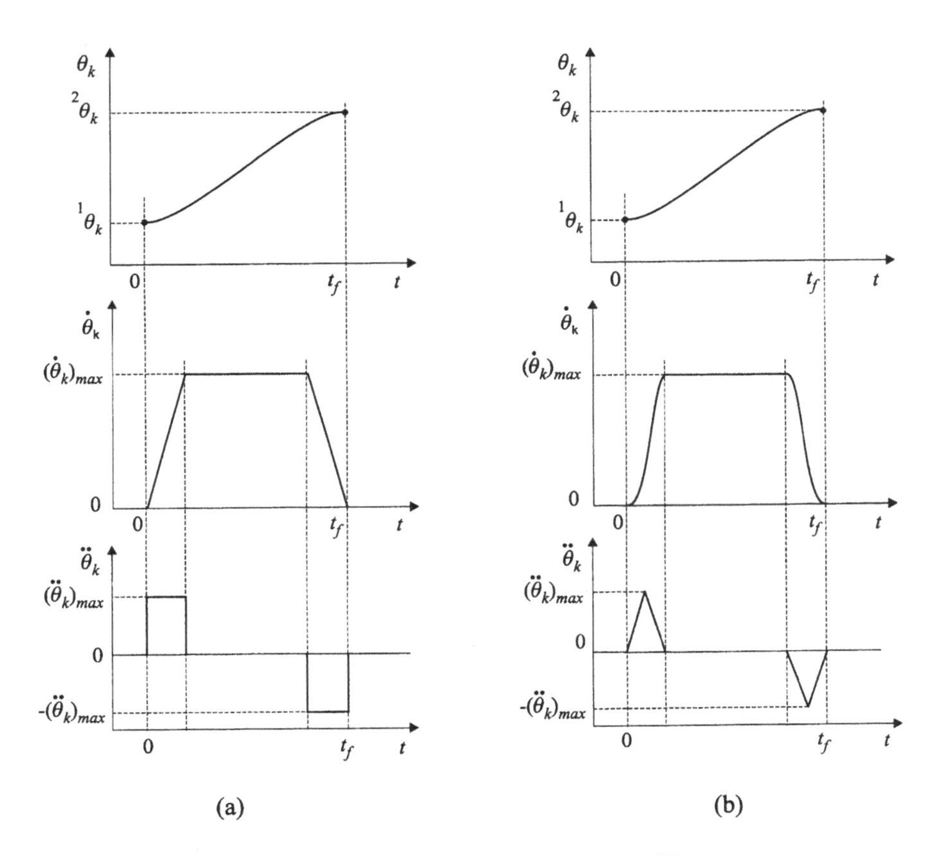

FIGURE 16 (a) Trapezoidal; (b) parabolic joint velocity profiles.

Once all joint trajectories have been planned, one may choose to carry out a computer simulation in order to determine whether the resultant PTP end-effector (Cartesian) path causes a collision. In case of a potential collision, we would need to select intermediate points to go around obstacles without actually stopping at these points. Such continuous PTP (CPTP) motion will be addressed below.

As noted above, PTP motion is normally carried at maximum joint speeds in order to minimize motion times and increase productivity. However, true minimum time PTP motion can only be obtained by considering the robot's dynamic model. Two such trajectory planning algorithms were developed in the early 1980s and have formed the basis of numerous other ones that followed them. The authors of these works were K. G. Shin and N. D. McKay from the University of Michigan at Ann Arbor and J. E. Bobrow, S. Dubowsky, and J. S. Gibson from the University of California at Irvine and Los Angeles and MIT, respectively. Both solution methods simultaneously determine the Cartesian end-effector path and corresponding joint trajectories that maximize joint torque utilization and thus minimize robot motion time subject to all robot kinematic and dynamic constraints.

Continuous Point-to-Point Motion: In CPTP motion, the robot end effector is required to move through all intermediate points (i.e., end-effector frames) without stopping and preferably following continuous joint velocity profiles. A common solution approach to this problem is to achieve velocity continuity at an intermediate point by accelerating/decelerating the joint motion prior to getting to that point and in the process, potentially, not pass through the point itself but only close to it. A preferred alternative strategy, however, would be the employment of spline curves for the individual segments of the joint trajectory.

Through an iterative process, one can fit cubic or quintic splines to all the trajectories of the robot's n joints, while satisfying displacement, velocity, and acceleration continuity at all the knots (Fig. 17). The overall motion time can be minimized through an iterative process or using parametric closed form equations, subject to all the joints' individual kinematic constraints (i.e., $\dot{\theta}_{max}$ and $\ddot{\theta}_{max}$).

Continuous Path Motion: In CP motion, the robot end-effector is required to follow a Cartesian path, normally with a constant speed (Fig. 14b). The motion (translation and rotation) of the end-effector frame is defined as a function of time. The solution of the (joint space) trajectory planning problem for CP motion requires discretization of the Cartesian path in terms of a set of points (frames) on this path separated by Cartesian distance and time. The robot can then be required to carry out a CPTP motion through these points, as described above, whose corresponding joint displacement values are determined by inverse kinematics.

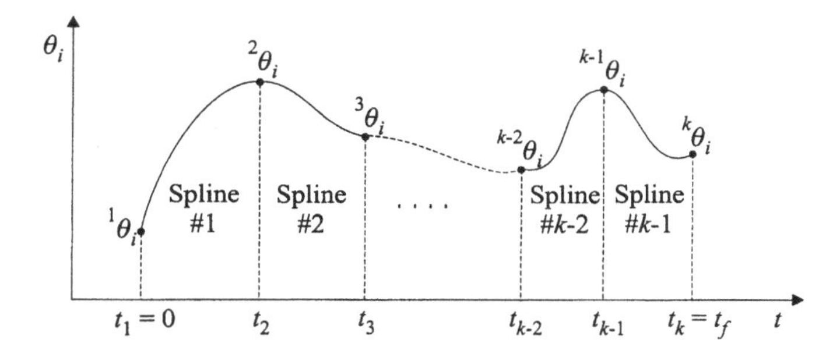

FIGURE 17 Spline fits for CPTP motion.

In CPTP motion, however, the end-effector can be forced to pass only through the selected set of points yielding a Cartesian path that approximates the desired path (e.g., a straight-line motion with constant end-effector orientation with respect to the workpiece). If the resultant path following errors in Cartesian space are greater than an acceptable threshold, we would have to increase the number of points selected to approximate the desired path.

As a generalization of CP motion for real-time implementation, one can simply use the inverse Jacobian matrix of the robot in determining joint velocities corresponding to the instantaneous end-effector velocity requirements (Eq. 14.5). This method, originally proposed by D. E. Whitney in the early 1970s, is commonly known as the resolved motion-rate-control method. If at any instant, the robot's dynamic capabilities cannot match the required end-effector motion requirements, a tracking error results.

Motion Control

Motion commands generated by a robot trajectory planning module are passed onto the individual joint controllers for their real-time implementation. As in NC machine tool control, the closed loop control of these (individual) manipulator joints yields the desired end-effector motion (Fig. 18). The majority of industrial robots employ DC (electric) servomotors for reliable displacement/velocity/acceleration/force control, though hydraulic drives have also been used in large load carrying applications (Chap. 12). The focus of this section is on the motion control of industrial robots that utilize DC servomotors.

The majority of industrial robots utilize PD or PID controllers, without relying on the existence of accurate manipulator dynamic models for trajectory tracking (Chap. 13). These controllers behave reasonably well

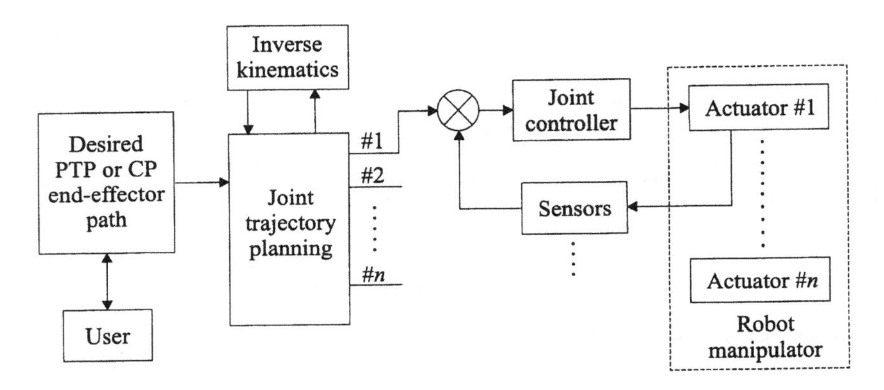

FIGURE 18 Robot motion control architecture.

for the execution of repetitive (constant load carrying and constant speed) motion tasks that do not involve (force-based) interactions with their environment. The joint acceleration commands in a PID controller can be simply calculated in the form

$$\ddot{\boldsymbol{\theta}} = \ddot{\boldsymbol{\theta}}_r + \mathbf{K}_d\dot{\mathbf{e}} + \mathbf{K}_p\mathbf{e} + \mathbf{K}_i \int \mathbf{e}\, dt \tag{14.9}$$

where the error vector term is $\mathbf{e} = \boldsymbol{\theta}_r - \boldsymbol{\theta}$ (i.e., the desired joint displacement value minus the current joint displacement value) and $\mathbf{K}_d$, $\mathbf{K}_p$, and $\mathbf{K}_i$ are the constant, diagonal (derivative, position, and integral) gain matrices, respectively.

Trajectory tracking can be significantly improved if the robot controller is provided with a reliable dynamic model. The computed torque (CT) technique is one of many control laws developed in the past two decades for the control of multilink manipulators using the robot dynamic model:

$$\boldsymbol{\tau} = \mathbf{M}(\boldsymbol{\theta})(\ddot{\boldsymbol{\theta}}_r - \mathbf{K}_d\dot{\mathbf{e}} - \mathbf{K}_p\mathbf{e}) + \mathbf{C}(\boldsymbol{\theta}, \dot{\boldsymbol{\theta}})\dot{\boldsymbol{\theta}} + \mathbf{G}(\boldsymbol{\theta}) \tag{14.10}$$

The above torque-control equation, based on the joint space dynamic model of the robot [Eq. (14.8)], can be converted into a Cartesian space form using the Jacobian matrix of the robot.

In the case of frequently varying task space operating conditions with which the above-mentioned PID or CT controllers cannot cope, users may choose to employ an adaptive control (AC) scheme. Such AC schema update the feedback gains according to the latest output measurements in joint space and/or task space coordinates.

All the above discussion focused on the position/velocity control of industrial robots carrying out simple motion tasks, such as painting, machine loading/unloading, or even spot welding. A variety of other industrial tasks, however, require the robot end-effector to exert controlled force on their environment (e.g., insertion, cutting). In order to cope with such phenomena, an industrial robot must be controlled using techniques such as impedance control or hybrid position/force control (Fig. 19). The former regulates the ratio of force to motion (i.e., mechanical impedance), while the latter decomposes the problem into two separate entities (i.e., force versus position control) and merges their solutions.

14.2.1 Robot Programming

The programming of industrial robots must be reviewed in the context of trajectory planning and control, as discussed above in Sec. 14.2.1. For PTP motion, the robot user aims at moving the manipulator from one point to another in the fastest possible manner with little regard to the actual path followed. For CP motion, on the other hand, a Cartesian path must be followed by the robot end-effector with a given velocity profile. Accordingly, one would expect to teach the robot the necessary points for PTP motion and the Cartesian path for CP motion, and instruct it, via a computer program, to execute the desired task.

In this section, robot programming will be addressed in three subsections: The first two subsections review the teaching and programming of robots in an on-line manner, which is valid for the vast majority of commercial robots, while the last subsection reviews the off-line programming of robot motions.

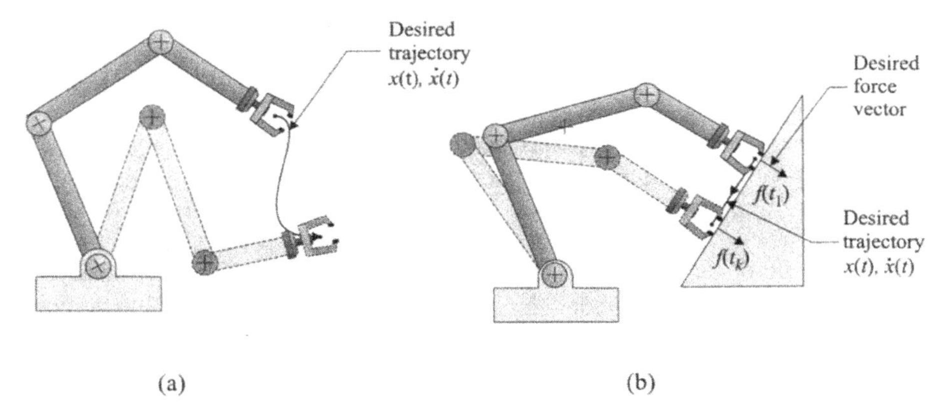

FIGURE 19 (a) Position control; (b) hybrid position/force control.

On-Line Teaching

The vast majority of commercial industrial robots are sold today without an accompanying kinematic model that would allow users to program the manipulator in an off-line manner. Furthermore, most industrial environments do not permit the specification of the exact locations (i.e., the reference coordinate frames) of the objects to be manipulated by the robot with respect to a global (world) coordinate frame. Thus the lack of a priori known precise (Cartesian) spatial relationships between a robot and objects necessitates the adaptation of teaching by showing (also known as lead-through) techniques. Naturally, such techniques are economically viable only for large batch sizes, where the cost of on-line programming (setup) of the industrial robot and, frequently, other production machines, can be divided and absorbed by the large number of (identical) parts. For small batch sizes or one-of-a-kind cost-effective manufacturing, we need to program all manufacturing equipment, including robots, in an off-line manner.

The most common method of teaching a fixed point (a Cartesian frame) is the use of a teach pendant (supplied with all commercial robots) (Fig. 20). A teach pendant allows the user to move the robot end-effector to any location within its workspace by moving the individual joints or by commanding the robot controller to move the end-effector in the Cartesian space. Cartesian space motion can be achieved either with respect to the robot's base frame or to its end-effector frame. Once a satisfactory location is obtained, the robot controller is asked to memorize this point. The majority of controllers memorize the joint displacement values corresponding to this point and not the Cartesian coordinates of the end-effector.

In regard to CP motion, the majority of commercial robots only allow straight line path following (with constant or varying end-effector orientation along the path) between two taught points; no other path geometries are permitted. Alternatively, users can chain link a large number of points for CPTP motion as an approximation of the continuous path.

For special purpose applications, such as spray painting, where one may need hundreds of points to approximate complex paths, some commercial companies provide users with a stripped-down version of the industrial robot for hand held lead-through teaching. The "slave" manipulator would be identical in mechanical configuration to its "master," except for the stripping of the high-ratio transmissions and other nonfunctional heavy components. Once the transmissions are removed, a human operator can physically hold the hand (or tool) of the robot and mimic the desired Cartesian space path, while the joint encoders/tachometers memorize the hundreds of points and time them.

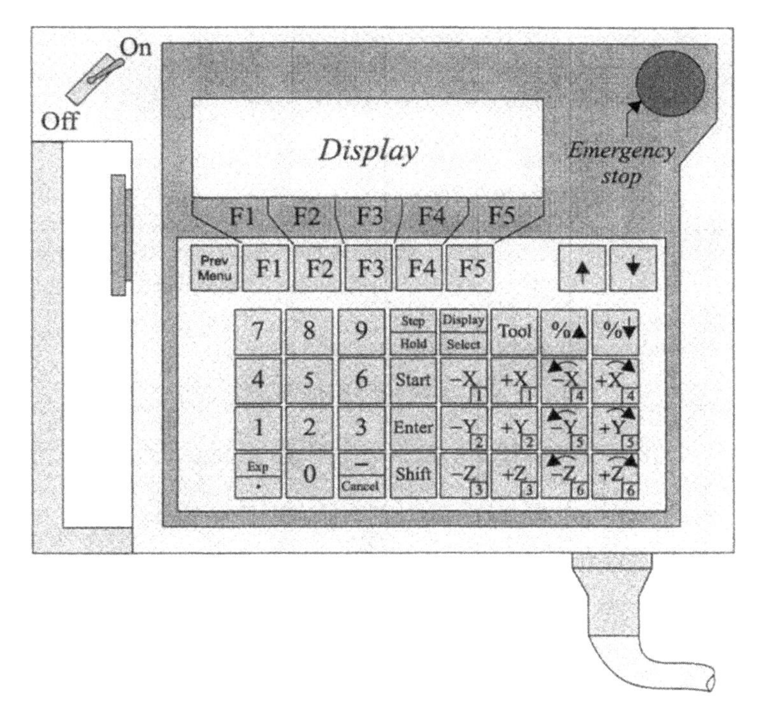

FIGURE 20 A GMF robot teach pendant.

Programming Languages

Industrial robots can be programmed to play back on-line taught tasks (PTP or CP motion). Unfortunately, over the past three decades, no single programming language has emerged and been adopted as a standard. Robot manufacturers today continue to market their hardware with accompanying proprietary software—operating system and programming language. As a consequence, there exist many robot programming languages, and industrial users must learn a variety of them if they own different makes of robots.

The majority of robot languages employ a limited number of commands (such as those of the APT language developed for machine tools): MOVE TO, APPROACH, OPEN GRIPPER, DEPART, etc. They also allow for sensory feedback using IF/THEN statements: for example, a robot end-effector can be instructed to move and grasp an object only after a positive sensory reading is obtained indicating object presence. At a higher level, a robot can also be instructed to follow a given trajectory subject to potential minor positional variations based on real-time sensory feedback: for example, in arc welding, seam tracking is achieved by receiving continuous distance measurements from a proximity sensor attached to the robot end-effector (Chaps. 12 and 13).

In all the above-mentioned cases, the robot is instructed to move between pretaught points or follow prespecified trajectories with or without sensory feedback. There can be no gross variations from planned Cartesian paths or the specified order of tasks. Computer programs can be directly input into the robot controller (by directly typing on the provided console) or, when available, prepared on an external personal computer and downloaded using a serial communication port.

Off-Line Programming

The challenge of flexible manufacturing, in robotic environments, can only be met through the use of off-line programming, that is, without interrupting the current operation of the manipulator. For highly structured environ-

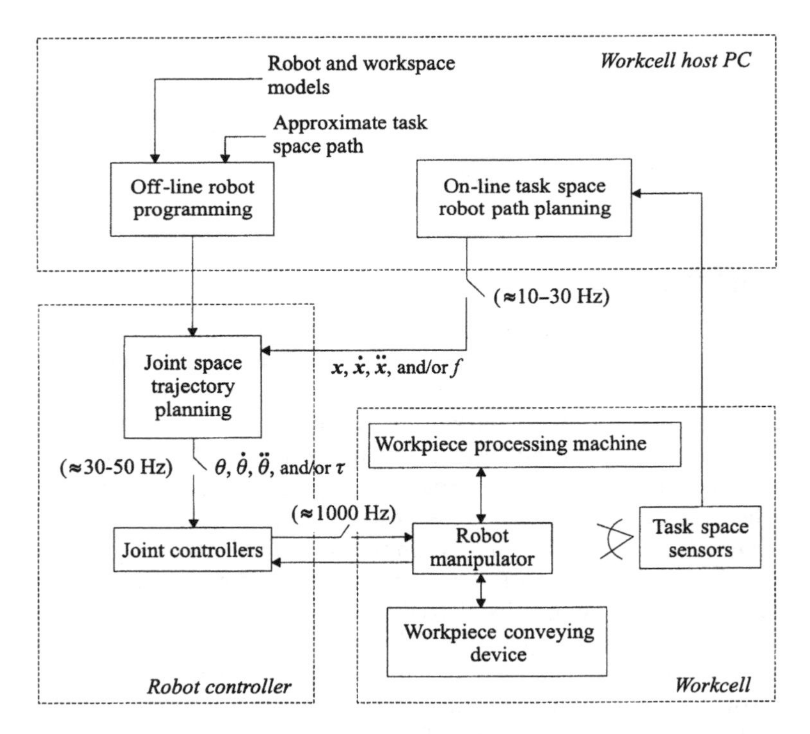

FIGURE 21 Robotic workcell planning/control.

ments, we must know the exact kinematic model of the robot as well as the geometric model of its environment in order to program off-line its motions. The commercial software CimStation Robotics, marketed by SILMA Inc. (a division of Adept Technology Inc.), allows users to create the CAD-based geometric model of their manufacturing environment and place in this environment the model of the industrial robot to be used (from an available database). Once all the motions are planned and graphically simulated, the user can ask the software for the automatic generation of the corresponding robot program (using a corresponding postprocessor, made available by SILMA) and download it to the robot's controller.

For manufacturing environments that are not highly structured (i.e., potential gross variations in object locations can be expected), no available robot programming language can be utilized for on-line or off-line programming of the industrial robot. In such cases, the robot must be programmed at the highest possible task level programming mode—for example, for the grasping of an object placed on a moving conveyor. This robot program must be supported by real-time information received from several sensors that monitor the robot's workspace and provide it with accurate and timely information regarding the object's latest state (position and velocity). Such languages and inference algorithms, although still in the research state, represent the near future of robotic implementation in loosely structured manufacturing environments. These programs/algorithms can also deal with inaccuracies in the robot's kinematic/dynamic models by directly feeding back information on the robot's end effector's (Cartesian-space) state in relation to the object's state (Fig. 21).

REVIEW QUESTIONS

1. Define the following terms as used for machine tools: numerical control (NC), computerized NC (CNC), and distributed NC (DNC).
2. Define point-to-point (PTP) motion and continuous path (CP) (contouring) motion in machining. Give specific examples.
3. Discuss adaptive control for machining. In your discussion make specific references to instrumentation and pattern recognition requirements.
4. Discuss multisensor integration and data fusion for *intelligent machining*.
5. Discuss *g*-code programming versus APT (or any other high-level language) programming of NC controllers.
6. Define the primary "words" used in *g*-code based programming.
7. Define computer-aided NC programming.
8. Discuss the objective in transforming motion commands between robot task space and joint space coordinate systems.

9. Describe the direct and inverse kinematic models for open chain serial robotic manipulators. How can one use the Denavit–Hartenberg (D-H) rigid body transformations in this context?
10. Why would one (time) differentiate the kinematic model of a robotic manipulator?
11. What is robot trajectory planning? Describe the need for using the manipulator's dynamic model in robot trajectory planning.
12. Define PTP motion and CP motion in robotics. How do these motion modes compare to their counterparts in machining? Explain similarities/differences.
13. What is trajectory tracking in robot motion control?
14. Compare robot on-line teaching to off-line programming. Discuss methods, feasibility, advantages/disadvantages, and so on.

DISCUSSION QUESTIONS

1. Discuss strategies for retrofitting an existing manufacturing enterprise with automation tools for material as well as information processing. Among others, consider issues such as buying turn-key solutions versus developing in-house solutions and carrying out consultations in a bottom-up approach, starting on the factory floor, versus a top-to-bottom approach, starting on the executive board of the company and progressing downward to the factory floor.
2. Job shops that produce one-of-a-kind products have been considered as the most difficult environments to automate, where a product can be manufactured within a few minutes or may require several days of fabrication. Discuss the role of computers in facilitating the transformation of such manual, skilled-labor dependent environments to intensive automation-based environments.
3. The factory of the future would be a totally networked enterprise. Information management in this enterprise would be a complex issue. In regards to planning, monitoring, and control, discuss the level of detail of information that the controllers (humans or computers) have to deal with in such environments. For example, some argue that in a hierarchical information management environment, activities are more of the planning type at the higher levels of the hierarchy and of the control type at the lower levels. It has also been argued that the level of detail significantly decreases as you ascend the enterprise ladder.
4. Machining centers with multitool turrets or other tool-changing mechanisms are designed to increase the operational flexibility of machine tools in terms of being able to carry out a variety of material removal

operations. Discuss the utilization of such machining centers as replacements for multiple single-objective machine tools.

5. When presented with a process planning problem for the machining of a nontrivial part, different (expert) machinists formulate different process plans. Naturally, only one of these plans is (time or cost) optimal. Considering this and other issues, compare manual (operator-based) machining versus NC-based machining, as enterprises are moving toward integrated and computerized manufacturing. Formulate at least one scenario where manual machining would be favorable.
6. An important beneficial feature of CNC machine tools is the potential of implementing an adaptive control strategy that would regulate the material removal (i.e., cutting) parameters in real time. Discuss the necessary conditions for such an implementation in terms of monitoring the material removal environment, signal processing, and decision making for making (real time) changes in input parameters.
7. Since most commercial CAD packages can automatically generate (controller specific) *g*-code programs for NC machines, discuss the value of learning a high-level NC programming language, such as APT and its variations, which would need to be subsequently also translated to *g*-code using a postprocessor (i.e., a compiler).
8. Computer-aided remote programming of NC machines is now possible through a factory's communications network. Most commercial CAD packages do allow users to plan cutting tool paths and automatically generate corresponding *g*-codes (for specific controllers) for such remote programming. However, one must note that none of these packages can optimize cutting parameters (e.g., depth of cut, feed rate). In the absence of such planners, discuss the logistics of generating tool paths automatically on a CAD workstation and the added value of this activity to the general computer-aided manufacturing planning process.
9. Tool wear can have a detrimental effect on satisfying the (geometric) dimensional specifications of a machined part, including its surface finish, especially for hard materials and complex three-dimensional surfaces. Discuss possible remedies to this problem in terms of on-line depth-of-cut compensation in turning, milling, and drilling. Address the issues of on-line sensory feedback (i.e., measurement of tool wear or object dimensions) and microscale depth-of-cut compensation using secondary (e.g., piezoceramic based) actuators (e.g., placed under the tool holder in turning).
10. The supervisory control of a manufacturing process relies on timely and accurate sensory feedback. However, owing to the difficult production conditions (high temperatures, high pressures, physical obstructions)

many production output parameters cannot be directly measured. Sensors instead observe and quantify certain physical phenomena (e.g., acoustic emissions) and relate these measurements to production output parameters (e.g., tool wear) that we desire to monitor. Discuss the following and other issues in the context of effective autonomous process control: the availability of effective signal-processing and pattern-recognition techniques, the use of multiple sensors to monitor one phenomenon (i.e., sensor fusion), the decoupling of information obtained by a sensor, whose outputs may have been influenced by several output parameters, the availability of models that could optimally change a machine input parameter in response to changes monitored in one or more output parameters.

11. In machining, tool change (because of wear) may constitute a significant part of setup time. This is especially true in multipoint cutting, such as milling, where all the inserts have to be replaced together. Almost a century-long work in the area of tool wear has yet to yield reliable models of the wear mechanisms, which would allow users to maximize utilization of the tools and thus minimize the number of tool changes. Discuss the use of a variety of sensors and pattern-recognition techniques for on-line intelligent machining in the absence of such models, or in support of approximate models.
12. Process planning in machining (in its limited definition) refers to the optimal selection of cutting parameters: the number of passes and tool paths for each pass, the depths of cut, the feed rates, the cutting velocities, and so on. It has been often advocated that computer algorithms be utilized for the search of the optimal parameter values. Although financially affordable in mass production environments, such (generative) programs may not be feasible in one-of-a-kind or small-batch production environments, where manufacturing times may be comparatively short. Discuss the utilization of group technology (GT)-based process planners in such computational-time limited production environments.
13. Analysis of a production process via computer-aided modeling and simulation can lead to an optimal process plan with significant savings in production time and cost. Discuss the issue of time and resources spent on obtaining an optimal plan and the actual (absolute) savings obtained due to this optimization, for example, spending several hours in planning to reduce production time from 2 minutes to 1 minute. Present your analysis as a comparison of one-of-a-kind production versus mass production.
14. NC machines can be efficiently programmed to execute a prepared process plan in terms of the relative motion of the tool and the

workpiece. Although this level of programmability does provide the users of such machines with automation and flexibility, the setup change requirements (e.g., workpiece fixturing) between products could negate these benefits. In this context, discuss effective ways of using NC machines in automated, flexible manufacturing environments.

15. Machining centers increase the automation/flexibility levels of machine tools by allowing the automatic change of cutting tools via turrets or tool magazines. Some machining centers also allow the off-line fixturing of workpieces onto standard pallets, which would minimize the on-line setup time (i.e., reduce downtime of the machine). While the machine is working on one part fixtured on Pallet 1, the next part can be fixtured on Pallet 2 and loaded onto the machine when it is has completed operating on the first part. Discuss the use of such universal machining centers versus the use of single-tool, single-pallet, uni-purpose machine tools.
16. Automation of materials processing or handling equipment has been often associated with increased product quality and reduced production cost. Discuss the specific benefits associated with automating material removal machines (e.g., lathes, milling machines) using NC and CNC technologies in comparison to manual machines, where the operator measures and sets the cutting parameters manually. In your discussion, also address the issues of one-of-a-kind production versus mass production and flexible manufacturing versus automated manufacturing.
17. The load carrying capacity of a spatial mechanism, for example an industrial manipulator, is a function of the path it follows and the velocity and acceleration profile of its end-effector along this path. One must consider the dynamics of the mechanism's motion when planning its end-effector's paths. This problem is complex in nature and thus rarely addressed for industrial robots. Most suppliers simply specify worst-case scenarios when defining industrial robot specifications in regard to achievable speeds and load carrying capacities. How would one deal with such a problem when integrating a robot into a manufacturing system?
18. The absence of accurate robot kinematic models, compounded with the absence of accurate geometric world models of their working environments, has often forced users to define Cartesian locations through manual teaching techniques (teleoperation). The robot end-effectors are moved to their desired destinations through teach pendants, and the controllers are required to memorize joint encoder readings at these locations. The use of such playback-based robot motion techniques thus forces objects always to be at their expected locations with very stringent tolerances. Discuss the potential of using

a variety of task-space sensors in controlling the robot motion that can lead to the relaxing of positioning requirements for objects that are static or those that are in motion.

19. Industrial robots have been often labeled as being deaf and blind operators with, furthermore, no tactile feedback detection capability. Discuss in general terms what would be the benefits of having a variety of visual and nonvisual sensors monitoring the robot's working environment and feeding back accurate and timely information to the motion controller of such manipulators.
20. Bin picking is a term used for the robotic grasping of a single component from an open bin that contains many randomly oriented, identical (and sometimes not identical) components. Discuss difficulties associated with such operations. Propose alternative solutions to robotic bin picking.
21. The necessary programming of robot task space locations by physically moving the robot's end-effector to these positions, while it is taken off the manufacturing line, has severely limited their use to mass production environments. That is, a although industrial robots provide a high level of automation, they cannot be time-efficiently programmed and used for one-of-a-kind or small batch productions. Discuss potential remedies that would allow robots to be programmed for their next task while they are performing their current task.
22. Most industrial robots need to be programmed using proprietary computer languages also developed by the makers of these robots (or their controllers). Discuss the potential negative impact such a diversification of programming languages can have on the decision-making process of purchasers/integrators/users of such machines.
23. Industrial robots have been often designed to replace the human operator in manufacturing settings. The past several decades have shown us, however, that there still exist significant gaps between humans' and robots' abilities, primarily because of the unavailability of artificial perception technologies. Compare humans to pertinent anthropomorphic robots in terms of the following and other issues: mechanical configuration and mobility, power source, workspace, payload capacity, accuracy, communications (wireless!), supervisory control ability, sensory perception, ability to process data, coping with uncertainties, and working in hazardous environments.
24. Human operators have been argued to be intelligent, autonomous, and flexible when compared to industrial robots. Discuss several manufacturing applications in which one would tend to use human operators as opposed to industrial robots (even those supported by a variety of sensors) in the context of these three properties.

25. The primary contributing factors to the achievement of high accuracies in machine tools are (1) the employment of high-precision and rigid linear actuators, (2) their Cartesian configuration, in which the linear stages are stacked on top of each other, thus avoiding significant inertia problems, and (3) the possibility of employing interferometers that can measure linear distances smaller than half a light wavelength. Discuss all three factors as you consider alternatives to the design of machine tools, for example, the use of rotary joint–based industrial robots.
26. Manufacturing systems, supported by computers, can be classified as manual versus automated and flexible (reconfigurable, reprogrammable, etc.) versus rigid. Discuss these classifications and elaborate on the intersections of their domains (e.g., manual and flexible, etc.). Note that each classification may have sublevels and subclassifications (i.e., different "levels of gray").
27. In the factory of the future, no unexpected machine breakdowns will be experienced! Such an environment, however, can only be achieved if a preventive maintenance program is implemented, in which all machines and tools are modeled (mathematically and/or using heuristics). These models would allow manufacturers to schedule maintenance operations as needed. Discuss the feasibility of implementing factory-wide preventive maintenance programs in the absence of our ability to model completely all existing physical phenomena, and furthermore in the lack of a large variety of sensors that can monitor the states of these machines and provide timely feedback to such models.

BIBLIOGRAPHY

Altintas, Yusuf. (2000). *Manufacturing Automation: Metal Cutting Mechanics, Machine Tool Vibrations, and CNC Design*. New York: Cambridge Press.

Amic, Peter J. (1997). *Computer Numerical Control Programming*. Upper Saddle River, NJ: Prentice Hall.

Angeles, Jorge. (1997). *Fundamentals of Robotic Mechanical Systems: Theory, Methods, and Algorithms*. New York: Springer-Verlag.

Armarego, E. J. A., Brown, R. H. (1969). *The Machining of Metals*. Englewood Cliffs, NJ: Prentice-Hall.

Asada, H., Slotine, J-J. E. (1986). *Robot Analysis and Control*. New York: John Wiley.

Bedworth, David D., Henderson, Mark R., Wolfe, Philip M. (1991). *Computer-Integrated Design and Manufacturing*. New York: McGraw-Hill.

Bobrow, J. E., Dubowsky, S., Gibson, J. S. (1985). Time-optimal control of robotic manipulators along specified paths. *International Journal of Robotics Research* 4(3):3–17.

Boothroyd, Geoffrey, Knight, Winston A. (1989). *Fundamentals of Machining and Machine Tools*. New York: Marcel Dekker.

Brooks, R. R., Iyengar, S. S. (1998). *Multi-Sensor Fusion: Fundamentals and Applications with Software*. Upper Saddle River, NJ: Prentice Hall.

Chang, Chao-Hwa, Melkanoff, Michel A. (1989). *NC Machine Programming and Software Design*. Englewood Cliffs, NJ: Prentice Hall.

Choi, G. S., Wang, Z., Dornfeld, D. A. (1991). Adaptive optimal control of machining process using neural networks. IEEE Proceedings of the International Conference on Robotics and Automation, Sacramento, CA (pp. 1567–1572).

Craig, John J. (1989). *Introduction to Robotics: Mechanics and Control*. Reading, MA: Addison-Wesley.

Croft, Elizabeth A. (1995). On-Line Planning for Robotic Interception of Moving Objects. Ph.D. diss., Department of Mechanical Engineering, University of Toronto, Toronto, Canada.

Croft, E. A., Fenton, R. G., Benhabib, B. (Apr. 1998). Optimal rendezvous-point selection for robotic interception of moving objects. *IEEE Transactions on Systems, Man, and Cybernetics,* Part B 28(2):192–204.

DeGarmo, E. Paul, Black, J. T., Kohser, Ronald A. (1997). *Materials and Processes in Manufacturing*. Upper Saddle River, NJ: Prentice Hall.

DeVries, Warren R. (1992). *Analysis of Material Removal Processes*. New York: Springer-Verlag.

Diei, E. N., Dornfeld, D. A. (August 1987). Acoustic emission sensing of tool wear in face milling. *Transactions of the ASME, Journal of Engineering for Industry* 109(3):234–240.

Dornfeld, D. A. (1991). Monitoring of the machining process by means of acoustic emission sensors. ASTM Symposium on Acoustic Emission: Current Practice and Future Directions, Charlotte, NC (pp. 328–344).

Doyle, Lawrence E., et al (1985). *Manufacturing Processes and Materials for Engineers*. Englewood Cliffs, NJ: Prentice-Hall.

Drozda, Thomas J., Wick Charles, eds. (1998). *Tool and Manufacturing Engineers Handbook*. Dearborn, MI: Society of Manufacturing Engineers.

Fu, K. S., Gonzalez, R. C., Lee, C. S. G. (1987). *Robotics: Control, Sensing, Vision, and Intelligence*. New York: McGraw-Hill.

Groover, Mikell P. (1996). *Fundamentals of Modern Manufacturing: Materials, Processes, and Systems*. Upper Saddle River, NJ: Prentice-Hall.

Hine, Charles R. (1970). *Machine Tools and Processes for Engineers*. New York, NY: McGraw-Hill.

Hsu, P. L., Fann, W. R. (Nov. 1996). Fuzzy adaptive control of machining processes with a self-learning algorithm. *Transactions of the ASME, Journal of Manufacturing Science and Engineering* 118(4):522–530.

Hujic, D., Croft, E. A., Zak, G., Fenton, R. G., Mills, J. K., Benhabib, B. (Sept. 1998). Time-optimal interception of moving objects—an active prediction, planning and execution system. *IEEE/ASME Trans. on Mechatronics* 3(3):225–239.

Joshi, Rajive, Sanderson, Arthur C. (1999). *Multisensor Fusion: A Minimal Representation Framework*. River Edge, NJ: World Scientific.

Kalpakjian, Serope, Schmid, Steven R. (2000). *Manufacturing Engineering and Technology*. Upper Saddle River, NJ: Prentice Hall.

Koivo, Antti J. (1989). *Fundamentals for Control of Robotic Manipulators*. New York: John Wiley.

Koren, Yoram. (1983). *Computer Control of Manufacturing Systems*. New York: McGraw-Hill.

Koren, Yoram. (1985). *Robotics for Engineers*. New York: McGraw-Hill.

Laumond, J.-P. (1998). *Robot Motion Planning and Control*. New York: Springer-Verlag.

Lewis, Frank L., Abdallah, C. T., Dawson, D. M. (1993). *Control of Robot Manipulators*. New York: Maxwell Macmillan International.

Mehrandezh, Mehran. (1999). Navigation-Guidance Based Robotic Interception of Moving Objects. Ph.D. diss., Department of Mechanical and Industrial Engineering, University of Toronto, Toronto, Canada.

Mehrandezh, M., Sela, M. N., Fenton, R. G., Benhabib, B. (1999). Robotic interception of moving objects using ideal proportional navigation guidance technique. *J. of Robotics and Autonomous Systems* 28:295–310.

Mongi, A. Abidi, Gonzalez, Ralph C. (1992). *Data Fusion in Robotics and Machine Intelligence*. Boston: Academic Press.

McMillan, Gregory K., ed. (1999). *Process/Industrial Instruments and Controls Handbook*. New York: McGraw-Hill.

Murray, Richard M., Li, Zexiang, Sastry, S. Shankar (1994). *A Mathematical Introduction to Robotic Manipulation*. Boca Raton, FL: CRC Press.

Nee, John G., ed. (1998). *Fundamentals of Tool Design*. Dearborn, MI: Society of Manufacturing Engineers.

Nof, Shimon Y., ed. (1999). *Handbook of Industrial Robotics*. New York: John Wiley.

Paul, Richard P. (1981). *Robot Manipulators: Mathematics, Programming, and Control*. Cambridge, MA: MIT Press.

Luo, Ren C., Kay, Michael G. (1995). *Multisensor Integration and Fusion for Intelligent Machines and Systems*. Norwood, NJ: Ablex.

Schey, John A. (1987). *Introduction to Manufacturing Processes*. New York: McGraw-Hill.

Sciavicco, L., Siciliano, Bruno (2000). *Modeling and Control of Robot Manipulators*. New York: Springer-Verlag.

Shiller, Z., Dubowsky, S. (1987). Acceleration map and its use in minimum-time motion planning of robotic manipulators. *ASME Proceedings of the International Computers in Engineering Conference,* New York (pp. 229–234).

Shin, K. C., McKay, N. D. (July 1985). Minimum-time control of robotic manipulators with geometric path constraints. *IEEE Transactions of Automatic Control* 30(6):531–541.

Siciliano, B., Valavanis, K. P., eds. (1998). *Control Problems in Robotics and Automation*. New York: Springer-Verlag.

Spong, Mark W., Vidyasagar, M. (1989). *Robot Dynamics and Control*. New York: John Wiley.

Spong, Mark W., Lewis, Frank L., Abdallah, Chauki T., eds. (1993). *Robot Control: Dynamics, Motion Planning and Analysis*. New York: IEEE Press.

Stephenson, David A., Agapiou, John S. (1997). *Metal Cutting Theory and Practice*. Marcel Dekker, New York.

Szafarczyk, Maciej, ed. (1994). *Automatic Supervision in Manufacturing*. New York: Springer-Verlag.

Tabarah, Edward (1993). Multi-Arm Robot Kinematics/Dynamics. Ph.D. diss., Department of Mechanical Engineering, University of Toronto, Toronto, Canada.

Tabarah, E., Benhabib, B., Fenton, R. G. (Dec. 1994). Motion planning for cooperative robotic systems performing contact operations. *ASME J. of Mechanical Design* 116(4):1177–1180.

Thyer, G. E. (1991). *Computer Numerical Control of Machine Tools*. New York: Industrial Press.

Tlusty, Jiri (2000). *Manufacturing Processes and Equipment*. Upper Saddle River, NJ: Prentice Hall.

Tsai, Lung-Wen (1999). *Robot Analysis: The Mechanics of Serial and Parallel Manipulators*. New York: John Wiley.

Valavanis, K., Saridis, George N. (1992). *Intelligent Robotic Systems: Theory, Design, and Applications*. Boston, MA: Kluwer.

Valentino, James (2000). *Introduction to Computer Numerical Control*. Upper Saddle River, NJ: Prentice Hall.

Vickers, G. W., Ly, M., Oetter, R. G. (1990). *Numerically Controlled Machine Tools*. New York: Ellis Horwood.

Whitney, D. E. (Dec. 1972). The mathematics of coordinate control of prosthetic arms and manipulators. *ASME Transactions, Journal of Dynamic Systems, Measurement, and Control*, 303–309.

Whitney, D. E. (1987). Historic perspective and state-of-the-art in robot force control. *International Journal of Robotics Research* 6(1):3–14.

Wright, P. K., Hansen, F. B., Pavlakos, E. (1990). Tool wear and failure monitoring on an open-architecture machine tool. Winter Annual Meeting of the American Society of Mechanical Engineers, Dallas, TX (pp. 211–228).

15

Supervisory Control of Manufacturing Systems

The focus of this chapter is the autonomous supervisory control of part flow within networked flexible manufacturing systems (FMSs). In manufacturing industries that employ FMSs, automation has significantly evolved since the introduction of computers onto factory floors. Today, in extensively networked environments, computers play the role of planners as well as that of high-level controllers. The preferred network architecture is a hierarchical one: in the context of production control, a hierarchical network of computers (distributed on the factory floor) have complete centralized control over the sets of devices within their domain, while receiving operational instructions from a computer placed above them in the hierarchical tree.

In a typical large manufacturing enterprise, there may be a number of FMSs, each comprising, in turn, a number of flexible manufacturing workcells (FMCs) (Fig. 1). These FMCs will be connected via (intercell) material handling systems such as automated guided vehicles (AGVs) and conveyors (Chap. 12).

FMCs have been, commonly configured for the fabrication and/or assembly of families of parts with similar processing requirements. A traditional FMC comprises a set of programmable manufacturing devices with their own controllers that are networked to the FMC's host computer for the downloading of production instructions (programs) as well as to a supervisory controller for the autonomous control of parts flow (Fig. 2).

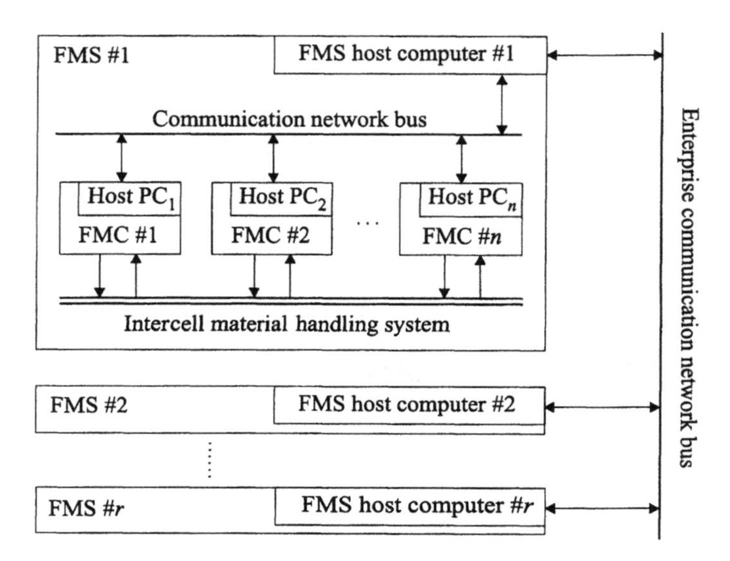

FIGURE 1 A networked manufacturing environment.

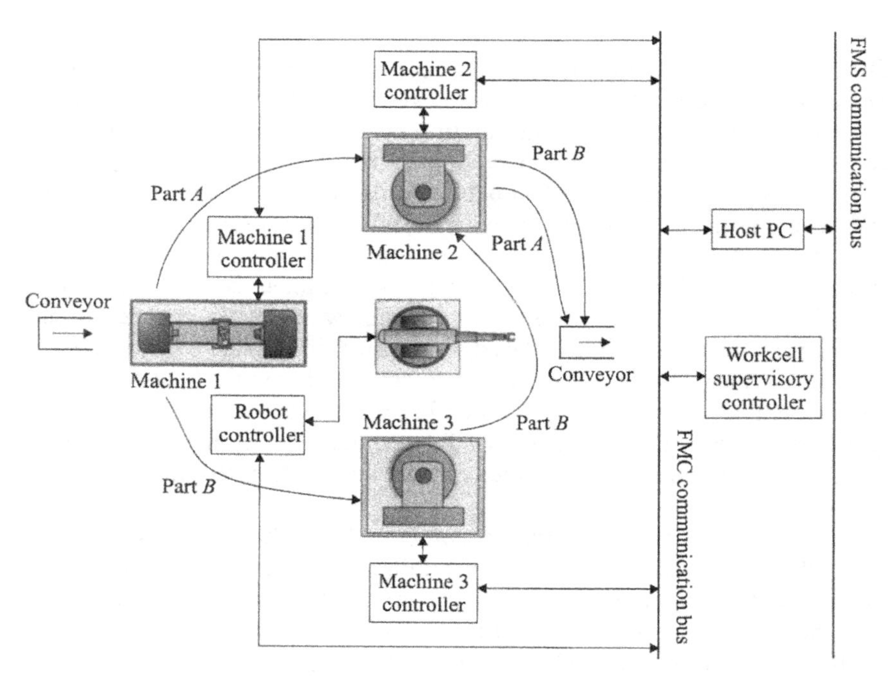

FIGURE 2 A flexible manufacturing workcell.

Although human operators have been traditionally used in the past century in the traffic control of part movements on the factory floor, personal computers (PCs) and programmable logic controllers (PLCs) have been replacing them since the early 1980s at a rapid pace. Such autonomous traffic controllers can be programmed with high-level instructions to make (correct) decisions in fractions of a second and communicate these decisions to the individual FMC devices with no delays. In turn, these devices can carry out their expected tasks as preprogrammed in their respective controllers, which will have been downloaded a priori or on-line from the host PC of the FMC. An FMC "supervisor" initiates/terminates device operations, though it does not interfere with the accomplishment of these tasks.

In contrast to time-driven (continuous variable) control of the individual devices in an FMC, the supervisory control of the FMC is event driven. The future actions of the FMC are solely dependent on the past events, as opposed to being clock driven. Thus manufacturing systems can be considered as discrete event systems (DESs) from a supervisory control perspective. DESs (also known as discrete event dynamic systems, DEDSs) evolve according to the (unpredictable) occurrence of events that are instantaneous, asynchronous, and nondeterministic.

The state of a DES changes in a deterministic manner based on the physical event that has just been observed, but the system overall is nondeterministic, since in any one state there may be several possible routes of actions ("enabled" events) that can take place. Nondeterminism implies that we may not know a priori which event (among the several possible) will take place, though once observed, this event can lead to only one future state of the DES (i.e., deterministic transition). For example, when a machine is working (state = Working), it may either complete its operation (event = Task completion) or break down (event = Failure), we do not know in advance which one will happen. However, we do know that the former will take the machine to its "Idle" state and the latter will take the machine to its "Down" state.

There exist three interested parties to this practical and very important manufacturing problem: users, industrial controller developers, and vendors and academic researchers. The users (customers) have been always interested in controllers that will improve productivity and impose minimal restrictions. Effective (supervisory) controllers are necessary for them to implement existing flexible manufacturing strategies. Industrial controller vendors have almost exclusively relied on the marketing of PLCs in the past two decades in response to the control needs of FMSs. Their efforts have largely concentrated on hardware improvements and better user interfaces, though continuously lagging behind developments by the PC industry by several years. The programming of PLCs must still be carried out in ad hoc manner (versus mathematical formalism), and thus it is prone to human error.

The academic community has spent the past two decades developing very effective formal control theories that are suitable for the supervisory control of manufacturing systems. Control strategies determined by invoking any one of these theories can be software coded and downloaded onto a PC or PLC for real-time (DES) control of limited-size FMCs. Naturally, although the successful control of such manufacturing systems have been shown in academic laboratory settings, appropriate software tools must be developed by current industrial controller developers/vendors prior to their adoption by the users (i.e., the manufacturing industry).

In this chapter, we will address two of the most successful DES control theories developed by the academic community: Ramadge–Wonham automata theory and Petri-nets theory. As proposed in Fig. 3, it is expected

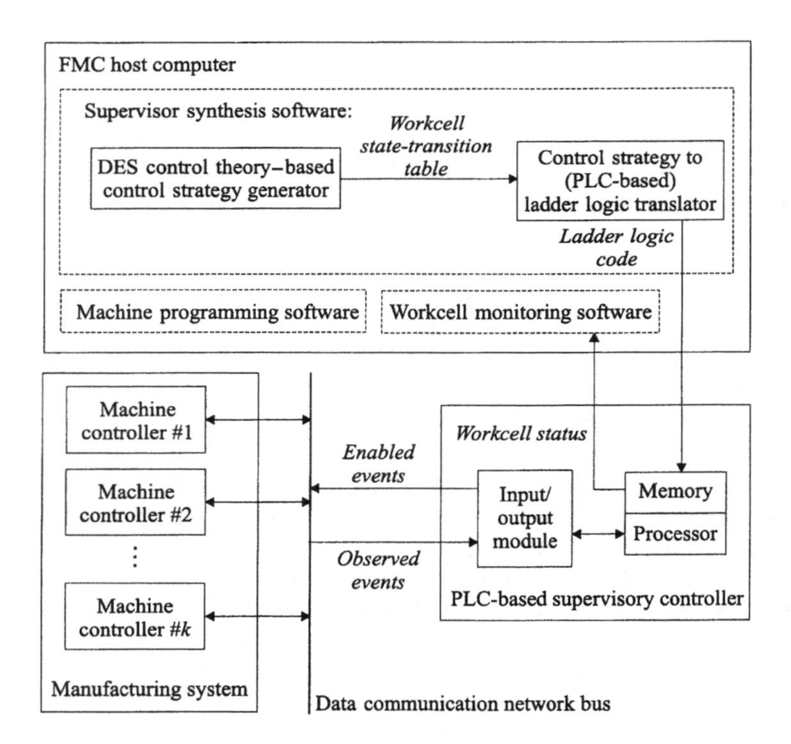

FIGURE 3 Software architecture for FMC control.

that in the future industrial users will employ such formal DES control theories in the supervisory control of their FMCs. The description of PLCs, used for the autonomous DES-based supervisory control of parts flow in FMCs, concludes this chapter. In Fig. 3 the term ladder logic refers to the programming language used by most current PLC vendors.

15.1 AUTOMATA THEORY FOR DISCRETE EVENT SYSTEM MODELING

Automata theory generally refers to the study of the dynamic behavior of information systems that can be described by a finite number of states and with discrete inputs and outputs. Although our focus in this chapter is on manufacturing systems, the field of automata theory was originally developed in response to the needs of computer science. It is of interest to note, however, that the first published work in the field of finite-state systems ("machines") by A. M. Turing in 1936 preceded all (digital) computers.

Significant advancements in the field of automata were reported in the 1950s and the early 1960s in the works of N. Chomsky, G. H. Mealy, and E. F. Moore. The application of automata theory to the supervisory control of manufacturing systems, though, was made possible only after the pioneering works of P. J. G. Ramadge and W. M. Wonham in the late 1980s (today known as the R–W theory). Thus, in this section, following a brief background review on the theories of languages and automata, we will present an overall description of the R–W theory.

15.1.1 Formal Languages and Finite Automata

Automata theory deals with systems whose dynamics is dependent on the occurrence of events that cause the system to change its state. Abstract algebra is an essential tool in the modeling and analysis of such DESs, in contrast to the use of differential calculus in time-varying systems.

Sets: A set is a collection of elements with a common property:

$$S = \{s \mid s \text{ has property } P\} \text{ or } \quad s \in S$$

Most common operations on sets include

Union (sum): $A \cup B = \{a \mid a \in A \text{ or } b \mid b \in B\}$.

Intersection: $A \cap B = \{a \mid a \in A \text{ and } b \mid b \in B\}$.

Cartesian product: $A \times B = \{(a, b) \mid a \in A, b \in B\}$.

For example, let $A = \{\alpha, \beta\}$ and $B = \{\gamma, \delta\}$, then

$$A \cup B = \{\alpha, \beta, \gamma, \delta\} \qquad A \cap B = \phi$$
$$A \times B = \{(\alpha, \gamma), (\alpha, \delta), (\beta, \gamma), (\beta, \delta)\}.$$

(The elements of $A \times B$ are termed as "ordered pairs").

Mapping: f: $A \rightarrow B$; the function, f, maps the elements of A into B.

For example

$$f(\alpha) = \delta \quad \text{and} \quad f(\beta) = \gamma \quad \alpha \in A \text{ and } \delta \in B.$$

Combinational logic: Logic elements can be used to perform logical operations on multiple inputs in order to yield a desired output. In binary-valued logic, the two most commonly used operations are AND and OR:

Input		Output, y				
x_1	x_2	AND	OR	NAND (not AND)	NOR (not OR)	Exclusive OR
0	0	0	0	1	1	0
0	1	0	1	1	0	1
1	0	0	1	1	0	1
1	1	1	1	0	0	0

The "not" operation, also known as the complementation, negates the value of the output (0 to 1, or 1 to 0). Although the above table only shows two input variables for clarity of discussion, there may be multiple input variables (≥ 2), on which the logical operations would be applied in the same manner.

Languages: In a DES, the set of all possible events can be considered as the alphabet, E, from which sequences of events, strings or words, can be generated. An (artificial) language is a collection set of strings (events). For example, for $E = \{\alpha, \beta, \gamma, \delta\}$, a language could be $L = \{\alpha\beta, \alpha\gamma\delta\}$.

Finite automata: A finite automaton comprises a finite set of states and a set of transitions (events) that occur according to the alphabet of the DES. Finite automata are also known in the literature as finite-state machines describing the dynamics of sequential machines (i.e., DESs). Automata are also considered as generators of languages according to well-defined rules. Formally, a finite-state automaton (FA) is defined by a quintuple,

$$\text{FA} = (S, E, f, s_0, F)$$

where S is a finite (nonempty) set of states, E is a finite (input) alphabet (events), f is a state-transition (mapping) function, f: $S \times E \rightarrow S$, s_0 is the initial state, $s_0 \in S$, and F is the set of final states, $F \subseteq S$.

For example, let us consider the finite automaton, M, shown in Fig. 4, where $S = \{s_0, s_1, s_2\}$, $E = \{0, 1\}$, $F = \{s_0\}$ and

$$f(s_0, 1) = s_2 \qquad f(s_1, 1) = s_0 \qquad f(s_2, 1) = s_1$$
$$f(s_0, 0) = s_1 \qquad f(s_1, 0) = s_2 \qquad f(s_2, 0) = s_0$$

In Fig. 4, the initial state is marked by an arrow labeled "start" and the final state is marked by two concentric circles. An input sequence (string) of $w = 000$ into M would yield the state s_0, $w = 00100$ would also yield s_0, etc.

A string w is said to be "accepted" by a FA, if $f(s_0, w) = p$, where $p \in F$. The language accepted by the FA, L(FA), is the set of all (accepted) strings satisfying this condition.

There exist two common finite-state machines with user-specified outputs at all of their states: Moore and Mealy machines. In Moore machines, the output at a specific state is defined regardless of how that state has been reached, while in Mealy machines, the output is dependent on the state as well as how it has been reached (i.e., the specific input transition to this state). Typical Moore and Mealy machines are given in Fig. 5a and Fig. 5b, respectively.

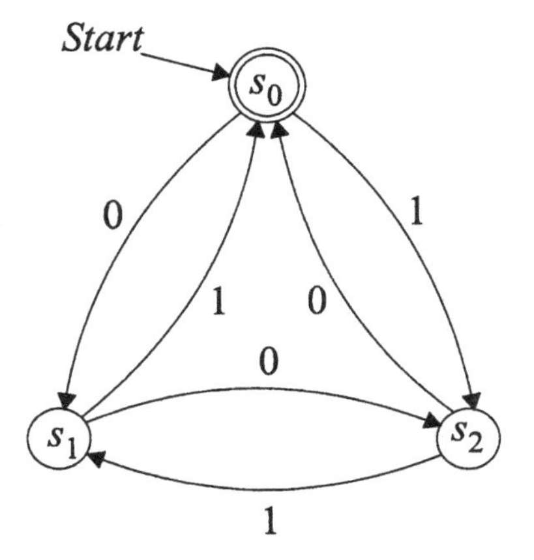

FIGURE 4 A finite-state automaton.

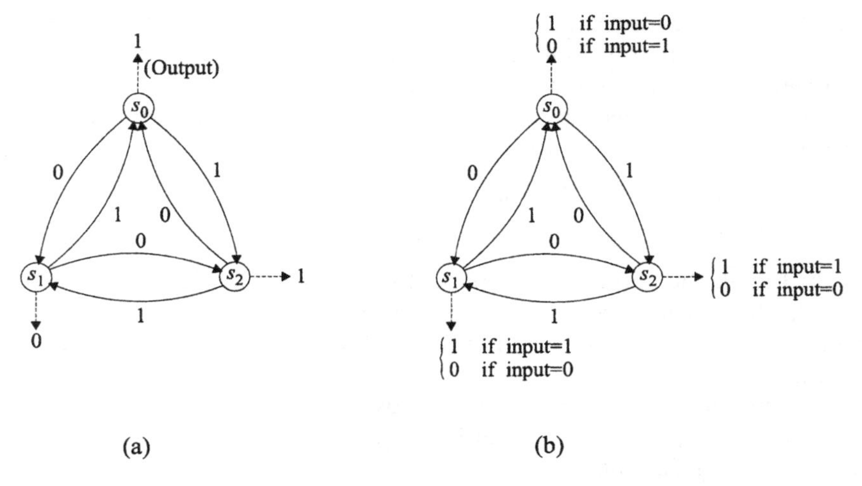

FIGURE 5 (a) A typical Moore machine; (b) a typical Mealy machine.

Formally, both Mealy and Moore machines are defined by a sextuple,

$$M = (S, E, O, f, g, s_0)$$

where S is a finite set of states, E is a finite input alphabet, O is a finite output alphabet, f is a state-transition function, g is output (mapping) function and s_0 is the initial state. In Mealy machines g is a function of the input as well, $g(s,e)$, $e \in E$. For example, in Fig. 5b, $g(s_0,1) = 0$, $g(s_0,0) = 1$, $g(s_1,1) = 1$, etc. Thus an input sequence of $w = 0011$ would yield an output of 1 in the Moore machine, while it would yield an output of 0 in the Mealy machine.

15.1.2 Ramadge–Wonham Supervisory Control Theory

Supervisory control of a DES, in the context of finite-state automata theory, can loosely be defined as the enablement (or disablement) of events at the latest reached state of the system. That is, a supervisor (a finite-state automaton) changes its state according to the latest event observed within the DES and informs the (controlled) DES what future events are enabled (or disabled). (Fig. 6). Naturally, only a subset of all events (defined in the alphabet, E) are controllable and only they can be enabled/disabled. For example, the start of an operation is a controllable event, whereas a breakdown event is uncontrollable by the supervisor.

The Ramadge–Wonham (R–W) controlled automata theory allows users to synthesize supervisors that are correct by construction. That is, all the system states within the supervisor are reachable through a

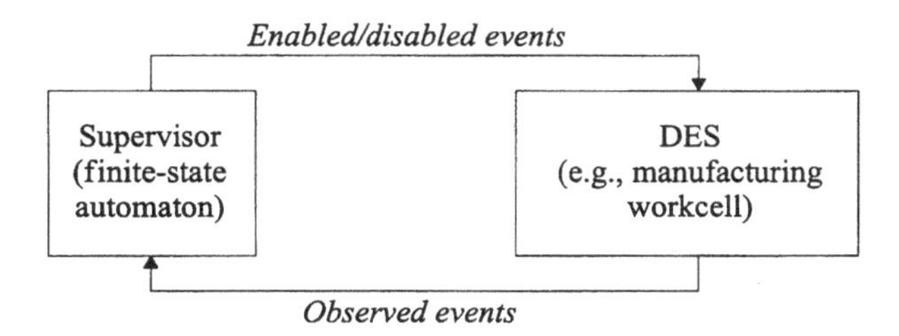

FIGURE 6 Supervisory control of a DES.

sequence of events (strings) included in the ("supremal-controllable") language of the automaton—a deadlock-free controller. Prior to the description of the controller synthesis process, the fundamentals of R–W (DES modeling) theory will be briefly described here. For consistency with the existing literature, the nomenclature introduced by Ramadge and Wonham will be utilized.

The R–W finite-state automaton, G, is defined by a quintuple,

$$G = (Q, \Sigma, \delta, q_0, Q_m)$$

where Q is the finite set of states, Σ is the finite alphabet of events, δ: $Q \times \Sigma \rightarrow Q$ is the (one-to-one mapping) function defining the transition between states according to observed events, q_0 is the initial state, $q_0 \in Q$, and $Q_m \subseteq Q$ is a subset of marker (completed task) states. A transition event is formally defined as a triple (q, σ, q'), where $\delta(\sigma,q) = q'$, for $\sigma \in \Sigma$ and q, $q' \in Q$.

The alphabet of events, Σ, is further partitioned into two disjoint subsets of controllable, Σ_c, and uncontrollable, Σ_u, subsets, where $\Sigma_{\hat{c}} \cup \Sigma_{\hat{u}}$. In an automaton, controllable events can be enabled (shown by a "tick" across the transition line in a directed graph), while uncontrollable events can be observed but not enabled or disabled. Fig. 7 illustrates a model of a machine with three states (idle, I, working, W, down, D) and four events (start to operate, α; finish, β; breakdown, λ; get repaired, μ), of which the breakdown and finish events are not controllable.

An automaton, G, is said to be nonblocking (deadlock free) if the language $L(M)$ includes the marked language accepted by M. The marked language, L_m, includes all strings that commence and terminate at the automaton's marker states (e.g., state I in Fig. 7). If the language, L, includes a string that leads to a nonmarker state with no controllable or uncontrollable event exiting it, then the DES is deadlocked at this state. Such (deadlock) states are labeled as not reachable and/or coreachable in R–W theory.

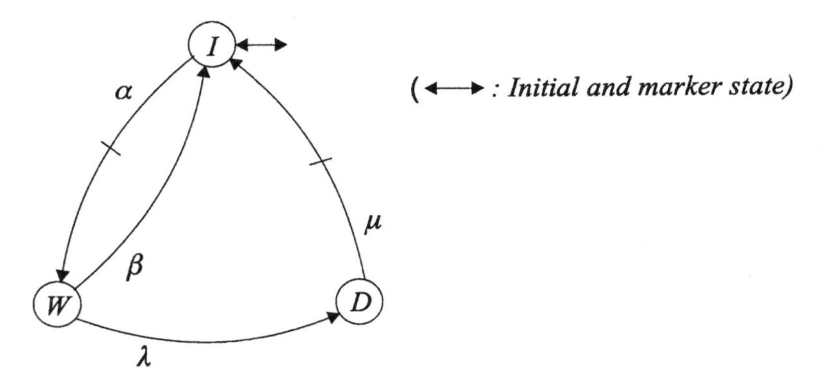

FIGURE 7 A (finite-state) automaton model for a machine.

The synthesis of a (controllable) supervisor is a two-step procedure: first, all the automata representing the individual machines of a DES are combined into one overall (uncontrolled) system automaton through a "shuffle" operation, while in parallel all automata representing the control specifications of this system are combined through a "meet" operation into one overall specifications automaton; second, the intersection of the languages of these two (system and specification) automata is obtained through a meet operation to determine the supremal- controllable language of the supervisor. This procedure is illustrated below through a simple manufacturing workcell example—two machines with a buffer of capacity one in between:

Shuffle operation: The shuffle operation (also known as the synchronous product) of two languages, $L_1||L_2$, yields a language comprising all possible interleavings of the strings of L_1 with those of L_2. The shuffled automaton of two machines, shown in Fig. 7, is given in Fig. 8. All shown system states (*II*, *WI*, *DI*, etc.) refer to the individual states of the two machines. For example, *IW* implies that the first machine, M_1, is idle, while M_2 is working. The indices of the events correspond to the machine numbers, $i = 1, 2$.

Meet operation: The meet operation applied on two languages yields their intersection, namely, a language comprising all the strings accepted by both their automata, $L = L(G_1) \cap L(G_2)$. As an example, the meet operation is applied herein on the (uncontrolled) system automaton shown in Fig. 8 and the control specification automaton shown in Fig. 9. This workcell specification does not allow M_1 to start operating unless the buffer, B, is already empty (preventing overflow) and does not allow M_2 to start operating unless the buffer contains a part that can be drawn by M_1

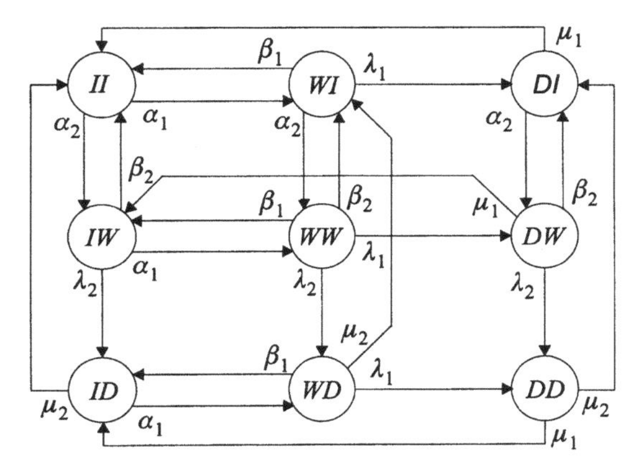

FIGURE 8 A shuffled automaton.

(preventing underflow). The resulting controllable supervisor for the specific M_1—B—M_2 DES is given in Fig. 10.

As shown in Fig. 10, the finite-state automaton (supervisor) of the overall manufacturing workcell, *SUP*, has 12 states and 25 transitions. The supervisor is nonblocking (deadlock-free) by construction. It enables controllable events and changes states by the observation of both controllable and uncontrollable events. A system state (label) in Fig. 10 is the

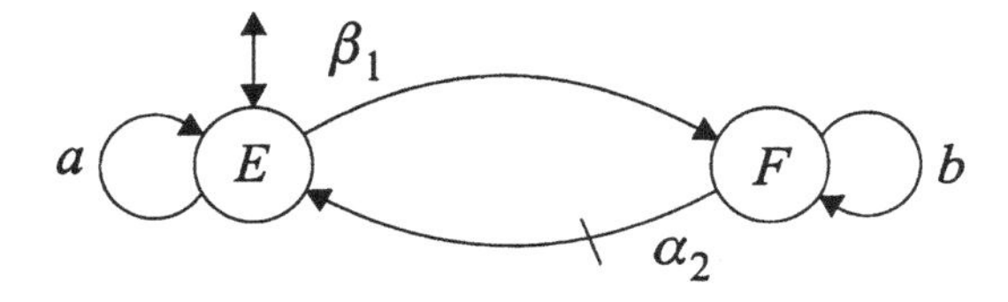

Self-loop-a: $\alpha_1, \lambda_1, \mu_1, \beta_2, \lambda_2, \mu_2$

Self-loop-b: $\lambda_1, \mu_1, \beta_2, \lambda_2, \mu_2$

E: Empty; *F: Full*

FIGURE 9 A control specification automaton, *B*.

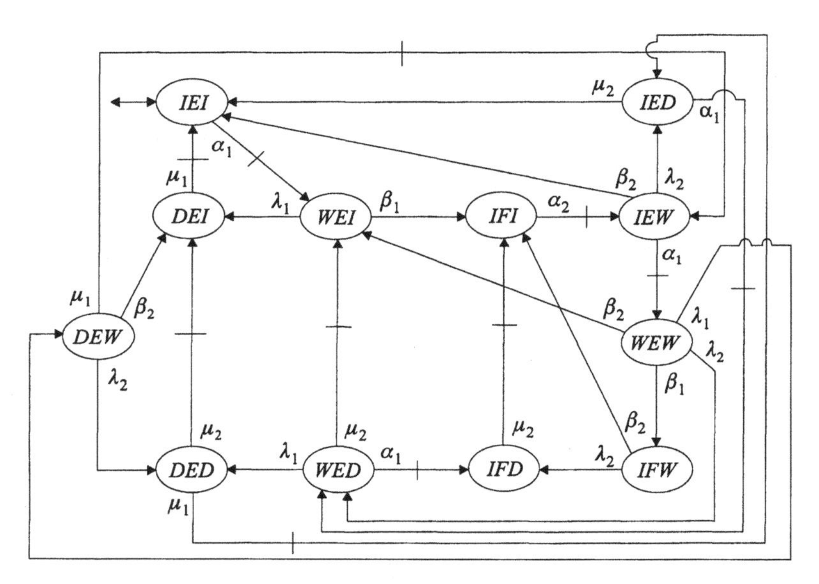

FIGURE 10 Supervisor, SUP, automaton.

concatenation of the individual states of the devices. For example, *IEI* refers to the machines, M_1 and M_2, being idle and the buffer, *B*, being empty.

15.2 PETRI NETS

Petri nets (PNs) provide engineers with a mathematical formalism for the modeling and analysis of DESs, such as manufacturing systems. They provide a simpler alternative to automata theory for the graphical representation of parts flow in a manufacturing system in terms of (system) states and transitions (events). (This graphical representation can be expressed by a set of linear algebraic equations.) However, the academic community has yet to illustrate clearly whether the formalism of PNs is superior to that of automata theory. In this section, we will only discuss the fundamentals of PNs and refrain from declaring a winner.

PNs were originally developed in the late 1950s and early 1960s by C. A. Petri. Petri's Ph.D. dissertation on the use of automata for the modeling and analysis of communications (events) within computer systems was published in 1962 in the Federal Republic of Germany. The use of PNs in manufacturing system modeling, however, started only in the early 1980s, coinciding with the start of the widespread use of computers in manufacturing planning and control activities.

Since the 1980s, significant advancements have been reported by the academic community in the use of PNs for queuing simulations (performance analysis), scheduling, and supervisory control of manufacturing systems using ordinary (event-based) PNs, timed PNs (stochastic or deterministic), and "colored" PNs, where colors (differentiators) are used for the modeling of a number of different parts within a PN. However, except for an isolated success in developing a PN-based programming language (GRAFCET) for sequential logic controllers, the implementation of PNs in industrial manufacturing environments has been sparse.

Our focus in this book will be on the modeling of manufacturing systems using deterministic (versus stochastic), nontime (versus timed), ordinary (versus colored) PNs. Furthermore, the emphasis will be on the potential use of PNs for the supervisory control of manufacturing systems (versus their performance evaluation).

15.2.1 Discrete Event System Modeling with Petri Nets

PNs allow engineers to model asynchronous (event-driven) manufacturing systems, with concurrent operations and shared resources, by formalizing precedence relations. A PN is a directed bipartite graph comprising nodes, places, and transitions joined by directed arcs. Places (states) are represented by circles and transitions (events) by bars/rectangles.

The dynamics of a PN is achieved by tokens that are moved from one place to another by a transition connecting them. A transition can be weighted to transfer multiple tokens at one instance. (For example, a transition can cause two tokens to leave a place, but arrive at the next place as only one token.) The marking of a PN is an n-component vectorial representation of the number of tokens stored in each of its places. An example PN with its initial marking, $m_{\hat{0}} = (3,1,1)$, is shown in Fig. 11. For ordinary PNs all the weights are equal to 1.

Formally, a marked PN can be represented by a quintuple,

$$PN = \{P, T, I, O, m_0\}$$

where $P = (p_1, p_2, \ldots, p_n)$ is a finite set of places, $T = (t_1, t_2, \ldots, t_p)$ is a finite set of transitions, I is an input function representing all directed arcs from P to T, $P \times T$, O is an output function representing all directed arcs from T to P, $T \times P$, and m_0 is the initial marking. Both I and O can be expressed as (incidence) matrices, whose elements are 0 or 1 for ordinary PNs representing the absence or presence of a joining arc, respectively.

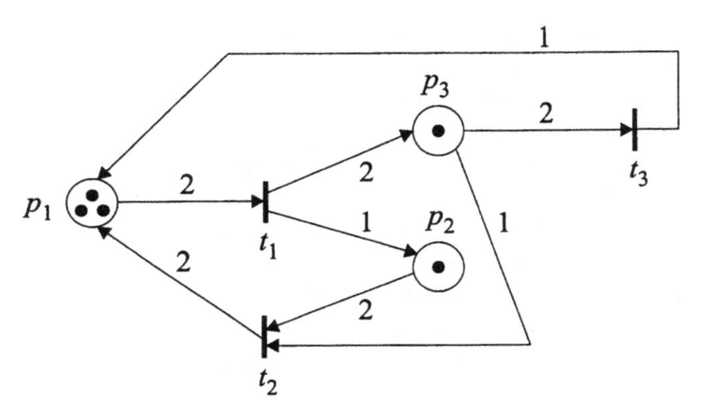

FIGURE 11 A marked Petri net.

A transition t is enabled if all places connected to it by input arcs contain tokens in numbers equal to or greater than the weights attached to the arcs. An event within the modeled system causes the corresponding transition to "fire." A fired transition causes transfer of tokens between places according to the specific weights. For example, in Fig. 11, the firing of transition t_1 yields the following marking: $m_1 = (1,2,3)$. A sequence of transitions, for example, $\sigma = \langle t_1, t_2, t_3, t_1 \rangle$, takes the same PN from its initial marking $m_0 = (3,1,1)$ to $m_4 = (2,1,2)$.

A transition without an input place is called a source and is always enabled. Similarly, a transition without an output place is called a sink that can be fired for the pure removal of tokens from the PN when enabled (Fig. 12). A self-loop is a circular representation of one place and one transition connected by an input as well as an output arc (Fig. 12). For example, a self-loop used in the modeling of a production machine would not allow the start of a new operation until the current operation is concluded.

Properties of PN Models

The properties of PNs can be classified as behavioral and structural. The former depend on the structure and the initial marking of the PN, while the latter depend only on the structure of the PN. Here we review several PN properties pertinent to manufacturing systems.

Reachability: A PN marking, m_k, (i.e., a specific system state) is said to be reachable if there exists a sequence of transitions, σ, that leads from m_0 to m_k. The (behavioral) reachability property of a PN can be analyzed by generating the corresponding reachability tree/graph, starting from the initial marking, m_0. In order to limit the size of the tree, markings (states),

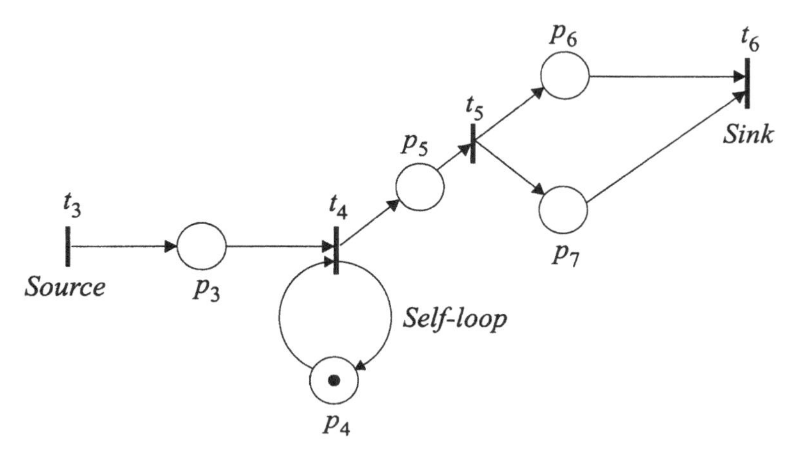

FIGURE 12 An example of an ordinary PN with a self-loop (all weights are 1 and thus not shown).

reached by (random) firing of transitions (events), that have already been noted as an earlier-encountered node on the tree branch from m_0, are labeled as old. No further transitions are fired from old markings. This elimination of duplicate markings result in a more compact coverability tree, which is equivalent to the reachability tree. In generating a reachability/coverability tree, one must note that a PN's marking can be changed by the simultaneous firing of multiple enabled transitions, as opposed to sequential firing.

Boundedness: Given the reachability set of all possible markings, a place, p_i, is l-bounded if it receives a maximum number of l tokens. The number l may or may not need to be a function of the initial marking. In manufacturing applications, boundedness can define the necessary capacity of a buffer or show its overflow. If the place examined is a machine, the term safeness is used to indicate a boundedness of $l = 1$, (i.e., only one operation at a time is allowed on that machine).

Liveness: A transition, t, is live if at any marking defined by the reachability tree there exits a sequence of subsequent transitions, σ, whose firing will lead to a marking that will reenable it. The PN is live as a whole if all of its transitions are live, i.e., the system is free of deadlock. A transition, t, is dead at a specific marking (also called dead marking) if there exits no subsequent sequence of transitions, σ, that will reenable it. A PN may have multiple dead markings, i.e., deadlock states. In the most common deadlock situations, called circular waiting, two or more processes, arranged in a circular closed-loop chain, each wait for resource availability next in the

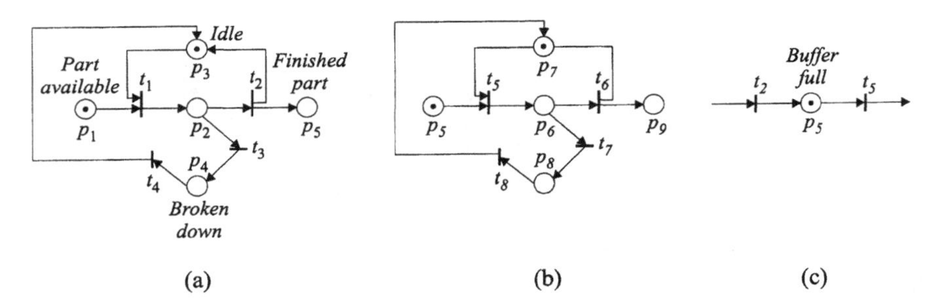

FIGURE 13 (a) PN model for M_1; (b) PN model for M_2; (c) PN model for B.

chain. A possible solution to such a practical problem is the utilization of buffers with sufficient storage capacity.

15.2.2 Synthesis of Petri Nets

The modeling of multiresource DESs, such as manufacturing workcells, can be carried out either by modeling the system as a whole or by modeling the individual resources first and then connecting them using a synthesis method. There are two primary PN synthesis methods: bottom-up and top-down.

A typical bottom-up approach would connect (live and bounded) multiple individual PNs into a larger system PN by merging common places into a new place. An alternative bottom-up approach would connect simple elementary paths shared by the individual PNs: for example, merging common paths terminated on both ends by a transition or by a place. A PN for a manufacturing line that comprises two machines (prone to failure), M_1 and M_2, and a buffer of size 1, B, that are combined in an M_1—B—M_2 configuration, whose PNs are given in Fig. 13, can be synthesized using a bottom-up approach as shown in Fig. 14.

In Figs. 13a and 13b, the PN model of the machine allows it to work if the machine was previously idle and a part is available (e.g., placed on its worktable). Once working, the machine can either finish its operation or break down. The machine returns to its idle state and the finished part is made available for the next resource/buffer/etc. after the machine is finished working. The reachability tree for such a machine model is given in Fig. 15. [As one can note, an external transition, t_e, making a part available to the machines (i.e., supplying a token to p_1 or p_5, respectively) is not included in the tree. Such a transition could happen only once the finished part is removed form the machine.]

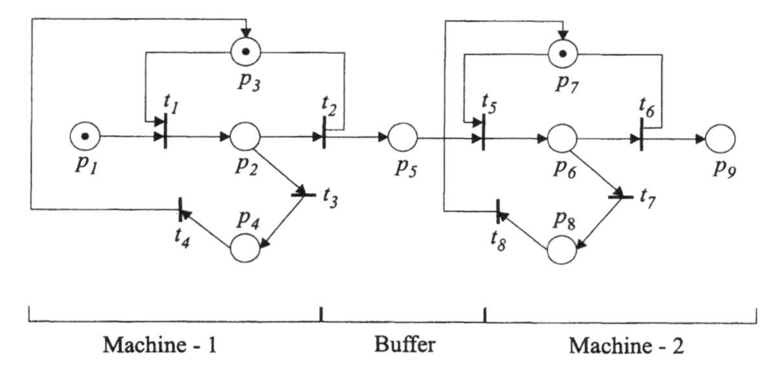

FIGURE 14 A PN for a M_1—B—M_2 manufacturing line.

In top-down synthesis techniques, the overall PN is developed in a gradual manner by (stepwise) refinement of places or transitions on an existing in-process PN model. That is, a more detailed submodel is inserted into the latest PN at hand as a block to replace a transition or a place. Unlike the bottom-up synthesis approaches, which provide users with flexible tools for the modular construction of large PNs, the top-down methods are more suitable for the minor refinement of already existing PNs, for example, the replacement of a resource or an operation.

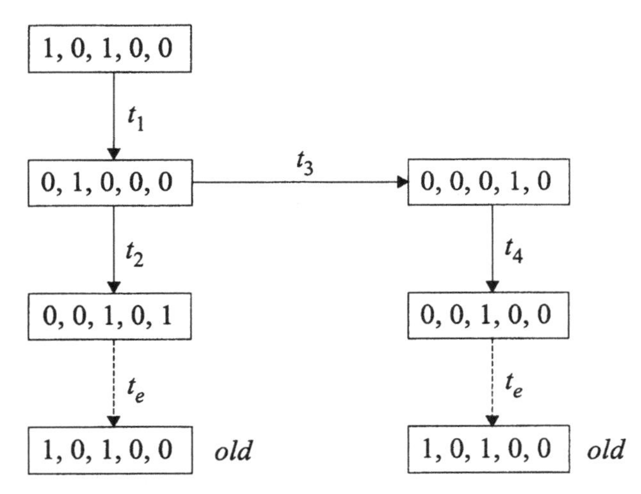

FIGURE 15 Reachability tree for M_1.

In regard to behavioral properties, although most synthesis methods (bottom-up or top-down) are expected to preserve liveness and boundedness, in practice it would be advisable to reexamine the properties of the resultant synthesized PN.

15.2.3 Supervisory Control Using Petri Nets

PNs have evolved since the early 1980s to cope with a variety of manufacturing constraints, such as modeling the flow of several different part types through the manufacturing system, sharing of resources, and so on. Although the primary utilization area of PNs has been in the simulation of manufacturing systems for performance analysis, there have also been supervisory control cases: for example, (1) the direct implementation of PNs through the use of a companion execution software and (2) the translation of PNs into sequential programming codes that can be implemented on a PLC or on an industrial PC with external I/O capability. The former method has also been known as playing the token game, i.e., keeping track of the locations of tokens in a PN as new transitions (events) occur in the corresponding physical system.

As one can infer, the synthesis of overall system PNs by combining the sub-PNs of their individual resources would yield very large nets, where places refer to the individual states of the resources as opposed to the overall state of the system, as would be the case with Ramadge–Wonham's supervisory control theory (Sec. 15.1.2). Thus, in any token game, one must keep track and examine all pertinent places for token movements. Transitions should be enabled based on firing rules, and such information must be effectively transmitted to individual device controllers. In PC-based control, the receipt of input signals from the manufacturing system, in regard to the actual occurrence of events, and the sending of output signals, in regard to the enablement of events, can only be achieved via multichannel I/O interface cards.

Among the efforts for generating a sequential programming code based on PNs, the work of a group of French academics and industrial participants in the mid 1970s stands out as unique. This programming standard, officially established in 1980, is today known as GRAFCET (graphe de commande étape transition). GRAFCET is a graphical programming tool directly derived from ordinary PNs for implementation on PLCs. The basic elements of GRAFCET are steps (places with capacity 1), transitions, and receptivities (logical conditions that need to be satisfied before a transition can fire). Directed arcs connect transitions and steps. The dynamics of the GRAFCET (net) is achieved by enabling transitions, whose

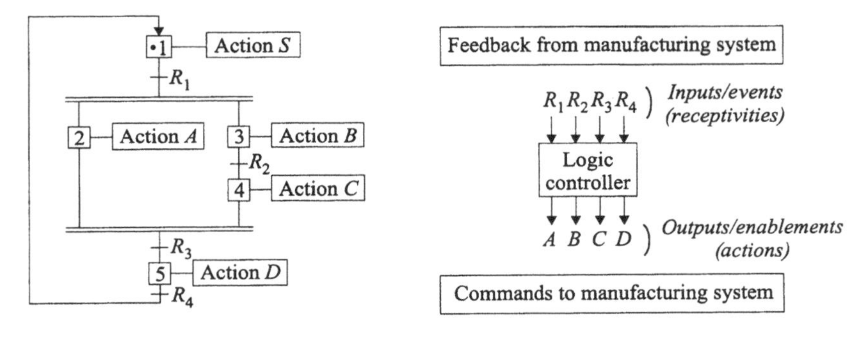

FIGURE 16 A GRAFCET example.

input steps satisfy the firing rules (i.e., have tokens in them), when the associated receptivities are true. A receptivity, R, may be an (external) event or a logic condition, or a combination of both. An example GRAFCET (net) is shown in Fig. 16.

As will be further discussed in subsection 15.3.2, PNs in general and GRAFCET in particular have led to the development of several industrial standards for PLC programming. The primary reason for this close relationship has been the similarities between PLC ladder logic coding and PNs in programming sequential systems via the use of logical expressions (AND, OR, NOR, etc.) that can be easily expressed in graphical form.

15.3 PROGRAMMABLE LOGIC CONTROLLERS

A PLC is a sequential controller that ensures (allows) the occurrence of events in a programmed sequence, through its output unit, based on feedback it receives from the system it is controlling, through its input unit. A control program stored in the memory of the PLC is continuously scanned (run in an endless loop) while examining all the inputs and "energizing" appropriate output ports (Fig. 17).

The first commercial PLC was developed and installed in 1969 at General Motors Hydra-Matic division by Modicon (Gould Electronics). The primary objective from GM's perspective was rapid retooling needed by product model changes. Electromechanical relays used on the factory floor to control the flow of parts prevented such rapid retooling owing to

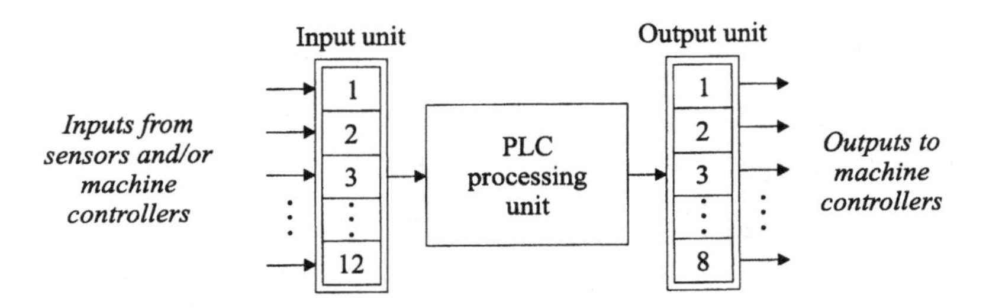

FIGURE 17 A typical PLC structure.

extensive (and expensive) rewiring. Thus GM engineers set the following specifications for the design of a sequential event controller:

Ease of on-the-floor reprogrammability
Modularity and ease of interface to factory devices
Ruggedness and cost effectiveness

All the above specifications were met by the original PLC designs, and their use became very widespread with the introduction of Intel's processors. Today the market for commercial PLCs is several billions of (US) dollars.

PLCs have long been accepted as industrial computers with efficient input/output (I/O) interface capability, and they competed successfully as an alternative to the use of PCs as process controllers. Recent advances in PLC technologies blur the differences even more in favor of PLCs. Today, PLCs, like PCs, can be networked (via Ethernet) to other remote PLCs and PCs, execute multiple programs (using coprocessors), communicate in digital and analog format, with a very large number of machines, and so on. Their modular structure also allows them efficient expandability to handle thousands of inputs and outputs, while in mini-PLC configuration (up to 24-32 I/O ports) they can be purchased for \$100 to \$150 (USA). PLCs' current primary weakness is their expected programmability using a low-level, device-specific language (i.e., lack of programmability by a high-level language) and difficulty of creating large programs that are verifiable, for example, for deadlocks.

15.3.1 PLC Hardware Structure

A PLC is an industrial computing device that continuously and sequentially checks its input ports to determine the most recent events that have occurred within the system it is controlling, and it activates (or deactivates) its output ports to allow (or disallow) other events to happen within the system (Fig. 17). The core unit of the PLC, as with any other computing

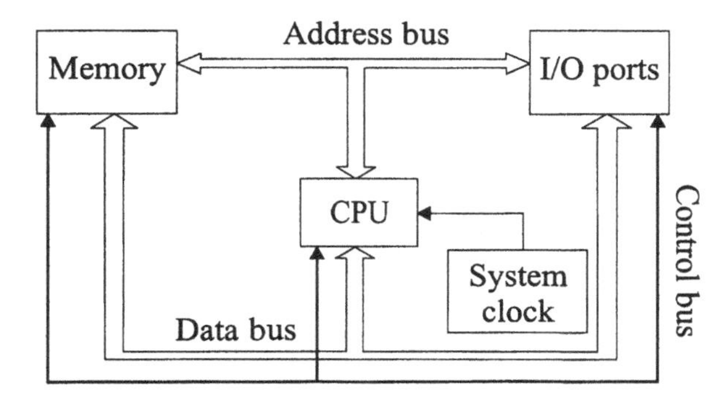

FIGURE 18 PLC communications.

device, is the central processing unit (CPU). The CPU controls all the operations within the PLC based on instructions specified by the user in the form of a program stored in its memory. The CPU communicates with the data- and program-memory modules and I/O units through the various (address, data, control) buses, according to its clock frequency (Fig. 18). Although CPU units used in PCs today can have a clock speed of above 2 GHz, most PLCs still use CPU units that are a few generations old (e.g., Intel's 80486) with clock speeds that are less than 200 MHz.

I/O System

The I/O system of a PLC is its interface (physical connection) to the entity it is controlling. The majority of input signals are received from a variety of sensors/switches and individual device controllers in the form of 5 V (TTL* level) or 24V DC and occasionally 100/120 V or 230/240 V AC. It is expected that all incoming (low- or high-voltage) signals are isolated through optical couplers in order to protect the PLC against voltage spikes. The output signals of the PLC also vary from 5 V (TTL level) DC to 230/240 V AC and are applied in the reverse order of the input signals. Every I/O point (port) is assigned a unique address that is utilized for its monitoring via the user supplied program. For example, Allen-Bradley's Series 5 PLCs denote addresses as I or O: two-digit rack number_one-digit module group number/two-digit port number: I:034/03, I:042/01, O:034/08, O:042/12, and so on.

As mentioned above, PLCs can be configured as single boxes that house all the logic and I/O units in one casing (with minimal variety on I/O

* Transistor–transistor logic (TTL).

signals) or as modular structures that allow users to choose from a variety of I/O modules (analog and discrete signal). The latter configuration is, naturally, flexible and expandable.

Memory Units

As do PCs, PLCs employ a variety of memory devices for the permanent or temporary storage of data and operating instructions. However, as with CPUs, these devices are generally several generations older than their counterparts in PCs in terms of capacity and speed. Most manufacturer-provided information, including the operating system of the PLC, is stored in a read-only memory (ROM) module. All user-provided programs and collected input data are stored in a random-access memory (RAM) module—such CMOS-based memory modules are battery supported and easily erasable and rewritable. Users can also choose to store certain programs and data on erasable programmable read-only memory (EPROM) modules for better protection. EPROMs can be completely erased through external intervention (e.g., using an ultraviolet light input through a window on the memory device) and reprogrammed for future read-only access cases. (Electrically erasable EPROMs are often called EEPROMSs.)

Almost all commercial PLCs provide users with a PC-based interface capability for their programming. Thus PLC programs can be developed on a host PC, stored on its hard drive, and downloaded to the PLC's RAM module when needed.

External Data Communications

Modern PLCs allow users to network their controllers for data communications between multiple PLCs as well as between PLCs and computers or other controllers on a factory floor. PLCs can be placed on local area networks (LANs) utilizing proprietary software/hardware (e.g., Allen-Bradley's Data Highway, Mitsubishi's Melsec-NET, and General Electric's Net Factory LAN) or Ethernet (nonproprietary network protocol and interface developed by Xerox, DEC, and Intel) (Fig. 19).

PLCs can also communicate between themselves and with other devices using serial communication interfaces. RS232 (also known as EIA 232) is the most commonly used serial interface standard, it uses a 25-pin connector—± 12 V signals indicate 0/1. Data transfer rates of up to 25 kilobaud (but typically only 9600 baud) can be achieved over short distances (less than 50 feet, 15 meters). The majority of PLCs also provide users with RS422 serial interface capability. RS422 can yield a transmission rate of up to 10 megabaud over a distance of up to 4,000 feet (1,200 meters). Other standards include the RS485 and the 20 mA current loops.

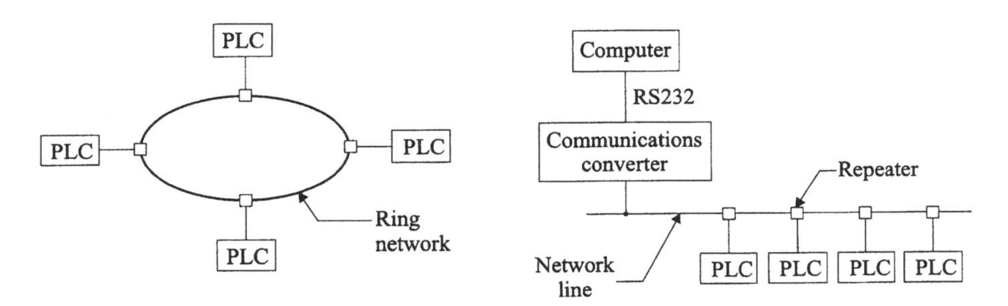

FIGURE 19 Networked PLCs.

15.3.2 PLC Programming

Ladder logic (LL) programming of PLCs is the current industrial standard, though instruction sets vary from one PLC make to another. LL programming is based on the logical representation of output decisions based on binary (0/1) values of the inputs. LL programs can be represented in a ladder diagram form or simply as sets of instructions. A rung typically is a combination of inputs that affect one output, though some commercial systems allow a rung to have multiple outputs.

Common logic gates (e.g., AND, OR, NOR) are utilized in the LL programming of PLCs. Fig. 20 illustrates typical logic symbols used in LL diagrams, while Table 1 lists some instructions used by commercial PLC manufactures.

A series of typical rungs are shown in Fig. 21. The first two correspond to examining multiple inputs and energizing the corresponding outputs, while the third illustrates a multiple output case.

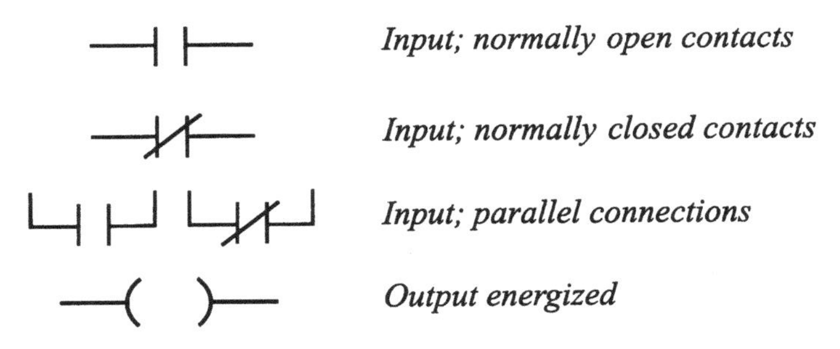

FIGURE 20 Ladder logic symbols.

TABLE 1 Ladder Logic Instructions

Action	Mitsubishi	Omron	Texas Instruments
Start a new rung	LD	LD	STR
Logical AND	AND	AND	AND
Logical OR	OR	OR	OR
Logical NOT	NOT	NOT	I
Output	OUT	OUT	OUT

As discussed above, a LL program loaded to the PLC's RAM module runs in an endless loop. During each scan cycle, the processor sequentially examines (reads) all the inputs and accordingly energizes or deenergizes the corresponding outputs. For example, in the third rung of Fig. 21, the output port with the address O001 is energized only if the PLC detects an "ON" (1) signal at input port address I003, while O002 is energized only if additionally the input port address I008 does not have an "ON" (1) value.

Timer and counter instructions, as they affect the outputs of a PLC, can also be programmed using LL. Timer instructions can be used to delay the activation of the output port, to deenergize it after a certain period of time and so on. Counters can perform similar tasks as timers by counting (up or down) the instances of signals generated to energize the output

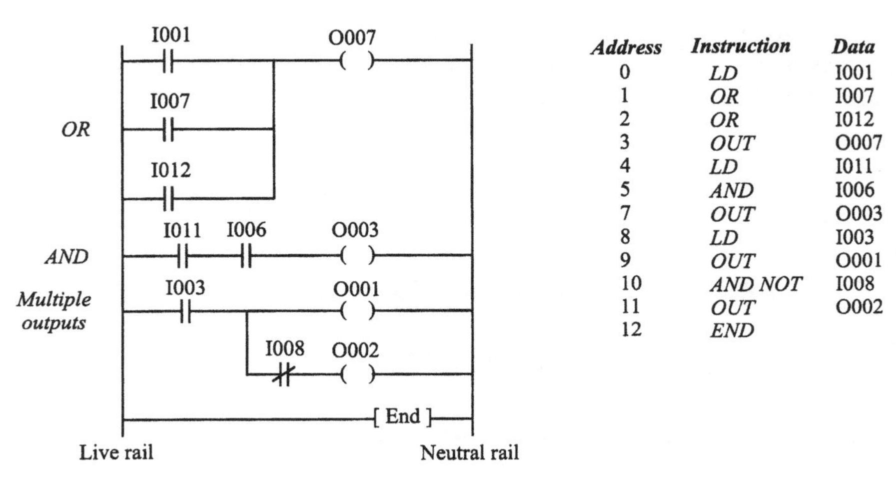

Address	*Instruction*	*Data*
0	*LD*	I001
1	*OR*	I007
2	*OR*	I012
3	*OUT*	O007
4	*LD*	I011
5	*AND*	I006
7	*OUT*	O003
8	*LD*	I003
9	*OUT*	O001
10	*AND NOT*	I008
11	*OUT*	O002
12	*END*	

FIGURE 21 Example ladder logic rungs.

port. Simple arithmetic functions (add, subtract, multiply, divide, square root, negate) can also be programmed using LL for data comparison instructions (e.g., equal, greater than or equal). Most PLCs also allow the creation and use of subroutines/procedures (external subprograms) within main programs.

The majority of PLC manufacturers today provide customers with interface software that runs on PCs for the writing of LL programs and their easy downloading to the PLCs. Although they differ in their graphical user interface capability, most software modules are very similar in aiding the programmer for syntax error detection and so on. One must realize, however, that having such an off-line programming tool does not guarantee the correct operation of the manufacturing system, for human errors in programming are normally detected only once the system is running and rarely ahead of time (i.e., off-line).

During the 1990s, two themes of research and development have been pursued by the industrial and academic communities. The former group have made some efforts in allowing users to program PLCs via high-level languages, while the academic community primarily concentrated on the automatic translation of supervisory controllers, developed via PN or automata tools, into LL programs.

The sequential function chart (SFC), made available to users by a number of major PLC manufacturers, is the most commonly used high-level language alternative to LL. This graphical sequencing language is defined within the IEC* 1131–3 standard and is derived from IEC 848 GRAFCET—a technique based on PN modeling of DESs.

The first revision of the IEC 1131-3 standard was published in 1993 for PLC programming that specifies the syntax, semantics, and display for several languages: LL, SFC, function block diagram (FBD), structured text (ST), and instruction list (IL). A PLC program can be built with any of these languages. Typically, such a program would consist of a network of functions and function blocks that are capable of exchanging data.

In particular, the SFC consists of steps linked with action blocks and transitions. Each step represents a particular state of the system that is being controlled. A transition causes the system to change states (steps). Steps are linked to action blocks that perform certain control actions. Steps and transitions can be arranged in series or in parallel (Fig. 22). Parallelism and other features of SFC allow the scanning of only the active states (steps) instead of sequential scanning of the entire logic, as is the case with LL programming.

*International Electrotechnical Commission.

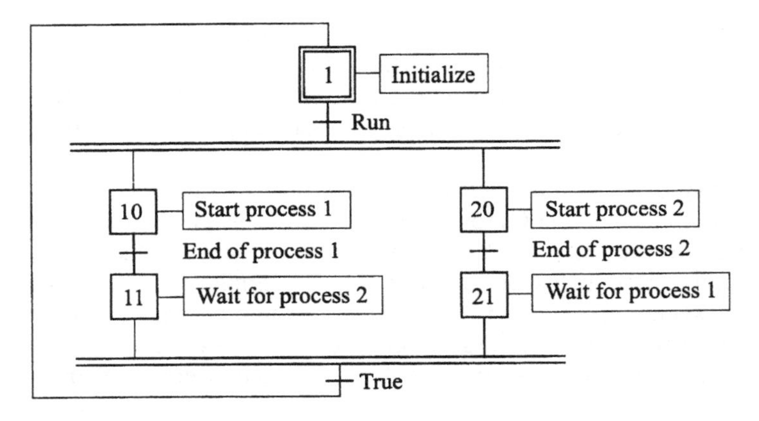

FIGURE 22 Sequential function chart programming.

REVIEW QUESTIONS

1. Define flexible manufacturing workcell (FMC) and discuss its autonomous supervisory control.
2. What is discrete event system (DES) control and how does it differ from time-driven control? Discuss *determinism* versus *nondeterminism* in the state transition of DES systems.
3. What is automata theory?
4. Define (artificial) language in the context of automata theory.
5. What is a finite-state automaton?
6. Define Moore (finite-state) machines versus Mealy machines.
7. What is the primary feature of Ramadge–Wonham (R–W) automata theory that makes it suitable to the supervisory control of manufacturing systems?
8. How do you obtain a R–W supervisor for an FMC?
9. What is a Petri net (PN) and how can it be utilized for the DES modeling of FMCs?
10. How is a transition enabled in PNs? What is the consequence of a fired transition?
11. Describe the following properties of PNs in terms of the control of FMCs: reachability, boundedness, and liveness. Furthermore, define a deadlock state.
12. How do you synthesize a PN supervisor for an FMC and how would you implement it?
13. What is a programmable logic controller (PLC) and how would it control an FMC?

14. Discuss the primary elements of a PLC and compare its architecture to that of a personal computer (PC). Furthermore, discuss PLC communications over an Ethernet-based network.
15. Discuss the programming of PLCs using ladder logic and other higher level programming languages.
16. How could one use automata or Petri net theory in the automatic generation of PLC programs versus their compilation manually by an experienced operator?

DISCUSSION QUESTIONS

1. Computers and other information management technologies have been commonly accepted as facilitators for the integration of various manufacturing activities. Define/discuss integrated manufacturing in the modern manufacturing enterprise and address the role of computers in this respect. Furthermore, discuss the use of intranets and extranets as they pertain to the linking of suppliers, manufacturers, and customers.
2. Manufacturing flexibility can be achieved on three levels: operational flexibility, tactical flexibility, and strategic flexibility. Discuss operational flexibility. Is automation a necessary or a desirable tool in achieving this level of flexibility?
3. The factory of the future will be a totally networked enterprise. Information management in this enterprise will be a complex issue. In regards to planning, monitoring, and control, discuss the level of detail of information that the controllers (humans or computers) would have to deal with in such environments. For example, some argue that in a hierarchical information management environment, activities are more of the planning type at the higher levels of the hierarchy and of the control type at the lower levels. It has also been argued that the level of details significantly decreases as you ascend the enterprise ladder.
4. Job shops that produce one-of-a-kind products have been considered the most difficult environments to automate, where a product can be manufactured within a few minutes or may require several days of fabrication. Discuss the role of computers in facilitating the transformation of such manual, skilled-labor-dependent environments to intensive automation-based environments.
5. Several fabrication/assembly machines can be physically or virtually brought together to yield a manufacturing workcell for the production of a family of parts. Discuss the advantages of adopting a cellular manufacturing strategy in contrast to having a departmentalized

strategy, i.e., having a turning department, a milling department, a grinding department, and so on. Among others, an important issue to consider is the supervisory control of such workcells.

6. In the factory of the future, it is envisioned that production and assembly workcells will be frequently reconfigured based on the latest manufacturing objectives without the actual physical relocation of their machines/resources. Discuss the supervisory control options that should be available to the users of such workcells, whose boundaries may exist only in the (computer's) virtual space.
7. The primary advantage of PLCs has been their robust modular design that allows large numbers of inputs and outputs to be directly connected to the controller. However, this advantage has also been discouraging users from implementing PLCs widely in flexible manufacturing workcells, who frequently opt for networkable industrial PCs. Compare PCs versus PLCs for the control of manufacturing systems that are configured for mass production versus those that are configured (and occasionally reconfigured) for small batch production.
8. In the factory of the future, it is expected that manufacturing workcells will operate autonomously under the control of an overall computer-based supervisor; all devices, including fixtures and jigs, will be equipped with their own (device) controllers that can effectively communicate with the workcell supervisor. Discuss difficulties in the integration and utilization of a large number of sensors and other communication components in remotely controlled (via two-way communications) FMCs.
9. Computer-based controllers (PC or PLC based) have been advocated as better supervisors than human controllers based on a number of factors including reaction time. Besides manufacturing environments, such controllers have been widely used in the health care, nuclear, aviation, security, and military industries. Discuss the advantages/disadvantages of using computer-based controllers in today's manufacturing industries. You may also extrapolate and discuss the role of such controllers in the factories of the future.
10. Discrete event system (DES)-based autonomous supervisory control of flexible manufacturing workcells require the planning of a new (software-based) controller, the reprogramming of the control hardware using this controller, and, frequently, rewiring of the communication lines between the control hardware and the input/output devices in the workcell every time the workcell is reconfigured to manufacture different products. All these three problematic issues and others discourage manufacturers from implementing computer-

based autonomous DES controllers. The manufacturers mostly continue to rely on limited feedback received from the workcell that is monitored by a human supervisor. Discuss remedies to these problems that would allow widespread use of autonomous computer-based supervisors.

11. Manufacturing systems, supported by computers, can be classified as manual versus automated and flexible (reconfigurable, reprogrammable, etc.) versus rigid. Discuss these classifications and elaborate on the intersections of their domains (e.g., manual and flexible, etc.). Note that each classification may have sublevels and subclassifications (i.e., different "levels of gray").
12. In the factory of the future, no unexpected machine breakdowns will be experienced! Such an environment, however, can only be achieved if a preventive maintenance program is implemented, in which all machines and tools are modeled (mathematically and/or using heuristics). These models would allow manufacturers to schedule maintenance operations as needed. Discuss the feasibility of implementing factory-wide preventive maintenance programs in the absence of our ability to model completely all existing physical phenomena and furthermore in the lack of a large variety of sensors that can monitor the states of these machines and provide timely feedback to such models.
13. Human factors (HF) studies encompass a range of issues spanning from ergonomics to human–machine (including human–software) interfaces. Discuss the role of HF in the autonomous factory of the future, where the impact of human operators is significantly diminished and emphasis is switched from operating machines to supervision, planning, and maintenance.

BIBLIOGRAPHY

AFCET (Association Française pour la Cybernétique Economique et Technique). (1977). Normalisation de la représentation du cahier des charges d'un automatisme logique. Journal Automatique et Informatique Industrielles. pp. 61–62, November–December.

Baker, A. D., Johnson, T. L., Kerpelman, D. I., Sutherland, H. A. (1987). GRAFCET and SFC as factory automation standards: advantages and limitations. Minneapolis, MN: Proceedings of the American Control Conference, pp. 1725–1730.

Batten, George (1994). *Programmable Controllers: Hardware, Software, and Applications.* New York: McGraw-Hill.

Boel, R., Stremersch G., eds. (2000). *Discrete Event Systems: Analysis and Control.* Boston, MA: Kluwer.

Booth, Taylor L. (1968). *Sequential Machines and Automata Theory.* New York: John Wiley.

Cao, X.-R., Ho, Y.-C. (June 1990). Models of discrete event dynamic systems. *IEEE Control Systems Magazine* 1(4):69–76.

Carrow, Robert S. (1998). *Soft Logic: A Guide to Using a PC as a Programmable Logic Controller.* New York: McGraw-Hill.

Cassandras, Christos G. (1993). *Discrete Event Systems: Modeling and Performance Analysis.* Homewood, IL: Aksen Associates.

Chomsky, N. (1956). Three models for the description of language. *IRE Transactions of Information Theory* 2(3):113–124.

Costanzo, Marco. (1997). *Programmable Logic Controllers: The Industrial Computer.* New York: Arnold.

Crispin, Alan J. (1997). *Programmable Logic Controllers and Their Engineering Applications.* New York: McGraw-Hill.

David, René, Alla, Hassane. (1992). *Petri Nets and GRAFCET: Tools for Modeling Discrete Event Systems.* New York: Prentice Hall.

Desrochers, A. Alan, Al-Jaar, Robert Y. (1994). *Applications of Petri Nets in Manufacturing Systems: Modeling, Control, and Performance Analysis.* Piscataway, NJ: IEEE Press.

Dicesare, F., et al. (1993). *Practice of Petri Nets in Manufacturing.* New York: Chapman and Hall.

Filer, Robert, Leinomen, George. (1992). *Programmable Controllers and Designing Sequential Logic.* New York: Saunders.

Ho, Yu-Chi, ed. (1992). *Discrete Event Dynamic Systems: Analyzing Complexity and Performance in the Modern World.* New York: Institute of Electrical and Electronics Engineers.

Hopcroft, John E. (1979). *Introduction to Automata Theory, Languages, and Computation.* Reading, MA: Addison-Wesley.

Hughes, Thomas A. (2001). *Programmable Controllers. Instrument Society of America.* NC: Research Triangle Park.

Johnson, David G. (1987). *Programmable Controllers for Factory Automation.* New York: Marcel Dekker.

Kain, Richard Y. (1972). *Automata Theory: Machines and Languages.* New York: McGraw-Hill.

Kohavi, Zvi (1970). *Switching and Finite Automata Theory.* New York: McGraw-Hill.

Lauzon, Stéphane C. (1995). An Implementation Methodology for the Supervisory Control of Manufacturing Workcells. M.A.Sc. thesis, Department of Mechanical Engineering, University of Toronto, Toronto, Canada.

Lauzon, S. C., Mills, J. K., Benhabib, B. (1997). An implementation methodology for the supervisory controller for flexible manufacturing workcells. *SME J. of Manufacturing Systems* 16(2):91–101.

Lewis, R. W. (1998). *Programming Industrial Control Systems Using IEC 1131-3.* Stevenage, UK: IEE Books.

Mealy, G. H. (1955). A method for synthesizing sequential circuits. *Bell System Technical Journal* 34(5):1045–1079.

Meduna, Alexander (2000). *Automata and Languages: Theory and Applications*. New York: Springer Verlag.

Moody, John O., Antsaklis, Panos J. (1998). *Supervisory Control of Discrete Event Systems Using Petri Nets*. Boston: Kluwer.

Moore, E. F. (1956). Gedanken-experiments on sequential machines. *Automata Studies, Annals of Mathematical Studies* (No. 34, pp. 129–153). Reading, MA: Princeton University Press.

Moore, Edward F., ed. (1964). *Sequential Machines: Selected Papers*. Reading, MA: Addison-Wesley.

Murata, T. (Apr. 1989). Petri nets: properties, analysis and applications. *Proceedings of the IEEE* 77(4):541–580.

Parr, E. A. (1999). *Programmable Controllers: An Engineer's Guide*. Boston: Newnes.

Petri, Carl Adam (1962). Kommunication mit Automaten (Communication with Automata). Schriften des IIM Nr. 3, Bonn: Inst. fur Instrumentelle Mathematik, Ph.D. diss.

Petruzella, Frank D. (1998). *Programmable Logic Controllers*. New York: Glenco.

Proth, Jean-Marie, Xie, Xiaolan (1996). *Petri Nets: A Tool for Design and Management of Manufacturing Systems*. New York: John Wiley.

Ramadge, P. J. G., Wonham, W. M. (Jan. 1989). Control of discrete event systems. *Proceedings of the IEEE* 77(1):81–98.

Ramirez-Serrano, Alejandro (2000). Extended Moore Automata for the Supervisory Control of Virtual Manufacturing Workcells. Ph.D. diss., Department of Mechanical and Industrial Engineering, University of Toronto, Toronto, Canada.

Ramirez, A., Zhu, S. C., Benhabib, B. (Aug. 2000). Moore automata for flexible routing and flow control in manufacturing workcells. *J. of Autonomous Robots* 9(1):59–69.

Ramirez, A., Benhabib, B. (Oct. 2000). Supervisory control of multi-workcell manufacturing systems with shared resources. *IEEE Transactions on Systems, Man and Cybernetics* 30(5):668–683.

Ramirez, A., Sriskandarajah, C., Benhabib, B. (Dec. 2000). Discrete-event-system modeling and control synthesis using extended Moore automata. *IEEE Transactions on Robotics and Automation* 16(6):807–823.

Rathmill, K. (1988). *Control and Programming in Advanced Manufacturing*. Bedford, UK: IFS.

Silva, M., Velilla, S. (1982). Programmable logic controllers and Petri nets: a comparative study. *Proceedings of the 3rd IFAC/IFIP Symposium on Software for Computer Control, Madrid, Spain, 29–34.*

Sobh, T. M., Benhabib, B. (June 1997). Discrete-event and hybrid systems in robotics and automation: an overview. *IEEE Robotics and Automation Magazine* 4(2):16–19.

Stenerson, Jon (1999). *Fundamentals of Programmable Logic Controllers, Sensors, and Communications*. Upper Saddle River, NJ: Prentice Hall.

Tzafestas, Spyros G. (1997). *Computer-Assisted Management and Control of Manufacturing Systems*. New York: Springer Verlag.

Turing, A. M. (1936–1937). On computable numbers with an application to the entscheidungs problems. *Proceedings of the London Mathematical Society* 43: 230–265 (correction ibid, Vol. 42, pp 544–546).

Viswanadham, N., Narahari, Y. (1992). *Performance Modeling of Automated Manufacturing Systems*. Englewood Cliffs, NJ: Prentice Hall.

Warnock, Ian G. (1988). *Programmable Controllers: Operation and Application*. Englewood Cliffs, NJ: Prentice-Hall.

Williams, Robert A. (1993). A DES-Based Hybrid Supervisory Control System for Manufacturing Systems. M.A.Sc. thesis, Department of Mechanical Engineering, University of Toronto, Toronto, Canada.

Williams, R. A., Benhabib, B., Smith, K. C. (1996). A DES-based supervisory control system for manufacturing systems. *SME J. of Manufacturing Systems* 15(2): 71–83.

Wonham, W. Murray (1997). *Notes on Control of Discrete-Event Systems*. University of Toronto: Department of Electrical and Computer Engineering.

Zhou, MengChu, DiCesare, Frank (1993). *Petri Net Synthesis for Discrete Event Control of Manufacturing Systems*. Boston: Kluwer.

Zhou, MengChu, ed. (1995). *Petri Nets in Flexible and Agile Automation*. Boston: Kluwer.

Zhou, MengChu, Venkatesh, Kurapati (1999). *Modeling, Simulation, and Control of Flexible Manufacturing Systems. A Petri Net Approach*. River Edge, NJ: World Scientific.

16

Control of Manufacturing Quality

The definition of quality has evolved over the past century from meeting the engineering specifications of the product (i.e., conformance), to surpassing the expectations of the customer (i.e., customer satisfaction). Quality has also been defined as a loss to customer in terms of deviation from the nominal value of the product characteristic, the farther the deviation the greater the loss.

The management of quality, according to J. M. Juran, can be carried out via three processes: planning, control, and improvement. *Quality planning* includes the following steps: identifying the customer's needs/expectations, designing a robust product with appropriate features to satisfy these needs, and establishing (manufacturing) processes capable of meeting the engineering specifications. *Quality control* refers to the establishment of closed loop control processes capable of measuring conformance (as compared to desired metrics) and varying production parameters, when necessary, to maintain steady-state control. *Quality improvement* requires an organization's management to maximize efforts for continued increase of product quality by setting higher standards and enabling employees to achieve them. A typical goal would be the minimization of variations in output parameters by increasing the capability of the process involved by either retrofitting existing machines or acquiring better machines. Among the three processes specified by Juran for quality management, the central issue addressed in this chapter is quality control with emphasis on on-line control (versus postprocess sampling): measurement technologies as well as statistical process control tools.

Cost of quality management has always been an obstacle to overcome in implementing effective quality control procedures. In response to this problem, management teams of manufacturing companies have experimented over the past several decades with techniques such as (on-line) statistical process control versus (postprocess) acceptance by sampling, versus 100% inspection/testing and so on. For example, it has been successfully argued that once a process reaches steady-state output in terms of conformance, it would be uneconomical to continue to measure on-line every product feature (i.e., 100% inspection), though a recent counterargument has been that latest technological innovation in measurement devices and computer-based analyzers do allow manufacturers to abandon all statistical approaches and instead carry out real-time quality control. Furthermore, it has been argued that new approaches to quality control must be developed for products with high customization levels achievable in flexible manufacturing environments.

No matter how great is the cost of quality control implementation engineers must consider the cost of manufacturing poor quality products. These lead to increased amounts of rejects and reworks and thus to higher production costs. Dissatisfaction causes customers to abandon their loyalty to the brand name and eventually leads to significant and rapid market-share loss for the company. Loyalty is more easily lost than it is gained. As will be discussed in greater detail later in this chapter, quality is commonly measured by customers as deviation from the expected nominal value. When

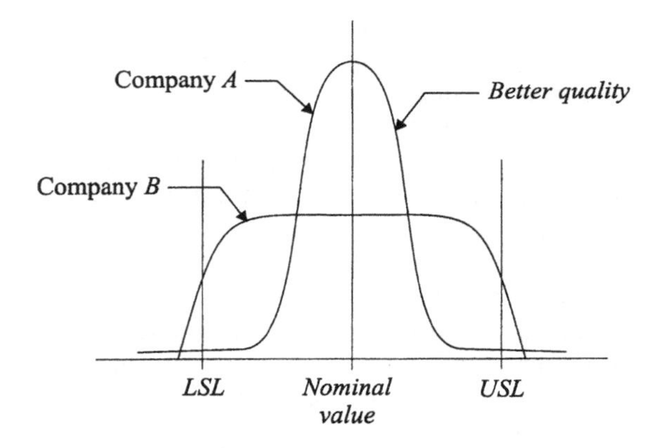

FIGURE 1 Quality control.

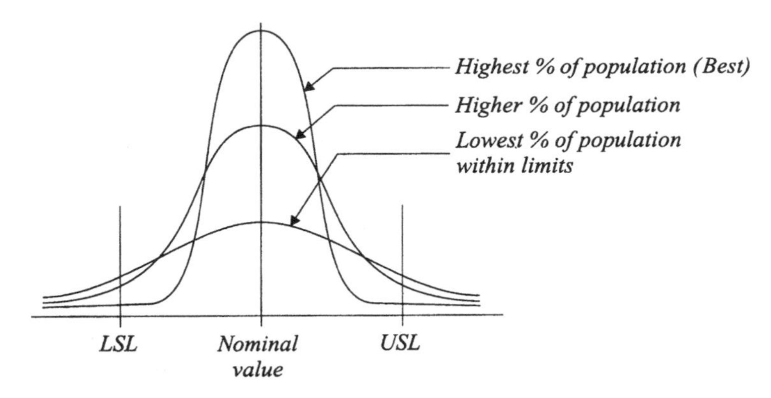

FIGURE 2 Variability about the nominal value.

two companies manufacture the same product, and equal percentages of their product populations fall within identical specifications (i.e., between LSL and USL: lower and upper specification limits, respectively), the company with the lower variation about the nominal value provides better customer satisfaction (Fig. 1). Naturally, a company with the lowest variation as well as the lowest percentage of the population of their products within their specification limits will have the best quality and the highest customer satisfaction (Fig. 2).

It has been erroneously argued that high-quality products can only be purchased at high prices. Such arguments have been put forward by companies who scrap their products that fall outside their specification limits and pass on this cost to the customers by increasing the price of their within-limits goods. In practice, price should only be proportional to the performance of a product and not to its quality. For example, a Mercedes-Benz car should deserve its higher price in comparison to a Honda or a Ford because of its higher performance with equivalent quality expectation by the customers.

16.1 MODERN HISTORY OF QUALITY MANAGEMENT

Quality management in the U.S.A. suffered a setback in the early 1900s with the introduction of F. W. Taylor's division-of-labor principle into (mass-production-based) manufacturing enterprises. Greater emphasis on productivity came at the expense of quality when workers on the factory floor lost ownership of their products. Quality control became a postprocess inspection task carried out by specialists in the quality-assurance department disconnected from the production floor.

The subsequent period of the 1920s to the 1940s was marked by the utilization of statistical tools in the quality control of mass produced goods. First came W. A. Shewart's process control charts [now known as statistical process control (SPC) charts] and then the acceptance by sampling system developed by H. F. Dodge and H. G. Romig (all from Bell Laboratories).

The 1950s were marked by the works of two modern pioneers of quality, W. E. Deming and J. M. Juran. Although both advocated continued reliance on statistical tools, their emphasis was on highlighting the responsibility of an organization's high-level management to quality planning, control, and improvement. Ironically, however, the management principles put forward by Deming and Juran were not widely implemented in the U.S.A. until the competitiveness of U.S. manufacturers was seriously threatened by the high-quality products imported from Japan in the late 1970s and the early 1980s. Two other modern pioneers that contributed to quality management in the U.S.A. have been A. V. Feigenbaum and P. Crosby.

Prior to the 1960s, products manufactured in Japan were plagued with many quality problems, and subsequently Japanese companies failed to penetrate the world markets. Behind the scenes, however, a massive quality improvement movement was taking place. Japanese companies were rapidly adopting the quality management principles introduced to them during the visits of Deming and Juran in the early 1950s as well as developing unique techniques locally. One such tool was K. Ishikawa's cause-and-effect diagram, also referred to as the fishbone diagram, which was developed in the early 1940s. The Ishikawa diagram identified possible causes for a process to go out of control and the effect of these causes (problems) on the process. Another tool was G. Taguchi's approach to building quality into the product at the design stage, that is, designing products with the highest possible quality by taking advantage of available statistical tools, such as design of experiments (Chap. 3).

In parallel to the development of the above-mentioned quality control and quality improvement tools, the management of many major Japanese organizations strongly emphasized company-wide efforts in establishing quality circles to determine the root causes of quality deficiencies and their elimination in a bottom-up approach, starting with the workers on the factory floor. The primary outcome of these efforts was the elimination of postprocess inspection and its replacement with the production of goods, with built-in quality, using processes that remained in control. Japanese companies implementing such quality-management systems (e.g., Sony, Toshiba, NEC, Toyota, Honda) rapidly gained large market shares during the 1970s to the 1990s.

In Europe, Germany has led the way in manufacturing products with high quality, primarily owing to the employment of a skilled and versatile labor force combined with an involved, quality-conscious management. Numerous German companies have employed statistical methods in quality control as early as in the 1910s, prior to Shewhart's work in the late 1920s. In the most of the 20th century, the "Made in Germany" designation on manufactured products became synonymous with the highest possible quality. In France and the United Kingdom, awareness for high quality has also had a long history, though, unlike in Germany, in these countries high quality implied high-priced products.

Participation in NATO (the North Atlantic Treaty Organization) further benefited the above-mentioned and other European countries in developing and utilizing common quality standards: in the beginning for military products but later for most commercial goods. The most prominent outcome of such cooperation is the quality management standard ISO-9000, which will be briefly discussed in Sec. 16.6.

16.2 INSPECTION FOR QUALITY CONTROL

Inspection has been loosely defined in the quality control literature as the evaluation of a product or a process with respect to its specifications—i.e., verification of conformance to requirements. The term testing has also been used in the literature interchangeably with the term inspection. Herein, testing refers solely to the verification of expected (designed) functionality of a product/process, whereas inspection further includes the evaluation of the functional/nonfunctional features. That is, testing is a subset of inspection.

The inspection process can include the measurement of variable-valued features or the verification of the presence or absence of features/parts on a product. Following an inspection process, the outcome of a measurement can be recorded as a numeric value to be used for process control or simply as satisfying a requirement (e.g., defective versus acceptable), i.e., as an attribute. Increasingly, with rapid advancements in instrumentation technologies, two significant trends have been developing in manufacturing quality control: (1) automated (versus manual) and (2) on-line (versus postprocess) inspection. The common objective to both trends may be defined as reliable and timely measurement of features for effective feedback-based process control (versus postmanufacturing product inspection).

Tolerances are utilized in the manufacturing industry to define acceptable deviations from a desired nominal value for a product/process feature. It has been convincingly argued that the smaller the deviation, the better the quality and thus the less the quality loss. Tolerances are design specifications, and the degree of satisfying such constraints is a direct function of the (statistical) capability of the process utilized to fabricate that product. For example, Process A used to fabricate a product (when "in control") can yield 99.9% of units within the desired tolerance limits, while Process B also used to fabricate the same product may yield only 98% of units within tolerance.

Prior to a brief review of different inspection strategies, one must note that the measurement instruments should have a resolution (i.e., the smallest unit value that can be measured) an order of magnitude better than the resolution used to specify the tolerances at hand. Furthermore, the repeatability of the measurement instruments (i.e., the measure of random errors in the output of the instrument, also known as precision) must also be an order of magnitude better than the resolution used to specify the tolerances at hand. For example, if the tolerance level is ± 0.01 mm, the measurement device should have a resolution and repeatability in the order of at least ± 0.001 mm.

16.2.1 Inspection Strategies

The term inspection has had a negative connotation in the past two decades owing to its erroneous classification as a postprocess, off-line product examination function based solely on statistical sampling. As discussed above, inspection should actually be seen solely as a conformance verification process, which can be applied based on different strategies—some better than others. However, certain conclusions always hold true: on-line (in-process) inspection is better than postprocess inspection 100% inspection is better than sampling, and process control (i.e., inspection at the source) is better than product inspection.

On-line inspection: It is desirable to measure product features while the product is being manufactured and to feed this information back to the process controller in an on-line manner. For example, an electro-optical system can be used to measure the diameter of a shaft, while it is being machined on a radial grinder, and to adjust the feed of the grinding wheel accordingly. However, most fabrication processes do not allow in-process measurement owing to difficult manufacturing conditions and/or the lack of reliable measurement instruments. In such cases, one may make intermittent (discrete) measurements, when possible, by stopping the process or waiting until the fabrication process is finished.

Sampling: If a product's features cannot be measured on-line, owing to technological or economic reasons, one must resort to statistical sampling inspection. The analysis of sample statistics must still be fed back to the process controller for potential adjustments to input variables to maintain in-control fabrication conditions. Sampling should only be used for processes that have already been verified to be in control and stable for an acceptable initial buildup period, during which 100% inspection may have been necessary regardless of economic considerations.

Source inspection: It has been successfully argued that quality can be better managed by carrying out inspection at the source of the manufacturing, that is, at the process level, as opposed to at (postprocess) product level. For fabrication, this would involve the employment of effective measurement instruments as part of the closed-loop process-control chain. For assembly, this would involve the use of devices and procedures that would prevent the assembly of wrong components and ensure the presence of all components and subassemblies—for example, using foolproofing concepts (*poka-yoke* in Japanese).

16.2.2 Measurement Techniques

Measurement is a quantification process used to assign a value to a product/process feature in comparison to a standard in a selected unit system (SI* metric versus English, U.S. customary measurement systems). The term metrology refers to the science of measurement in terms of the instrumentation and the interpretation of measurements. The latter requires a total identification of sources of errors that would affect the measurements. It is expected that all measurement devices will be calibrated via standards that have at least an order of magnitude better precision (repeatability). Good calibration minimizes the potential of having (nonrandom) systematic errors present during the measurement process. However, one cannot avoid the presence of (noise-based) random errors; one can only reduce their impact by (1) repeating the measurement several times and employing a software/hardware filter (e.g., the median filter) and (2) maintaining a measurement environment that is not very sensitive (i.e., robust) to external disturbances.

As will be discussed in the next subsections, variability in a process' output, assuming an ideal device calibration, is attributed to the presence of random mechanisms causing (random) errors. As introduced above, this random variability is called repeatability, while accuracy represents the totality of systematic (calibration) errors and random errors. Under ideal conditions, accuracy would be equal to repeatability.

Since the objective of the measurement process is to check the conformance of a product/process to specifications, the repeatability of the measurement instrument should be at least an order of magnitude better than the repeatability of the production process. Thus random errors in measuring the variability of the output can be assumed to be attributable

* Système International.

primarily to the capability (i.e., variance) of the production device and not to the measurement instrument. As will be discussed in Sec. 16.3, the behavior of random errors can be expressed by using a probability function.

In Chap. 13, a variety of measurement instruments were discussed as a prelude to manufacturing process control, which includes control of quality. Thus in this section, we will narrow our attention to a few additional measurement techniques to complement those presented in Chapter 13.

Mechanical Measurement of Length

Length is one of the seven fundamental units of measurement—the others are mass, time, electric current, temperature, light intensity, and amount of matter. It is commonly measured using simple yet accurate manual (mechanical) devices on all factory floors worldwide. The vernier caliper is frequently used to measure length (diameter, width, height, etc.) up to 300 to 400 mm (app. 12 to 14 in.) with resolutions as low as 0.02 mm (or 0.001 in.). A micrometer can be used for higher resolution measurements, though at the expense of operational range (frequently less than 25 mm), yielding resolutions as low as 0.002 mm (or 0.0001 in.). Micrometers can be configured to measure both external and internal dimensions (e.g., micrometer plug gages).

Coordinate measuring machines (CMMs) are typically numerical control (NC) electromechanical systems that can be used for dimensional inspection of complex 3-D-geometry product surfaces. They utilize a contact probe for determining the x, y, z coordinates of a point (on the product's surface) relative to a reference point on the product inspected. The mechanical architecture of a CMM resembles a 3-degree-of-freedom (Cartesian) gantry-type robot (Chap. 12), where the probe (i.e., end-effector) is displaced by three linear (orthogonal) actuators (Fig. 3). Some CMMs can have up to five degrees of freedom for increased probing accuracy on curved surfaces.

Mechanical-probe-based CMMs can have an operating volume of up to 1 m^3, though at the expense of repeatability (e.g., 0.005 mm). There also exist a variety of optical-probe-based (noncontact) CMMs, which increase the productivity of such machines to carry out inspection tasks. However, mostly, CMMs are expensive machines suitable for the inspection of small batch or one-of-a-kind, high-precision products. Owing to their slow processing times, they are rarely employed in an on-line mode on factory floors.

Surface finish is an important length metric that has to be considered in discrete part manufacturing. Besides checking for surface defects (e.g., cracks, marks), engineers must also verify that a product's surface roughness satisfies the design specification. Stylus instruments have been commonly

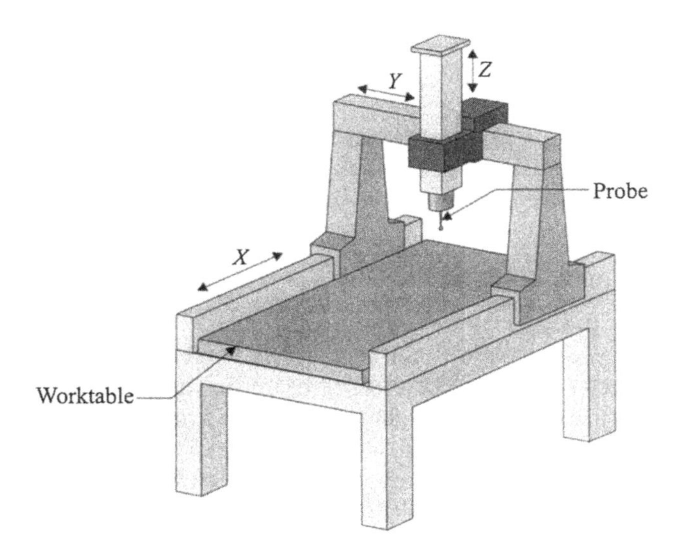

FIGURE 3 A coordinate measuring machine architecture.

utilized to quantify surface roughness: typically, a diamond-tip stylus is trailed along the surface and its vertical displacement is recorded. The roughness of the surface is defined as an average deviation from the mean value of the vertical displacement measurements (Fig. 4),

$$R_a = \frac{1}{L}\int_0^L |y(x)|\,dx \tag{16.1}$$

where L is the sampling length.

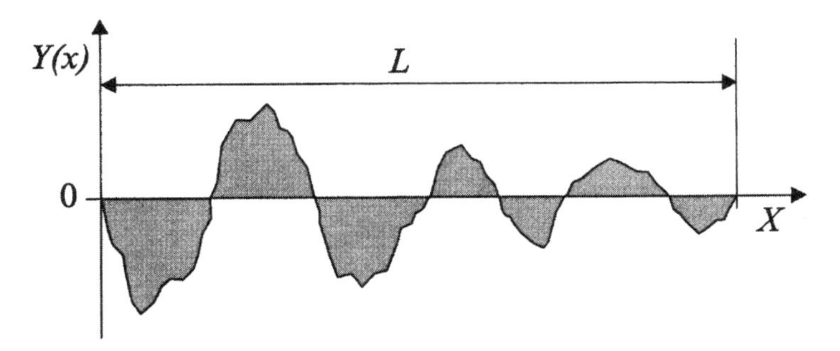

FIGURE 4 Surface profile.

Mechanical systems such as the stylus instrument can measure roughness in the order of thousandths of a millimeter (or microinches). However, it should be noted that, despite the minimum force applied on the stylus tip, a trace might be left on the surface owing to the minute diameter of the diamond tip. Thus for surface roughness measurements that require higher precisions, an interferometry-based device can be used for nondestructive inspection.

Electro-Optical Measurement of Length

A variety of electro-optical distance/orientation measurement devices have been discussed in Chap. 13 and thus will not be addressed here in any great detail. These devices can be categorized as focused beam (i.e., use of a laser light) or as visual (i.e., use of a CCD camera) inspection systems. The former systems are highly accurate and in the case of interferometers can provide resolutions as low as half a light wavelength or better. The latter (camera-based) systems are quite susceptible to environmental disturbances (e.g., changes in lighting conditions) and are also restricted by the resolution of the (light receiving) diodes. Thus, for high-resolution systems, CCD camera–based inspection systems should be coupled to high-resolution optical microscopes.

For surface roughness measurement, interferometric optical profilometers can be used for the inspection of highly smooth surfaces in a scale of

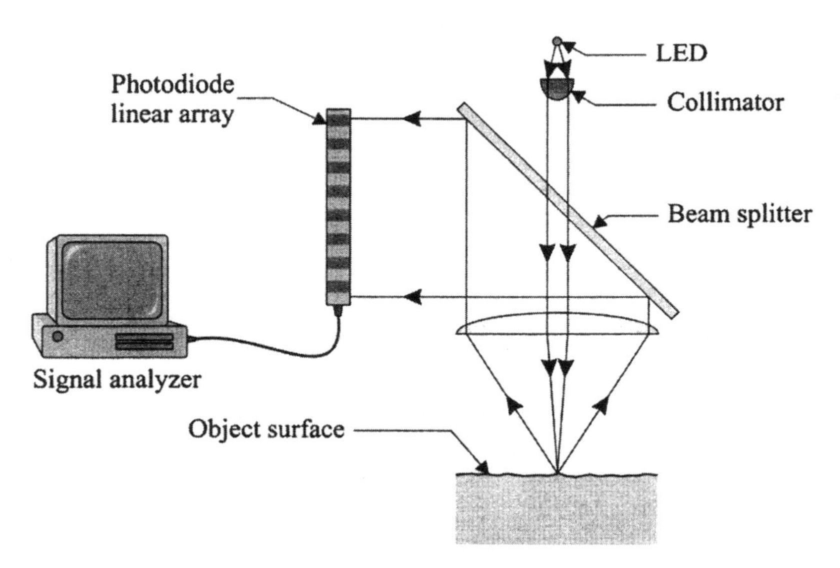

FIGURE 5 An optical surface roughness inspection instrument.

nanometers, such as optical lenses and metal gages used to calibrate other measurement instruments. In the case of intermediate microroughness products, one can utilize a light scattering technique, in a scale of better than micrometers: such devices correlate the intensity of specularly reflected light to surface roughness (R_a). Smoother surfaces have a higher fraction of the incident light reflected in the specular mode (versus diffusive) with a clear Gaussian distribution. Such a commercially available (Rodenstock) surface-roughness-inspection instrument is shown in Fig. 5.

X-Ray Inspection

Electromagnetic radiation (x rays or gamma rays) can be effectively utilized for the inspection of a variety of (primarily metal) products in on-line or off-line mode. Measurements of features are based on the amount of radiation absorbed by the product subjected to (in-line) radiation. The intensity of radiation and exposure times are dictated by material properties (i.e., attenuation coefficient). The amount of absorbed radiation can be correlated to the thickness of the product (in-line with the radiation rays) and thus be used for thickness measurement or detection of defects.

In the most common transmissive x-ray radiographic systems, the radiation source is placed on one side of the product, while a detector (e.g., x-ray film, fluorescent screen) is placed on the opposite side (Fig. 6). In cases where one cannot access both sides of a product, the x-ray system can be used in a backscattering configuration: the detector, placed near the emitter, measures the intensity of radiation scattered back by the product. The thicker the product, the higher the level of backscatter will be.

Computed tomography (CT) is a radiographic system capable of yielding cross-sectional images of products whose internal features we wish

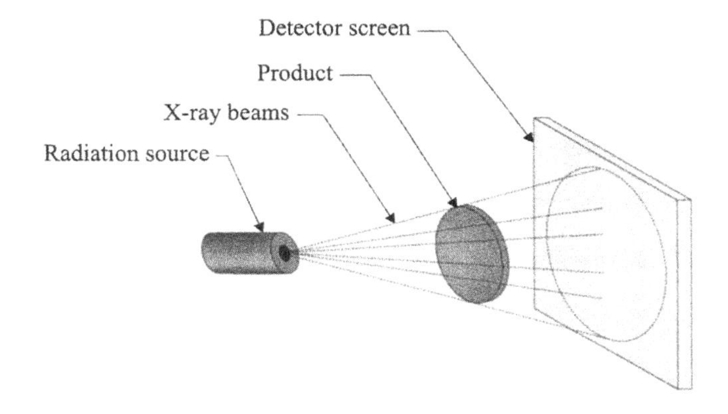

FIGURE 6 Transmissive x-ray imaging.

to examine. CT machines typically utilize a fan-beam-type x-ray source and a detector array (placed on opposite sides of a product) rotating synchronously around an axis through the product (Fig. 7). A series of x-ray images (up to 1,000) that are collected after a complete 360° rotation around the product are then reconstructed into a cross-sectional 2-D image via mathematical tools. Through an (orthogonal) translation along the rotational axis, several 2-D cross-sectional images can be collected and utilized for 3-D (volumetric) reconstruction. One must note, however, that CT is primarily useful for product geometries with low aspect ratios—i.e., nonplanar. Furthermore, even with today's available computing power, CT-based image analysis may consume large amounts of time unacceptable for on-line inspection.

X-ray laminography is a variant of the CT system developed for the inspection of high-aspect-ratio products. A cross-sectional image of the product is acquired by focusing on a plane of interest, while defocusing the planes above and below via blurring of features outside the plane of interest (i.e., reducing their overall contrast effect). This laminographic effect of blurring into the background is achieved though a synchronized rotational motion of the x-ray source and the detector, where any point in the desired focal plane is always projected onto the same point in the image (Fig. 8). During the rotation of the source and detector a number of images are taken and subsequently superimposed. Features on the focal plane maintain their sharpness (since they always occupy the same location in every image and

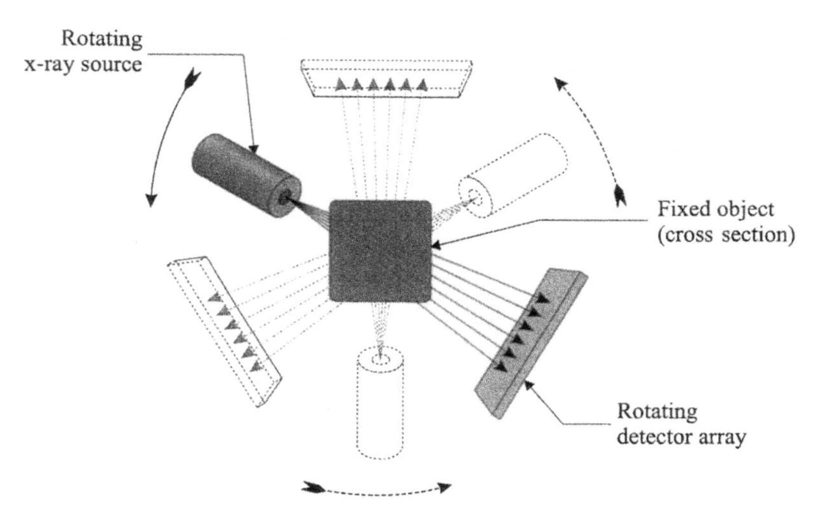

FIGURE 7 Computed tomography.

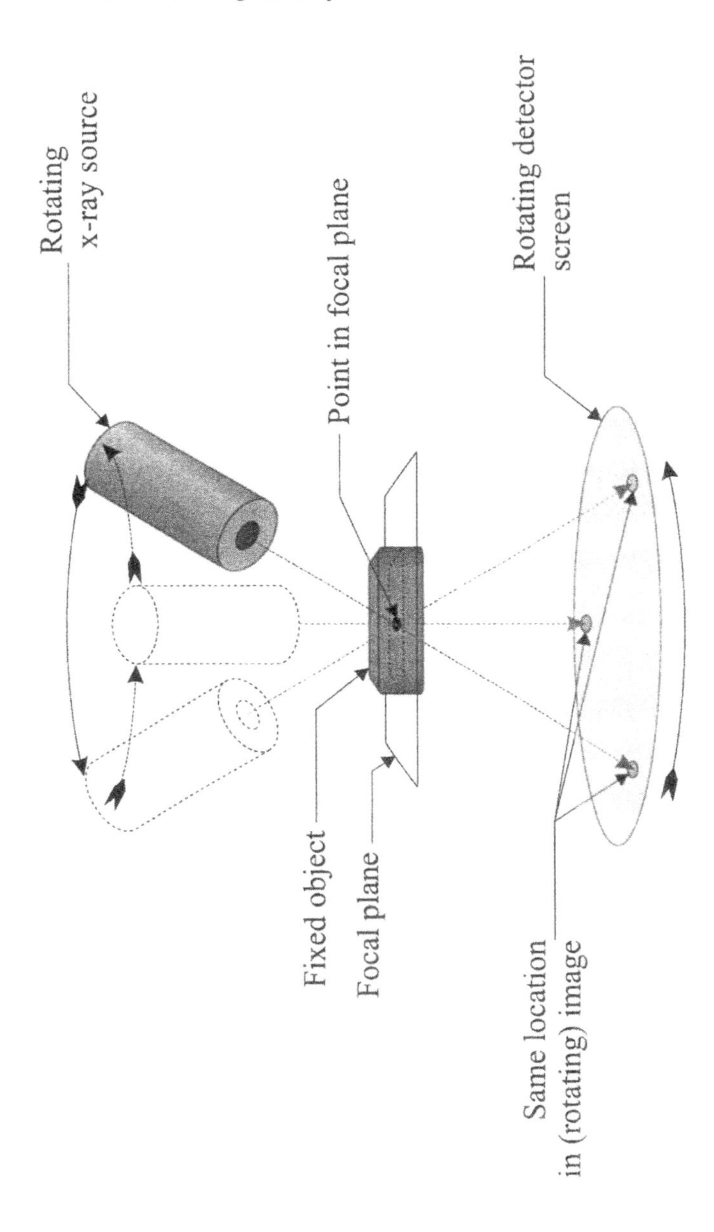

FIGURE 8 X-ray laminography.

yield perfect overlapping), while out-of-plane features get blurred into a (gray) background (since they never occupy the same location in every image).

As in CT systems, different 2-D cross-sectional images, obtained by translating the product in an orthogonal direction, can be used to reconstruct a 3-D representation of the product. However, one must first overcome the blurring effect generated by the laminographic movement of the source–detector pair.

In all x-ray radiography systems, transmissive, CT, and laminography, mirrors can be used to reflect the image formed on a phosphor screen onto a visible-light CCD array camera for the automatic analysis of measurement data.

16.3 BASICS IN PROBABILITY AND STATISTICS THEORIES

Statistics theory is concerned primarily with the collection, analysis, and interpretation of experimental data. The term experiment is a generic reference to any process whose (random) outcome is measured for future planning and/or control activities. Probability theory, on the other hand, is concerned with the classification/representation of outcomes of random experiments. It attempts to quantify the chance of occurrence of an event. The term event is reserved to represent a subset of a sample space (the complete set of all possible outcomes of a random experiment).

The study of risk in modern times can be traced to the Renaissance period in Europe, when the mathematicians of the time, such as B. Pascal in the mid 1600s, were challenged by noble gamblers to study the games of chance. In 1730, A. de Moivre suggested that a common probability distribution takes the form of a bell curve. Next came D. Bernoulli's work on discrete probability distributions and T. Bayes' work on fusing past and current data for more effective inference, both in the mid-1700s. In the early part of the 1800s, C. F. Gauss further enforced the existence of a bell curve distribution based on his extensive measurements of astronomical orbits. He observed that repeated measurements of a variable yield values with a given variance about a mean value of the variable. Today, the bell-curve distribution is often called the Gaussian probability distribution (or the "normal" distribution).

16.3.1 Normal Distribution

Probability distributions can be classified as discrete or continuous. The former type is used for the analysis of experiments that have a finite number of outcomes (e.g., operational versus defective), while the latter type is used for experiments that have an infinite number of outcomes (e.g., weight, length, life). Both types have a number of different distributions within their own class: for example, binomial versus Poisson for discrete and Gaussian (normal) versus gamma for continuous probability distributions. In this chapter, since our focus is on the statistical quality control of manufacturing processes whose outputs represent continuous metrics, only the normal distribution is reviewed.

In practical terms, the variance of a process output (for a fixed set of input control parameters) can be viewed as random noise superimposed on a desired signal. For a perfectly calibrated system (with no systematic, nonrandom errors), the variance in the output can be seen as a result of random noise present in the environment and that cannot be eliminated. This noise, ε, would commonly have a normal distribution with a given variance, $\sigma^2 \neq 0$, and zero mean, $\mu = 0$, value (Fig. 9).

For the case where the desired output signal, μ ($\neq 0$), is superimposed with normally distributed noise, represented by the variance, σ^2, the random measurements of the output variable, X, are represented by the probability distribution function

$$f(x) = \frac{1}{\sigma\sqrt{2\pi}} \exp\left(-\frac{1}{2} \left[\frac{x-\mu}{\sigma} \right]^2 \right) \qquad -\infty < x < \infty \tag{16.2}$$

where the variable, X, is of the continuous type and $f(x) \geq 0$. One must note that, although, for a specific variable value, x_0, the corresponding $f(x_0)$ value is nonzero, the actual probability of this measurement value to occur in practice is near zero [i.e., $P(X = x_0) \approx 0$]. This is true because there exist

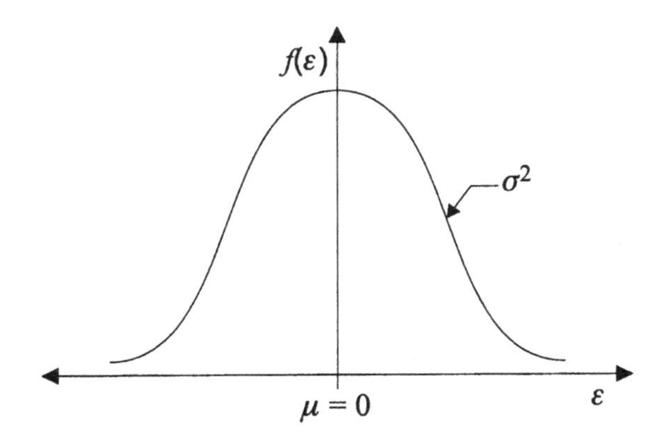

FIGURE 9 Normally distributed noise.

infinite possible of outcomes to the experiment, Eq. (16.2), where each outcome has a near zero probability of occurrence. Therefore, in continuous probability distributions, the probability of occurrence is specified for a range of measurements, as opposed to for a specific outcome.

The probability of X being in a given range $[x_1, x_2]$ is defined by the integral of the probability function (Fig. 10):

$$P(x_1 < X < x_2) = \int_{x_1}^{x_2} f(x)\, dx \tag{16.3}$$

The lack of computers and hand-held electronic calculators prior to the 1950s, which could have been used for the calculation of integrals [such as the one in Eq. (16.1)], led to the normalization of the Gaussian distribution with respect to (μ_x, σ_x) and generation of quick-reference lookup tables. The normalization was achieved by using the transformation variable

$$Z = \frac{X - \mu_x}{\sigma_x} \tag{16.4}$$

where $P(x_1 < X < x_2) = P(z_1 < Z < z_2)$. The Z-distribution is characterized by $\mu_z = 0$ and $\sigma_z{}^2 = 1$ (Fig. 11).

Evaluation of the integral in Eq. (16.3), for a normal distribution, Eq. (16.2), yields the probability values commonly referred to in engineering measurements:

$$P(\mu_x - \sigma_x < X < \mu_x + \sigma_x) = P(-1 < Z < 1) \cong 68.26\%$$
$$P(\mu_x - 2\sigma_x < X < \mu_x + 2\sigma_x) = P(-2 < Z < 2) \cong 95.44\%$$
$$P(\mu_x - 3\sigma_x < X < \mu_x + 3\sigma_x) = P(-3 < Z < 3) \cong 99.74\%$$

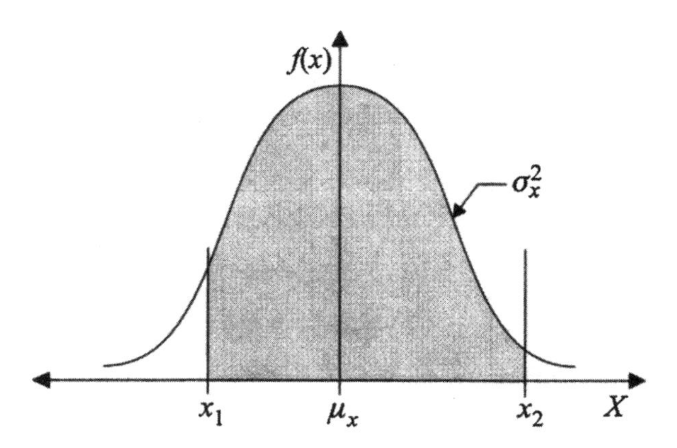

FIGURE 10 Probability of $(x_1 < X < x_2)$.

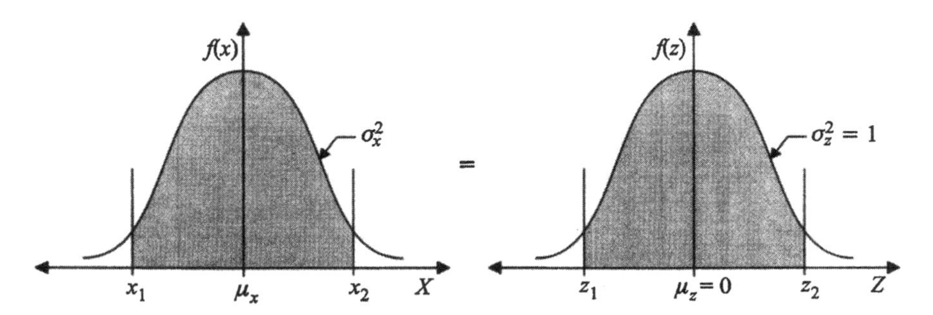

FIGURE 11 Equivalence of probability distributions.

16.3.2 Sampling in the Normal Distribution

As discussed in Sec. 16.3.1, if one knows the two metrics (statistics) (μ_x, σ_x) of a normally distributed population of measurements, the probability of a random outcome to be in the range [x_1, x_2] can be calculated using Eq. (16.3). However, in practice, the statistics (μ_x, σ_x) are not readily available, but must instead be approximated by sampling. Based on a sample drawn from the infinite-size population, one would estimate the upper and lower limits for the true (μ_x, σ_x) values at some confidence level. The first step in understanding this estimation process, however, is the analysis of the sampling process.

For a normally distributed population of measurements (i.e., random outcomes of an experiment), samples of size n are characterized by the metrics sample mean, $\overline{X}$, and sample variance, S^2:

$$\overline{X} = \frac{1}{n}\sum_{i=1}^{n} x_i \qquad \text{and} \qquad S^2 = \frac{1}{n-1}\sum_{i=1}^{n}(x_i - \overline{X})^2 \tag{16.5}$$

Furthermore, it can be shown that the distribution of sample means is characterized by a normal distribution and the distribution of sample variances can be defined by a chi-squared distribution.

Distribution of Sample Means

The mean values of samples of size n, $\overline{X}$, drawn from a population of normally distributed individual x_i values, $i = 1$ to n, also has a normal distribution with the statistics

$$\mu_{\overline{x}} = \mu_x \qquad \text{and} \qquad \sigma_{\overline{x}}^2 = \frac{1}{n}\sigma_x^2 \tag{16.6}$$

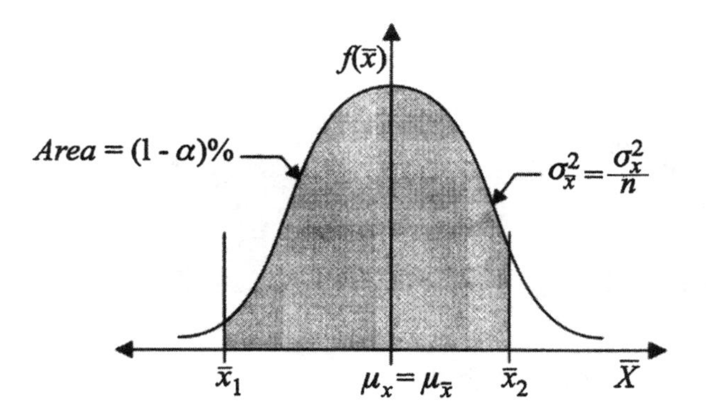

FIGURE 12 Sample mean distribution.

Therefore, based on Eqs. (16.2), (16.3), and (16.6), one can calculate the probability of a randomly drawn sample of size n to have a mean value in the range $[\bar{x}_1, \bar{x}_2]$ (Fig. 12):

$$P(\bar{x}_1 < \bar{X} < \bar{x}_2) = \int_{\bar{x}_1}^{\bar{x}_2} f(\bar{x}) d\bar{x} \tag{16.7}$$

As an example, let us consider that a machine is set to produce resistors of a nominal resistance value equal to 2 ohms. Based on the process capability of the machine, one assumes that a normally distributed noise affects the output of this machine, where $\mu_\varepsilon = 0$ and $\sigma_\varepsilon^2 = 0.01$. Analysis of this population's statistics indicates that a randomly chosen resistor has a resistance value, X, in the range 1.743 to 2.257 ohms with 95% certainty (probability). Furthermore, the analysis also indicates that a future randomly chosen sample of $n = 30$ resistors would have a mean value, $\bar{X}$, in the range 1.953 to 2.047 ohms with 95% probability, since $\mu_{\bar{x}} = 2$ and $\sigma_{\bar{x}}^2 = 0.002$.

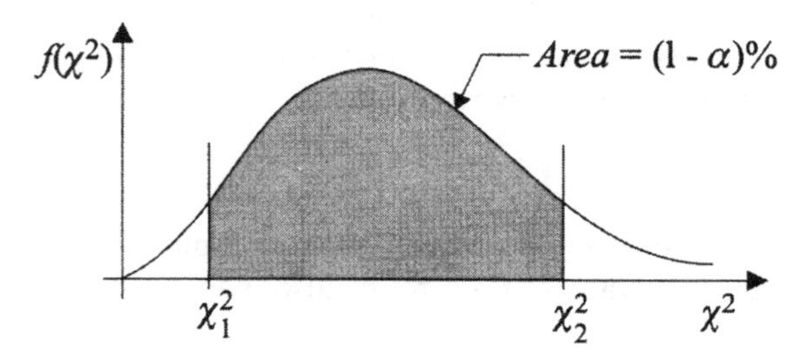

FIGURE 13 Chi-square distribution.

Distribution of Sample Variances

The variance values of samples of size n, S^2, drawn from a population of normally distributed individual x_i values, $i=1$ to n, would have a chi-squared (χ^2) distribution (Fig. 13),

$$f(u) = \left(2^{k/2}\,\Gamma\left(\frac{k}{2}\right)\right)^{-1} u^{(k/2)-1}e^{-u/2} \qquad u > 0 \tag{16.8}$$

where the variable u refers to χ^2, Γ is the gamma function, and $k=n-1$ The variance, S^2, is expressed as a function of the χ^2 variable in Eq. (16.8) as follows,

$$\chi^2 = \frac{kS^2}{\sigma_x^2} \tag{16.9}$$

Based on the integration of Eq. (16.8), between the two limits $[\chi_1^2, \chi_2^2]$, one can calculate the probability of a randomly drawn sample of size n to have a variance value in the range $[s_1^2, s_2^2]$, where the conversion from S^2 to χ^2 is achieved via Eq. (16.9),

$$P(s_1^2 < S^2 < s_2^2) = P(\chi_1^2 < \chi^2 < \chi_2^2) = \int_{\chi_1^2}^{\chi_2^2} f(\chi^2)d\chi^2 \tag{16.10}$$

For the above-considered example of a population of resistors with a normal distribution, $\mu_x=2$ ohms and $\sigma_x^2=0.01$, let us consider drawing a random sample of size $n=30$. Based on Eqs. (16.8) to (16.10), it can be shown that the variance of this sample would be in the range $(0.0055 < S^2 < 0.0158)$ with 95% certainty (probability), where $\chi_1^2=16.047$ and $\chi_2^2=45.722$.

16.3.3 Estimation of Population Statistics

In Sec. 16.3.2 above, the behavior of sample statistics was discussed while assuming that the population statistics, (μ_x, σ_x), are known. In practice, however, the population statistics are not known and must be estimated using the statistics of one or more randomly drawn samples. The outcome of this estimation process is a range for the population mean and a range for the population variance: $[\mu_L, \mu_U]$ and $[\sigma_L^2, \sigma_U^2]$ for a $(1-\alpha)$% confidence level, where $(1-\alpha)$ is the area under the distribution curves between the two limits (Figs. 12 and 13).

For example, let us consider a randomly chosen sample of size $n=30$, whose statistics are $\bar{x}=1.98$ ohms and $s^2=0.012$. It can be shown that, based on these sample statistics, for a confidence level of 95%, the estimated ranges of the population statistics would be

$$1.9285 < \mu_x < 2.0315 \qquad \text{and} \qquad 0.0076 < \sigma_x^2 < 0.0217$$

We are only 95% confident that the above ranges are valid. There exist a 5% chance that the sample drawn may not have its statistics within the 95% confidence-level limits set about the true (μ_x, σ_x) values of the population, thus yielding invalid range estimates for the population statistics.

Although we can estimate ranges (i.e., confidence intervals) for both μ_x and σ_x as will be discussed later in this chapter, most quality-control procedures only recommend the use of a (large) sample's statistics as the population statistics: $\mu_x \cong \overline{X}$ and $\sigma_x^2 \cong S^2$.

16.4 PROCESS CAPABILITY

Specification limits (or tolerance limits) define conformance boundaries for a product's characteristics as specified or dictated by design requirements. Such limits must be carefully defined as constraints to be satisfied and not arbitrarily chosen. One must remember that the tighter the tolerance limits, the higher the cost of achieving them.

The problem of satisfying the specification limits can be formulated as a typical optimization problem, where the objective function to be minimized is the deviation of an individual product characteristic value, X, from the desired nominal value, μ_x,

$$\begin{aligned} &\text{Min}(X - \mu_x) \\ &\text{subject to} \qquad LSL_x \leq X \leq USL_x \end{aligned} \tag{16.11}$$

where LSL_x and USL_x are the specification limits on X.

The quality (or cost) ramifications of satisfying the specification limits of a product characteristic have commonly been addressed by evaluating the capability of a process to satisfy these limits. All process capability indices used by the manufacturing industry attempt to quantify the variance of the process output with respect to the engineering-defined range of specification limits. Assuming a normal distribution of output values about a desired mean value, most indices employ a variance range of $\pm 3\sigma_x$. The commonly used C_{pk} index is

$$C_{pk} = \min\left[\frac{\mu_x - LSL_x}{3\sigma_x}, \frac{USL_x - \mu_x}{3\sigma_x}\right] \tag{16.12}$$

As an example, let us consider the production of resistors by a machine whose population variance is defined as $\sigma_x^2 = 0.01$. If the specification limits on the desired $\mu_x = 2$ ohms were to be set as $LSL_x = 1.85$ ohms and $USL_x = 2.3$ ohms (note the unequal limits), then the C_{pk} index would assume a value of 0.5. The higher the value of C_{pk}, the lower the percentage of products outside the specification limits. For this example, using Eq. (16.3), one can determine that about 6.8% of products fall outside the specification limits.

In practice, as mentioned above, the population statistics (μ_x, σ_x^2) are not available. A simple approach to coping with this problem is to use a sample's statistics, ($\overline{X}$, S^2), as approximations to population statistics: that is, use $\mu_x \cong \overline{X}$ and $\sigma_x \equiv S$ in Eq. (16.12). A $\hat{C}_{pk}$ value obtained using these approximated values can be called a middle-of-the-road or a liberal estimate of the true index value. Alternatively, one can calculate a range for $\hat{C}_{pk}$ using the two ranges determined for the population statistics for a certain $(1-\alpha)\%$ confidence level (Fig. 14),

$$\frac{\mu_L - LSL_x}{3\sigma_U} < \hat{C}_{pk} < \frac{USL_x - \mu_U}{3\sigma_U} \tag{16.13}$$

where $[\mu_L, \mu_U]$ and $[\sigma_L^2, \sigma_U^2]$ are the lower and upper limits of the population's statistics calculated from a sample's statistics.

The process capability index is a simple measure of variance normalized with respect to the product specification limits. A process could be quite capable of manufacturing one product (e.g., $C_{pk} \geq 1.5$), while labeled a poor process for another product (e.g., $C_{pk} \leq 0.5$), while having the same variance, σ_x^2, during the manufacturing of both products.

For a perfectly calibrated process, the variance is a result of random-error mechanisms. Thus, when faced with a process capability problem, the manufacturing engineer must cope with a common practical dilemma in

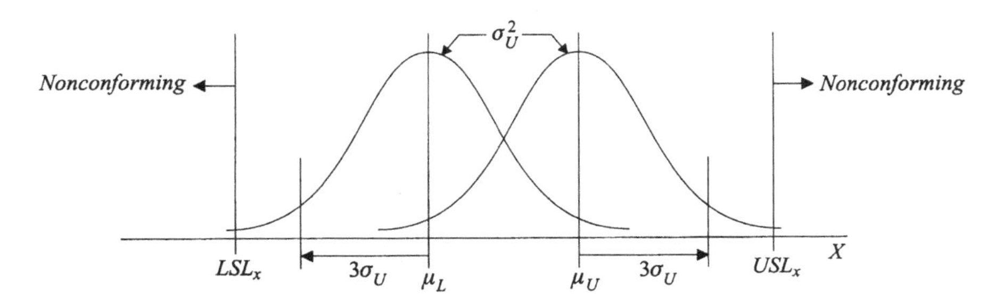

FIGURE 14 Approximation of process capability range.

order to provide the customers with products within the specification limits: (1) Acquire and utilize a new machine/process that can yield an acceptable C_{pk} value, or (2) inspect all the products and scrap those outside the specification limits. The latter is a short-term solution and should be employed only for unusual and very infrequent product orders. The former solution must be chosen if the C_{pk} index is frequently lower than the acceptable value.

16.5 STATISTICAL PROCESS CONTROL

All manufacturing processes must be controlled in a closed loop mode. The input variables of a process should be adjusted in real time in response to unacceptable deviations of the output variables from their nominal values. In the context of statistical process control (SPC), the question that needs to be answered is, Is the deviation from the nominal, $\Delta X = |\mu_x - X|$, statistically significant to require intervention in the input variable, ΔY? The common answer is that we should change the input variable value of a process only in response to an assignable cause other than a random mechanism. The SPC concept is discussed below first via an example.

Let us consider a bottle-filling operation: bottles of a soft-drink company are required to be filled on average with 330 mL liquid, (i.e., $\mu_x = 330$ mL). The machine used for this process is controlled via a timer that regulates the flow of liquid into each bottle. For a constant filling rate of 150 mL/sec, the timer is set to keep the liquid flow on for 2.2 seconds, corresponding to $\mu_x = 330$ mL per bottle. This (calibrated) machine is known to have a timing variance that translates into an output variance of $\sigma_x^2 = 4$ (i.e., variance in volume of liquid filled) per bottle. Thus, assuming a normal distribution, one can conclude that the output of this machine, X, is expected to have the population statistics of $\mu_x =$ 330 mL and $\sigma_x^2 = 4$.

In order to provide (closed loop) feedback control to this process, the bottles are weighed after they have been filled and the amount of volume is calculated. It is assumed that variations in (empty) bottle weights are negligible compared to variations in liquid volumes. As discussed above, the objective of SPC is to determine whether variations in output are due to only a random mechanism or to other assignable causes as well. The SPC process advocates monitoring variations through sampling, $\overline{X}$. It can be shown that variations due to assignable causes can be better detected by examining sample statistics, $(\overline{X}, S)$ versus individual X values.

In the bottle-filing example, it can be shown that [via Eqs. (16.6) and (16.7)] samples of size $n = 5$ have their sample mean values in the range

$327.3 < \overline{X} < 332.7$ in 99.74% of cases. Thus, in the absence of any identifiable trend, if 99.74% of the samples collected at regular intervals have their means in the above range, while the sample variances also satisfy a random behavior requirement, one can choose not to vary the input variable (i.e., the timer value of 2.2 sec). This SPC-based control example is schematically illustrated in Fig. 15. The specific details of calculating the appropriate control limits and examining the sampling data for determining random versus assignable-cause-based variance are discussed in the subsections below.

It is important to note that control limits for SPC define only statistical capability limits of the specific process considered. They define population percentages expected to be within statistical limits about the mean value of the population (i.e., upper and lower control limits, *UCL* and *LCL*, respectively). For example, when sampling a process output that has a normal distribution, one can expect 99.74% of sample means to be within the range $\mu_{\overline{X}} - 3\sigma_{\overline{X}} \leq \overline{X} \leq \mu_{\overline{X}} + 3\sigma_{\overline{X}}$, where $\mu_{\overline{X}} = \mu_x$ and $\sigma_{\overline{X}} = \sigma_x/\sqrt{n}$.

SPC limits are not specification limits (tolerances), which are specified by product designers/engineers regardless of the process variance that defines the statistical control limits. A process can be perfectly in control (operating subject to statistical random errors) while yielding a large percentage of defects: this phenomenon indicates a poor process capability and not any control problems.

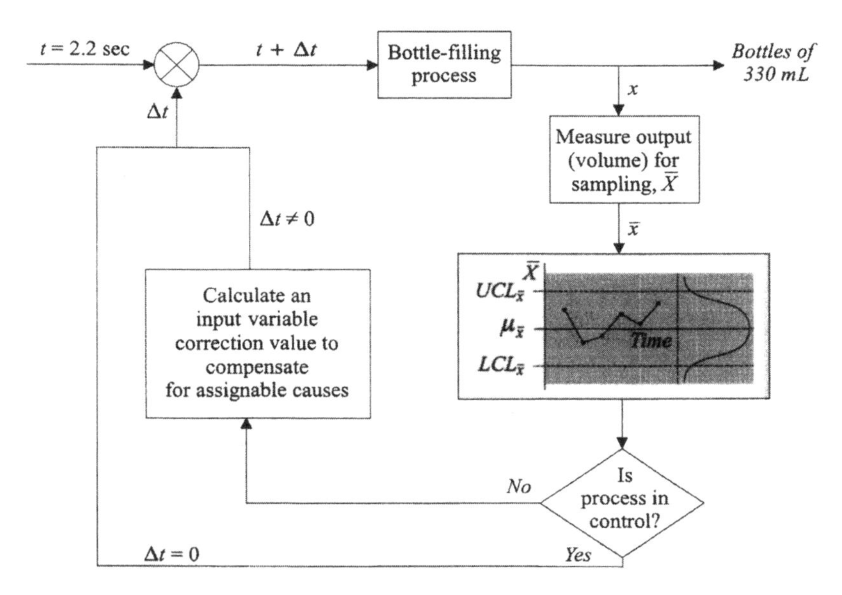

FIGURE 15 Statistical process control example.

16.5.1 $\overline{X}$-*R* Control Charts for Variable Data

SPC can be carried out by monitoring the magnitude of (continuous) variable output data or by monitoring the (binary) conformity of a product to desired specification. The former is commonly called the SPC of variable data, while the latter is called the SPC of attribute data. The focus of this section is only on the determination of control charts for variable data.

The most commonly used control charts for variable data are the $\overline{X}$-R charts (R: range), which must be utilized as a pair. The $\overline{X}$ chart is used to track potential deviations from the desired population mean value, μ_x, while the R-chart is used to track potential changes in process variance, σ_x^2. A process is in control if both charts indicate compliance.

Control charts provide a graphical user interface for the tracking of the process output with respect to its expected statistical behavior. A control chart is a two-dimensional plot of a sample statistic versus time. Compliance with process control requirements can be checked by verifying that an expected proportion of sample statistics are within the control limits as well as by monitoring trends.

In an $\overline{X}$ chart, the SPC control limits should be calculated for a certain statistical range utilizing the true mean value for the population of measurements, which we assume to have a normal distribution. In practice, this range is, typically, set to include the 99.74% of the sample means, though it could be set for any other percentage:

$$LCL_{\bar{x}} = \mu_x - \frac{3\sigma_x}{\sqrt{n}} \quad \text{and} \quad UCL_{\bar{x}} = \mu_x + \frac{3\sigma_x}{\sqrt{n}} \tag{16.14}$$

where n is the size of the samples collected.

In the absence of true (μ_x, σ_x) values known to us, we must use approximations. Historically, these population statistics have been approximated by

$$\mu_x = \mu_{\bar{x}} \equiv \overline{\overline{X}} \quad \text{and} \quad \sigma_x \cong \frac{\overline{R}}{d_2} \tag{16.15}$$

where

$$\overline{\overline{X}} = \frac{1}{k}\sum_{j=1}^{k}\bar{x}_j \quad \text{and} \quad \overline{R} = \frac{1}{k}\sum_{j=1}^{k}R_j = \frac{1}{k}\sum_{j=1}^{k}(x_{\max} - x_{\min})_j$$

Above, $\bar{x}_j$ is the mean value of the jth sample, $j = 1$ to k (k is usually about 20), collected prior to starting the SPC process; R_j is the range of all x_i, $i = 1$

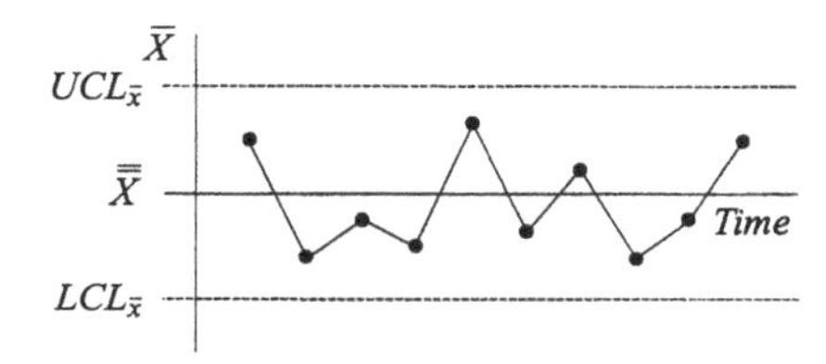

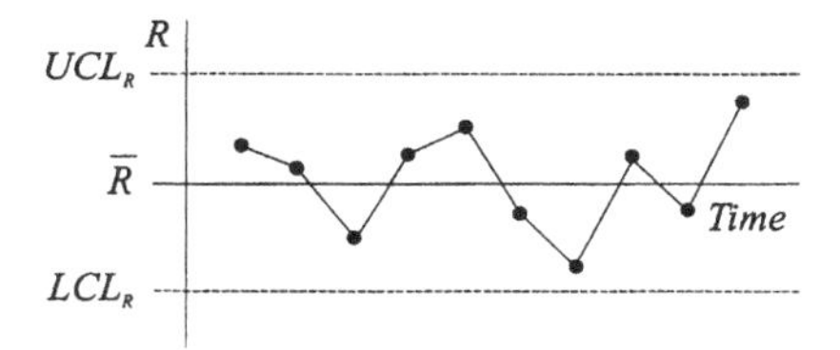

FIGURE 16 $\overline{X}$-R charts.

to n (n is usually 4 to 6), measurements within this sample; and, d_2 is a correction factor, whose value is a function of the samples' size, n.

The limits for the $\overline{X}$ and R charts based on the above approximations are, then, defined as (Fig. 16)

$$LCL_{\bar{x}} = \overline{\overline{X}} - 3\frac{\overline{R}}{d_2\sqrt{n}} \quad \text{and} \quad UCL_{\bar{x}} = \overline{\overline{X}} + 3\frac{\overline{R}}{d_2\sqrt{n}} \tag{16.16a}$$

$$LCL_R = \overline{R} - 3\frac{d_3}{d_2}\overline{R} \quad \text{and} \quad UCL_R = \overline{R} + 3\frac{d_3}{d_2}\overline{R} \tag{16.16b}$$

where d_3 is a correction factor whose value is a function of the samples' size, n. One must note that LCL_R cannot have a value below zero.

The use of a range of measurements, R, as opposed to sample variance, S, can be attributed to the absence of portable electronic calculators (or personal computers) on the factory floors of the first half of the 20th century. (In the next subsection, the use of $\overline{X}$-S charts will be reviewed.)

The approximation of σ_x in Eq. (16.16a) is an acceptable solution to the unavailability of population variance. However, the approximations used in the definition of the R chart may not be acceptable to some. As discussed in Sec. 16.3.2, sample variances S, follow a chi-squared distribution. However, the distribution used for the R values in Eq. (16.6b) is Gaussian, with some correction factors (d_2, d_3). This approximation may yield unacceptable conclusions at the extremes (near control limit values).

16.5.2 $\overline{X}$-S Control Charts for Variable Data

The pair of $\overline{X}$-S charts may also be utilized for SPC purposes. As for $\overline{X}$-R charts, it has been customarily assumed that both sample mean and sample variance values have normal distributions. In the absence of true μ_x and σ_x

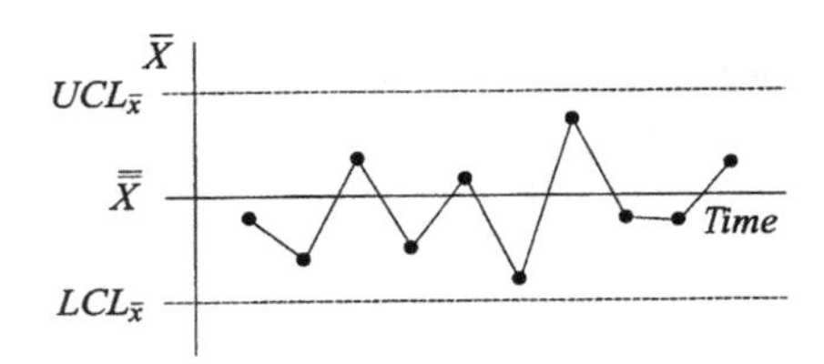

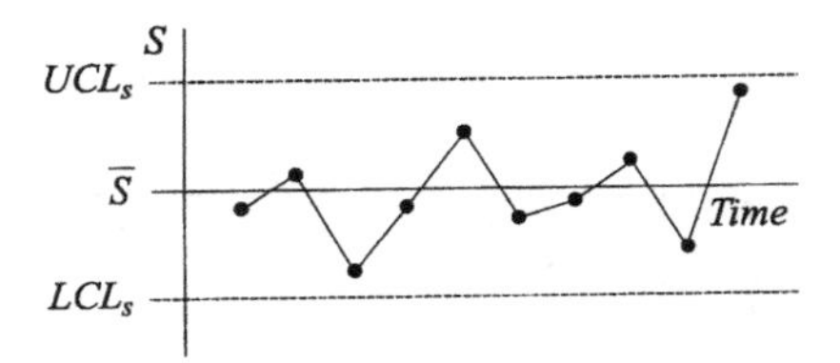

FIGURE 17 $\overline{X}$-S charts.

values, the following approximations have been used to define the control limits for 99.74% population ranges (Fig. 17):

$$LCL_{\overline{x}} = \overline{\overline{X}} - 3\frac{\overline{S}}{c_4\sqrt{n}} \quad \text{and} \quad UCL_{\overline{x}} = \overline{\overline{X}} + 3\frac{\overline{S}}{c_4\sqrt{n}} \tag{16.17a}$$

$$LCL_s = \overline{S} - 3\frac{\overline{S}\sqrt{1-c_4^2}}{c_4} \quad \text{and} \quad UCL_s = \overline{S} + 3\frac{\overline{S}\sqrt{1-c_4^2}}{c_4} \tag{16.17b}$$

where $\overline{S}$ is the mean of the k samples' standard deviations used to calculate the control limits (each sample of size n), and c_4 is a correction factor used to compensate for the normal distribution approximation of the actual chi-squared distribution of the sample variances. As for R charts, LCL_s cannot have a value below zero.

16.5.3 Implementation and Interpretation of Control Charts

SPC is based on monitoring output data and providing feedback information to the process controller. The monitoring and analysis of output data is carried out through sampling theory. Once the type of control charts is chosen ($\overline{X}$-R versus $\overline{X}$-S), the next step is gathering data for the calculation of control limits: for most applications, it is recommended that 20 to 30 samples (k = 20 to 30), each of the same size n = 4 to 6, be collected and approximations for population statistics be established. These are

For $\overline{X}$-R charts,

$$\mu_{\overline{x}} \cong \overline{\overline{X}} \quad \text{and} \quad \sigma_{\overline{x}} \cong \frac{\overline{R}}{d_2\sqrt{n}}$$

$$\mu_R \cong \overline{R} \quad \text{and} \quad \sigma_R \cong \frac{\overline{R}d_3}{d_2}$$

For $\overline{X}-S$ charts;

$$\mu_{\bar{x}} \cong \overline{\overline{X}} \quad \text{and} \quad \sigma_{\bar{x}} \cong \frac{\overline{S}}{c_4\sqrt{n}}$$

$$\mu_s \cong \overline{S} \quad \text{and} \quad \sigma_s \cong \overline{S}\frac{\sqrt{1-c_4^2}}{c_4}$$

The control limits for the SPC charts need to be calculated from the desired population range to be considered. In Secs. 16.5.1 and 16.5.2, the 99.74% range was utilized in the derivation of these limits. Once the control limits have been established from preliminary training data, the process can be started and its output monitored by frequent sampling: The exact frequency of data collection is a function of the reliability of the process and the cost of data collection.

Interpretation of data gathered is simply the application of probability theory: First, the limits chosen dictate what percentage of data could be allowed to fall outside the limits (e.g., 0.26% for the 99.74% limits used in this section). Second, if we assume a reasonable approximation of the center lines (i.e., $\mu_{\bar{x}}$, μ_R, and μ_s) the data points should be equally distributed on both sides of the center lines. We can claim that the probability of having two consecutive points on one side of the center line is equal to (50% × 50% =) 25%, for three consecutive points it is 12.5%, and so forth. Therefore rules can be established to monitor out-of-the-ordinary sequences of occurrences, e.g., 7 out of 7 consecutive, 10 out of 11 consecutive, 14 out of 17 consecutive points one side of the centerline could be considered as indicating a process going out of control, since each case would approximately have a 1% chance of occurrence.

Other symptoms of a process potentially going out of control include cyclic behaviors, high ratios of data points near the control limits (in contrast to the expected normal behavior of having most points around the centerlines), sudden spikes, trends of points showing steady increase/ decrease in values, and so on.

16.6 ISO 9000

ISO 9000 is a family of standards on quality management systems and related supporting standards on terminology and specific tools. In ISO 9000, quality refers to "all product features that are required by the customer." Quality management refers to "all actions that an organization must carry out to ensure that its products conform to the customer's requirements: quality assurance, quality control, and quality improvement." Quality assurance encompasses all planned activities required to provide adequate confidence that a product/process fulfils the quality requirements, such as documenting plans and specifications, reporting results, and so on. Quality control encompasses all operational procedures necessary to fulfill the quality requirements, such as measuring conformity in real time using appropriate sensors and providing feedback to the process controller. Quality improvement encompasses all actions that yield beneficial changes in quality performance, such as reducing the spread of variations in a manufacturing processes, or reducing failure rates.

ISO 9000 dictates the way an organization carries out its work and not directly the result of this work. It is about processes and not products. It specifies generic requirements for compliance, as opposed to most other standards that specify technical engineering specifications or other precise criteria to be used consistently as rules or guidelines to ensure that products, processes, and services are fit for their purposes.

16.6.1 ISO 9000:1994

One of the original standards established for quality management programs was issued by the U.S. Department of Defense, MIL Q 9858A, in 1959. This standard was followed by NATO's Allied Quality Assurance Publication (AQAP-1) in 1968 and by the U.K. Ministry of Defense standard Def Stan 05-08 in 1970. By the mid-1980s, many countries had developed quality management standards that no longer heavily targeted military products as did their predecessors during the period 1960 to 1980. In 1987, the first international quality management standard was issued by the International Organization for Standardization (ISO). However, most of the countries involved (over 25) in the development of this standard adopted national equivalents, as opposed to the original ISO standard. The subsequently published ISO 9000 series (family) of standards in 1994 (known as ISO 9000:1994) were, however, more successfully adopted by the participating countries.

The ISO 9000:1994 family of standards allowed organizations to choose one of three standards, tailored for specific quality management system applications, ISO 9001, ISO 9002, and ISO 9003 for registration. All organizations, however, were encouraged to implement the fourth standard, ISO 9004, which stated the exact quality management requirements that would lead to certification under one of the three quality assurance standards, ISO 9001, ISO 9002, or ISO 9003. The primary members of the 1994 version of the ISO 9000 family were

ISO 8402:1994: Quality management and quality assurance vocabulary
ISO 9000:1994: Guidelines for the selection and application of ISO 9001, ISO 9002, and ISO 9003

ISO 9001:1994: Model for quality assurance in design, development, production, installation, and servicing

ISO 9002:1994: Model for quality assurance in production, installation, and servicing

ISO 9003:1994: Model for quality assurance in final inspection and test

ISO 9004:1994: Guidelines for quality management and quality system elements

Despite exact equivalence in content, different countries still adopted their own coding and some varied application procedures of the ISO 9000:1994 family of standards: In the U.S.A., ANSI/ASQ Q9000 was issued by the American National Standards Institute (ANSI); in Canada, CAN/CSA-ISO 9000 was issued by the Canadian Standards Association (CSA); in the U.K., BS 5750; in France, NF-EN 29000; in Germany, DIN ISO 9000; and, in Japan, JIS Z 9900. These standards have been used extensively as the basis for independent quality system certification for over 400,000 organizations worldwide.

The ISO 9000:1994 family of standards contained a common set of 20 principles/requirements to be complied with for certification:

1. *Management responsibility*: The organization's management shall define and document its policy for quality and provide adequate resources for its implementation.
2. *Quality system*: The organization shall establish a quality system to ensure that products conform to specified requirements, including preparation of quality control plans, identification of measurement techniques and tools, and so on.
3. *Contract review.*
4. *Design control*: The organization shall establish and maintain procedures to control the design of products to ensure that requirements are met.
5. *Document and data control.*
6. *Purchasing (evaluation of subcontractors).*
7. *Control of customer supplied product (storage and maintenance).*
8. *Product identification and traceability.*
9. *Process control*: The organization shall ensure that production, installation, and servicing processes that affect quality are carried out under controlled conditions.
10. *Inspection and testing (procedures).*
11. *Control of inspection, measuring, and test equipment.*
12. *Inspection and test status (results).*
13. *Control of nonconforming product.*
14. *Corrective and preventive action.*

15. *Handling, storage, packaging, preservation, and delivery of products.*
16. *Control of quality records.*
17. *Internal quality audits.*
18. *Training*: The organization shall provide training to all personnel performing operations affecting quality and verify qualifications on the basis of education, training, and/or experience.
19. *Servicing.*
20. *Statistical techniques*: The organization shall identify the need for statistical techniques required for quality control of processes and products.

16.6.2 ISO 9000:2000

Since ISO protocol requires that all standards be reviewed at least every five years to determine whether they should be confirmed, revised, or withdrawn, the 1994 versions of the ISO 9000 family of standards were revised by the ISO's Technical Committee (TC) 176 in 2000. The original ISO 9000 family (developed during the late 1980s and the mid 1990s) contained more than twenty standards and documents. This proliferation of standards was a concern to many users and customers. The latest revisions of the core series standards in the ISO 9000 family, ISO 9000:2000, were published on December 15, 2000. ISO 8402 and part of the content of ISO 9000 were merged into a new ISO 9000:2000 standard. The earlier three (quality assurance) standards, ISO 9001, ISO 9002, and ISO 9003, were integrated into the new ISO 9001:2000. ISO 9004:1994, though maintaining its code, was also substantially revised.

As of 2001, ISO 9000 certification is to be achieved only through adhering to ISO 9001:2000, the practices described in ISO 9004:2000 may then be implemented to make the quality management system effective in achieving the quality assurance goals. ISO 9001:2000 and ISO 9004:2000 have been formatted as a consistent pair of standards to facilitate their use. Organizations must upgrade their quality management systems to meet the requirements of ISO 9001:2000 by December 15, 2003, in order to maintain an accredited certificate.

Currently, the primary ISO 9000 family standards are

ISO 9000:2000: Quality management systems—fundamentals and vocabulary.

ISO 9001:2000: Quality management systems—requirements. This is the requirement standard needed to assess an organization's ability to meet customer and applicable regulatory requirements.

ISO 9004:2000: Quality management systems—guidelines for performance improvements: This standard provides guidance for continual improvement of the quality management system.

ISO 10007:1995: Quality management—guidelines for configuration management. This standard provides guidance to ensure that a complex product continues to function when components are changed individually.

ISO/DIS 10012: Quality assurance requirements for measuring equipment—Part 1. Metrological confirmation system for measuring equipment. This standard provides guidance on the main features of a calibration system to ensure that measurements are made with the intended accuracy.

ISO 10012-2:1997: Quality assurance for measuring equipment—Part 2. Guidelines for control of measurement of processes. This standard provides supplementary guidance on the application of statistical process control when it is appropriate for achieving the objectives of Part 1.

ISO/TS 16949:1999: Quality systems—automotive suppliers—particular requirements for the application of ISO 9001:1994. This standard provides sector specific guidance to the application of ISO 9001 in the automotive industry.

ISO 19011: Guidelines on quality and/or environmental management systems auditing.

As mentioned above, the revised ISO 9001 and 9004 constitute a consistent pair of standards. Their structure and sequence are identical in order to facilitate an easy transition between them. Although they are standalone standards, their new structures promote enhanced synergy between the two. It is also intended that the ISO 9000 standards have global applicability and be used as a natural stepping stone towards total quality management (TQM).

The revised ISO 9000:2000 series standards are based on eight quality management principles that provide management with a framework to guide their organization towards improved performance.

Customer focus: Organizations should strive to exceed current and future customer expectations. (Improved customer loyalty leads to repeat business.)

Leadership: Organizations should encourage leaders to create an internal environment in which people can become fully involved in achieving the organization's vision. (Providing people with the required resources, training, and freedom to act with responsibility and accountability.)

Involvement of people: People at all levels of an organization should be fully involved. (Motivated, committed, and involved people lead to innovation and creativity.)

Process approach: A desired outcome can be achieved more efficiently when activities and resources are managed as a process. (Focused and prioritized improvement opportunities.)

System approach to management: Identifying, understanding, and managing interrelated processes as a system contributes to the organization's effectiveness and efficiency in achieving its objectives. (Integration and alignment of processes that will best achieve the desired results.)

Continual improvement: Continual improvement of the organization's overall performance should be a permanent objective of the organization. (Flexibility to react quickly to opportunities.)

Factual approach to decision making: Effective decisions can only be carried out based on the (factual) analysis of data and information. (Ensuring that data and information are sufficiently accurate and reliable and analyzing data and information using valid methods.)

Mutually beneficial supplier relationships: An organization and its suppliers are interdependent, and an effective relationship will enhance their competitiveness. (Flexibility and speed of joint response to changing customer expectations can optimize costs and resources.)

The revision of the ISO quality management system standards (yielding ISO 9000:2000), while retaining the essence of the original requirements, has repositioned the 20 elements of the ISO 9001:1994 and the guidelines of ISO 9004:1994 into the following eight classes:

1. Scope
2. Normative references
3. Terms and definitions
4. Quality management system
   - 4.1. General requirements
   - 4.2. Documentation requirements
5. Management responsibility
   - 5.1. Management commitment
   - 5.2. Customer focus
   - 5.3. Quality policy
   - 5.4. Planning
   - 5.5. Responsibility, authority, and communication
   - 5.6. Management review
6. Resource management
   - 6.1. Provision of resources
   - 6.2. Human resources
   - 6.3. Infrastructure
   - 6.4. Work environment

7. Product realization
   7.1. Planning of product realization
   7.2. Customer-related processes
   7.3. Design and development
   7.4. Purchasing
   7.5. Production and service provision
   7.6. Control of monitoring and measuring devices
8. Measurement, analysis, and improvement
   8.1. General
   8.2. Monitoring and measurement
   8.3. Control of nonconforming product
   8.4. Analysis of data
   8.5. Improvement

REVIEW QUESTIONS

1. Define quality and quality management.
2. Should the cost of quality management be added to the cost of the product or should it be recovered through increased market share?
3. Consider two makes of (electrical) batteries, A and B. The population of batteries of Make A has a mean life longer than that of Make B. However, the (life) variance of Make A is significantly larger than that of Make B. Discuss the following two issues: quality versus performance and the pricing of the two makes.
4. Define inspection versus testing.
5. Define destructive versus nondestructive inspection/testing.
6. Define accuracy versus repeatability (also known as precision).
7. Discuss on-line versus postprocess inspection.
8. Discuss 100% versus sampling-based inspection.
9. Discuss the use of coordinate measuring machines (CMMs) for inspection purposes.
10. Discuss the need for x-ray-based inspection in manufacturing. Elaborate on x-ray-based inspection for mass production versus for one-of-a-kind production.
11. What is computed tomography (CT)?
12. Can product life be represented using a Gaussian (normal) probability distribution? Explain.
13. Define population statistics versus sample statistics. Discuss the estimation of population statistics using finite-size sample statistics.
14. What is process capability? Discuss its use in product design.

15. What is statistical process control (SPC)? Can a process in total (statistical) control yield defective products (i.e., with feature values that are outside the product's specification limits)?
16. Compare the use of $\overline{X}$ charts versus $\overline{X}$ charts in SPC.
17. Provide a step-by-step SPC implementation procedure.
18. What is the primary purpose of ISO 9000?

DISCUSSION QUESTIONS

1. Computers and other information management technologies have been commonly accepted as facilitators for the integration of various manufacturing activities. Define/discuss integrated manufacturing in the modern manufacturing enterprise and address the role of computers, especially in the context of quality management.
2. Discuss the concept of progressively increasing *cost of changes* to a product as it moves from the design stage to full production and distribution. How could you minimize necessary changes to a product, especially for those that have very short development cycles, such as portable communication devices?
3. Information collected on failed products may provide valuable information to manufacturers for immediate corrective actions on the design and manufacturing current and/or future lines of products. Discuss how would you collect and analyze product failure (or survival) data for industries such as passenger vehicles, children's toys, and computer software.
4. The performance of a multicomponent product or system (e.g., the force required to close a car door) would be significantly improved as the dimensional parameters of the individual components approach their respective nominal values. In order to address this issue, some designers tend to narrow the acceptable ranges of these parameters (i.e., select stringent specification limits, tolerances) without any regard to the capability of the manufacturing processes to be used in fabricating the individual components. Discuss the above issue of tolerance specification in the profitable production of multicomponent products that will meet customer (quality) expectations.
5. Nondestructive quality control techniques are widely utilized in the manufacturing industry. Discuss the need for destructive testing in terms of government regulations, lack of reliable nondestructive testing techniques, testing time and cost, and so on. Use exemplary products during your discussion and state features that would be tested.

6. Go–no-go gages/setups/etc. have long been used in mass-production environments to ensure that every single part shipped to a customer meets the engineering specifications. Discuss if such techniques may contribute to the quality control of the manufacturing process even though they do not provide much feedback on the statistical behavior of the process. Under what conditions could go–no-go quality checks be useful or necessary?
7. Tool wear can have a detrimental effect on satisfying the (geometric) dimensional specifications of a machined part, including its surface finish, especially for hard materials and complex three-dimensional surfaces. Discuss possible remedies to this problem in terms of on-line depth-of-cut compensation in turning, milling, and drilling. Address the issues of on-line sensory feedback (i.e., measurement of tool wear or object dimensions) and microscale depth-of-cut compensation using secondary (e.g., piezoceramic based) actuators (e.g., placed under the tool holder in turning).
8. In mass-production environments, it is a common practice to have 100% inspection until the manufacturing process reaches a stable state, and then to employ statistical control methods to maintain the highest possible quality levels. Discuss a comparable viable quality-assurance strategy for one-of-a-kind or small-batch-size manufacturing.
9. SPC was developed as a monitoring tool that can identify problematic trends in production that may lead to quality problems. SPC can be considered as a "virtual sensor." Discuss the use of SPC in closed loop feedback control of fabrication processes, where manufacturing parameters are adjusted in an adaptive mode in response to the output of the SPC "sensor."
10. SPC is a process monitoring technique, whose objective is to ensure that the process is performing to its utmost capability defined by a statistical variation index. A fully calibrated process that is in control may, however, produce a large percentage of defectives, whose engineering specifications are outside those defined by its design. Although the process is in control, it is incapable of meeting the stringent engineering specifications. Compare SPC limits to engineering specification limits in elaborating on the above scenario. Discuss approaches to supplying a customer with a desired threshold percentage of parts that meet the engineering specifications, even when faced with a process capability problem in which the machine/system cannot meet this threshold percentage requirement.
11. Quality improvement is a manufacturing strategy that should be adopted by all enterprises; that is, although quality control is a primary concern for any manufacturing company, engineers should attempt to

improve quality: In statistical terms, all variances should be minimized, and furthermore where applicable the mean values should be increased (e.g., product life, strength) or decreased (e.g., weight) appropriately. Discuss the quality improvement issue and suggest ways of achieving continual improvements. Discuss also whether companies should concentrate on gaining market share through improved product performance or/and quality or only through cost/price.

12. The achievement of product specifications can be significantly improved with the availability of sensors that can provide the manufacturing process with feedback information while the fabrication of the product is ongoing. Discuss the role of postproduction quality control techniques in such environments, i.e., as complementing online quality-control strategies.
13. Production machines' (statistical) capability in terms of providing different levels of precision must be considered at the design stage of the product. Discuss the impact of this data on the decision-making process during the product development stage with respect to the following scenarios and others: proceed with the design of a product whose several components might have to be contract manufactured owing to the absence of economic in-house manufacturing capability; adopt a strategy of producing many components, using the ones that meet specifications and scrapping the rest; design, produce, and market products that only fractionally meet the design specifications; purchase better machines.
14. The factory of the future will be a totally networked enterprise. Information management in this enterprise will be complex. In regards to planning, monitoring, and control, discuss the level of detail of information that the controllers (humans or computers) would have to deal with in such environments. For example, some argue that in a hierarchical information management environment, activities are more of the planning type at the higher levels of the hierarchy and more of the control type at the lower levels. It has also been argued that the level of details significantly decreases as you ascend the enterprise ladder.
15. In the factory of the future, no unexpected machine breakdowns will be experienced! Such an environment, however, can only be achieved if a preventive maintenance program is implemented, in which all machines and tools are modeled (mathematically and/or using heuristics). These models would allow manufacturers to schedule maintenance operations as needed. Discuss the feasibility of implementing factory-wide preventive maintenance programs in the absence of our ability to model completely all existing physical phenomena and

furthermore our lack of a large variety of sensors that can monitor the states of these machines and provide timely feedback to such models.

BIBLIOGRAPHY

Batchelor, Bruce G., Hill, Denys A., Hodgson, David C. (1985). *Automated Visual Inspection*. Kempston, UK: IFS.

Benhabib, B., Charette, C., Smith, K. C., Yip, A. M. (1990). Automatic visual inspection of printed-circuit-boards—an experimental system. *IASTED, Int. J. of Robotics and Automation* 5(2):49–58.

Bennett, J. M. (May-June 1985). Comparison of techniques for measuring the roughness of optical surface. *Optical Engineering* 24(3):380–387.

Bernstein, Peter L. (1996). *Against the Gods: The Remarkable Story of Risk*. New York: John Wiley.

Besterfield, Dale H. (2001). *Quality Control*. Upper Saddle River, NJ: Prentice Hall.

Brodmann, R., Thurn, G. (1984). An optical instrument for measuring the surface roughness in production control. *CIRP Annals* 33:403–412.

Brodmann, R., Gerstorfer, O., Thurn, G. (May-June 1985). Optical roughness measuring instrument for fine-machined surfaces. *Optical Engineering* 24(3):408–413.

Cielo, Paolo G. (1988). *Optical Techniques for Industrial Inspection*. Boston: Academic Press.

Cleaver, Thomas G., Michels, James C., Jr., Dennis, Lloyd A. (1991). *Appearance Inspection of Finished Surfaces*. Milwaukee, WI: ASQC Quality Press.

Crosby, Philip B. (1996). *Quality is Still Free: Making Quality Certain in Uncertain Times*. New York: McGraw-Hill.

Demant, Christian, Streicher-Abel, Bernd, Waszkewitz, Peter (1999). *Industrial Image Processing: Visual Quality Control in Manufacturing*. New York: Springer-Verlag.

Deming, W. Edwards (1952). *Elementary Principles of the Statistical Control of Quality: A Series of Lectures*. Tokyo, Japan: Nippon Kagaku Gijutsu Remmei.

Deming, W. Edwards (2000). *Out of The Crisis*. Cambridge, MA: MIT Press.

DeVor, Richard E., Chang, Tsong-How, Sutherland, John W. (1992). *Statistical Quality Design and Control: Contemporary Concepts and Methods*. New York: Maxwell Macmillan International.

Dodge, Harold F., Romig, Harry G. (1951). *Sampling Inspection Tables: Single and Double Sampling*. New York: John Wiley.

Feigenbaum, Armand V. (1991). *Total Quality Control*. New York: McGraw-Hill.

Goetsch, David L., Davis, Stanley B. (1998). *Understanding and Implementing ISO 9000 and ISO Standards*. Upper Saddle River, NJ: Prentice Hall.

Grant, Eugene L., Leavenworth, Richard S. (1996). *Statistical Quality Control*. New York, NY: McGraw-Hill.

Groover, Mikell P. (2001). *Automation, Production Systems, and Computer Integrated Manufacturing*. Englewood Cliffs, NJ: Prentice-Hall.

Gryna, Frank M. (2001). *Quality Planning and Analysis: From Product Development Through Use*. New York: McGraw-Hill.

Harrington, H. J., Mathers, Dwayne D. (1997). *ISO 9000 and Beyond: From Compliance to Performance Improvement*. New York: McGraw-Hill.
Hoyle, David (1999). *ISO 9000 Quality Systems Handbook*. Boston: Butterworth-Heinemann.
International Organization for Standardization. http://www.iso.ch/iso/en/iso9000-14000/ iso9000/iso9000index.html, 2001.
Ishikawa, Kaoru (1982). *Guide to Quality Control*. Tokyo, Japan: Asian Productivity Organization.
Ishikawa, Kaoru (1985). *What is Total Quality Control?* Upper Saddle River, NJ: Prentice-Hall.
Johnson, Perry L. (2000). *ISO 9000: The Year 2000 and Beyond*. New York: McGraw-Hill.
Joseph M., Juran, A. Blanton Godfrey, eds. (1998). *Juran's Quality Handbook*. New York: McGraw Hill.
Juran, Joseph M. (1995). *A History of Managing for Quality*. Milwaukee: Quality Press.
Juran, Joseph M. (1989). *Juran on Leadership for Quality: An Executive Handbook*. New York: Free Press.
Kak, Avinash C., Slaney, Malcolm (1988). *Principles of Computerized Tomographic Imaging*. New York.
IEEE PressKalender, Willi A. (2000). *Computed Tomography: Fundamentals, System Technology, Image Quality, Applications*. Munich, Germany: MCD Verlag.
Kennedy, Clifford W., Hoffman, Edward G., Bond, Steven D. (1987). *Inspection and Gaging*. New York: Industrial Press.
Landry, Pierre D., et al., eds. (2000). *The ISO 9000 Essentials: A Practical Handbook for Implementing the ISO 9000 Standards*. Etobicoke, Ontario, Canada: Canadian Standards Association.
Maldague, Xavier (1993). *Nondestructive Evaluation of Materials by Infrared Thermography*. New York: Springer-Verlag.
Mitra, Amitava (1998). *Fundamentals of Quality Control and Improvement*. Upper Saddle River, NJ: Prentice Hall.
Montgomery, Douglas C. (1996). *Introduction to Statistical Quality Control*. New York: John Wiley.
Peach, Robert W. (1997). *The ISO 9000 Handbook*. Chicago: Irwin Professional.
Pyzdek, Thomas (1999). *Quality Engineering Handbook*. New York: Marcel Dekker.
Robinson, Stanley L., Miller, Richard K. (1989). *Automated Inspection and Quality Assurance*. Milwaukee, WI: ASQC Quality Press.
Rooks, Stephen M. (1993). The Development of an Inspection Process for Solder-Ball Connect Technology Using Scanned-Beam X-Ray Laminography: M.A.Sc. thesis, Department of Mechanical Engineering, University of Toronto, Toronto, Canada.
Rooks, S., Benhabib, B., Smith, K. C. (Dec. 1995). Development of an inspection process for ball-grid-array technology using scanned-beam x-ray laminography. *IEEE Transactions on Components, Hybrids, and Manufacturing Technology* 18(4):851–861.

Shewhart, Walter Andrew (1931). *Economic Control of Quality of Manufactured Product*. New York: Van Nostrand.
Shingo, Shigeo (1986). *Zero Quality Control: Source Inspection and the Poka-Yoke System*. Stamford, CT: Productivity Press.
Society of Automotive Engineers (1999). *Quality Systems, Aerospace, Model for Quality Assurance in Design, Development, Production, Installation and Servicing*. Warrendale, PA: SAE International.
Taguchi, Gen'ichi, Elsayed, A. Elsayed, Hsiang, Thomas C. (1989). *Quality Engineering in Production Systems*. New York: McGraw-Hill.
Tricker, Raymond L. (1997). *Quality and Standards in Electronics*. Boston: Newnes.
Zuech, Nello, ed. (1987). *Gaging with Vision Systems*. Dearborn, MI: SME Publications.

Index

CPSIA information can be obtained
at www.ICGtesting.com
Printed in the USA
LVHW081107131218
599554LV00003BB/27/P

9 780824 742737